Intensive Livestock Production in Transition

Barbara Grabkowsky • Thomas Blaha
Editors

Intensive Livestock Production in Transition

Analyses, Concepts and Strategies for Sustainability Transformation of the Livestock Value Chain

Springer

Editors
Barbara Grabkowsky
Center for Sustainability Transformation
Research in AgriFood Systems, Vechta
Institute of Sustainability Transformation
in Rural Areas
University of Vechta
Vechta, Germany

Thomas Blaha
Former Head of the Field Station
for Epidemiology (Bakum)
University of Veterinary Medicine Hannover
Hannover, Germany

ISBN 978-3-031-97871-5 ISBN 978-3-031-97872-2 (eBook)
https://doi.org/10.1007/978-3-031-97872-2

This Springer imprint is published by the registered company Springer Nature Switzerland AG
The registered company address is: Gewerbestrasse 11, 6330 Cham, Switzerland

Preface

The "Great Transformation" of the Agri-Food System: A Shift in Values, Sustainability Imperatives and Geopolitical Uncertainty

We live in an era of profound transformation in which climate change, biodiversity loss, geopolitical shocks and economic volatility reshape global systems. These interlocking pressures expose vulnerabilities in our financial and agricultural models and make sustainability an imperative. The transition is complex: it involves conflicts of societal values, competing interests of involved stakeholders, unavoidable trade-offs and manifold uncertainties. The agri-food system sits at the intersection of planetary limits and food security, amid rising social expectations and difficult policy choices.

Agriculture has long been a pillar of stability, yet today it is entwined with geopolitical disruption, market volatility and shifting societal expectations. The war in Ukraine, trade frictions and inflation have revealed the fragility of supply chains and fuelled debates on food security—covering availability, access and utilisation, and the reliability of supply over time—and on food sovereignty—the governance claim that societies should have agency over how food is produced, by whom and under which standards—alongside questions of strategic autonomy.

In the light of this, *Intensive Livestock Production in Transition* presents analyses, concepts and strategies for the sustainability transformation of the livestock value chain. In this book, we use 'transition' to denote the ongoing change process and 'transformation' for the resulting structural reconfiguration towards sustainability.

Livestock Production and Food Systems at a Crossroads

Livestock farming is at the centre of this transformation. It underpins food security and rural economies while contributing to climate change, deforestation, biodiversity loss and water scarcity. The concept of Rockström's planetary boundaries—ecological limits within which humanity can safely operate—underscores the need to reduce agriculture's environmental footprint while safeguarding economic viability and social acceptance.

Reform pressures arise from regulation, consumer preferences, technological change and global competition. Path dependencies, economic constraints and cultural factors render the shift anything but simple. The core challenge is to strike a balance between sustainability and competitiveness without compromising food security, rural livelihoods or the resilience of agri-food systems.

Regions of High Livestock Concentration

Over recent decades, several regions around the world have built highly specialised, intensive livestock systems with integrated feed, slaughter and processing industries, supported by dense supplier networks. These clusters have underpinned efficiency and growth, yet they have also become focal points for debates on sustainability, animal welfare and environmental impact as regulation tightens, ecological constraints intensify and consumer preferences shift.

A case in point is the Oldenburger Münsterland in northwest Germany—one of Europe's most dynamic agri-food clusters, comparable to North Brabant in the Netherlands, Catalonia in Spain and Brittany in France. The region combines deep value-chain integration, close research–industry linkages and a pronounced entrepreneurial culture with an unusually dense institutional landscape.

Precisely because the full value chain and its institutions are well integrated—and owing to this integration, collaboration and entrepreneurial drive—the Oldenburger Münsterland is exceptionally well placed to serve as a credible locus for the evolution of livestock systems. If the region succeeds in advancing its sustainability and animal-welfare standards while maintaining its productivity edge, the resulting market-tested operating models will provide transferable reference designs. In peer clusters with comparable institutional density and collaboration—such as North Brabant or Brittany—these designs can be adapted to local conditions and scaled.

Our Perspective—and Why This Book

As editors, our vantage point stems from sustained, hands-on work with farmers, veterinarians, processors, retailers, standard-setters and civil-society organisations—in the Oldenburger Münsterland, in comparable clusters across Europe and internationally. Over the years, we have worked on antimicrobial stewardship, animal health, food safety and animal welfare, and on transition strategies for resilient value chains with partners across the sector.

Thomas Blaha, Professor of Epidemiology and former head of the Field Station for Epidemiology at the University of Veterinary Medicine Hannover (TiHo), brings deep experience spanning surveillance and risk assessment of farm-to-fork food safety and antimicrobial-resistance mitigation through to applied improvements in animal welfare. **Barbara Grabkowsky**'s path led from university research to seven years in industry and, since 2017, to directing trafo:agrar, a transdisciplinary platform that convenes science, practice, policy and civil society to co-create and test operating models and methods in living-lab settings.

This combined perspective shapes how we frame this book; it has grown out of these efforts. With our joint, sustained on-the-ground engagement in the region for many years, we have witnessed and supported this evolution, and we continue to do so.

Many chapters are authored by long-time colleagues and international partners from the University of Minnesota (United States), the University of British Columbia (Canada), IFIP—Institut du Porc (France) and Aarhus University (Denmark). Further contributions come from scholars affiliated with the trafo:agrar consortium—among them the Georg-August University of Göttingen, the University of Veterinary Medicine Hannover (TiHo) and the University of Vechta—as well as from Bielefeld University, Freie Universität Berlin, Humboldt-Universität zu Berlin, the University of Osnabrück, Hochschule Osnabrück—University of Applied Sciences, the Technical University of Braunschweig, the Technical University of Munich (TUM), the University of Hildesheim, the German Research Center for Artificial Intelligence (DFKI), Osnabrück, the Friedrich-Loeffler-Institut (Federal Research Institute for Animal Health), the Johann Heinrich von Thünen Institute, Braunschweig, the Leibniz Institute of Agricultural Development in Transition Economies (IAMO), Weihenstephan-Triesdorf University of Applied Sciences (HSWT) and Zoo Arche Warder e.V. Perspectives from practice and industry also inform the volume, including Big Dutchman AG (Calveslage), FutterAllianz Nordwest GmbH & Co. KG (Schneiderkrug), Opteinics™ (Chemovator GmbH) (Mannheim) and the Agrar- und Ernährungsforum Nord-West e.V. (AEF). Our sincere thanks go to all chapter authors and contributors; their expertise and collaborative spirit have shaped this volume.

Together, these contributions reflect the diversity of perspectives needed to navigate systemic change.

Purpose of This Book

This book aims to:

- Diagnose system drivers and constraints across ecology, economics and risk assurance—covering biophysical limits and footprints, market structures and the safeguards provided by food safety and biosecurity
- Design and evaluate solution pathways for practice—ranging from circularity and upcycling to digital technologies, entrepreneurial approaches and future husbandry structures
- Integrate the societal dimension—ethical discourse, education, animal health and welfare as prerequisites, and consumer behaviour and labelling—to align transformation with public expectations

- Enable change through governance—policy instruments and multi-stakeholder processes, comparative international perspectives and an overarching transformation framework

Contributions span animal health and welfare, economic and policy frameworks, environmental challenges (including climate impact and resource efficiency), societal expectations and consumer behaviour, and technological innovation for sustainable practice.

This is not a one-size-fits-all blueprint. It is a pragmatic guide—a compass for understanding and navigating complexity. It explains why progress takes time, why conflicts emerge and why emotions run high—because, at its core, this transformation touches livelihoods, cultural identities and humanity's fundamental need for food security.

This book aligns with the European Commission's Transition pathway for the EU agri-food industrial ecosystem and extends beyond Europe to foster a global dialogue on sustainable livestock systems.

Our sincere thanks go to Vera Schockemöhle, whose outstanding support—coordinating author communication, maintaining a structured filing system, and tirelessly managing documents—greatly facilitated this book.

We hope this book will inspire researchers, policymakers, industry leaders and practitioners worldwide, and help to shape agri-food systems that are not only sustainable and economically viable but also globally just and resilient.

Vechta, Germany Barbara Grabkowsky
Hannover, Germany Thomas Blaha

Contents

About the Editors and Contributors

About the Editors

Barbara Grabkowsky is Director of the "Center for Sustainability Transformation in Agrifood Systems" at the University of Vechta, Germany. She holds a master degree in Environmental Sciences and a doctoral degree in Biosecurity and Risk Assessment in livestock production systems.

Her profile combines academic expertise with several years of international leadership experience in the value chain for animal-based food, where she was responsible for sustainability, food safety, and cross-sector collaboration. This dual perspective enables her to connect scientific and entrepreneurial thinking—a key asset for advancing sustainability transformations.

Dr. Grabkowsky follows a systemic and transdisciplinary approach, integrating ecological, political, social, and economic dimensions. Her work focuses on the transformation of intensive agricultural regions, resilient agrifood systems, and educational formats that support change management and transformative learning. A central aim of her work is to foster "out-of-the-box-thinking" for developing holistic, practicable solutions for multidimensional challenges.

She has contributed to expert groups at the European Food Safety Authority (EFSA) and the Food and Agriculture Organization of the United Nations (FAO). She is also active in national boards and advisory bodies with a focus on water protection, nature conservation, and sustainability education—and contributes to the discourse on the future of agriculture and food systems as a lecturer, speaker, and author.

Thomas Blaha is Professor Emeritus at the University of Veterinary Medicine Hannover Foundation, Germany, where he held the Chair of Epidemiology at the Field Station in Bakum. Before that, he served as a university professor at the University of Minnesota (USA). Since 2007, he has been Chairman of the Veterinary Association for Animal Protection (TVT) and is deeply committed to advancing animal welfare and health in national and international contexts. He is the author and editor of more than 250 scientific publications and 3 books on animal health and welfare.

He is a diplomate of the European College of Veterinary Public Health (ECVPH) as well as a founding member and diplomate of the European College of Porcine

Health Management (ECPHM). In addition, he has been an active member of the Animal Welfare Working Group (AWWG) of the Federation of Veterinarians of Europe (FVE) for 5 years, and he served on several ad-hoc working groups of the European Food Safety Authority (EFSA) in Parma. He is also a former president of the International Society of Animal Hygiene.

Prof. Blaha's long-standing international experience, his extensive professional network, and his profound expertise make him a pioneering voice in the field. His holistic thinking has significantly influenced how sustainable and ethically responsible livestock systems can be designed in light of societal transformation processes. He continues to serve as a thought leader in developing future-proof approaches to animal husbandry that align with broader sustainability goals.

Contributors

Brianne A. Altmann Institute of Animal Hygiene, Animal Welfare and Farm Animal Behaviour, University of Veterinary Medicine Hannover, Hannover, Germany

Marketing for Food and Agricultural Products, Department of Agricultural Economics and Rural Development, University of Goettingen, Goettingen, Germany

Karen Arcia Department of Agricultural Economics and Rural Development, University of Göttingen, Göttingen, Germany

Alfons Balmann Leibniz Institute of Agricultural Development in Transition Economies (IAMO), Halle (Saale), Germany

Thomas Blaha Former Head of the Field Station for Epidemiology (Bakum), University of Veterinary Medicine Hannover, Hannover, Germany

Stephan Brosig Leibniz Institute of Agricultural Development in Transition Economies (IAMO), Halle (Saale), Germany

Michael Bucherer Institute of Construction Design, Industrial and Health Care Building, Technical University of Braunschweig, Braunschweig, Germany

Franz J. Conraths Friedrich-Loeffler-Institut, Federal Research Institute for Animal Health, Greifswald-Insel Riems, Germany

Kai Frölich Zoo, Ark Warder e.V., Warder, Germany, and Department of Biology, Institute of Biology and Chemistry, University of Hildesheim, Hildesheim, Germany

Julia Gickel WING (Science for Innovative and Sustainable Poultry Farming), University of Veterinary Medicine Hannover, Foundation, Hannover, Germany

Barbara Grabkowsky Center for Sustainability Transformation Research in AgriFood Systems, Vechta Institute of Sustainability Transformation in Rural Areas, University of Vechta, Vechta, Germany

Harald Grethe Humboldt-Universität zu, Berlin, Germany

Klaus G. Grunert Aarhus University, Aarhus, Denmark

Johannes Halbe Institute of Geography and Institute of Environmental Systems Research, University of Osnabrück, Osnabrück, Germany

Jantje Halberstadt Economics of Sustainability, Institute of Sustainability Transformation in Rural Areas, University of Vechta, Vechta, Germany

Sebastian Hartstang University of Veterinary Medicine Hannover, Hannover, Germany

Franziska Hauff Leibniz Institute of Agricultural Development in Transition Economies (IAMO), Halle (Saale), Germany

Joachim Hertzberg Institute of Computer Science, Osnabrück University, German Research Center for Artificial Intelligence (DFKI), Cooperative and Autonomous Systems Group, Osnabrück, Germany

Jan Holzhausen Institute of Construction Design, Industrial and Health Care Building, Technical University of Braunschweig, Braunschweig, Germany

Elisa Husson IFIP-institut du porc, Pacé, France

Johannes Isselstein Division of Grassland Science, Department of Crop Sciences, Georg-August-University Göttingen, Göttingen, Germany

Campus Center of Biodiversity and Sustainable Land Use, University of Göttingen, Göttingen, Germany

Manfred Kayser Division of Grassland Science, Department of Crop Sciences, Georg-August-University Göttingen, Göttingen, Germany

GeoLab, University of Vechta, Vechta, Germany

Sophia Kollhoff Economics of Sustainability, Institute of Sustainability Transformation in Rural Areas, University of Vechta, Vechta, Germany

Martin Komainda Division of Grassland Science, Department of Crop Sciences, Georg-August-University Göttingen, Göttingen, Germany

Sarah Kühl Agricultural Marketing, Department of Sustainable Agriculture and Energy Systems, University of Applied Sciences Weihenstephan-Triesdorf, Freising, Germany

Peter Kunzmann University of Veterinary Medicine Hannover, Hannover, Germany

Diana Meemken Working Group Meat Hygiene, Institute of Food Safety and Food Hygiene, School of Veterinary Medicine, Freie Universität, Berlin, Germany

Clara Mehlhose Marketing for Food and Agricultural Products, Department of Agricultural Economics and Rural Development, University of Goettingen, Goettingen, Germany

Mariann Molnár Animal Welfare Program, Faculty of Land and Food Systems, University of British Columbia, Vancouver, BC, Canada

Daniel Mörlein Division for Quality of Animal Products, Department of Animal Sciences, University of Goettingen, Goettingen, Germany

Henning Müller German Research Center for Artificial Intelligence (DFKI), Cooperative and Autonomous Systems, Osnabrück, Germany

Andrea Nelke University of Veterinary Medicine Hannover, Hannover, Germany

Claudia Pahl-Wostl Institute of Geography and Institute of Environmental Systems Research, University of Osnabrück, Osnabrück, Germany

Carolina Probst Friedrich-Loeffler-Institut, Federal Research Institute for Animal Health, Greifswald, Germany

Marco Rieckmann Faculty I, Department of Education, Working Group Higher Education Development, University of Vechta, Vechta, Germany

Christine Roguet IFIP-institut du porc, Pacé, France

Irene Rosique-Conesa Opteinics™, Chemovator GmbH, Mannheim, Germany

Gerald Shurson Department of Animal Science, College of Food, Agriculture, and Natural Resource Sciences, University of Minnesota, St. Paul, MN, USA

Achim Spiller Marketing for Food and Agricultural Products, Department of Agricultural Economics and Rural Development, Georg-August-Universität Göttingen, Göttingen, Germany

Annika Thies Institute of Market Analysis, Johann Heinrich von Thuenen Institute, Braunschweig, Germany

Caroline Timme FutterAllianz Nordwest GmbH & Co. KG, Schneiderkrug, Germany

Christian Visscher Institute for Animal Nutrition, University of Veterinary Medicine Hannover, Foundation, Hannover, Germany

Stephan von Cramon-Taubadel Department of Agricultural Economics and Rural Development, University of Göttingen, Göttingen, Germany

Lina Sofie von Fricken Big Dutchman International GmbH, Vechta, Germany

Tabea Weber Bielefeld University, Bielefeld, Germany

Wilhelm Windisch Technical University Munich (TUM), TUM School of Life Sciences, Munich, Germany

Jasmin Zöllmer Humboldt-Universität zu, Berlin, Germany

Nana Zubek Department of Agricultural Economics, Faculty of Agricultural Sciences and Landscape Architecture, Hochschule Osnabrück – University of Applied Sciences, Osnabrück, Germany

1 Introduction

Thomas Blaha, Barbara Grabkowsky, and Gerald Shurson

Food security is essential for human life and has played a critical role in sustaining human societies for millennia. Food insecurity, defined as a lack of access to nutritious and affordable food, has historically led to devastating consequences such as the Great Famine in Ireland (1845–1852) and the famine in Ukraine (1932–1933) that caused unnecessary loss of human life, mass migration, and profoundly changed the cultural, political, and demographic landscapes of these countries. However, food insecurity continues to exist today in low- and middle-income countries where it leads to political unrest, migration, poor physical and mental health, child development problems, disease, and unnecessary suffering. Food insecurity and "food deserts" also exist in rural and urban areas in affluent countries in Europe and the United States that disproportionately affect black, brown, and Indigenous populations due to income inequality and inadequate social foundations despite abundant food supplies being produced (Smets et al. 2022; Karpyn et al. 2020). In 1996, the United Nations Food and Agriculture Organization (FAO) defined food security as follows: "Food security exists when all people, at all times, have physical and economic access to sufficient, safe and nutritious food that meets their dietary needs

T. Blaha (✉)
Former Head of the Field Station for Epidemiology (Bakum), University of Veterinary Medicine Hannover, Hannover, Germany
e-mail: thomas.blaha.ir@tiho-hannover.de

B. Grabkowsky
Center for Sustainability Transformation Research in AgriFood Systems, Vechta Institute of Sustainability Transformation in Rural Areas, University of Vechta, Vechta, Germany
e-mail: barbara.grabkowsky@uni-vechta.de

G. Shurson
Department of Animal Science, College of Food, Agriculture, and Natural Resource Sciences, University of Minnesota, St. Paul, MN, USA
e-mail: shurs001@umn.edu

B. Grabkowsky, T. Blaha (eds.), *Intensive Livestock Production in Transition*,
https://doi.org/10.1007/978-3-031-97872-2_1

and food preferences for an active and healthy life". Considering all the technological innovations humans have created, including space travel to the moon, it is astonishing that global food security for all has never been achieved, and it continues to be one of our greatest challenges. In fact, the 11th Global Food Security Index Report published in 2022 showed a decline in global food security for the third year in a row based on food availability, affordability, quality and safety, and sustainability and adaptation among 113 countries (GFSI 2022).

Historically, agriculture and food production have undergone significant transformations in response to food insecurity. Following World War II, hunger and malnutrition were widespread for most people. The limited amounts or absence of high-quality animal proteins in the daily human diet was deemed detrimental to the quality of life (Leroy et al. 2023). Much of this nutritional deficit was attributed to the dominance of small-scale farming systems, which were prevalent but were not capable of producing enough quality food, especially meat and meat products, to meet the nutritional needs during this post-war period. The rapid development of industrialisation and urbanisation further limited the availability of animal-derived foods because the proportion of farmers producing food declined dramatically relative to the increase in the number of food consumers. Furthermore, the low productivity and output of food produced per farmer was not sufficient to supply the increasing demand for food in the rapidly growing urban centres. Therefore, the logical and socially acceptable solution at that time was to promote and subsidise the development of larger farms and production units, especially for livestock production. Beginning around 1947 and continuing through many years of the Cold War in Europe, the "Western Bloc" provided significant government funding to support research and extension education that was targeted towards helping farmers learn and implement new technologies that enabled them to become more efficient and profitable and increase the size of their farms. In contrast, a different model emerged in the "Eastern Bloc" where the creation of larger and more efficient production units was achieved by expropriating private farmers by using forced collectivisation, which also led to extremely large food animal production units that were planned and ordered to be constructed by communist party leaders. In both government models, animal breeding programmes were successfully implemented to improve productivity and efficiency (i.e. increased daily weight gain, eggs per hen per year, milk yield per cow, and feed conversion ratio) of food animal production.

In the 1960s, 1970s, and 1980s, in the West and the East, livestock productivity and the amount of meat, milk, and eggs produced on livestock farms increased dramatically. As a result, the abundant supply of foods of animal origin became a staple in most people's daily diets in Europe. This achievement was a direct result of publicly funded animal production and health research, which was widely viewed by society as a great success story until the mid-1990s. However, in recent decades, increasing income inequality, along with other factors, has led to a divergence of food availability, affordability, and choices among consumers who have become disconnected from farmers and farming practices and are more concerned about how their food is produced.

By the late twentieth century, feelings of dislike and distrust of modern food production systems and practices began to spread among consumers in well-nourished societies. These new perspectives were particularly evident in the highly intensive innovation clusters of food animal production in the Northwest region of Germany, the Brittany region in France, North Brabant in the Netherlands, Emilia-Romagna in Italy, and Catalonia in Spain. Terms like "factory farming" became widely used to stigmatise intensive animal farming systems as being cruel to animals by animal rights groups, which led to animal welfare becoming a major topic of widespread public concern. In addition, environmental concerns associated with the production and management of large amounts of manure produced from large intensive animal production facilities were soon added to the list of criticisms because of gas and odour emissions and the eutrophication of groundwater and surface water with nitrate and phosphorus. Increasing awareness and public concern about climate change and global warming led to regulatory pressure to limit imports and use of soybean meal from Brazil due to devastating environmental impacts caused by the extensive destruction of the Amazon rainforest to allow the expansion of soybean and cattle production in Brazil. Public opinions about intensive animal agriculture have continued to evolve towards the need to reduce meat consumption per capita because of the contribution of animal agriculture, especially beef production, towards methane and greenhouse gas emissions, which contribute not only to global warming but also to the detrimental effects on human health caused by excessive consumption.

Feeding the global population remains an intricate challenge due to multiple interacting factors:

- The *agricultural sector* is highly *dynamic* (subject to changing economic, climatic, and technological factors).
- *Conflicting priorities* exist between global trade policies, local food security, and sustainability goals.
- Food systems are *vulnerable* to extreme weather events, geopolitical conflicts, and economic crises.
- Inequitable food production and distribution conditions persist due to *subsidies, trade tariffs, and economic disparities.*

Global agriculture and food production significantly contribute to the exceedance of six out of nine planetary boundaries: climate change, biodiversity loss, land system change, freshwater depletion, excess nitrogen and phosphorus emissions, and novel entity pollution (including microplastics and per- and polyfluorinated alkyl substances (PFAS), often referred to as "forever chemicals"). These boundaries are approaching irreversible thresholds that could threaten the stability of life on Earth (Richardson et al. 2023). Even more striking is the fact that about 33–40% of the food produced globally is lost or wasted and not consumed (Wang et al. 2023), which results in wasting 38% of total energy consumption and 30% of land use (FAO 2019), along with 25% of freshwater (FAO 2023). About 50% of total greenhouse gas (GHG) emissions from the global agriculture and food production system have been attributed to food loss and waste, with 30% of these emissions produced from end-of-life disposal (Zhu et al. 2023).

The extent and causes of food loss and waste differ significantly between low-income and high-income countries. Likewise, there are major differences in the productivity and efficiency of food production systems between these two types of economies that have evolved in response to different cultural, economic, and climatic circumstances and require context-specific modifications (Henchion et al. 2021).

In low-income countries, productivity and efficiency of food production are so low that, together with high pre-consumer food losses, along with inequitable and unjust food availability and distribution, nearly 800 million people continue to be hungry and undernourished due to the inability to supply all people with adequate amounts of nutritious, healthy, and safe food. This has been a recurring problem for many decades. One of the United Nations Millennium Goals for the period of 2000 to 2015 was to eradicate extreme poverty, hunger, and malnutrition (United Nations 2015), but we are moving further away from achieving all these goals by 2030 (FAO, IFAD, UNICEF, WFP and WHO 2024). In 2024, nearly one-third (29.6%) of the world's population faced moderate to severe food insecurity, and up to 783 million people suffer from hunger (UNEP 2024). These startling numbers of undernourished people do not include the unknown number of malnourished people who consume sufficient calories but are deficient in protein and micronutrients from the foods they consume.

In high-income countries, the development and implementation of state-of-the-art intensive agriculture and food production technologies have resulted in highly productive and efficient food production systems to provide an abundant and affordable food supply. Thus, intensive farming systems are often considered to be global models for feeding future generations, but they have serious flaws that must be addressed. Among these challenges is high animal density, which has led to animal welfare concerns, vulnerability to disease transmission, antibiotic use, and detrimental environmental impacts from manure production. Food insecurity is also a significant problem in high-income countries despite the abundance of food.

In fact, food insecurity coexists with high obesity rates, which predispose populations to numerous adverse health effects and place a burden on public healthcare systems (Frongillo and Bernal 2014). Like low-income countries, excessively large amounts of food loss and waste also occur in high-income countries but at different stages of the food supply chains and for different reasons (e.g. exaggerated "cosmetic" standards for fruits and vegetables, expiration date labelling in supermarkets, unsold restaurant food, and household food waste). However, because of the affluence of many people in high-income countries and their ability to make food choices based on how their food is produced, animal-derived foods are often criticised for not being healthy, environmentally sustainable, or produced under unacceptable animal welfare conditions. As a result, there are common challenges as well as major differences between low-income and high-income countries in achieving food security for all. Therefore, the ultimate question is, "How should food be produced?"

Our agri-food world is broken, and we must repair it. The environmental, social, and economic costs of food animal production are too high to sustain the planet and human societies for future generations. A new blueprint for a sustainable and resilient food system is needed to replace the current model that is destroying our environment by exploiting natural resources, diminishing the biodiversity of ecosystems, accelerating climate change, and polluting the Earth beyond its carrying capacity. We also need to redesign our social foundation to overcome the many problems associated with the wide inequality gaps and remove the unjust and inequitable conditions preventing food security for all. Achieving these goals will not be easy, and public discourse about how food should be produced and the types of food that should be consumed has become highly contentious in high-income societies. The many themes and chapters included in this book provide multiple perspectives on transforming our current unsustainable food animal production systems into sustainable and resilient frameworks that are based on a balanced, systems approach that strives to achieve food security for all and simultaneously protects the environment, increases the wellbeing of all food producing animals, maintains biodiversity of ecosystems, minimises global warming and climate change, and prevents further exploitation of natural resources beyond planetary boundaries.

Achieving a sustainable and resilient food system requires empowering people to drive the necessary changes, including addressing social sustainability challenges. Intersectional environmentalism advocates for protecting both people and the planet by recognising links between environmental degradation and social injustice (Thomas 2022). In addition, ethical aspects have also gradually become part of the public discourse regarding the future of intensive food animal production because of profound changes in social values and increasing knowledge about the adverse side effects of exploiting animals and the environment. Climate and environmental protection have become increasingly important societal issues (Wang-Erlandsson et al. 2022) and have shifted from an anthropocentric perspective to a pathocentric perspective and even a biocentric perspective (Selter and Salloch 2023), demanding a fundamental shift in attitudes towards animals, soil, water, and air. Sustainability now requires balancing the needs of humans, animals, plants, and the ecosystem, which all must also be respected and met. This development towards a careful, sustainable use of natural resources has been proposed as the basis for a future-proof and sustainable economy and food system (Richardson et al. 2023; WEF 2025).

There are numerous challenges to transforming our currently unsustainable intensive food animal production systems into circular, regenerative, sustainable, and resilient systems. Some of the key challenges are as follows:

(a) *Multi-dimensional Complexity*. Our current agri-food system is extremely complex, with many stakeholders upstream and downstream in supply chains that will be either positively or negatively affected by any change. A major transformation of the magnitude required makes it a classical "wicked" problem, as defined by Rittel and Webber (1973) for multi-dimensional, complex, and unsolvable problems. Instead of a solution, only continuous interventions in small steps are realistic. This approach is not at all acceptable for many fundamental critics. There are many trade-offs rather than "wrong or right" decisions that must be publicly debated to find context-specific agreements and governance of how we produce and consume food of animal origin. This will require an organised discourse where everyone has a voice and is transdisciplinary, representing multi-stakeholder groups.

(b) *"Systems Thinking" Required.* The current system of livestock production is not only in itself highly complex with many "internal" and "external" interactions and feedback loops with other areas of agriculture, food systems, industrial activities, and natural ecosystems. The transformation of intensive food animal production cannot take place in an isolated "bubble" that includes only those directly involved in food animal production. Many aspects of the transformation will have many positive and negative impacts on farmers, agribusinesses, and the general economy, and result in many unintended consequences which must be foreseen, managed, and mitigated. Therefore, governments need to use a systems approach and frameworks that include internal and external interactions and feedback loops to make meaningful changes to achieve greater food security and more sustainable agri-food systems.

(c) *Understanding and Managing Trade-offs.* There are many differences in perspectives, conflicting goals and targets, and trade-offs associated with implementing transformative practices. For example, the implementation of some practices that improve sustainability will increase production costs and lead to increased food prices that ultimately will make food unaffordable for people with low incomes. Another example of trade-offs that must be considered is the use of extensive, outdoor animal production systems which are generally less efficient in resource utilisation, and more of a concern for biosecurity to prevent disease transmission and food safety concerns than indoor, intensive production systems. Although conflicts of business interest can theoretically be managed through government regulations, differences in values and objectives need to be thoughtfully balanced to achieve long-term sustainability. Changes made to reduce the environmental footprint of food production, must not compromise advances in food safety, curbing antimicrobial resistance, preventing and minimising zoonotic risks to human health, and the increasing understanding and emerging efforts to promote and achieve One Health.

(d) *The Role of Consumers.* Consumers are one of the major drivers of transforming agricultural and food production practices. Food preferences and price points for food purchases are as varied as the many types of consumers in the market. Some consumers believe that organic, local, or natural foods are more nutritious, safer, ethically produced, and environmentally sustainable, and they are willing and capable of paying more for these types of foods, while the majority of consumers simply want safe and high-quality foods at the lowest price possible. Other consumers choose to follow vegan, vegetarian, or flexitarian lifestyles and food choices for various reasons, including religious beliefs, personal health concerns, or finding some types of animal-derived foods unappealing. As a result, the wide variety of food preferences among consumers creates many niches, increases the complexity of food production systems, and makes it difficult for producers to change production practices in response to clear market demand signals. Furthermore, the "Consumer-Citizen Gap" indicates that people make food purchasing decisions differently as price-sensitive consumers than as accountable citizens (Busch and Spiller 2018). Consumers are not a consistent "body", have no organisational structure, cannot be directed, but can only be nudged into changing purchasing habits. However, even if there is a consensus in society on which directions consumers should be nudged to follow, there will always be stakeholders that will try to hinder certain attempts and nudge consumers in a direction that is favourable for them. Consumer buying decisions are also dynamic, where economic conditions can dramatically alter buying behaviour, causing them to change their preference for types of food products, the amount purchased, and the price they are willing to pay for those goods.

(e) *Sorting Facts from Fiction.* Universal access to accurate information by everyone has never been greater. Unfortunately, there are very few "guard rails" to help people distinguish between facts and misleading and inaccurate information. As a result, strong beliefs are created and propagated through the Internet and social media regarding various viewpoints about how food should be produced and the reasons for choosing the types of food to be eaten. This makes an objective, fact-based public discussion about food production system sustainable transformations more difficult. The Internet has accelerated the distribution of irrational and erroneous ideas that are propagated as absolute truths. Simultaneously, effective communication of science and research findings plays a decreasing role in our information-flooded world. The sustainable transformation of our food production system must include different perspectives from stakeholders and society in general, but it must be science-based and factual to have a meaningful chance of success.

(f) *We Need Farmers.* The vehement protests of farmers throughout the European Union in early 2024 were a clear indication that farmers feel marginalised, ignored, and disregarded especially by the urban population. As a result, politicians started to make concessions in partial support of their demands. The recently enacted carbon tax on farms in Denmark was a result of a compromise towards a system that appears to make sense for all. This is an example of the urgent need for governments to not impose environmental sustainability regulations without obtaining perspectives and reasonable considerations from the farming community. Farmers must be embraced and included in the public discourse and decision-making processes to create a supportive framework that enables them to work under good conditions, earn a reasonable income, and enjoy a good quality of life.

This book explores the multi-faceted transformation of intensive food animal production systems into sustainable, resilient, and ethically sound frameworks. It is a thoughtful collection of chapters written by international experts in their respective disciplines that explore multiple themes and support the transformation towards a more sustainable and animal welfare-friendly production system that is efficient enough to guarantee food security for all while meeting the highest food safety standards and protecting human, animal, and environmental health. It is intended to provide numerous scientifically based perspectives that can be used to facilitate holistic, balanced, and functional sustainable food animal production systems.

References

Busch G, Spiller A (2018) Consumer acceptance of livestock farming around the globe. Anim Front 8(1):1–3. https://doi.org/10.1093/af/vfx005

FAO (2019) The State of Food and Agriculture 2019, Moving Forward on Food Loss and Waste Reduction, Food and Agriculture Organization of the United Nations, Rome. https://www.fao.org/3/ca6030en/ca6030en.pdf

FAO (2023) The State of Food and Agriculture 2023 – revealing the true cost of food to transform agrifood systems, Food and Agriculture Organization of the United Nations, Rome. https://doi.org/10.4060/cc7724en

FAO, IFAD, UNICEF, WFP and WHO (2024) The State of Food Security and Nutrition in the World 2024 – financing to end hunger, food insecurity and malnutrition in all its forms. Rome. https://doi.org/10.4060/cd1254en

Frongillo EA, Bernal J (2014) Understanding the coexistence of food insecurity and obesity. Curr Pediatr Rep 2:284–290. https://doi.org/10.1007/s40124-014-0056-6

Global Food Security Index (2022) Economist impact. https://impact.economist.com/sustainability/project/food-security-index/reports/Economist_Impact_GFSI_2022_Global_Report_Sep_2022.pdf

Henchion M, Moloney AP, Hyland J, Zimmerman J, McCarthy S (2021) Review: trends for meat, milk and egg consumption for the next decades and the role played by livestock systems in the global production of proteins. Animal 15(Suppl. 1):100287. https://doi.org/10.1016/j.animal.2021.100287

Karpyn AE, Riser D, Tracy T, Wang R, Shen Y (2020) The changing landscape of food deserts. UNSCN Nutr 44:46–53. Available at: https://pmc.ncbi.nlm.nih.gov/articles/PMC7299236/

Leroy F, Smith NK, Adesogan AT, Beal T, Iannotti L, Moughan PJ, Mann N (2023) The role of meat in the human diet: evolutionary aspects and nutritional value. Anim Front 13:11–18. https://doi.org/10.1093/af/vfac093

Richardson K, Steffen W, Lucht W, Bendtsen J et al (2023) Earth beyond six of nine planetary boundaries. Sci Adv 9:eadh2458. https://doi.org/10.1126/sciadv.adh2458

Rittel HWJ, Webber MM (1973) Dilemmas in a general theory of planning. Pol Sci 4(2):155–169. https://doi.org/10.1007/BF01405730

Selter F, Salloch S (2023) Whose health and which health? Two theoretical flaws in the One Health paradigm. Bioethics 37(7):674–682. https://doi.org/10.1111/bioe.13192

Smets V, Cant J, Vandevijvere S (2022) The changing landscape of food deserts and swamps over more than a decade in Flanders, Belgium. Int J Environ Res Public Health 19:13854. https://doi.org/10.3390/ijerph192113854

Thomas L (2022) The intersectional environmentalist: how to dismantle Systems of Oppression to protect people + planet. Hatchette Book Group, Inc, New York. ISBN: 1800812868, 9781800812864

UNEP (2024) Food Waste Index Report 2024. https://wedocs.unep.org/20.500.11822/45230

United Nations (2015) The Millennium Development Goals Report 2015. ISBN: 9789211013207, 9211013208

Wang-Erlandsson L, Tobian A, Ent R, Fetzer I, te Wierik S, Porkka M, Staal A, Jaramillo F, Dahlmann H, Singh C, Greve P, Gerten D, Keys P, Gleeson T, Cornell S, Steffen W, Bai X, Rockström J (2022) A planetary boundary for green water. Nature Reviews Earth & Environment. https://doi.org/10.1038/s43017-022-00287-8

Wang X, Dou Z, Feng S, Zhang Y, Ma L, Zou C et al (2023) Global food nutrients analysis reveals alarming gaps and daunting challenges. Nat Food 4:1007–1017. https://doi.org/10.1038/s43016-023-00851-5

World Economic Forum (2025) The global risks report 2025. 20th Edition. Insight report. Available at: https://reports.weforum.org/docs/WEF_Global_Risks_Report_2025.pdf

Zhu J, Luo Z, Sun T, Li W, Zhou W, Wang X et al (2023) Cradle-to-grave emissions from food loss and waste represent half of total greenhouse gas emissions from food systems. Nat Food 4:247–256. https://doi.org/10.1038/s43016-023-00710-3

Changing the Goal from Maximising "Financial" to Optimising "Sustainable" Return on Investment for the Future of Livestock and Food Production, People, Ecosystems, and the Planet

2

Gerald Shurson

1 The Current State of Our World

We did not create planet Earth, but somehow, we have forgotten that all life, including ours, is entirely dependent on the many resources that it provides. Unlike our Indigenous communities, we have become disconnected from nature. We have not respected the Earth by protecting its life-giving resources to ensure that we have clean air to breathe and clean water to drink. Instead, we have treated the Earth and exploited its resources much like how we operate machines in a factory, with a belief that we can somehow control and manipulate nature for our constant benefit. However, nature continues to prove us wrong. We have created an industrialised economy that is dependent on perpetual gross domestic product (GDP) economic growth, consumption of fossil fuels, and a "take, make, consume, and waste" society. Emissions from our addiction to burning fossil fuels as an unsustainable energy source have led to irreversible climate change. Our excessive materials consumption has also led us to the depletion of natural resources and pollution that has exceeded the Earth's carrying capacity and ability to regenerate. Our human-induced activities are threatening the continuation of life on Earth.

At the core of our current unlimited GDP growth and fossil fuel–dependent economy are a few multinational companies that have enormous wealth, power, and influence which they are using to control governments and avoid deviating from this economic model at any cost. This has led to a "tug-of-war" between capitalism and climate change policies, which have been thoughtfully discussed in recent books by Kate Raworth (Doughnut Economics; Raworth 2017), Jason Hickel (Less Is More; Hickel 2022), and Naomi Kline (This Changes Everything; Klein 2014) among

G. Shurson (✉)
Department of Animal Science, College of Food, Agriculture, and Natural Resource Sciences, University of Minnesota, St. Paul, MN, USA
e-mail: shurs001@umn.edu

B. Grabkowsky, T. Blaha (eds.), *Intensive Livestock Production in Transition*,
https://doi.org/10.1007/978-3-031-97872-2_2

others. Raworth (2017) and Klein (2014) have taught us that we need to remember and recognize that there are limits to Earth's resources and we need to respect them. They have also taught us that we need to stop treating nature as "ecosystem services" and once again, become agrarians and better connected with nature where we only take what we need and put back what we don't use to regenerate for the future. By increasing investments in renewable solar and wind energy, we reconnect with nature in a positive, nondestructive way, and reduce the abusive treatment of the Earth caused by emissions from fossil fuels and excessive extraction of resources that continue to dominate global government and corporate business decision-making. Unfortunately, the actions required to prevent further climate change and environmental degradation are contrary to the unregulated, perpetual GDP economic growth that our current capitalistic model needs to sustain itself. Klein (2014) reminded us that capitalism is organized with the imperative of constant expansion and growth leading to continually increasing levels of industrial production and consumption which is measured by GDP. The purpose of this economic mindset is to extract and accumulate increasing amounts of profit and wealth in every industry, in every sector, and in every national economy to grow all the time with no identifiable end point. Monasterolo (2020) suggested that the global financial system could help achieve climate targets if investments were aligned with sustainability, but climate-related financial risks are difficult to overcome using traditional economic and financial risk assessment approaches. Therefore, one of our greatest grand challenges is to resolve this "tug-of-war" between maximising financial return on investment and GDP growth versus optimizing sustainable return on investment to support climate change action, ecosystem health, and the creation of a safer, more equitable, and just operating space for humanity and the future of life on planet Earth.

Science has enabled the creation of technology that has increased the well-being of the lives of billions of people. Today, most people are healthier because of modern medicine, have access to abundant and affordable food because of intensive agriculture production systems, and are more educated, peaceful, and materially comfortable than at any time in history. Although the quality of life and life expectancy for billions of people has improved, especially in rich and powerful regions such as North America and Europe, we still live in a world that fails to meet the basic food, healthcare, employment, and equitable living condition needs of billions of people. Climate change is making this worse (Lenton et al. 2023). Without drastic changes in our growth-dependent economic model and reducing our reliance on fossil fuels, continued human population growth will make living conditions worse and further contribute toward exceeding the Earth's carrying capacity. Emerging technologies including artificial intelligence, quantum computing, robotics, and synthetic biology (genetic sequencing, modification, and synthesis) have the potential to dramatically improve our lives even more, in previously unimaginable ways. However, if these emerging powerful technologies are abused by bad actors, catastrophic consequences could occur. Suleyman and Bhaskar (2023) warn us that this "coming wave" of new technologies is going to require everyone working together along a narrow path that requires developing a governance system that will allow human societies to enjoy the enormous benefits of these technologies while controlling and preventing intentional abuse, and unintentional and unexpected consequences that could occur.

Our current economic system and governance models are not working. The fossil fuels industry, banks, and corporations have been given enormous power and control of our economy. Essential services like energy and water that are part of the public good have been privatized in most countries. Trust in democratic governments has declined dramatically and has led to an increase in populism, authoritarianism, and political instability. Increasing income inequality and concentration of wealth; inflation; stagnant education levels; declining working-age populations, public goods, and services; as well as increasing social inequalities have caused enormous strain on societies. All of these dynamics of our modern world have led to the tipping points of ecological, social, and financial collapse that must be addressed immediately (Klein 2014).

In a unified attempt to provide a global framework for improving environmental, social, and economic sustainability, the United Nations created 17 Sustainable Development Goals (SDGs) in 2012 to help guide the redesign of economies and reverse the devastation to societies and the environment that our current GDP economic growth model has created. Unfortunately, we have not implemented climate change actions fast enough, nor with enough impact to meet climate pledges (IPCC 2023). Continuing down the path of "business as usual" is not an option. Continuing to follow our current path with no comprehensive strategy, no clear deadlines, and a lack of courage, determination, and persistence to transform our current economic and governance system into a better one that overcomes these challenges will have devastating consequences. We need to become more impatient with the slow pace of making transformative changes.

Winning this "tug-of-war" is about much more than increasing investment in environmentally friendly practices and enacting government regulations to achieve sustainability goals. We are addicted to unsustainable bad behaviors and need to retrain our minds to think differently about the world we live in. All of us should be advocating for holistic approaches and implementing actions that not only reduce the abuse and restore Earth's valuable resources but also transform our economies so that we can rebuild a world that will enhance our social system and the human experience. We are at a critical point in history where we have a unique opportunity and an imminent need to create a new economy, energy system, democracy, and relationship to the planet and nature and to each other, while also achieving public governance of land, water, and food resources to create a thriving global society for all that is respectful of human and Indigenous rights and dignity (Fig. 2.1). In fact, Klein (2014) suggested that climate change could be the catalyzing force for positive change by rebuilding and reviving local economies, reclaiming democracies from corrosive corporate influence, blocking harmful new trade deals and rewriting old ones, investing in deteriorated public infrastructure such as mass transit and affordable housing, reclaiming ownership of essential public services like energy and water, improving agricultural systems to become healthier and regenerative, opening borders to migrants whose displacement is linked to climate change, and respecting Indigenous land rights to help end gross inequalities. This means that if

Fig. 2.1 This graphic illustrates the current global "tug-of-war" between continuing on an environmentally destructive and unsustainable path driven by GDP economic growth and a fossil fuel–dependent global economy (right side) versus a sustainable, renewable energy-based, degrowth, and "doughnut" economics-based economy (left side) that can enable 9+ billion people (surrounding the earth) to flourish while not exceeding the Earth's capacity to provide life-giving resources. The outcome of this "tug-of-war" (hands and arms pulling the rope in opposite directions symbolizing the current conflict between maintaining the status quo versus adopting a global sustainable food, agriculture, and livestock system) depends on our courage to change our addictive behaviors that have created our current unsustainable environmental, economic, and social crises and embrace a new economic model and sustainable way of thinking that includes achieving Sustainable Development Goals and One Health (center panel) of our global food, agriculture, and livestock production systems (center pull out view depicting a redesigned sustainable agriculture and food production system utilizing renewable wind and solar energy and agroecology practices)

we are cognizant and thoughtful about how we transform our global food systems, while identifying and considering the tradeoffs and potential unintended consequences associated with choices that are made, there is a tremendous opportunity to build a nontoxic, resilient economy where all basic human needs are met and opportunities to thrive exist.

1.1 We Need a New Economic Model and Sustainable Way of Thinking

Sustainable development has been defined as the harmonization of economic, social, and environmental components for the benefit of current generations without compromising the capabilities and opportunities of future generations (Neumayer 2003). New economic models, such as doughnut economics, have been proposed as a framework for beginning this transformation (Raworth 2017). This model consists of two concentric circles (doughnut shape) which include a social foundation to ensure that no one is denied access to the essentials of life, and an ecological ceiling to ensure that humanity does not exceed the planetary boundaries (regenerative capacity) that protect the Earth's life support systems (Rockström et al. 2023). Raworth (2017) proposed seven ways of thinking like a twenty-first-century economist that can help lead to a world where everyone prospers while living within the planetary boundaries of life's essential resources including (1) changing the goal, (2) seeing the big picture, (3) nurturing human nature, (4) getting savvy with systems, (5) designing to distribute, (6) creating to regenerate, and (7) being agnostic about growth.

The first step is to change the goal from economic GDP growth, as the primary measure of progress, to the goal of meeting the basic needs and rights of every person within the means of our planetary resources. In fact, Hickel (2022) proposed "degrowth" as an alternative economic model to refocus our economic system on more self-sufficiency. To do this, we need to become "systems thinkers" and begin seeing the big picture (Meadows 2008). This is not easy because it is much easier to think in a linear, reductionist, "cause and effect" way than to try to understand the many elements, interconnections, feedback loops, functions, and purposes that make up systems. The doughnut economics model places the economy embedded within society, and society under the Earth's umbrella of natural resources in a closed loop system where energy from the sun flows to it and materials cycle within it in a circular economy. The Earth is not only a source of finite resources but is also a depository for waste created by human activity, which has resulted in near depletion and degradation of natural resources and exceedance of the earth's regenerative and absorptive capacity (Raworth 2017).

We also need to find a way to create a more dynamic balance between environmental, social, and economic sustainability. This involves nurturing human nature (Raworth 2017). We all want to live in a world where we can thrive rather than survive, but our ability to thrive depends on social capital which is created by how we build norms, rules, and relationships that allow us to cooperate and depend on each other to meet our fundamental social needs of participation, leisure, protection, and belonging. The amount of trust and reciprocity that we create in our social interactions with others makes us better people, but our economy plays a major role in our relationships, public spirit, and distribution of wealth. Many of us need to develop a new way of thinking and behaving as individuals if we want to live in a society that allows us to thrive rather than survive. This means that we need to find ways to get better connected with nature, protect our environment, and improve our social relationships and communities by trusting, reciprocating, and cooperating to do things that we cannot do alone (Raworth 2017). To integrate the social and environmental aspects within the doughnut economy, a new design is needed for distributing income and wealth to overcome the problems caused by inequality and improve societal welfare and democracy, while minimizing ecological degradation and economic instability (Raworth 2017). Our economy must also become more circular, where we create food and products in a recycled system that enables the regeneration of resources (Raworth 2017). Other researchers have provided a more detailed framework for creating cost-effective circular economy pathways for agricultural, energy, transportation, waste management, and other industries using life cycle assessment (Yang et al. 2023a). Therefore, it doesn't matter if we are dealing with biological materials or other materials, our new economy should reward the use of renewable materials to produce and consume food, goods, and services, and minimize losses of matter and heat through processes of recycling, reusing, repairing, and regeneration.

Hickel (2022) proposed that if material consumption is reduced to sustainable levels in high-income nations, it would reduce pressure on ecosystems by reducing deforestation, habitat destruction, and loss of biodiversity. Reducing material consumption would also reduce energy use derived from fossil fuels and potentially accelerate the transition to renewable energy sources. Furthermore, a reduction in total energy consumption would result in the need for fewer solar panels, wind turbines, and batteries, which would reduce pressure on the need to extract required minerals from the Earth, and the negative impacts that this practice has on ecosystems of communities in these locations. In fact, Hickel (2022) suggested that reducing material and energy consumption can actually improve people's lives if we change the economic goal toward satisfying human needs rather than accumulating profit and wealth, which could also eliminate huge amounts of waste. Therefore, he proposed (1) ending planned obsolescence of goods, (2) ending advertising, (3) shifting ownership to usership, (4) ending food waste, (5) scaling down ecologically destructive industries such as fossil fuels, and (6) decommodification of public goods and expand the commons, which are all viable, potential solutions to achieve a healthy, more sustainable world. Achieving these outcomes requires following two key principles to stay in balance with the living world: (1) never extract more than ecosystems can generate and (2) never waste or pollute more than ecosystems can safely absorb. This requires putting limits on resource use and waste (Hickel 2022).

1.2 Environmental Regulations: Are We Following the Rules and Keeping Promises?

The current pace of reduction of greenhouse gas (GHG) emissions is far too slow, and we did not meet the global goal of preventing a 1.5 °C rise in temperature in 2023 (IPCC 2023). The difference between a global warming increase of 1.5 °C and 2.0 °C means that hundreds of millions of more people's lives are at risk from extreme weather events, and some ecosystems will be destroyed (https://www.cfr.org/backgrounder/paris-global-climate-change-agreements). A significant part of the slow action among all countries is due to the international "tug-of-war" occurring between countries due to the debate on "fairness," and the many questions that involve taking responsibility and keeping promises for taking meaningful climate action (Hormio 2023). Who is responsible? Who pays? Who is capable of paying? Who is affected the most? What is each country's fair share? What are the consequences of noncompliance? Some argue that the greatest GHG-emitting countries should be required to invest more financial resources and be expected to have more aggressive climate reduction agendas than others. In this case, of the estimated 53.8 billion tonnes of GHG produced in the world in 2022, China (29.2%) would be expected to invest the most and have the most aggressive climate action plan, followed by the United States (11.2%), India (7.3%), European Union (6.7%), Russia (4.8%), and Brazil (2.4%), which represent 61.6% of global GHG emissions, 50.1% of global population, 61.2% of GDP, and 63.4% of global fossil fuel consumption (Crippa et al. 2023). Wealthy nations in the European Union and the United States have made investments and implemented actions that have reduced emissions compared with China and India, which are the world's most populous nations and emerging economies, and where emissions are still increasing (IPCC 2023). However, when expressed on a per capita basis, the average United States citizen produced 17.9 tonnes of CO_2 (carbon dioxide) equivalent per year, similar to Russian citizens (18.0 tonnes), compared with 11.0 tonnes of CO_2 equiv. for citizens in China, with even lesser amounts for citizens in the European Union (8.1 tonnes), Brazil (6.1 tonnes), and India (2.8 tonnes), suggesting that the United States and Russia should be expected to invest more in climate change reduction practices than the other four greatest emitters (Crippa et al. 2023). Furthermore, governments from low-income countries believe that high-income countries have greater historical responsibility for climate change because they built their economies and wealth by creating this problem. Climate change will have more devastating impacts on low-income countries compared with high-income countries because they often have greater social problems, such as poverty and hunger, and will require greater investment and effort to build supply chain and climate resilience. Therefore, governments of low-income countries argue that they should be given the opportunity to develop industry and subsequent pollution levels to achieve growth and income that wealthy, high-emitting countries have already achieved.

Regardless of the metrics used, all countries have a collective responsibility as contributors to climate change to mitigate further emissions (Hormio 2023). Although all countries need to act on climate change, they have different histories and capabilities. The United States and European Union are frequently targeted as high emitters that have historically been responsible for current global warming and are now considered to be "in debt." India is on the other end of the spectrum because it has had much less historical contribution to the climate crisis and also has fewer resources and capabilities to contribute toward future mitigation. However, there are many ways to determine a country's fair share of emissions reductions and there is no single universal answer that can be implemented. Hickel (2020) proposed using an equality-based attribution approach to quantify national responsibility of CO_2 emissions in excess of planetary boundaries based on per capita access to the atmospheric commons. Using this approach, the United States, European Union, Russia, Japan, and Canada are responsible for 85% of excess global CO_2 emissions and would be expected to contribute the most to mitigation, with most countries in the Global South, including China and India, emitting within their boundary "fair shares" and would be expected to contribute less (Hickel 2020).

Ultimately, this unresolved international debate will require diplomacy for developing, implementing, and complying with government policies and regulations. This international debate should also be occurring for land use, water resources, and environmental impacts of N and P flows through global trade. However, because no country has met its climate pledge, it is reasonable to question the effectiveness of current climate regulations and government oversight. Sustainability governance regulations are emerging and will be important for providing direction and oversight of climate actions and natural resource use in the future. Three major programs have been established including the International Sustainability Standards Board (ISSB), US Security Exchange Commission (SEC) Directive, and the European Union Corporate Sustainability Reporting Directive (CSRD). The ISSB was established in 2020 at COP26 in Glasgow, Scotland, to address the fragmentation of voluntary, sustainability-related standards and requirements that add cost, complexity, and risk to companies and investors (https://www.ifrs.org/groups/international-sustainability-standards-board/). The objectives of ISSB are to (1) develop global standards and baselines for sustainability disclosures, (2) meet information needs of investors, (3) enable companies to provide comprehensive information to global capital markets, and (4) facilitate interoperability with disclosures that are specific to jurisdictions and aimed at broader groups of stakeholders. The US SEC (https://www.sec.gov/securities-topics/climate-esg) has proposed rule changes requiring companies to include certain climate-related disclosures in their registration statements and reports to provide investors with information related to climate risks affecting companies they own. The EU CSRD (https://finance.ec.europa.eu/capital-markets-union-and-financial-markets/company-reporting-and-auditing/company-reporting/corporate-sustainability-reporting_en) requires all large (more than 500 employees) public interest companies and listed companies to disclose information on the risks and opportunities from social and environmental issues and the impact of their activities on people and the environment. Specifically, new CSRD rules require large companies to publish information related to environmental matters, social matters and treatment of employees, respect for human rights, anti-corruption and bribery, and diversity of age, gender, educational, and professional background on company boards. Although these sustainability governance regulations are an initial step toward holding companies accountable for implementing sustainability measures, greater international governance coordination and efforts are needed in the global agricultural and food system.

1.3 Alternative Climate Mitigation Models Are Expensive, Unpredictable, and Dangerous

We have procrastinated too long, and the more we continue to avoid implementing meaningful changes to slow the rise in global temperature causing climate change, the worse the problems will become. In fact, some scientists and experts believe we have already waited too long (World Meteorological Organization 2023). We have at least two options for taking action to prevent further escalation of global warming. The first option is to directly target the root causes of climate change and loss of ecosystem health, along with addressing the many social well-being problems facing societies today. This can be accomplished by doing the difficult work of implementing doughnut economic and degrowth models to achieve greater long-term sustainability. The other option is to attempt to implement a "quick fix" of geoengineering, which has been proposed by supporters of the current perpetual GDP growth economic model and the fossil fuel industry. Geoengineering technologies were created in response to the need to find immediate solutions to "fix" our climate change problem without changing human behaviors that can reduce extraction, use, and emissions from fossil fuels. This option would involve investing billions of dollars in unproven, high-risk, and potentially dangerous geoengineering and carbon capture technologies with unpredictable and unknown consequences if they were to be implemented (Keith 2003; Reynolds 2019; Sovacool 2021). These technologies include emitting sulfur in atmospheric space to cool temperatures, cloud seeding to fix droughts, ocean fertilization to cope with acidification, and carbon-sucking machines to pull carbon from the air and transport it thousands of miles through pipelines to store it underground (Baur et al. 2023; Sovacool 2021; Zhang et al. 2015). Compared with creating a circular economy, carbon removal technologies are prohibitively expensive and range from $100 to $1200 US dollars per ton of CO_2 sequestered (Yang et al. 2023a). In contrast, shifting to an agricultural model based on agroecological practices would safely sequester large amounts of carbon, reduce emissions, and increase food security (Hawes et al. 2021) without the high-risk uncertainties. Therefore, adopting any of these geoengineering technologies would further limit our ability to change current behaviors and achieve a state of ecosystem and planetary health that requires no further intervention (Klein 2014). Implementing geoengineering "fixes" would lead to an outcome of never living on a healthy planet again because the Earth would be constantly reliant on the use of these technologies for the rest of its existence, and the potentially life-threatening unknown interactions and unintended consequences (Klein 2014). Let's choose the first option and redesign our agriculture and food systems model.

2 Agriculture in Transition

Agriculture is the foundation of our global food system and it affects people and the health of the planet in a multitude of ways. Schneider et al. (2023) developed a holistic food system framework for monitoring progress toward achieving health and sustainability goals of food system transformation that involve five major themes including (1) diet, nutrition, and health; (2) environment, natural resources, and production; (3) livelihoods, poverty, and equity; (4) governance; and (5) resilience, which were divided into 50 different indicator domains. Using this framework, they reported that every country evaluated had positive health and sustainability outcomes in some aspects of their food systems, with some indicators being independent of national income, and that each country aspires to achieve healthy, sustainable, and just food systems (Schneider et al. 2023).

However, another "tug-of-war" occurring in the agriculture and food sectors is between "doing business as usual," or at a minimum, "doing what pays," versus proactively implementing more aggressive climate smart and sustainable practices. We need to make a rapid behavioral transition from a mindset of focusing strictly on maximizing return on financial investment at any cost to an attitude of wanting to optimize sustainability return on investment. This requires taking actions that shift the balance toward improving environmental, social, and economic sustainability that benefits "us," rather than the current short-sighted focus of maximizing return on financial investment that benefits "me," with minimal interest or intent on contributing toward improving the environment and society. This requires reversing the "profits before people" culture that is inherent in numerous corporations, companies, and organisations around the world.

The OECD-FAO (2023) global agricultural outlook for the next 10 years (2023–2032) highlighted several trends relative to the future sustainability and resiliency of livestock, poultry, food production systems, people, ecosystems, and the planet, which are listed here and discussed in greater detail throughout this chapter:

- Economic growth is expected to be about 2.6% due to a decrease in China's population.
- Energy prices are expected to gradually increase up to 2032.
- Demand for biofuels (ethanol and biodiesel) is expected to slowly increase, with increased use of agricultural crops as feedstocks in India and Indonesia due to increased transport fuel use and increased biofuel blending requirements.
- Demand for biofuels is expected to decline in the European Union due to reduced transport fuel use and a shift toward using other renewable fuels.
- Overall global use of sugarcane and vegetable oils for biofuel production is expected to increase while maize use is expected to decrease.
- The Russian war against Ukraine has created uncertainties in global food, energy, and input prices.
- Growth in total agriculture production is expected to increase by 1.1% annually, mainly in middle- and low-income countries, assuming wide access to inputs and despite minimal change in technology investment, infrastructure, and training from previous years.
- Global consumption of food calories is projected to increase by 1.3% annually, but less than previously estimated due to a decrease in global population and per capita income growth.

- Global increases in crop production are expected to be driven primarily by increased yields (79%), followed by cropland expansion (15%), and increased cropping intensity (6%).
- Mineral fertilizer prices will likely increase, which will directly increase crop production costs and indirectly increase compound feed prices for swine and poultry, resulting in increased food prices for consumers.
- Rapid expansion and intensification of livestock production in low- and middle-income countries is expected to increase demand for animal feed.
- Slower growth in livestock and poultry production is expected in high- and some upper-middle-income countries, including China, and improved feed efficiency will likely reduce feed demand.
- The 1.3% projected annual growth in livestock, poultry, and fish production will be driven mainly by improvements in animal productivity, herd management, and feed utilization.
- Poultry production is expected to account for about half of the growth in meat production.
- Pig production is projected to increase if recovery from African swine fever is successful in East Asia.
- Milk production is expected to sharply increase due to increased growth and demand in India and Pakistan.
- Greenhouse gas (GHG) emissions from agriculture are expected to increase by 7.6%, which is less than the previous decade, but greater effort is needed to implement effective mitigation practices, especially in livestock production, which accounts for 80% of total GHG emissions from agriculture.
- More frequent and extreme weather events will require agricultural production systems to develop adaptation strategies for greater resilience.
- Global trade rebounded from disruptions during the COVID-19 pandemic and is expected to increase with increasing production if transparency and the rules-based multilateral trading system continue to function well.

2.1 Is Global Trade an Asset or Liability for Sustainability?

Globalization of economies and agriculture is driven by international trade and investment that connects people, businesses, and governments around the world (Robinson 2018). In 1995, the World Trade Organization was formed to provide international oversight for global trade, which is when the conflict between policing climate change and global trade began (Klein 2014). International trade of agricultural products is rarely considered when identifying major contributors to climate change, environmental pollution, and depletion or disruption of natural resource flows. The Russian invasion of Ukraine has been a stark reminder of the inherent dependencies on global trade for food security, the vulnerabilities and capabilities of countries to absorb supply chain disruptions and depletion, and the need to reevaluate open trade and the socioeconomic value of agriculture for achieving greater stability of food security in vulnerable countries (Hassen and Bilali 2022;

Dyson et al. 2023; Lin et al. 2023; Hellegers 2022). Global trade of agricultural products has increased by an average of 6% annually from 2000 to 2016, and 72% of total agricultural products exported were produced in ten countries or regions (i.e., European Union, United States, Brazil, China, Canada, Indonesia, Thailand, India, Australia, and Mexico) in 2018 (Balogh and Jámbor 2020). International trade has had a positive impact on reducing inequality by 25–33% around the world relative to the distribution of where food is produced (Carr et al. 2016). However, if international trade is going to continue to enable inclusive and sustainable production in food-deficit, low-income countries, policies and investments must focus on (1) diversifying production and markets to increase resilience from external disruptors, (2) increasing competitiveness and improving market access for local farmers, and (3) developing collective international action to incorporate externalities in international trade (Van Berkum 2021).

Limited studies have been conducted to assess the environmental impacts of international trade of agricultural and food products, but results from most of these studies indicate that international trade and trade liberalization have detrimental effects on the environment. Global trade increases pollution and degradation of natural resources caused by the associated increases in production, resource trade, transboundary pollution, transportation, and relocation of businesses to other countries to avoid environmental regulations (Balogh and Jámbor 2020). Global trade flows of agricultural products have adversely affected land systems that impact food and nutrient availability, natural habitat conversion, biodiversity loss, and ecosystem carbon storage (Kastner et al. 2021). The negative effects of international food trade on biodiversity loss associated with land use have resulted in an estimated loss of 17% in total species, with exports from Indonesia to the United States and China having the greatest impact (Chaudhary and Kastner 2016). Although global food production relies on unsustainable irrigation practices that deplete freshwater supplies, disrupt environmental flows, and impair aquatic ecosystems, the effects of trading agricultural products internationally are seldom recognised as an important contributor to water scarcity and sustainability considerations (Rosa et al. 2019). About 52% of global irrigation is unsustainable, 15% is virtually exported, and about 33% is associated with demand from export markets in several countries, which indicates that government policies are needed to achieve water and food security goals in the coming decades (Rosa et al. 2019). Understanding the impact of water usage on crop production can improve governance and increase the efficiency of water allocation and use (Falsetti et al. 2020). Global prices of wheat, maize, soybeans, and potatoes typically account for the quantities of water used in their production, but there are minimal, if any, embedded water use-price links for food products such as coffee, cocoa beans, tea, and vanilla (Falsetti et al. 2020).

Carbon dioxide emissions from production of globally traded goods and services are a significant contributor to the total CO_2 emissions, which doubled from 1996 to 2011, and shifted from Europe to Asia (Zhang et al. 2019a). Because global trade is a major contributor to total carbon emissions, it has been a contentious part of the discussion regarding which countries should accept greater responsibility for reducing carbon emissions to meet climate targets (Zhang et al. 2019a). Nitrogen (N) and phosphorus (P) flows and mass balance are also significantly affected by global trade of agricultural products (Chen et al. 2023). Population density, amount of animal-sourced food consumed, crop and animal productivity, N management, and manure recycling are the major factors that determine the relationships between trade, N balance, and utilization efficiency of imports, as well as exported food and feed (Wang et al. 2022a). Global trade of agricultural products has also contributed to changes in the global P cycle, which increased eightfold from 1961 to 2011, with 44% of global P traded in animal feed, 28% of P attributed to human food, and 28% of P in crops produced for other uses (Nesme et al. 2018). These results indicate that global trade policies need to be revisited to assess the environmental consequences of agricultural product flows as part of the much-needed agricultural and food system transformation.

Social sustainability is another important component of global agricultural trade that is rarely considered or evaluated, especially ethics, fairness, local resource scarcity, and fragility of ecosystems (Dalin and Rodríguez-Iturbe 2016). Global trade of agricultural products also has indirect effects on farmer displacement on marginal lands, which causes deforestation and soil erosion, encourages increased land use, and shifts crop production systems toward export crops in developing countries. Furthermore, international trade has led to overconsumption of foods relative to defined health and sustainable diet thresholds, indicating that agriculture imports are misaligned with dietary health and sustainability outcomes and suggests that current trade policies need to be modified to avoid these outcomes in the future (Mapes et al. 2022).

2.2 Does Foreign Land Ownership Contribute to Sustainability?

Agricultural land is an essential resource of our global food system but the "tug-of-war" over ownership or control of how land is used has profound effects on environmental, social, and economic sustainability. Foreign investment in agricultural land, sometimes called "land grabs," has become a controversial geopolitical subject in many countries and can affect local employment, change production systems, and increase food insecurity (Robinson 2018). Globally, about 90 million hectares of agricultural land have been acquired by foreign investors during the past two decades (Müller et al. 2021). Due to increasing global demand for agricultural products, these investments are considered to provide a means for acquiring modern agricultural technology that improves crop yields and subsequently provides economic development in target countries (D'Odorico et al. 2017). Most of the large-scale land acquisitions have occurred in countries and regions of sub-Saharan Africa, Asia, eastern Europe, and Latin America that want to increase profits by replacing traditional local farming systems with intensive, high-yielding commodity crop production (D'Odorico and Rulli 2013). However, the displacement and shift away from production of local staple crops threatens local food security because essential dietary nutrients are diverted away from local communities and toward the export market (Müller et al. 2021; Dogan 2022). Large-scale foreign land ownership often adversely alters the management of all types of landscapes that local farmers and communities depend upon, such as access and use of water resources, types of land cover, and often the production of nonfood crops (D'Odorico et al. 2017). In addition, land and water "grabbing" are a threat to sustainable development and negatively affect all 17 United Nations SDGs because of their detrimental effects on local communities, traditional land users, vulnerable Indigenous people, and the environment (Dell'Angelo et al. 2017). In most developing countries, governance of people and community access to land, water, fisheries, and forests is based on unwritten customs and practices rather than written policies and laws, which consequently can disrupt the social structure of local communities and create food insecurity and poverty under scenarios of foreign land ownership (Dogan 2022). Therefore, there is a significant need to develop and implement national and international governance measures that require responsible and sustainable investment in land and agricultural production systems in developing countries (Santangelo 2018; Dogan 2022).

2.3 Is the Production of Biofuels Part of the Sustainability Solution?

Another controversial use of agricultural land has been for the production of feedstocks, such as grains and oilseeds, for the production of biofuels to partially replace the use of fossil fuels as energy sources. This has led to a recurring "food versus fuel" debate that ebbs and flows depending on changing economic conditions and government policies. Because of the global significance and implications of the "food versus fuel" tug-of-war in the sustainable transformation of our energy, food production, and livestock systems, the current economy policies and subsidies that support the biofuels industry require a thorough review and rethinking.

Global energy consumption continues to increase despite indisputable scientific evidence that it is the greatest contributor to climate change. Global CO_2 emissions from energy consumption increased 0.8% in 2022 compared with 2021 and represent 87% of total global emissions (Energy Institute 2023). Significant increases (14.7%) in the production and use of renewable energy sources (i.e., solar, wind, ethanol, biodiesel) to replace fossil fuels are occurring in many countries, with the greatest increases in solar (24.9%) and wind (13.5%) energy compared with only a 3.4% increase in ethanol and biodiesel use worldwide (Energy Institute 2023). Therefore, despite the popularity of using ethanol and biodiesel as renewable energy sources, the magnitude of their contributions toward reducing GHG emissions from total energy consumption is minimal. Furthermore, although the increased production and use of all renewable energy sources are having a measurable impact on reducing CO_2 emissions, their impact is muted due to the continually escalating total global energy use. In fact, despite the clear and enormous detrimental impact of burning fossil fuels on climate change, fossil fuels continue to be the predominant form of energy among all major biofuel producing countries (i.e., United

States, Brazil, European Union, China). These facts lead to key questions that are at the core of the food versus fuel debate. Is the use of agricultural land and water to produce crops for biofuels a sustainable use of increasingly scarce natural resources? Is the production of biofuels environmentally sustainable? Should current government policies that are subsidizing and encouraging farmers to produce increasing amounts of feedstocks for biofuel production be channeled differently and toward production practices that have a greater impact on improving food security, sustainability, and reducing GHG emissions?

Only an average of about 5% ethanol is blended with gasoline globally (Ramsey et al. 2023). However, there is a wide disparity of ethanol use between Brazil and the United States. Brazil is the world's third largest consumer of gasoline and uses ethanol directly as fuel or at very high blending rates with gasoline compared with the United States, which consumes the most gasoline globally but only uses relatively low (10–15%) ethanol blend rates for most of the gasoline supply, and minimal amounts of 85% ethanol blends (Ramsey et al. 2023). Similarly, 2–20% of biodiesel and renewable diesel is blended with petroleum diesel, which represents a relatively small proportion of total energy consumption worldwide (OECD-FAO 2023). According to US Energy Information Administration, liquid biofuels (ethanol and biodiesel) contributed only 3% to total energy production and consumption worldwide in 2021 (https://www.eia.gov/international/data/world/biofuels/biofuels-consumption). Is this an appropriate sustainable return on investment? Why not reduce energy consumption along with using less environmentally invasive forms of renewable energy, such as wind and solar, instead of continuing to exploit land, water, and biodiversity, while relying on increasing amounts of fossil fuel derived N fertilizer and finite reserves of P fertilizer, to expand crop production for biofuel production?

Although it may seem that investing in biofuel production is a climate-smart use of natural resources and agricultural commodities, a closer look at the upstream and downstream effects of biofuel production does not indicate a good environmental sustainability return on investment. Agriculture is heavily dependent on energy from many sources, but unlike other industries, it also provides the feedstocks to produce biofuels. According to OECD/FAO (2023), about 60% of ethanol is produced from maize, 23% from sugarcane, 7% from molasses, 3% from wheat, and the remainder from other grains, cassava, or sugar beets, while about 70% of biodiesel is derived from palm oil (29%), soybean oil (23%), rapeseed oil (14%), and used cooking oils (25%). The United States is the largest producer of ethanol (48%) and the second largest producer of biodiesel (18%), while the European Union produces the most (32%) biodiesel globally (OECD/FAO 2023). These statistics are significant because about 40% of the annual maize crop produced in the United States is used to produce ethanol (Ramsey et al. 2023), and about 46% of the soybean oil produced in the United States is used to produce renewable diesel and other biofuels (https://www.ers.usda.gov/data-products/u-s-bioenergy-statistics/). Rapeseed oil, palm oil, and used cooking oil are used as feedstocks for biodiesel production in the European Union (OECD/FAO 2023). During the 1990s, when United States maize production exceeded demand, building a bioethanol industry was considered to be an appropriate way to utilise the excess maize, increase the price and economic value of maize production for farmers with the help of government subsidies, replace methyl tert-butyl ether (MTBE) as a safer oxygenation agent in gasoline, and reduce carbon emissions by blending ethanol with gasoline (Shurson 2017). However, the economic incentives to expand United States ethanol production led to increased maize production that required more land, more fertilizer, and more water. A similar scenario is now occurring with the rapid growth of renewable diesel industry in the United States, which is creating increased demand for fats and oils and encouraging more land use for soybean production (https://www.eia.gov/biofuels/renewable/capacity/).

Significant amounts of the land, water, fertilizer, and energy resources are used to produce the grain and oilseed feedstocks used in biofuel production, which must also be considered in the "tug-of-war" between their relatively small impact on global climate change mitigation, compared with their much greater magnitude of long-term impacts on land, water, and fertilizer use, and other environmental consequences associated with their production. We are not seeing the big picture of these broader environmental impacts from converting crops to biofuels. An extensive global infrastructure has been built for producing biofuels that require large amounts of valuable agricultural resources, without considering the unintended consequences and downstream effects on other equally important environmental considerations, including land use change and associated biodiversity loss, production and emissions of large amounts of CO_2 from ethanol fermentation, and inefficient nitrogen (N) and phosphorus (P) use from feeding maize coproducts to livestock and poultry.

Although the low ethanol blends with gasoline in the United States fuel market have contributed to a 544MMT reduction of CO_2 equivalent emissions from 2005 to 2019 (Lee et al. 2021b), estimates of the carbon intensity of maize ethanol production in the United States have often been underestimated (Spawn-Lee et al. 2021). Smith et al. (2017) compared CO_2 emissions associated with using maize for ethanol, beef, pork, and broiler production using a county-based, spatially explicit analysis of maize supply chain flows in the United States. Results from this analysis showed that total CO_2 emissions from maize used in ethanol production were 1.7–3.3 times greater than emissions from maize attributed to beef, pork, and broiler production. This is a remarkable finding because of the general prevailing assumption that maize used for ethanol production is good for the environment, and maize fed to animals is bad for the environment. In fact, results from several studies indicate that maize used to produce ethanol also has greater negative impacts on irrigated water use, water scarcity, and direct land use change than the many of environmental footprints associated with maize use as animal feed in livestock and poultry production.

Smith et al. (2017) compared embedded irrigation water use in maize production and supply chain flows between ethanol, beef, pork, and broiler sectors in the United States and reported that about 2.8 times more irrigated water use was attributed to maize supplied for ethanol production than for maize used in diets fed to animals in the pork and broiler industries. Using the same commodity flow model, Brauman et al. (2020) linked spatially distributed water resource use for maize and soybean production with individual ethanol and meat processing facilities to show unique water scarcity footprints and water risks inherent in ethanol production and meat supply chains in the United States. Results showed that 39% of irrigated maize used for ethanol production was attributed to five companies, which indicates significant risk based on the need to control access to scarce water resources and the potential for increased supply costs (Brauman et al. 2020).

The United States ethanol industry, along with the broiler industry, is also associated with the greatest overall amount of direct land use change from maize cropland expansion compared with the amount of land use to produce maize attributed to beef and pork production (Pelton et al. 2021). Similarly, soybean oil processing and the broiler industries were associated with the greatest amount of land use change associated with soybean cropland expansion among all commodity groups (Pelton et al. 2021). Like maize production, the increased land use change from the expansion of soybean production to meet oil demands for biodiesel and renewable diesel production is detrimental to the environment. In the United States and Europe, farmers were once required to "set aside" a certain percentage of their total planted acreage and devote it to approved conservation uses. Piñeiro et al. (2009) indicated that agricultural land set aside for conservation can be a better climate investment than producing maize for ethanol production because the increased demand for maize has contributed to the loss of Conservation Reserve Program (CRP) land, which are typically marginal lands that result in lower yields in the United States. Because much of the CRP land provides important habitat for grassland species (Weins et al. 2011), conversion of these lands to crop production results in less biodiversity and abundance of vertebrates in these habitats (Fletcher Jr. et al. 2010). In fact, Groom et al. (2008) suggested that using maize as a feedstock for ethanol production is the worst choice among the alternative feedstocks due to the consequential effects on carbon balance, land use, and biodiversity loss. Furthermore, landscapes used to produce first-generation crops, such as maize and soybeans that provide starch, sugar, and oil for use in biofuels production, have greater negative effects on biodiversity loss than second-generation crops that provide lignocellulose, wood-based substrates and residues as feedstocks for biofuel production (Núñez-Regueiro et al. 2021). Therefore, devoting large amounts of arable land for the purpose of producing crops for biofuel production not only displaces land available for producing crops consumed directly by humans and animals but it is also detrimental to biodiversity and ecosystem health around the world.

Proponents of the use of maize for ethanol production remind us that the residual biomass after the starch is fermented to ethanol and CO_2 is recycled into animal feed in the form of distiller grains (DG) coproducts and is a positive contribution to a circular agri-food economy. The United States ethanol industry produces about 37 million tons of DG coproducts annually. About one-third of this total is exported for use in animal feeds in more than 50 countries around the world, with the remaining 25 million tons consumed by beef cattle (45%), dairy cattle (33%), swine (15%), and poultry (6%) domestically (Renewable Fuels Association 2023). Although DG coproducts have become major components of livestock and poultry diets globally, they contribute to reduced N and P utilization efficiency in animal production systems. Feeding DG coproducts to livestock and poultry results in 1.1 million tons of total P excreted in manure annually in the United States, which is enough P to displace 12.5% of total inorganic P fertilizer used in crop production each year (Ruffatto et al. 2022). In fact, the amount of P in DG coproducts transferred from ethanol facilities to animal feeding operations was nearly two times that of the P in human waste before treatment, which indicates that maize ethanol biorefineries are a key leverage point for improving national P management in the United States (Ruffatto et al. 2023). High dietary inclusion rates and suboptimal protein quality of maize coproducts also cause increased N excretion in manure compared with feeding conventional diets without DG for all species (Lim et al. 2023). Furthermore, the life cycle assessment (LCA) of swine feeding programs indicates that feeding diets containing DG to growing-finishing pigs resulted in greater climate change, acidification, marine and freshwater eutrophication, water use, and fossil resource use, but less land use, compared with feeding maize-soybean meal diets (Yang et al. 2023b). Therefore, the use of maize coproducts in circular livestock systems does not contribute toward reducing the environmental footprint of livestock and poultry production systems except for land use.

Crude glycerol is the primary by-product comprising 10% of biodiesel production, but unlike DG coproducts, it has low economic value due to several impurities (Ayoub and Abdullah 2012). About 1.2 million tons of crude glycerol is produced annually in the United States (Ayoub and Abdullah 2012), and this amount is expected to increase because of the expanding renewable diesel industry. However, because crude glycerol contains several impurities including salt, methanol, and free fatty acids, high purification costs, and a saturated market, its use in a circular economy has been economically challenging (Chozhavendhan et al. 2018). Despite the relatively high energy value of crude glycerol in swine and poultry diets, variable composition, high salt concentrations, and feed safety concerns involving methanol content have prevented its use in animal feeds (Kerr et al. 2011). Furthermore, the international crude glycerol market is unstable and is dependent on demand for conventional and new value-added chemical uses (Ayoub and Abdullah 2012). Potential uses of unrefined crude glycerol have been proposed and include production of various chemical products, hydrogen, fuel additives, development of fuel cells, ethanol or methanol production, animal feed, co-digestion for biogas production, and as a source of organic carbon for denitrification of wastewater (Leoneti et al. 2012). However, these potential uses require more development before they can become viable contributors to a circular bioeconomy. Therefore, crude glycerol as a by-product from biodiesel production does not contribute toward reducing the environmental footprint of agriculture and food production.

Another important part of the environmental cost of bioethanol production is the enormous amount of CO_2 emissions that occur during the fermentation process of converting starch from maize into ethanol and DG coproducts. This CO_2 is not captured to prevent release into the atmosphere and, therefore, contributes toward increasing rather than reducing GHG emissions associated with ethanol production. Of the 60 billion liters of ethanol that were produced in 220 ethanol facilities in the

United States in 2018, almost all of the 50 million tons of CO_2 generated during the fermentation process was emitted into the atmosphere (Lee et al. 2021a). In general, about 8.6 kg of CO_2 is produced per bushel (25.4 kg) of maize that is fermented to produce ethanol. In fact, 100 g of glucose from maize or sugarcane feedstocks will produce about the same amount of ethanol (51.1 g) as CO_2 (48.9 g; Badger 2002). Unfortunately, CO_2 emissions generated during ethanol production have historically been ignored as part of the discussion and consideration of the benefits of using ethanol in gasoline to reduce carbon emissions. Most ethanol production companies in the United States do not capture CO_2 emitted from their facilities, but recently, a few new businesses have unsuccessfully attempted to obtain environmental permits to collect CO_2 emissions from several ethanol facilities, transport it via pipelines across five Midwestern intensive agriculture states, and bury the CO_2 deep underground. This is a similar concept used by some petroleum companies, where CO_2 emissions are collected from production facilities, but the CO_2 is injected underground to release more oil and gas between rocks. During the past few decades, carbon capture technologies have emerged and have been promoted as a way to combat climate change. These technologies are being developed using federal tax incentives and billions of United States government dollars. It is understandable that carbon capture technology, transport pipelines, and sequestration underground would be supported by the renewable fuels industry and farming organisations to reduce the carbon intensity score to meet environmental impact requirements in some markets. It is also understandable that this type of approach for mitigating carbon emissions is opposed by many landowners and environmental groups who question its safety and effectiveness. Current costs are \$500 to \$600/ton to remove CO_2 by direct air capture, and facilities using this technology can obtain a tax credit of \$50/ton, which may increase to \$180 per ton in the future depending on political motivations. For now, this tug-of-war has led to the denial of permits to build pipelines for the purpose of collecting, transporting, and burying CO_2 from many ethanol facilities in these five Midwestern states. Again, we need to ask the question "Does this economic model provide an appropriate sustainable return on investment for using maize to produce bioethanol based on the environmental costs versus the benefits associated with it?"

Lastly, climate change is often considered to be a consequence of a market failure that can be corrected by assigning a price to carbon to penalize carbon-intensive activities and the social costs associated with it (Rosenbloom et al. 2020). The creation of carbon markets and carbon trading is another financial mechanism that enables emitters to pay non-emitters or low emitters to absorb their emissions and continue with "business as usual." Nothing changes. There is minimal, if any, net reduction in overall emissions, just financial transactions to redistribute carbon emissions ownership (Pearse and Böhm 2014). Furthermore, trading of carbon credits results in socially unjust consequences and inequities along with emissions of co-pollutants that affect people and places in ways not accounted for by this market model (Lejano et al. 2020). Therefore, carbon pricing is not an adequate approach to mitigate climate change (Rosenbloom et al. 2020). Although, Pearse and Böhm (2014) described ten reasons why carbon markets will not bring about substantial reductions in carbon emissions, Miltenberger et al. (2021) argued that despite the current flaws of voluntary carbon markets, they may be a first step for enabling the transition to reorganise global economies by internalizing market-based climate action solutions.

3 It All Comes Down to Us: Our Addiction to Bad Behavior and Whether We Have the Courage to Change

We are addicted to bad behaviors that continue to prevent us from making meaningful changes to redesign and improve the world we live in. We are trapped in a tug-of-war between the mindset of needing unregulated, perpetual economic growth focused on maximizing financial return on investment versus actions to address climate change, sustainability, One Health, and improve human well-being and justice for the future of life on planet Earth. Some critics of sustainability have suggested that the term is “too boring” to gain public attention, “too vague” to provide guidance, or “too late” to address the world’s environmental problems (Dernbach and Cheever 2015). Other critics argue that economic growth and technological innovations will solve our environmental problems (Asafu-Adjaye et al. 2015), while others believe that sustainability is being used as a means to promote socialist and globalist agendas (Fajack 2014). There is also extensive “greenwashing” in all sectors of the economy where sustainability language is used by companies and businesses to make unsubstantiated or misleading claims of using environmentally friendly practices that are really just a disguise for “business as usual” (Ramus and Montiel 2005; Cherry and Sneirson 2012; Lyon and Montgomery 2015). However, it is abundantly clear that our current economic model is not providing the basic needs of the vast majority of people in the world. Climate change, poverty, social inequities, and biodiversity loss have continued to threaten life on Earth despite efforts to implement sustainability practices around the world (IPCC 2023; IPBES 2019). This indicates that we need to change our behavior and economic model. Can we change our goal to focus on upending social inequalities and poverty to improve the well-being of all of human society while preserving and restoring the wealth of our planetary resources?

It is inherently difficult to change our current thinking and understanding of the numerous and complex sustainability challenges and potential solutions toward a more holistic worldview. Thinking about sustainability requires a major shift to a new, complex, multidimensional paradigm that requires learning new things and exploring unfamiliar areas. To do this, Loring (2020) proposed the use of threshold concepts, which can be used as a starting point to provide learners with new ways of understanding the world. Threshold concepts can cause a significant shift in learner understanding; enable remembering a new perspective; expose previously hidden connections and help explain them in new ways; define limits of a perspective; and stimulate debate because they deviate from assumptions and reject or redraw boundaries of an existing paradigm (Meyer and Land 2005, 2006). Five complementary and linked threshold concepts have been proposed for sustainability thinking which include complexity, collaborative institutions, ways of knowing, no panaceas, and adaptation (Loring 2020). Using these threshold concepts to develop a new sustainable way of thinking will enable us to modify our actions to potentially reduce the abuse and exploitation of Earth's valuable resources, and restore them to safe operating spaces, while also transforming our economy to allow us to rebuild a world that will enhance the human experience.

Rates of adoption of sustainability practices vary among different countries, sectors, commodity groups, stages of supply chains, product and service providers, competitors, and farmers. Farmers are generally very risk averse and are reluctant to adopt new technology unless they are convinced there is a low risk of failure and significant potential for economic benefit. Much of the relative risk aversion among different types of people can be described as categories of early adopters (those who make things happen), early and late majority adopters (those who watch what happens), and laggards (those who wonder what happened). In the globally competitive business world driven by profitability and financial return on investment, decisions are usually made to meet financial targets to satisfy shareholders rather than the broader good to society. Corporations were not designed to be sustainable, and any investments made in sustainability must be justified to shareholders. Likewise, human lifestyles were not designed to be sustainable, and humans generally do not change behavior unless they are forced to do so.

Therefore, if we are going to be successful in transitioning to a new economy and a new sustainable agricultural and food system, we need to reframe our thinking to find a balance between our current industrial-based food systems and ways of reconnecting with nature to create more nature-based food systems (Loring 2023). Our current industrial-based food system is dominated by the goals of standardisation, consolidation, profit, growth, and increased material use and efficiency (Giampietro 2023; Müller et al. 2021; Thompson 2017). These goals are often inconsistent with social sustainability goals focused on diversity, equity and inclusion, anti-racism, indigenisation, and sovereignty (Carlisle 2022; Figueroa-Helland et al. 2018; Thompson 2017). Loring (2023) suggested that to increase the rate of implementing climate smart and sustainable practices in agriculture and food systems, we need to replace old industrial terminology of scalability, standardization, efficiency, and control, with nature-based terms of regenerative design, Indigenous stewardship, adaptation, and diversity when framing and redesigning our food system, which is summarised in Table 2.1.

3.1 We Urgently Need Human Resources with Knowledge and Experience in Sustainability

Although we have known for several decades that dramatic changes were needed in society, in businesses, and in agricultural and food systems to avert climate change and resource depletion, we have not adequately educated and prepared decision-makers in government, agribusiness, and the farming community to understand the complexities, tradeoffs, interrelationships, unintended consequences, and potential

Table 2.1 Comparative summary of differences of key features between industrial- and ecosystem-based food systems (Loring 2023)

Features of industrial systems	Features of ecosystems
Centralization, specialization, and control of inputs, processes, and environment to produce the same things consistently and efficiently over time	Not uniform nor standardized; order is determined by localised processes of learning, adaptation, and emergent self-organization; problems are solved by trial and error; diversification and coalescence into new ecological communities increases productivity and resilience; always in flux to fill niches, respond to change, and cycle energy from usable to useless forms
Closed or specialized circular systems	Not circular, closed, or specialized but cyclical, open, and perpetually evolving at multiple scales
Change is driven by innovation to create new ideas, improvements, or deviations from existing norms and practices that emerge out of necessity or by originality which occurs by systemic learning; fitness of the system is improved by accelerating or creating new niches for growth and development	Change driven by adaptation, which is based on systemic learning by biological organisms and ecosystems to accelerate or create new niches for growth and development and improve the fitness of the system
Innovation adoption occurs through processes where technology is offered by creators and is gradually applied over time by new practitioners depending on economic incentives, imperfect markets, barriers to learning, and risk aversion	Adaptations spread because of the need for increased survival and well-being of communities, which are not inherently helpful or necessarily in the right direction, and can become maladaptive and die off or be abandoned; future relevance is always in question, being tested, and renegotiated
Scalability is the ability of a system to handle an increasing amount of work, determines whether it gains or loses efficiency, creates uniformity in isolation from externalities, and prioritises outcomes across places	Fitness of an adaptation is not determined by scale, but rather by how well it coexists relative to the diversity around it in the ecosystem; coexistence is interactive, interdependent, and focused on individual solutions, as well as synergies and relationships among diverse collections
Uniformity to achieve efficiency and stability is done by developing highly standardized monocultural production systems using large plots of land and large equipment to produce a few staple crops to keep land prices and food prices stable, and can be consumed all year through global supply chains; the consequences are decreased nutritional value, depleted soil fertility, increased chemical inputs, loss of biodiversity, and decline in prosperity and well-being of people in the low- and middle-income countries	Diversity maximizes productivity, not uniformity; uniformity and stability are not naturally occurring in living systems but rather involve a series of cycles on multiple scales of fast and slow release and renewal, growth and decline, and predator-prey cycles in ecological communities with continual movements of matter and energy from simple to complex forms; uniformity and standardization are dead ends in evolution

(continued)

Table 2.1 (continued)

Features of industrial systems	Features of ecosystems
Stability is the goal for achieving resilience based on consistency and a static level of production every year	Resilience is the ability to recover rapidly from disturbances and is dependent on diversity and achieving a more dynamic form of stability over time
Science-driven technologies without diligent consideration of potential unintended consequences	Rejection of extractive and oppressive systems used to develop and implement technologies
Efficiency is achieved by using scalability and uniformity to predictably produce more with less resources in linear, additive production systems; behavior and performance of the system can be predicted by analysing the individual parts	Generativity is the capacity to grow, self-organise, and create new forms by the amount taken and given back to the system to flourish; outcomes in complex systems are not linear and additive but circular, emergent, and novel
Disconnected with natural cycles and human and ecosystem communities	Regenerative and agroecological design and indigenous stewardship confer human resilience to the rest of the wild community

benefits of implementing more sustainable practices. Because of the immense urgency to address climate change, inexperienced leaders have been rather suddenly expected to adapt to the rapidly changing business and sustainability environment. There is an incredibly high demand for students trained in environmental, social, and governance (ESG) standards, which are used by companies to manage their environmental and social impacts. Unfortunately, university training and degree programs have been slow to adapt to these educational needs, and continue to produce graduates who are too specialized and do not have the systems thinking skills necessary to meet these new needs of businesses in the agriculture and food sectors. As a result, most chief sustainability officers evolve from various types of specialised training and technical backgrounds and must rapidly learn to become systems thinkers and knowledgeable in many new domains. These business leaders must shift their focus to learning everything nonfinancial including a new vocabulary; sustainability metrics; understanding current and emerging regulatory requirements; navigating political division between centralization vs. decentralisation; becoming aware of existing and new sustainability technologies and programs; assessing the relative maturity of sustainable practice adoption and compliance with other partners upstream and downstream in supply chains; and attempting to rapidly increase involvement in secondary markets, emerging opportunities, and alliances. As a result, these decision-makers are often overwhelmed with the need to monitor and understand numerous social and environmental factors, summarised in Table 2.2, to guide the businesses they serve. Therefore, the rate, direction, and magnitude of improvements in sustainability of agricultural and food systems have been, and will continue to depend on the desire, training, and experience of business leaders and decision-makers at all levels of the supply chains including farmers, agri-business advisors and professionals, and government policymakers.

Table 2.2 Examples of the variety of topic domains that are part of decision-making of successful chief sustainability officers in agricultural and food businesses

Global interconnectivity	Corporate purpose
Global conflicts	Climate change
Human migration	Employees and equity
Democracy under threat	Human rights
Income inequality	Land and soil management
Artificial intelligence	Deforestation
Synthetic biology	Ecosystem health
Regulations	Regenerative agriculture
Activists	Water quality and availability
Supply chains	Oceans
Input sourcing	Waste
Government policies	One Health

3.2 We Must Harness the Insights Gained from Indigenous Knowledge

Indigenous people have discovered, shared, and applied information and knowledge about nature, interactions with biophysical and ecological processes, landscapes, ecosystems, and species for millions of years before scientific inquiry was introduced. Years of observing flourishing natural ecosystems led Indigenous peoples to develop permaculture, which is an approach to land management and agricultural ecosystems intended to be sustainable and self-sufficient (Hernandez 2022). However, Bill Mollison (Permaculture: A Designer's Manual, 1997) is wrongly credited as the pioneer and the Western scientist who introduced permaculture to global agriculture (Hernandez 2022). Indigenous knowledge has made significant contributions to the scientific literature to improve our understanding of ecology, evolution, physiology, and applied ecology, and has important distinctions from "Western" science in motivation and approach, but it also shares similar and complementary conceptual themes that are beneficial (Jessen et al. 2021). Indigenous knowledge is a critical component of sustainable transformation of our global food system, but it has been marginalized in science, practice, and government policy (Vijayan et al. 2022). Furthermore, Indigenous knowledge of permaculture and agroecology is often ignored or dismissed in educational programs (Massicotte and Kelly-Bisson 2019). Due to the limited impact that current climate change mitigation, as well as water, land, and biodiversity management strategies have had, we must begin applying Indigenous knowledge and technologies that can provide important solutions for overcoming these challenges (Imoro et al. 2021). Therefore, there is tremendous benefit to engage Indigenous communities and utilise their novel insights, historical observations, interactions, and experimentation in redesigning a more sustainable food system.

Social sustainability of Indigenous people must also become an urgent priority. Indigenous people live on all human-inhabited continents, and their territories comprise a significant proportion of the Earth's surface (IPBES 2019). However, indigenous populations inhabit a remarkably small proportion of land that they previously occupied because governments have forced them out of ancestral lands to live on constrained and controlled reservations (Virtanen et al. 2020). The process of genocide of Indigenous people began more than five centuries ago in the Americas and has spread globally. Current international, national, and local governing bodies and laws continue to fail Indigenous people and all life around the world (McGregor et al. 2020). Treaty violations, battles for food and water security, and protection of culturally and religiously sacred sites have resulted in Indigenous resistance to government and corporate invasion of their lands for extractive energy projects (Hernandez et al. 2022), and consequently have violated every dimension of social sustainability. It is encouraging that reparations are beginning to emerge (Gilio-Whitaker 2019), and restoration strategies are being widely deployed by environmental nongovernment organisations and government agencies (Grenz and Armstrong 2023) that can lead to Indigenous environmental justice.

Climate change is projected to have greater detrimental effects on Indigenous populations than on non-Indigenous populations (Vinyeta and Lynn 2013). In the United States, traditional ecological knowledge has led Native American tribes to implement culturally appropriate strategies to adapt to climate change (Vinyeta and Lynn 2013). Traditional ecological knowledge systems of Indigenous people have also enabled conservation of about 80% of the biodiversity in the world (Nitah 2021). In addition, recycling and reuse practices among Indigenous people in northern Russia ensure that nothing goes to waste (Siragusa and Arzyutov 2020). These culturally and historically embedded behaviors and practices can serve as an excellent model for guiding the development and implementation of new sustainable practices in agriculture. Working with Indigenous people and their knowledge is essential for assessing and building linkages between nature and people (Hill et al. 2020), and Vijayan et al. (2022) proposed seven approaches for including Indigenous food systems knowledge in global sustainability transformations. Indigenous people define the main dimensions of sustainability to be context-based relationality involving community-based governance, education, language, quality of life and health, and communal recognition of certain nonhumans as givers of life (Fernández-Llamazares and Virtanen 2020; Virtanen et al. 2020). Part of the context-based relationality is the capacity of a community to create and maintain communal existence through management of local natural resources in a way that ensures survival and connectedness of members of the community with their environment (Kealiikanakaoleohaililan and Giardina 2016). Although these core components are similar to the UN Sustainable Development Goals, they are not the same due to conceptual and cultural differences among people and communities. However, these practices match the principles of doughnut economics perfectly.

3.3 Who Will Produce Our Food?

Farmers and farm workers are the essential foundation of our global food system. Small farms (20 hectares or less) produce more than 75% of most food commodities in sub-Saharan Africa, southeast Asia, south Asia, and China, but much larger (> 50 hectares) and more intensive farms dominate production in North America, South America, Australia, and New Zealand which contribute 75–100% of all cereal grains, livestock, and fruit production to the global economy (Herrero et al. 2017). There is a wide range in farm size and types across various geographic regions. In general, as farm size increases, the diversity of agriculture and nutrient production decreases, but some areas of the world with greater agricultural diversity produce more nutrients regardless of farm size (Herrero et al. 2017). Achieving economic viability of farming is an ongoing global problem but it is essential for sustainable agriculture and food systems. Giller et al. (2021a) identified several challenges that will determine the direction of evolution of farming systems which include (1) changes in patterns of land ownership, rental, and exchange, (2) declining employment and rural depopulation, (3) implementation of more environmentally sustainable practices, (4) becoming less dependent on external inputs, (5) development of more regionalised production systems, (6) increased resilience by moving away from specialization and moving toward recoupling arable land and livestock farming, (7) slowing or reversing the rate and scale of farm expansion, and (8) decreasingly limited possibilities for smallholders to significantly contribute to agricultural growth. As a result, different types of farmers are emerging in our global food system that include (1) adaptive farmers, (2) corporate farmers, (3) intensive farmers, (4) patrimonial farmers, (5) controlled environment farmers, (6) cell farmers, (7) social care farmers, (8) lifestyle farmers, (9) regenerative farmers, (10) urban farmers, (11) serious hobby farmers, and (12) community-provisioning farmers (Table 2.3; Bock et al. 2020). These different types of farmers have very diverse ideas and goals of what farming should be, and consequently they influence government policy, supply chains, and contribute toward offering a wider variety of food choices to consumers.

3.3.1 We Need to Stop Exploiting Essential Immigrant Workers

Legal and illegal immigrant workers are another essential human resource on crop and livestock farms. These workers play critical roles as hired farm workers, truck drivers, employees in meatpacking plants and dairy processing facilities, and constitute a high proportion of the workforce in high-income countries in Europe as well as in the United States. The declining number of farmers and farm workers is a major social and economic sustainability concern among countries in the European Union (Nori 2017). More than 1.3 million native European Union farmers and farm workers left agriculture between 2011 and 2017 (Palumbo et al. 2022). During this same time period, the number of people migrating and seeking asylum and safety increased in all European Union countries but varied among them (Henrekson et al. 2019). Therefore, the reduction in native farmers and farm workers was partially offset by the entry of 585,000 European Union citizen migrants and 837,000 non-European Union migrants into agriculture (Palumbo et al. 2022). However, the integration of immigrants into the labor market creates national and local socioeconomic cohesion problems in all countries, including Sweden, which has one of the most extensive labor market integration policies and structures for accommodating large

Table 2.3 Emerging types of farmers with different perspectives of how food should be produced (Bock et al. 2020)

Farmer type	Characteristics
Adaptive farmer	Make best use of all farm resources to provide a diverse range of goods and services based on demand
Corporate farmer	Trained agronomists and professional managers with diverse skills in agricultural production who manage a portion of production supply needed by large food corporations
Intensive farmer	Independent, innovative, environmentally conscious precision farmer focused on using the best available technologies and processes to efficiently produce commodity products for the region and world
Patrimonial farmer	Do not have the motivation or means to change business models and practices to adapt to new demands; may have limited access to knowledge, financing, or alternative supply chains; or lack employment opportunities outside of agriculture resulting in low expectations and a desire to maintain traditional activities
Controlled environment farmer	Independent, indoor farmer in urban, peri-urban, and some rural areas where there is high consumer demand throughout the year for locally sourced foods that are part of healthy diets and contribute to circularity and food security
Cell farmer	Cell farmers use advanced technology to grow animal-derived muscle cellsor microorganisms in cell cultures using bioreactors that require constant innovation and risk to contribute to food security but avoid animal welfare concerns
Social care farmer	Social farmers feel a strong social responsibility, enjoy working with people, and combine interests in agriculture, nature, and environmentally sustainable approaches to contribute to human well-being
Lifestyle farmer	Working-aged people and retirees who move to rural areas for a lifestyle change away from declining quality of living in an urban environment and to improve personal well-being
Regenerative farmer	Farmers who are motivated by a holistic view of reconnecting humans and agriculture as part of natural ecosystems and prioritize environmental sustainability outcomes over productivity
Urban farmer	Farmers of varied backgrounds and motivations ranging from newcomers, trained farmers who desire living in the city, or part of city networks of local food producers interested in contributing to food security, healthy diets, and improving city green spaces, micro-climates, biodiversity, organic recycling, and social initiatives
Serious hobby farmer	Farmers who enjoy working with soil, plants, and animals without expecting income from their activities and are motivated by a farm inheritance, keeping traditions, need for better work-life balance, and experimenting with different practices for improvement
Community provisioning farmer	Citizen farmers in urban and rural areas who are motivated produce a large amount of healthy food for their own consumption

numbers of immigrants into its workforce among all countries in the European Union (Barth and Zalkat 2023). Quickly gaining employment in their new country of residence is one of the best ways for migrants to achieve socioeconomic integration (Auer et al. 2018). Fortunately, immigrant agricultural workers have been valued as important sustainability assets for agriculture and rural communities in the European Union (Lähdesmäki and Suutari 2020), and agricultural employers eagerly hire immigrant workers to fill this labor deficit (Findlay and McCollum 2013).

The United States has some of the largest and most labor-intensive farms in the world, and hired labor represents about 60% of employment on these farms (Martin 2015). Foreign-born workers are an essential part of the labor force in United States agriculture, especially in the livestock and dairy industries. Workers in the food animal industry include those who work on ranches and farms as breeders and animal caretakers, in meatpacking facilities that slaughter and process meat, as trucker drivers who transport animals and products, and as dairy workers who are animal caretakers or process milk for distribution. The lack of enough workers to meet demand in the meat and dairy industries in the United States is a significant sustainability challenge. The main causes of this labor shortage are an increasing number of farmers and farm workers reaching retirement age and reduced labor participation rates since the COVID-19 pandemic (American Immigration Council 2022). The American Immigration Council (2022) summarised the current status of immigrant labor on livestock and dairy farms where (1) nearly 30% of mostly US-born (89%) workers on livestock farms will reach retirement age in the next decade, (2) the increased demand for labor has led to a 33.7% increase in wages (2019 to 2022) that employers must pay to attract workers, (3) increased wages have resulted in a 4.5 to 5.5% increase in consumer meat prices and a 6 to 7% increase in dairy prices, (4) between 20% and 51% (Adcock et al. 2015) of workers on dairy farms were immigrants, (5) 20% of workers on livestock farms were immigrants, (6) 45% of workers in meat packing plants were immigrants, and (7) 26% of transportation workers in the meat industry are foreign-born. These statistics indicate the essential role of immigrant workers in ensuring the sustainability of livestock and poultry production in the United States.

However, despite the essentiality of immigrant labor in agriculture and livestock systems in the United States, the sustainability of this labor force is threatened by immigration reforms that make it more difficult for migrants to enter the United States and gain employment on farms (Martin 2015). The United States is a nation of immigrants, but it has struggled for decades to find solutions to manage illegal immigration (Martin and Midgley 2006). Immigration policy designed to limit legal and illegal immigrants into the United States has been used to convince citizens that it will benefit domestic workers when there is little evidence to indicate that this has occurred or will occur in the future (Champlin and Hake 2006). Furthermore, the current focus on illegal immigration and border enforcement reduces bargaining power for immigrant workers, which provides direct economic benefits for industries such as meat packing and processing that freely hire unauthorised immigrants (Champlin and Hake 2006). More than 50% of hired workers on farms in the United States have been unauthorised since the mid-1990s, and almost all new migrant farm workers are unauthorized (Martin 2015). Employers on farms that rely on immigrant workers want new government programs that make it easier to hire foreign workers. However, there must also be better recruitment, housing, and wage protection for immigrant workers to ensure that they are not exploited, while also ensuring they are not preferred over US-born workers (Martin 2015). Without significant immigration reform, the amount of labor needed to support specialty crop, livestock, and dairy production will not be met; domestic supply of food will be reduced causing increased reliance on imports; consumer food prices will increase; other jobs in sales, marketing, and transportation of food supply chains will be affected; and the global competitiveness of United States agriculture will be reduced (Mercier 2014). Therefore, major efforts are needed to improve social sustainability of immigrant farm workers in global food animal production, transport, and processing systems in the United States.

3.3.2 Forced Labor Is Modern Slavery That Must End

Elimination of forced labor in food systems is a high priority among Sustainable Development Goals. Forced labor, also known as modern slavery and severe labor exploitation, is a significant problem involving more than 50 million people around the world who work in labor-intensive sectors such as agriculture in low-, middle-, and high-income countries (Kunz et al. 2023). This is a lesser-known, hidden issue within global food supply chains that rarely is exposed when consumers make food purchasing decisions about how their food was produced, and is in direct contrast with consumer expectations for meeting animal welfare conditions in the production of meat, milk, and eggs. Forced labor involves situations where people are coerced to work using violence or intimidation, accumulated debt, retention of identity papers, or threats of being reported to immigration authorities (International Labour Organization 2014). Illegal labor trafficking operations continue to thrive in the United States despite the passage of laws intended to prevent them (Childress et al. 2023). In fact, meat, poultry, and egg supply chains in the United States had the highest risk of forced labor among all food commodities, and 62% of forced labor risk in the land-based food supply was attributed to the domestic production or processing sectors (Blackstone et al. 2023). Another major labor concern in agriculture in the United States is the unregulated involvement of youth in farming operations. In the United States, youth of all ages are legally permitted to work in agriculture, but not in other industries, which is a result of an outdated law and has been attributed to more than half of youth work-related fatalities occurring in agriculture (Iannacci-Manasia 2023).

3.3.3 More Veterinarians Are Needed

Veterinarians are another essential group of professionals in the global food animal industry. They are responsible for monitoring, preparedness, and mitigation of emerging and foreign animal disease outbreaks; prevention and treatment of existing animal diseases; ensuring animal welfare, food safety, and quality along the entire food chain; and addressing public health concerns involving zoonosis and food security (Navarre et al. 2020). However, recruitment, training, and retention of food animal veterinarians is a major concern in the United States as well as in other countries. Extremely high student debt; challenges living in rural communities such as lack of social and cultural opportunities and lack of access to jobs for spouses and childcare; long workdays; and high on-call demands are some of the factors that are causing inadequate numbers of food animal veterinarians working in rural communities in the United States (Navarre et al. 2020). The multiple factors that limit recruitment and retention of food animal veterinarians make it difficult to implement changes and the much-needed multifaceted solutions to ensure that enough adequately trained public and private practice veterinarians will meet current and future demands (Navarre et al. 2020). This is another important social sustainability problem that needs to be addressed to achieve sustainable food, agriculture, and livestock production systems.

3.3.4 Quality of Life and Work of Farmers and Workers Must Be Improved in the Food Animal Industry

One of the seven principles in doughnut economics is to nurture human nature (Raworth 2017). People want choices and the opportunity to thrive rather than simply survive. While most of the 17 United Nations SDGs involve important components of social sustainability, these social aspects are rarely considered or included as prominently as environmental sustainability indicators of livestock production systems. Furthermore, although a healthy environment is a human right, the social sciences have generally been ignored as contributors to potential solutions for climate change mitigation and achieving environmental sustainability. However, identifying alternatives to overcome the inequitable social structures that contribute to our current high carbon footprint must be included in climate mitigation strategies (Klinenberg et al. 2020).

Social sustainability involves the equity, justice, and quality of life obtained from social relationships. The sustainability pillar of the United National SDGs is less structured than the environmental and economic pillars, but success in one area of sustainability reinforces success in the others. For example, adoption of sustainable agriculture practices promotes long-term ecological health and economic benefits, while also contributing to vibrant communities and satisfying the livelihoods of farmers and other members of the food production system. Social sustainability involves two major considerations which include (1) the response to external societal demands (i.e., animal welfare, environmental sustainability practices) and the inclusion of farmers and the positive role of farming in society, along with (2) the response to internal conditions of farmer well-being and quality of life of farmers and farm workers.

External social well-being and connections occur at different levels or scales of social relationships (Table 2.4). Internal dimensions of social sustainability of livestock farmers and workers involve many different domains related to well-being and quality of life including (1) physical and mental health, (2) meeting social, cultural, and spiritual needs, (3) equitable treatment, and (4) adequate income. Brennan et al. (2023) suggested multiple criteria for assessing internal social sustainability of livestock farmers and farm workers, which are also applicable to other workers in industries with supporting roles including trucking, meatpacking, and dairy processing industries (Table 2.5).

The internal social pillar is also important to the overall sustainability of farming because of the need to create social conditions that attract new farmers, farm

Table 2.4 External social sustainability requirements of farmers at multiple scales (Guptill 2021)

Scale	Social requirements
Personal and household	Basic needs are met
	Protect and improve health and well-being
	Sustain cherished cultural values and practices
	Feel a sense of autonomy to act creatively according to their social and environmental ethics
	Feel a sense of personal fulfillment from farming
Farm—relationships among farmers, employees, interns, landowners, and service providers	Safety and comfort of working conditions
	Respectful treatment of employees
	Intentional learning and professional development
	Daily communication among farmers and employees and to identify and resolve issues
	Access to needed resource and information to enhance sustainable farming
Local community—relationships between farmers and communities, including nonfarm people and those with shared cultural identities	Connections between people on the farm and people in the local community
	Farmers' and employees' sense of belonging to the local community
	Economic and ecological vitality of the community
	Capacity of the community to cooperate for mutual benefit
Agri-food network—relationships among farmers, customers, suppliers, lenders, and service providers	Fair and transparent negotiations among suppliers, lenders, contractors, and buyers
	Mutually supportive relationships between farms and their customers
	Mutually beneficial relationships among farms
	Equitable and beneficial access to agricultural services and technical assistance
	Farmers' sense of sharing risks and rewards with others
Society at large—farmers' influence on public policy and society's views of agriculture, food, and sustainability	Society's awareness of farmer's contributions
	Consideration of agriculture and food system issues in public sector decision-making
	Equitable access to public resources

Table 2.5 Criteria for assessing internal social sustainability of livestock farmers and workers (Brennan et al. 2023)

Job satisfaction	Connectedness with their job
	Do they like their job?
	Individualized aspects of jobs
Safety and security	Enabling good work
	Training and opportunities
	Benefits
	Autonomy
	Responsibilities
Aspiration reality	Level of achievement and ambition
Working conditions	Hours of work
	Rest periods
	Work schedules
	Remuneration
	Physical conditions
	Mental demands
Decent working environment	Opportunity for productive work
	Provides fair income
	Security in the workplace
	Social protection for all
	Potential for personal development and social integration
	Freedom to express concerns, organize, and participate in decisions that affect their lives
	Equality of opportunity and treatment for all men and women
Quality of life	Perception of their position in the context of culture and value systems they live in and relation with their goals, expectations, standards, and concerns
Intrinsic benefits	Pleasure
	Meaning
	Self-accomplishment
	Relations to nature
Immaterial benefits or rewards	Lifestyle
	Opportunities to grow and evolve
	Opportunities to train
	Opportunities to take on new responsibilities and careers
	Contracts that are inclusive, clear, and fair

workers, and young people to enter the farming professions; retain existing farmers and farm workers; and minimize early exit from farming that can occur from physical and mental health problems to overcome the decreasing number of workers in livestock farming (Dedieu et al. 2022; Santhanam et al. 2021). Overcoming these challenges requires incorporating measures of farmers' well-being, satisfaction with work, and quality of life as essential components of sustainability goals (Sabillón et al. 2022). To do this, sociostructural trends (e.g., automation, expansion) in farming and their consequences on the skills, stress, and flexibility of farmers need to be considered. Public policies also need to be evaluated and structured to improve work conditions and limit unsatisfactory situations for farmers and farm workers. Creating awareness and opportunities for individuals interested in farming that match their personal and career goals with different farming models, such as those described in Table 2.3, could also increase work satisfaction. Unlike economic and environmental sustainability measures, social sustainability measures are a combination of objective and subjective indicators. Breslow et al. (2018) proposed the "4 C's" of social sustainability to include (1) connections to others and to nature, (2) capabilities to evolve and choose, (3) conditions of living standards, and (4) cross-cutting issues of resilience and security.

Immigrant workers are essential contributors to the sustainability of livestock farms, trucking, meatpacking, and dairy processing operations. However, the oppression and exploitation of these workers are extensive and continue to worsen in many countries around the world. There are many legal, social, economic, and cultural factors that lead to exploitation, low wages, and conditions of vulnerability of migrant workers in the agri-food sector in Europe (Palumbo et al. 2022). Most migrant workers work long hours for low pay, are exposed to toxic pesticides and herbicides, endure extreme heat in summer and cold in winter, live in degrading and unsanitary conditions, and have limited access to basic services (Palumbo et al. 2022). The employment of inexpensive and exploitable migrant workers has contributed to the survival and sustainability of agri-food companies as supply chains have become increasingly integrated (Palumbo et al. 2022). As a result, a system has been created that promotes worker dependency on employers, confines the migrant workforce to specific sectors, facilitates continuous replacement, and results in farm owners and companies profiting from social, legal, and economic vulnerability of migrant workers (Palumbo et al. 2022).

Like in the European Union, immigrant workers in the United States are a vulnerable worker population that receives lower wages and works longer hours in jobs that are hazardous to their health, and often in worse conditions that involve human rights violations, abuse, human trafficking, and violence, compared with nonmigrant workers (Moyce and Schenker 2018). Immigrant workers are often provided minimal occupational safety training and protection which can lead to poor health, workplace injuries, and fatalities. A survey of Latino immigrant cattle feed yard workers conducted by Ramos et al. (2018) indicated that 40% of workers had not received health or safety training from their current employer, and only 13% had ever participated in any formal safety and health training program despite more than 90% of them indicating interest in receiving training. Results from a more extensive survey that was conducted to assess the health, safety, and job satisfaction among Latino dairy farm workers in the United States, indicated a need for improvement in the physical environment, equipment issues, the social-psychological environment, management, and organisational factors (Menger et al. 2016). Health-related problems among immigrant workers are also a result of language and cultural barriers, limited access to health care, documentation status, and the political situation in the host country (Moyce and Schenker 2018).

Slaughterhouses are a major sector of the livestock industry that has a long history of devaluing and exploiting immigrant workers, people of color, and women, along with promoting racism and xenophobia to support capitalism and the accumulation of wealth (Nibert 2014). Current economic policies, politics, free trade agreements, and the location of facilities in states that prevent or disable organized labor have created an unsustainable scenario for continuing to rely on immigrant labor to meet this critical need in the food system (Nibert 2014). Meatpacking plants have restructured labor in rural towns and communities by becoming the largest employer but offering poor jobs with low wages, high injury rates, and a grueling work pace and schedule with limited, if any, involvement from organized labor unions (Haedicke 2013). The social structure and stability of rural communities with small populations can be quickly transformed with an influx of immigrant workers, which often occurs at locations of meatpacking facilities. Steps can be taken to ease this transformation by (1) establishing planning committees and initiating community conversations about potential changes, (2) assisting local school districts to prepare for and accommodate ethnically diverse students entering classrooms, (3) actively recruiting language translators to help at schools and places of employment, (4) encouraging local business to provide goods and services representing cultural preference, (5) training and employing social workers capable of identifying acculturative stress and use culturally appropriate intervention, (6) providing business incentives to encourage diversification of local employment opportunities and community development, and (7) establishing cross-community networks to share experiences and learn to enhance community transformation (Dalla et al. 2005). Helping new immigrant farmers become familiar with agricultural service providers, such as government agencies (i.e., USDA), extension service, and agricultural nongovernment organizations, can be an important approach for accelerating their professional development and social well-being (Lewis 2009).

4 Accurate Information Is Essential for Making Informed Decisions

We not only need a socially sustainable agricultural workforce, but we also need better educated "systems thinkers" to provide leadership and make informed decisions when implementing sustainability practices on farms and various sectors of food supply chains. However, this requires having adequate, relevant, high-quality, accurate data for generating accurate and useful information to make thoughtful, high-impact decisions. We cannot have accurate information without access to accurate data. Therefore, data collection, analysis, ownership, and information management play a critical and essential role in the overall sustainability framework of agriculture, food, people, ecosystems, and the planet, but it is often underappreciated. We need answers to numerous questions to improve data acquisition and use to make fact-based sustainability decisions. How can relevant data be collected? Who owns the data? Can data be shared, and if so, at what cost? How do we evaluate and manage data quality? How much data granularity or aggregation is needed? How is the integrity of data maintained to avoid bias that leads to inaccurate or misleading information that could unfairly provide a market advantage? What are the benefits and limitations of using public versus private data? How do we capture the benefits of Big Data and the use of artificial intelligence (AI)? Will there be a role for quantum computing in agriculture and food systems in the future?

Finding answers to these questions is critical for successful adoption of sustainability practices in agriculture and the global food system. In fact, agriculture is in the initial stages of a digital revolution that includes satellites, drones, Big Data, AI, robotics, and quantum computing that will enable the production of more food produced on less land with fewer resources and less environmental footprint if we can overcome the many barriers that currently exist (Weersink et al. 2018). We also need to be conscious of the potential differences in interpreting data and information at various scales and system boundaries to identify potential unintended consequences resulting from cascading effects within the system or supply chain. The previous discussion on whether the production and use of biofuels is a good environmental sustainability strategy is a good example of the need to understand upstream and downstream effects in supply chains and systems when evaluating the overall impact of a practice on sustainability.

Data collection to establish benchmarks and monitor progress toward achieving sustainability goals begins at the farm level and must be connected to each segment throughout entire supply chains. However, farmers are concerned about many aspects of collection, sharing, and unauthorized use of their data with third parties. Legal agreements are often ambiguous and may not comply with data privacy practices. A summary of data collection, ownership, sharing, and management concerns, along with some potential solutions, is summarised in (Table 2.6).

4.1 How Do We Know If Information Is Accurate?

Communication of accurate scientific information among citizens, farmers, agribusiness professionals, scientists, and policymakers is essential for making informed decisions and assessing whether any progress is being made toward achieving sustainability goals. Therefore, evaluating the accuracy of information is a major determinant of our ability to successfully transform economies and agri-food systems to become sustainable. Furthermore, if farmers are to implement climate mitigation and sustainability practices, they need to be aware of the technologies that will enable them to do so, which requires communication of accurate information from multiple sources (Fierros-González and López-Feldman 2021). Unfortunately, government policy has led to reduced public investment in agricultural research (Pardey et al. 2006), which has increased privatization of agricultural advisory services (Swanson and Davis 2014). As a result, economies and politics have become poorly aligned with the behavioral, social, and psychological changes necessary to effectively transition to more environmentally friendly agricultural and food systems (Burton 2004). In addition, extensive amounts of misinformation about climate change and sustainable agriculture have been undermining the role of research, science, and fact-based decision-making. Increasing amounts of misinformation, driven by social media and online search engines, are continuing to have

Table 2.6 Summary of challenges and some potential solutions for data and information management in agriculture and livestock systems

Farmers have many concerns regarding access, sharing, and use of agricultural data; fragmented and unclear data governance diminishes the willingness of farmers to share data which can reduce availability and accessibility of agriculture data for policymaking, research for developing new innovations, and developing and adopting digital solutions and services for farmers; policy makers must develop agriculture data regulations that provide a balance that protects privacy, confidentiality, and economic interests of farmers while also making it possible to use data for research that leads to innovations and digital solutions for farmers	Jouanjean et al. (2020)
No laws regulate data usage in the US dairy industry despite increased dependency and monetization of data; some codes of practices have been published but are legally vulnerable; a "farmers bill of rights" was proposed to standardize and promote equitable data governance and regulations among farmers, industry professionals, and companies	Cue et al. (2021)
Results from a US survey of dairy farmers and non-farmers indicated that data sharing is a valuable activity, farm data should be owned by the farmer, incentives could motivate continued or greater data sharing; lack of awareness of data collection protocols, frequent lack of data sharing agreements, concerns about lack of control once data are accessed by others, along with ownership, confidentiality, security, and lack of integration need to be addressed	Fadul-Pacheco et al. (2022)
Agricultural technology and data services are providing an increasing amount of decision-making tools and services for farmers that require very little analytic skill or active participation by farmers but raise questions about who owns the data, how data can be used, implications for new market entrants, and future competition and industry dynamics such as commodity market performance; farmers may be worried about the wrong things such as data use for speculation rather than competition among agriculture data services and reductions in valuable information flows that may impact future competitive market outcomes	Sykuta (2016)
Australian farmers are reluctant to share data due to lack of transparency and clarity regarding data ownership, portability, privacy, trust, and liability in commercial relationships associated with smart farming; lack of trust between farmers and third parties that collect, aggregate, and share farmer data is the primary concern; changes in social relationships, education, and awareness of agriculture data collection and governance are needed to build trust for adopting smart farming systems	Wiseman et al. (2019)

Sensors, drones, and other data collection and analysis tools collect large amounts of farm data; farmers are concerned about unauthorized access, collection, and sharing of data with third parties by agriculture technology providers; lack of adoption of best practices and security of farm data, as well as ambiguity of legal aspects of data collection, processing and sharing create uncertainty of data privacy practices among farmers; farmers are reluctant to adopt new technologies that can be detrimental to stakeholders, government, and the public; protecting data privacy and respecting farmer rights should be collectively addressed by farmers, governments, agriculture technology providers, and supply chain stakeholders	Kaur et al. (2022)
Potential benefits of new innovations from public-private research relationships using big data may not be realized without addressing data privacy and security concerns; well-defined data privacy standards along with using an economically informed, high-, medium-, and low-risk tolerance data security approach can be an effective strategy for meaningful data governance	Wilgenbusch et al. (2022)
Strategies to minimize pest and disease risks of plants and animals in global supply chains are dependent on research using highly spatial and temporal data in epidemiological models to accurately assess and predict biotic risks; data privacy concerns among data owners and collectors have limited ability of researchers to access complex, multisource, multiscale data on pathogens, hosts, and environments; design and use of innovative data sharing and analysis platforms can address these data privacy challenges	Senay et al. (2022)
Agricultural food and products are vulnerable to spoilage, loss of quality and nutritional value, adulteration, and safety concerns that require traceability due to increased globalization and international trade; traceability requires robust data collection related to products during the entire life cycle, information management, and communications; blockchain technology provides traceability by providing applications that enable monitoring, observing, and managing all transactions among stakeholders; Internet of Things technology provides approaches for verifying all transactions which are recorded and stored in centralized databases	Kaur et al. (2022)
Artificial intelligence (AI), big data, and global food safety data can be used to predict short-term food safety risks in food that can initiate early mitigation actions, but long-term predictions are needed to design mitigation strategies that reduce the number of food safety recalls	Stoitsis et al. (2023)
AI applications in agriculture can support precision farming, reduce labor requirements, conserve water from irrigation systems, improve fertilizer and water use to support soil and plant health, and provide timely disease detection in crops to reduce pesticide use and other chemicals; the main challenges for achieving these benefits are the need to create and improve datasets and develop AI-based algorithms that represent natural environments and are computationally efficient	Singh and Kaur et al. (2022)

detrimental effects on democratic governments and public health (Aslett et al. 2023). In the United States, misinformation about science is often driven by political views, social networks, and an individual's ability and motivation to identify falsehoods (Scheufele and Krause 2019). Emotions can also be potent, pervasive, predictable, sometimes harmful, and sometimes beneficial factors associated with information acceptance or rejection (Lerner et al. 2015). The widespread availability and use of social and online media networks, which have limited accountability, have led to intentional and unintentional spread of false, inaccurate, or misleading information that is readily believed, but there is no holistic framework in place to prevent this from happening (Chowdhury et al. 2023). Even more concerning is that people conducting online searches to evaluate the truthfulness of false news articles are more likely to believe false information because they are likely to experience data and information voids or information spaces with corroborating evidence from low-quality sources (Aslett et al. 2023). Because of the detrimental impact of misinformation in society, media literacy programs need to be developed to help users sort fact from fiction, and search engines need to be modified to minimise the propagation of false information (Aslett et al. 2023).

4.2 How Do We Measure Sustainability in Agriculture?

In addition to the challenges of accurate data and information management, there are also many challenges for measuring all the dimensions of agricultural sustainability. Transforming to sustainable agricultural and food systems involves synergies and trade-offs among the three pillars of environmental, social, and economic sustainability goals. Synergies and tradeoffs also exist between increased mechanisation and the role of human labor, especially in low- and middle-income countries, that must be considered among all sustainability pillars (Daum 2023). Despite their importance, Robling et al. (2023) observed that 20 sustainability indicators for 12 of 20 sustainability themes were difficult to measure due to lack of scope, temporary data availability, or additional data requirements. These researchers concluded that there are trade-offs between data availability, comprehensibility, and analytical validity for some indicators. The Food and Agriculture Organization of the United Nations has supported efforts to develop and implement a composite indicator (SDG 2.4.1) that measures the percentage of agricultural area under productive and sustainable agriculture for individual farms, which includes 11 specific sub-indicators of economic, environmental, and social sustainability (Gennari and Navarro 2019). Similarly, Castillo-Diaz et al. (2023) developed a standardised, quantitative, composite sustainability indicator from a system of circular economy indices for the European agri-food system and showed that overall sustainability can be estimated using ten indicators. Wiśniewska (2011) acknowledged that a holistic approach that accounts for interactions between the farming sector, national economy, society, and natural environment is needed. He proposed that sustainable agriculture must intersect social, economic, institutional, and ecosystem sustainability subsystems that represent the level of justice, efficiency, competency, and condition of the natural environment. However, none of these approaches and schemes include connections with One Health, regenerative agriculture, and ecology.

Ecosystem health is an integrated, transdisciplinary framework that includes environmental, social, and economic sustainability considerations along with all of the components from Sustainable Development Goals, One Health, regenerative agriculture, and agroecology. Therefore, ecosystem health should be used as a more comprehensive, holistic approach for measuring, assessing, and achieving sustainability of our agriculture and food system at multiple dimensions. Ecosystem health also aligns with the doughnut economics model and principles because it includes human elements embedded within ecosystems in which we interact and depend. Although ecosystem health frameworks have been proposed, and multiple indicators and metrics have been defined, this "systems" approach has rarely been used or discussed relative to its benefits and limitations for assessing, monitoring, and achieving sustainability, and resiliency of livestock production systems. Most of the sustainability assessments in agriculture and food systems have focused on LCA of environmental impacts of agricultural and food production systems inputs and outputs without directly considering social and economic sustainability and One Health implications. Although LCA is becoming widely used to estimate environmental impacts in livestock systems and products, LCA methods lack accuracy, robustness, and consistency in assessing sustainability across livestock systems and products (Goglio et al. 2023). Despite standardised guidelines for determining LCAs in agriculture, there are also issues with data quality, differences in systems boundaries, and allocation methods (i.e., economic, energy, mass) used that can influence results and interpretation. Improved LCA estimates for livestock systems could be achieved by including data for food, feed, fuel, and biomaterial competition; crop-livestock interactions and the circular economy; biodiversity; animal welfare; nutrition; and GHG emissions (Goglio et al. 2023). Although livestock are an important part of the regenerative agriculture framework to restore and prevent further degradation of ecosystems and improve their health, they play a minor role (Giller et al. 2021b; Newton et al. 2020; Lal 2020; Tittonell et al. 2022). Re-wilding is another restorative action that has been proposed to improve ecosystem health, but it has significant implications for wildlife-livestock interactions related to disease transmission, emergence or reemergence of new pathogens, and predator management (Egoh et al. 2021; Jepson et al. 2018; Jørgensen 2015; Schmitz et al. 2023). Therefore, although an ecosystem health framework can provide a more holistic approach toward the management of natural systems and resources to achieve overall sustainability and One Health goals, guidance and standardised methodologies for data collection and metrics are needed.

5 We Need to Create a Permaculture-Agroecosystem-Precision Farming-Regenerative Agriculture-Modern Digital Agriculture Nexus

Permaculture (i.e. “permanent agriculture”) is a design framework of best practices for creating and managing sustainable and resilient agroecosystems (Krebs and Bach 2018). Permaculture originated from Indigenous knowledge that was co-opted in the 1970s by Bill Mollison and David Holmgren, based on Mollison’s experiences living with Indigenous people from Tanzania, and involves the development of agricultural ecosystems that are self-sustainable and self-sufficient without depending on supplemental factors to support them (Hernandez 2022). Specifically, Holmgren (2002) defined permaculture as the “conscious design and maintenance of agriculturally productive ecosystems which have the diversity, stability, and resilience of natural ecosystems.” The main idea of permaculture is that energy-consuming industrial technologies used in agriculture that cause pollution can be reduced or replaced by intensive use of biological resources based on thoughtful, holistic designs that mimic natural ecosystems (Scialabba 2022). The three core components of permaculture systems are (1) care for the earth, (2) care for people, and (3) setting limits of consumption and reproduction while redistributing the surplus (Krebs and Bach 2018). These core components are consistent with the core principles of doughnut economics proposed by Raworth (2017). Compared with current monoculture systems, permaculture involves using integrated land management practices to improve soil and water quality along with integrated pest management and crop rotations to diversify ecosystems (McLennon et al. 2021). Livestock and poultry production is often included in permaculture systems as a means of ensuring efficiency and ecological integrity by recycling nutrients and optimizing grazing management of pasture lands (Scialabba 2022). Furthermore, integrating permaculture with regenerative agriculture, precision agriculture, and modern digital agriculture can improve decision support systems to enable sustainable crop production that reduces the environmental footprint and improves global food security (McLennon et al. 2021). The permaculture framework is also compatible with management processes for developing business strategies that integrate ecological and social aspects for achieving sustainability of organizations (Akhtar et al. 2015).

Permaculture is the most widely practiced form of agroecology, and these concepts are often used interchangeably to describe an ethical framework and principles for designing diverse and sustainable systems to meet the needs of various cultural and ecological scenarios (Hathaway 2016). The 12 design principles of permaculture systems include (1) observe and interact; (2) capture and store energy, nutrients, and water; (3) obtain a yield; (4) apply self-regulation and accept feedback; (5) use and value renewable resources such as sunlight and rainwater; (6) produce no waste; (7) use nature’s patterns as a template for design; (8) integrate rather than segregate; (9) use small and slow solution; (10) use and value diversity; (11) value marginal spaces; and (12) creatively respond to changes in ecological systems (Hathaway 2016). Agroecology principles provide a transdisciplinary, interdisciplinary, and holistic framework for addressing social, cultural, political, environmental, and economic issues to transform food systems to become more resilient, equitable, and sustainable (Wezel et al. 2020). The Food and Agriculture Organization of the United Nations (FAO, 2018a) proposed ten elements of agroecology (i.e., diversity, co-creation of knowledge, synergies, efficiency, recycling, resilience, human and social values, culture and food transitions, responsible governance, and circular and solidarity economy) and 13 agroecological principles that correspond to these elements which include (1) recycling, (2) input reduction, (3) soil health, (4) animal health, (5) biodiversity, (6) synergy, (7) economic diversification, (8) co-creation and sharing of knowledge, (9) social values and diets, (10) fairness, (11) connectivity, (12) land and natural resource governance, and (13) participation (Wezel et al. 2020). Following these principles, five progressive levels of transition toward sustainable food systems have been defined as (1) increasing efficiency of input use and reducing costly, scarce, or environmentally damaging inputs, (2) substituting conventional inputs and practices with agroecological alternatives, (3) redesigning agroecosystems, (4) reconnecting consumers and producers through the development of alternative food networks, and (5) building a new global food system on based on participation, localness, fairness, and justice (Wezel et al. 2020). These principles are also essential for redesigning and transforming intensive livestock systems into sustainable and more resilient components of our global food system.

Precision livestock farming (PLF) is an emerging, innovative, multidisciplinary, and intensive management approach using principles of process engineering, data science, computer science, advances in animal science, veterinary medicine, agricultural engineering, and environmental science to automatically and continuously monitor animal health and performance in real-time for improved efficiency and productivity of animal production systems (Berckmans 2017; Pomar et al. 2019; Jiang et al. 2023). Therefore, the main purpose of PLF is to manage any variable in real time that disrupts the production process and warns the farmer that a problem has been detected (Tzanidakis et al. 2023). Precision livestock feeding is an essential component of PLF that enables feeding individual animals customized diets that meet their daily nutrient requirements and optimizes nutrient utilization efficiency to reduce manure nutrient excretion and environmental impacts and improve profitability (Pomar et al. 2019). Although PLF is being used in all types of livestock and poultry production systems, it has been most extensively implemented on dairy farms where animal sensors, production data, and external data are integrated for early disease detection and management decision-making (Kleen and Guatteo 2023). Precision livestock farming technologies that are used in pig production enable early warnings of problems and better decision-making to improve pig performance, welfare, and farm sustainability, but adequate training of farmers is necessary to optimize their use (Vranken and Berckmans 2017). Several studies with growing-finishing pigs have shown that using precision feeding techniques for individual growing-finishing pigs is effective in reducing nitrogen and phosphorus excretion by 20–30%, as well as other environmental impacts (climate change, eutrophication, and acidification potential) by 3–6%, and feed cost by 10% without affecting growth performance and carcass composition (Andretta et al. 2014, 2016,

2018). Coupling PLF with precision crop farming in mixed agricultural production systems offers even greater opportunities to improve productivity and efficiency of resource use, profitability, and reduce environmental impacts of food production (Monteiro et al. 2021).

European pig, poultry, and dairy farmers who receive adequate technical support from providers have a positive view of using PLF technology, which enables them to recognize problems much earlier than conventional methods, but many farmers acknowledge that personal contact with animals cannot be replaced with video monitoring (Hartung et al. 2017), and farmers have ethical responsibilities to livestock in their care (Werkheiser 2018). Using PLF technology can change the human-animal relationship, and veterinarians will need to adapt and play an active role in further development and implementation of these technologies on farms (Kleen and Guatteo 2023). Although the use of PLF has been promoted to improve animal welfare, Tuyttens et al. (2022) identified 12 potential threats to animal welfare which involve direct harm and indirect harm via the end user, changes in housing and management, and ethical stagnation or degradation. Precision livestock farming equipment is expensive, timely and adequate maintenance may not occur from some technology companies, and farmers want more experience and testing of technology on demonstration farms before implementing it on their own farms (Hartung et al. 2017). European consumers believe that the use of PLF technologies will improve animal health and welfare, reduce environmental impacts, and increase transparency of livestock and poultry production but worry that it will lead to more industrialized livestock farming, make technology and data more vulnerable to misuse and cybercrime, and that information will not be adequately communicated to consumers (Krampe et al. 2021). Therefore, effective communication of using PLF technology is needed for transparency among various social groups including farmers, consumers, regulators, and the food industry (Elliott and Werkheiser 2023).

5.1 Are Local, Organic, Plant-Based, and Cultured Food Production Systems Part of the Sustainability Solution?

If we are going to transform our food system to better serve society within planetary boundaries, we need to consider how it will affect the connections among people, where and how their food is produced, and who has control and can participate in how food is produced and distributed (Loring 2023). This involves understanding the expectations, perceptions, and preferences among different types of consumers for how their food is produced, and the potential synergies and incompatibilities of food production systems for achieving environmental, social, and economic sustainability goals. Locally produced, organic, plant-based meat alternatives, and emerging cultured meat foods are popular food choices among some, more affluent consumers, but are these types of foods more sustainable?

Many governments and community organisations have promoted local food systems for several years based on claims of (1) increased consumer access to healthy food, (2) consumer willingness to pay more for locally sourced food versus nonlocal food, (3) increased social recognition of farmers, (4) increased economic benefits to local farmers, (5) increased community social connectedness, (6) benefits to the local economy, (7) encouraging environmentally friendly production practices, and (8) contributing toward mitigating climate change (Enthoven and Van den Broeck 2021). However, the impact of local food systems on social, economic, and environmental sustainability varies between types of supply chains, products, and countries (Enthoven and Van den Broeck 2021). In general, producing and consuming local foods has minimal benefit for reducing the carbon footprint of food production compared with emissions associated with different types of food produced and consumed (Ritchie 2020). Overall, animal-based foods tend to have a greater GHG footprint than plant-based foods, with land use and farm-level emissions representing about 80% of total emissions for most foods (Ritchie 2020). Because food transport accounts for only about 6% of total emissions, there is minimal advantage of focusing on consuming locally produced foods (Ritchie 2020). In fact, locally produced food does not necessarily infer greater environmental, social, or economic sustainability and supply chain resiliency than nonlocal food systems even though they may contribute to rural development and enhance local community connectedness (Stein and Santini 2021).

Organic agricultural production and food systems, which represent only about 1% of global agricultural land, are also often perceived as being more sustainable than conventional farming systems (Meemken and Qaim 2018). However, although organic farming may be less polluting than conventional farming per unit of land, when sustainability factors are evaluated per unit of output, organic farming results in lower crop yields, loss of natural habitats, and increased costs and food prices, making food less affordable for poor consumers in low income countries (Tuomisto et al. 2012; Meemken and Qaim 2018). Organic farms generally have soils with greater organic matter content and use less energy, but they also have greater nitrogen leaching from soil, nitrous oxide and ammonia emissions, land use, eutrophication potential, and acidification potential per unit of product output than conventional systems (Tuomisto et al. 2012). Therefore, although organic farming is not a viable solution for increasing food security and sustainability, some elements can provide environmental benefits if they are integrated into conventional crop production systems to improve soil quality, recycle nutrients, and protect biodiversity (Tuomisto et al. 2012; Meemken and Qaim 2018). Gaudaré et al. (2021) compared organic and conventional production systems for dairy cattle, pigs, and poultry, and reported that organic systems resulted in 12% less animal productivity, and an overall average of 14% less feed utilization efficiency, but provided 46% less human-food versus animal-feed competition for organic dairy cattle systems. Van Wagenberg et al. (2017) conducted a more comprehensive review of numerous studies to compare economics, productivity, environmental impacts, animal welfare, and public health differences for dairy cattle, beef cattle, pigs, broilers, and laying hens in conventional and organic production systems in Europe, North America, and New Zealand. They reported that conventional systems had the advantage of less labor requirements per unit of product produced, lower income risk per animal, greater animal productivity per unit of time, greater reproduction rate and feed conversion efficiency, less land use, generally less acidification and eutrophication potential per unit of product produced, equal or better udder health for cows, and equal or less microbial contamination than organic systems. However, organic systems had the advantages of more income per animal or full-time employee, less impact on biodiversity, and equal or less likelihood of bacterial antibiotic resistance. Therefore, like crop production systems, the productivity of food animal production systems is superior in conventional systems, but the integration of some aspects of organic systems into conventional systems would enhance overall sustainability.

Results from a global meta-analysis of LCAs of 742 agricultural systems and more than 90 unique foods produced in high-input systems showed that organic production systems require more land, cause more eutrophication, use less energy, and have similar GHG emissions than conventional systems (Clark and Tilman 2017). These authors concluded that shifting diets toward low environmental impact plant-based foods along with increasing efficiency of agricultural input use would provide greater environmental benefits than switching from conventional to organic production systems. Furthermore, Clark et al. (2022) developed an approach to estimate environmental impacts of 57,000 food products using publicly available data, combined with a nutrition and health scoring system (NutriScore) used in many countries, to provide integrated nutrition and environmental sustainability information that can be used to guide consumers and policymakers make more informed decisions about food choices.

Implementing more sustainable production practices on farms and along food supply chains often requires significant financial investment and increased costs, which can threaten economic sustainability of farmers unless these costs can be absorbed through prices paid by consumers or by other segments of supply chains. Therefore, consumer willingness to pay for food that is produced using practices that reduce the carbon and environmental footprint and enable achieving other sustainability measures is a frequent topic of debate and cause for slow adoption of sustainable practices in our global food system. For example, Denver et al. (2023) conducted a survey of consumers in China, Denmark, Germany, and the United Kingdom to assess their willingness to pay for pork produced under improved welfare conditions, decreased use of antibiotics, safety from *Salmonella* and foodborne bacterial contamination, and avoidance of climate impacts from using pig feed containing ingredients produced in deforested rain forest regions. Results of this survey indicated that while most consumers in all countries were willing to pay a premium price for more sustainable pork, climate impact was the least important of the choices provided because consumers believed that improvements in climate impacts could be achieved by other means, unlike animal welfare and food safety (Denver et al. 2023). Results from another survey of conflicting choices of pig farming and pig meat characteristics among German consumers showed that personal health benefits were a higher priority than animal welfare considerations, while environmental sustainability benefits ranked the lowest, even though consumer preferences varied based on psychographic and socio-demographic factors (Klink-Lehmann et al. 2023).

Bryant et al. (2019) surveyed consumers in the United States, India, and China regarding their perceptions and acceptance of plant-based meat alternatives and cultured meat from animal cells (also referred to as in vitro meat, lab meat, synthetic meat, factory-grown meat, or clean meat) and found high consumer acceptance in all three countries, especially in China and India. In contrast, Weinrich et al. (2020) reported that German consumers were only moderately accepting of cultured meat. Cultured meat is an emerging technology that has been promoted as being more environmentally sustainable and animal welfare friendly than conventional meat production (Sinke et al. 2023). Cultured meat production is energy-intensive, but it is more efficient in converting crops into meat than poultry production, and as a result, contributes to less nitrogen emissions due to the lack of manure produced compared with animal-derived meat produced in conventional systems (Sinke et al. 2023). Newman et al. (2023) suggested that although emerging cell culture technologies have great potential and provide economic opportunities to produce protein with a lower environmental footprint than conventional food animal production, there is a need for regulation and development of government policies to ensure that the environmental and social benefits provided by these technologies are achieved.

Plant-based meat alternatives have become commercially available to consumers who are generally affluent vegetarians or flexitarians (person who is primarily a vegetarian but occasionally eats meat or fish) who want to minimise or stop eating meat because of environmental or animal welfare concerns associated with conventional livestock and poultry production, but still want to enjoy the eating experience of meat consumption (Caldwell and Mills 2023). Although these plant-based food products provide the benefits of eliminating animal welfare concerns and reducing carbon emissions and land use, several challenges need to be overcome to provide an organoleptic eating experience and nutritional profile at a similar price compared with conventional meat sources (Caldwell and Mills 2023). Therefore, these plant-based meat alternatives are not a long-term sustainability solution for most of the global population of consumers.

6 Are Livestock and Poultry Production Systems Assets or Liabilities for Sustainability?

Food animal production has been an essential and integral part of human societies for many centuries. However, one of the most widely debated topics involving food security and sustainability is whether food animal production should continue to be a part of our global food system. Much of this controversy was elevated when the EAT-Lancet Commission proposed a planetary healthy diet that requires substantially restricting the global consumption of highly processed and animal-sourced foods because of detrimental effects on human health and the environment (Willett et al. 2019). However, implementing this recommendation will be extremely difficult because dietary preferences are inherently personal and local, and require engaging policymakers and actors at all levels of the food system to find an appropriate balance between health, environment, culture, and economy (Biesbroek et al.

2023). Furthermore, although the consumption of relatively high amounts of animal-sourced foods is associated with several forms of chronic diseases in high-income Western countries, minimal amounts of meat products are consumed by millions of malnourished poor people worldwide (Leroy et al. 2022). The authors of the EAT-Lancet study did not consider the significant contributions of highly bioavailable forms of essential vitamins and minerals provided by animal-sourced foods, which are globally scarce (Beal et al. 2023). About 2 billion people do not consume enough vitamins and minerals to meet their daily requirements; more than 150 million children under age 5 experience nutritional deficiencies causing stunted growth; and millions of people have impaired cognitive development because of malnutrition which could be prevented by consumption of adequate amounts of animal-sourced foods (Adesogan et al. 2020). Therefore, increasing the proportion of animal-sourced foods in diets for these malnourished populations was recommended to meet the recommended daily intake of six vitamins and minerals that are globally deficient (Beal et al. 2023).

Another limitation of the EAT-Lancet study was that it focused only on the role of animal-sourced foods on human health and environmental sustainability and overestimated the environmental impacts of different types of livestock and poultry production systems without accounting for their highly variable contributions around the world. Critics argue that food animal production systems should be eliminated due to the diversion of natural resources away from plant-based food production, their inefficient use of nitrogen and phosphorus to produce edible food products (Gerber et al. 2014), significant contributions to greenhouse gas emissions and global warming (Xu et al. 2021), and negative dietary effects from excess consumption on human health in affluent societies (Neuenschwander et al. 2023). However, Leroy et al. (2022) reminded us that although several agricultural practices associated with food animal production, such as intensive crop production for feed use, deforestation, and water pollution, are detrimental and require mitigation, these impacts are not directly caused by livestock farming. They also remind us that the impacts of livestock production on land use, water wastage, and GHG emissions are highly contextual, and estimations are inaccurate due to a reductionist use of metrics.

However, focusing only on human health, climate change, and the environmental impacts of livestock and poultry production is short-sighted and very risky because reducing or eliminating food animal production would have much greater socioeconomic losses than the potential environmental benefits that may be achieved (Lee et al. 2021a, 2021b). Livestock and poultry production systems provide numerous food and nonfood contributions to society in a wide range of types of production systems around the world (Table 2.7). In fact, well-managed livestock production systems can contribute to 12 of the 17 Sustainable Development Goals and can have several positive effects on ecosystem health and circular food systems. Another valid criticism of the EAT-Lancet analysis was that the contributions of livestock and animal-sourced foods toward reducing poverty, improving gender equality, improving livelihoods, increasing food security, and improving health were not considered but are all essential components of sustainability (Adesogan et al. 2020).

Table 2.7 The sustainability benefits and essentiality of livestock and poultry in global societies (Alders et al. 2021)

Provide food—meat, milk, and eggs
Food security
High-quality, nutrient-dense protein sources
Provide livelihoods and income
Livelihoods for about 1 billion smallholder farmers
Contributes 40% of agricultural GDP
Contributes up to >33% of household incomes
Provide draft power and transport for smallholders in low-income countries
Source of natural hides and fiber for clothing and other consumer goods
Major contributor to circular food systems
Consume biomass (crop by-products, residues, and food waste) unfit for direct human Consumption
Utilize forage from grasslands to produce meat and milk
Slaughter by-products recycled into livestock, poultry, and companion animal feed
Manure provides nitrogen, phosphorus, and organic matter to enhance soil fertility
Anaerobic digesters produce biogas from manure
Manure can serve as substrate for insect production
Contribute to ecosystem health
Transform inedible biomass into nutritious food for humans
Grazing, browsing, and production of dung and urine enhance ecosystem health
Change landscape locations in response to resource availability and weather patterns
Land cover and biodiversity are increased in well-managed systems
Carbon-neutral, profitable livestock systems can be achieved with correct balance of trees, grassland, and livestock stocking density
Sustainable livestock farming can limit soil erosion and water evaporation
Provide habitats for many types of invertebrate and vertebrate species
Grazing reduces fuel loads to minimise prevalence and severity of wildfires
Social benefits
Women empowerment in low-income countries who manage poultry and small ruminants
Certain types of livestock and poultry are essential for religious feasts, sacrifices, ceremonies
Beneficial for mental health including reducing stress and anxiety, social support, facilitate interpersonal relations, essential for spiritual health in some religions, and provide meaningful daytime occupation
Human medical/physical health benefits
Production of immunoglobulins in anti-sera
Source of insulin for treating diabetes
Source of xenografts such as heart valves and skin grafts
Serve as models for vaccine development and testing
Source of therapeutics for infectious diseases
Provide treatments for metabolic, neoplastic, and genetic disorders

Livestock and poultry production is essential for economic development in Africa and Asia because animals are an asset and means of storing wealth, provide a source of protein and micronutrients to ensure a healthy population, serve as a pathway for women empowerment and gender equality (Baltenweck et al. 2020), and provide a means for poverty reduction and economic development (Upton 2004). Therefore, food animal production systems are an essential part of ecosystems and human society that cannot be removed or replaced without detrimental consequences. However, we need to develop a more holistic and multidimensional sustainability assessment to develop and implement balanced solutions that reduce the environmental footprint of food animal production while maintaining its essential role in human societies and ecosystems.

6.1 What Are the Ecosystem Health Costs of Livestock and Poultry Production Systems?

Despite the many benefits that livestock provide human society and ecosystems, they are also significant contributors toward climate change, ecosystem health degradation, and exceedance of several planetary boundaries. An initial report by the Food and Agriculture Organization of the United Nations (Steinfeld et al. 2006) provided a summary of the major environmental impacts of livestock systems:

1. Intensification and industrialisation of food animal production is causing increased competition for scarce land, water, and other essential natural resources as well as increasing waste and pollution.
2. Livestock production utilises 70% of all agricultural land (30% of land surface on Earth) and contributes to deforestation, overgrazing, soil erosion, and nutrient pollution.
3. Significant amounts of methane and nitrous oxide are produced from livestock systems which contribute to climate change, and ammonia emissions contribute to acid rain and acidification of ecosystems.
4. Food animal production requires large amounts of freshwater which contributes to shortages and depletion of surface and groundwater resources, and is a major contributor to water pollution, eutrophication of ecosystems, human health problems, and antibiotic resistance.
5. Livestock production systems have contributed to biodiversity loss in 306 of 825 terrestrial ecosystem regions due to their direct and indirect impacts on climate change; air and water pollution; land degradation and deforestation on habitat loss; fertilizer, pesticide, and herbicide use for crop production; and antibiotic and hormone use.

Therefore, transforming and redesigning livestock systems for the twenty-first century must result in significant improvements in these key areas along with addressing several other One Health factors and reducing reliance on fossil fuel use (Table 2.8). Low carbon livestock production can be achieved by (1) increasing the

Table 2.8 Ecosystem costs that must be addressed to improve sustainable return on investment in global food animal production

Ecosystem health indicator	Main effects	References
Greenhouse gas emissions	Global livestock production contributes 11–17% of total global GHG emissions primarily from feed and manure	Steinfeld et al. (2006), Herrero et al. (2011), Gerber et al. (2013), Glatzle (2014), Poore and Nemecek (2018), Xu et al. (2021), Chang et al. (2021), Hayek and Miller (2021), and Twine (2021)
Disrupted nitrogen (N) flows	Global soil N surplus from crop-livestock production systems increased 3.8-fold from 1950 to 2000 and is projected to increase 23% from 2000 to 2050; livestock N emissions represent about 35% of total N emissions from agriculture and 29–34% of total N emissions from the global economy; livestock N emissions represent 39% of total nitrates (NO_3-), which cause surface and ground water pollution and biodiversity loss, 60% of total ammonia (NH_3), 23% of total nitrogen oxides (NO_x), which contribute to air pollution, human health risks, eutrophication, and acidification, and 32% nitrous oxide (N_2O), which is the most potent GHG; feed production and manure management systems are the major sources of N emissions; N utilization efficiency of converting dietary N to edible outputs at the farm level ranges from 5% to 45% among cattle, swine, and poultry; international trade of meat, feed ingredients, and fertilizer alters N flows and utilization efficiency	Bouwman et al. (2013), Uwizeye et al. (2020), and Gerber et al. (2014)
Disrupted phosphorus (P) flows	About 50% of total P inputs in global agriculture systems accumulated in soils from 2002 to 2010, with the remainder lost to water bodies, but 32% of cropland and 43% of pastures in specific regions had P deficits; globally, soil P surplus from crop-livestock production systems increased 5.5-fold from 1950 to 2000 and is projected to increase 54% during 2000 to 2050; about 90% of global rock phosphate reserves are concentrated in 5 countries; P utilization efficiency of converting dietary P to edible outputs at the farm level ranges from 21% to 74% among cattle, swine, and poultry; international trade of meat, feed ingredients, and fertilizer alters P flows and utilization efficiency	Lun et al. (2018), Bouwman et al. (2013), Gerber et al. (2014), and Nesme et al. (2018)

(continued)

Table 2.8 (continued)

Ecosystem health indicator	Main effects	References
Loss of biodiversity	Livestock systems contribute directly or indirectly to 5 main factors that cause global biodiversity loss including habitat change, pollution, climate change, invasive species, and over-exploitation; positive impacts include habitat restoration, habitat creation and maintenance, landscape connectivity, beneficial practices, nutrient cycling, carbon storage, and control of invasive species and can hold ecosystems together; negative impacts include habitat degradation, habitat destruction and occupation, landscape fragmentation, detrimental practices, nutrient pollution, ecotoxicity, greenhouse gas emissions, disease spread, invasive species spread, and competition with large mammals; recommended indicators to assess biodiversity impacts include habitat protection, habitat change, wildlife conservation, invasive alien species, pollution and aquatic biodiversity, off-farm feed, and landscape-scale conservation	FAO (2019, 2020)
Freshwater use	Agriculture uses about 70% of freshwater withdrawals with 20% used in livestock feed production and < 1% used in animal consumption, cooling, and other direct uses; feed production is the greatest contributor to water footprint (WF) which decreases with improved feed conversion and increased level of animal production intensification; WF is greater for meat than milk or egg production and is much greater for beef than for other major meat species; feed production uses 4387 km^3 of blue and green water annually which represents 41% of total agriculture water use; animal protein produced per m^3 of water consumed varies greatly between livestock types, geographic regions, and production systems; intensification can improve water use efficiency for swine and poultry but may not improve efficiency for ruminants	FAO (2018b), Ibidhi and Ben Salem (2020), Heinke et al. (2020), and Weindl et al. (2017a)
Freshwater eutrophication	78% of global ocean and freshwater eutrophication is caused by N and P derived from agriculture	Poore and Nemecek (2018)

(continued)

Table 2.8 (continued)

Ecosystem health indicator	Main effects	References
Land occupation and management	80% of overall land use is attributed to global livestock production, which consumes 58% of plant biomass; land use for animal production modifies local biodiversity, soils, nutrient cycling, regional light/radiation reflection and hydrology, and global changes in greenhouse gas and aerosol concentrations; pasture is the largest land cover representing 22% to 26% of ice-free land surface on earth; livestock contribute to overgrazing, deforestation, compaction or crusting of soils, rangeland soil degradation, excessive soil nutrients from mismanagement, and excessive application of manure	Weindl et al. (2017b), Phelps and Kaplan (2017), and Scanes (2018)
Air quality	80% of air quality-related deaths in the United States are directly attributable to animal production and indirectly from growing animal feed; carbon dioxide, ammonia, hydrogen sulfide, and other reduced sulfide compounds, and water vapor are the major gases produced; ammonia cause respiratory health problems while hydrogen sulfide can be lethal to animals and humans; particulate matter can adsorb gases, odorous compounds, and microorganisms; offensive odors include hydrogen sulfide, volatile organic compounds, amines, aromatics, and sulfur compounds and can be disruptive to surrounding communities and adversely affect human health; zoonotic airborne pathogens can be transmitted from animal facilities to humans	Domingo et al. (2021), Cambra-Lopez et al. (2010), and Scanes (2018)
Antibiotic-resistant bacteria	Use of antibiotics in livestock production has resulted in antibiotic residues in animal-derived foods and antibiotic-resistant bacteria that are widely disseminated in the environment through animal manure and easily transmitted to humans via food chains; farming intensification can increase antibiotic resistance bacterial genotypes and phenotypes in feces of livestock; livestock and poultry, manure, and run-off water from production facilities can be a source of antibiotic-resistant bacteria	Wang et al. (2023a), Manyi-Loh et al. (2018), Tang et al. (2017), and Scanes (2018)

(continued)

Table 2.8 (continued)

Ecosystem health indicator	Main effects	References
Synthetic androgens, estrogens, progestogen, and ractopamine	Hormone contamination of aquatic systems can cause reproductive impairment of aquatic organisms; androgens, estrogens, and progestogen implants are used in beef and dairy cattle in some countries; manure excretion and run-off water from cattle facilities where these compounds are used can cause ecotoxicity; progesterone appears to be the main concern; grazing cattle and applying manure to frozen fields is greatest risk for elevated hormones in runoff; leaching potential of trenbolone acetate peaks at 5–30 days post-implantation; concentrations of ractopamine can be 3 to –4 times greater than trenbolone acetate and melengesterol acetate in feces and runoff water can contain gram quantities of ractopamine	Jones et al. (2014), Scanes (2018), Havens et al. (2020), and Challis et al. (2021)
Zoonotic disease transmission	Animal manure and mortalities can contain several human pathogenic viruses, bacteria, and protozoa; particulate matter in air can contain pathogenic organisms	Cambra-Lopez et al. (2010) and Scanes (2018)
Fossil fuel use	Amount of fossil fuels burned varies among animal species and type of animal product produced; about 30% of global energy consumption is in the agri-food chain and about 9% is used on farms; 50% of on-farm fossil fuel use is attributed to production of seed, herbicides, pesticides, nitrogen fertilizer, and operating farm machinery to produce feed crops; on-farm energy use for heating, cooling, and ventilation is much greater for intensive livestock farms than for smaller scale, extensive or pasture-based livestock farms	Steinfeld et al. (2006) and FAO (2012)

efficiency of animal production and resource use, (2) increasing recycling and minimizing nutrient losses in a circular bioeconomy, (3) using nature-based solutions to decrease carbon footprints, (4) accounting for alternatives when striving for healthy and sustainable diets, and (5) developing government policies to drive change (FAO 2020). In fact, GHG emissions from livestock and poultry production systems can be reduced by as much as 30% by implementing improved nutritional, genetic, health, and management strategies (Gerber et al. 2013).

Limited efforts have been devoted toward understanding the impacts of animal health on sustainability (Perry et al. 2018). Climate change has increased the occurrence of diseases in some agricultural systems, but predictive models need to be developed that integrate ecophysiology and community ecology data involving the host-pathogen interactions under various climatic conditions (Altizer et al. 2013). Little is known about the role of infectious agents in climate change, but Ezenwa et al. (2020) proposed that climate change not only increases the prevalence of infectious animal diseases, but pathogens and parasite infections in ruminants also increase methane emissions creating an ongoing climate-disease cycle. However, Charlier and Morgan (2021) criticised the Ezenwa et al. (2020) analysis as an oversimplification that requires more empirical data and more rigorous modeling because the impacts of climate change on disease are nonlinear, and a more holistic environmental impact assessment is needed. Collaborations between animal health experts and emissions modelers resulted in identifying 56 metrics related to feeding that were effective using farm-scale models for estimating livestock disease impacts on GHG emissions (Kipling et al. 2021). Capper (2023) used simple deterministic models to estimate the impact of several globally important vaccine-preventable diseases on productivity and GHG emissions in intensive dairy, extensive beef, commercial swine, and backyard poultry production systems. Results of this study showed that diseases that increase mortality (e.g., anthrax, classical swine fever, and high pathogenicity avian influenza) cause greater increases in GHG emissions per unit of meat or milk production than diseases that mainly affect yields or reproduction (e.g., brucellosis, low pathogenicity avian influenza). Their results also showed that disease prevalence also affects the magnitude of GHG emissions, where a reduction in the prevalence of porcine reproductive and respiratory syndrome in swine, foot and mouth disease in cattle, and avian infectious bronchitis in poultry reduced GHG intensities. Therefore, the inclusion of animal disease prevalence and its effects on growth, reproductive performance, and mortality should be a significant part of assessing the sustainability of food animal production systems.

6.2 Livestock Systems Are Essential Components of a Circular Economy and Food System

One of the seven principles of doughnut economics is to create to regenerate (Raworth 2017). This means that we need to continue to strive to create a circular economy and food system that recaptures, reuses, and recycles nutrients from food loss and waste for new productive purposes, which is an essential role that livestock and poultry provide in sustainable, resilient food production systems. Yang et al. (2023a) summarized the multiple benefits provided by implementing strategies in a circular bioeconomy which include:

- Using ecological waste management systems that maximise economic efficiency, reduce resource consumption, and prevent environmental pollution.
- Reducing carbon emissions from industry, waste disposal, energy use, building construction, and transportation.
- Improving air and water quality, reducing energy and natural resource consumption, and effectively disposing of solid and toxic waste.
- Reducing nonrenewable resources in food production, increasing externally recyclable resources, and recovering internally generated waste.
- Shortening supply chains and reducing the scale of food consumption by improving efficiency of supply chains.

- Providing additional waste treatment methods for food waste management and increasing the value of by-products.

Feed production contributes 50–85% of climate change impact, 64–97% of eutrophication potential, 70–96% of energy use, and about 100% of land occupation in monogastric animal production systems (Garcia-Launay et al. 2018). Therefore, because feed ingredients used in animal feed production have a large influence on environmental impact, one of the most effective strategies to reduce the environmental impact of livestock and poultry production is to use multiobjective diet formulation approaches that include feed ingredients that have low environmental impacts based on LCA methodologies (Mackenzie et al. 2016; Garcia-Launay et al. 2018; de Quelen et al. 2021; Méda et al. 2021; Soleimani and Gilbert 2021). Life cycle assessment is a compilation and evaluation of the inputs, outputs, and environmental impacts of a system used to produce a product throughout its life cycle (Van Middelaar et al. 2019). Standardised methodology and guidelines have been established for determining various LCA environmental impact indicators (LEAP 2015), but most of the limited number of LCA feed ingredient databases include feed ingredients used in the European Union which are not applicable to other geographic regions around the world. However, an extensive database of feed ingredients with 18 LCA environmental indicator variables has been developed and is available from the Global Feed LCA Institute (GFLI; https://tools.blonkconsultants.nl/tool/16/), which not only includes estimates for specific sources of individual ingredients on global warming potential with and without land use change, embedded water use and land use, but also includes estimates on acidification potential, eutrophication potential, ecotoxicity, ozone formation and depletion, fine particulate matter, mineral and fossil resource use, fine particulate matter formation, and human health indicators. Therefore, sourcing and using feed ingredients with low environmental impacts in animal diets can now be done to significantly reduce the environmental burden of livestock and poultry production systems associated with feeding programs in major food animal production countries.

6.2.1 Recycling Agro-industrial By-products into Animal Feed

One of the most important roles of livestock and poultry in circular, sustainable, and resilient food systems is to utilise nutrients from a wide variety of by-products, biomass, and food loss and waste sources in feeds to produce nutrient-rich animal-sourced foods, and reduce the environmental burden of crop production and use in animal feeds (Uwizeye et al. 2016; Shurson and Urriola 2022; Shurson 2020). The feed industry was created in response to the need for recovering, redirecting, and repurposing valuable nutrients in by-products produced by the grain milling, meat packing, and milk processing industries in animal feeds to comply with laws prohibiting the discarding of these waste streams into water resources and causing environmental pollution (Coffey et al. 2016). However, there continues to be a tremendous need to use more of these by-products and wasted materials in animal feeds to meet the increasing demand for feedstuffs caused by the expansion of animal production systems around the world (Kim et al. 2019). The use of grains and oilseeds to produce biofuels (i.e., ethanol, biodiesel) has been a controversial issue regarding its climate change mitigation benefits relative to the broader impacts on sustainability of agriculture and food production (Shurson 2017). Distillers grains coproducts derived from the ethanol industry generally contain high amounts of dietary fiber and protein with amino acid imbalances, which results in suboptimal nutrient use efficiencies in swine and poultry diets (Jang et al. 2021). However, bioprocessing using fungal fermentation, is an emerging technology that can enhance the feeding value of high fiber by-products from the biofuel industry (e.g., wet maize distillers grains with solubles, soybean hulls (Barnharst et al. 2021; Sun et al. 2021a, b, c), and food waste (Sun et al. 2023a, b, 2024) to reduce their environmental footprint in animal feeds.

6.2.2 Upcycling Food Loss and Waste to Animal Feed

Food loss and waste is a major global problem that contributes to food insecurity and inefficient use of water, land, carbon, nitrogen, and phosphorus, and causes significant economic losses. Globally, about 33–40% (1.6–2.5 billion tons) of food produced for humans is not consumed annually (Wang et al. 2023b; World Wildlife Fund 2021). As a result, food loss and waste is attributed to 30% of agricultural land use, 20% of freshwater use, and 38% of total energy consumption that is wasted, and contributes to 8% of anthropogenic greenhouse gas emissions (FAO 2019). Unfortunately, the use of food waste in animal feed has been minimal relative to the amount produced in most countries because of government regulations restricting or prohibiting its use in animal feed due to biosafety concerns, and a lack of infrastructure and economic incentives (Shurson 2020; Shurson et al. 2023; Dou et al. 2021; Pinotti and Ottoboni 2021). Historical association of transmission of animal diseases from contaminated uncooked or inadequately heat-processed food waste used in animal feed has been the main reason for restricted or prohibited use of animal-derived materials in animal feed in many countries (Shurson 2020; Shurson et al. 2022). However, use of adequate thermal processing inactivates all of the viable disease-causing agents (parasites, bacteria, viruses), except prions that may be present in ruminant-derived tissues, and African swine fever virus that may be present in contaminated pig tissues, in these specific types of food waste materials (Shurson 2020).

Shurson et al. (2023) proposed rethinking current laws and government regulations restricting the use of food loss and waste in animal feeds because of the enormous environmental benefits that upcycling food waste to animal feed could provide. Uwizeye et al. (2019) estimated that feeding food waste to pigs in industrial production systems would increase nitrogen use efficiency and reduce nitrogen losses from feed, resulting in saving the equivalent of 31 million tons of soybeans and 20 million tons of grains that would spare 16 million hectares of arable land globally. Greater impacts can be achieved by considering the use of food waste in poultry and ruminant diets (Dou et al. 2022). Several sources of thermally processed pre-consumer food waste have greater or similar energy and nutritional content compared with conventional feed ingredients such as maize and soybean meal in swine diets (Fung et al. 2019a, 2019b). Directing food loss and waste streams toward animal feed not only enhances nitrogen (Shurson and Kerr 2023; Kerr et al. 2019; Uwizeye et al. 2019) and phosphorus recovery, and improves nutrient use efficiency in swine and poultry diets, but it is a more environmentally sustainable option than diverting these waste materials toward biogas production, composting, and landfills (Padeyanda et al. 2016; Zu Ermgassen et al. 2016; Salemdeeb et al. 2017; Shurson 2020).

6.2.3 Upcycling Agri-food Wastewater Nutrients via Microalgae Production

Another untapped opportunity to upcycle nutrients from waste streams into animal feed is to produce microalgae from agri-food wastewater. Microalgae grow rapidly and can efficiently utilize various chemical forms of carbon, nitrogen, and other nutrients in wastewater to potentially produce an economical source of protein and oil for use in animal feeds and human diets if large-scale commercial cultivation systems can be developed. Selected microalgae strains can efficiently and economically use nutrients in dairy wastewater (Lu et al. 2016) and meat processing (Lu et al. 2015) streams to remove nitrogen and increase biomass yield, protein concentration, and oil content. Microalgae by-products may also provide dietary health benefits in monogastric animals (Urriola et al. 2018).

6.2.4 Upcycling Animal Manure and Food Waste Nutrients via Insect Production

There is considerable interest in producing insects from nutrients present in animal manure and food waste for subsequent use as nutrient-rich supplements in animal feeds. Several commercial companies have developed insect production systems for five insect species that upcycle nutrients from waste streams and concentrate protein and lipids in their biomass (Makkar et al. 2014). In addition, several bioactive compounds have been identified in insect meal that may enhance immune responses when fed to animals (Sogari et al. 2019). However, despite insect meal being promoted as an environmentally sustainable feed ingredient for use in food animal production, very few studies have focused on determining LCA environmental impacts (DeFoliart and Paoletti 2005; Oonincx and de Boer 2012; Van Zanten et al. 2015; Salomone et al. 2017).

6.3 How Do We Measure Economic Sustainability in Livestock Systems in a Redesigned Global Economy?

Economics is the main cause of our inability to overcome the many environmental and social sustainability challenges described throughout this chapter. Although economic viability is essential for households, farms, businesses, and communities to thrive, our current capitalistic economic model has created an economic and financial infrastructure that is not sustainable. Following the leadership of Raworth (2017), Klein (2014), and Hickel (2022) described throughout this chapter, we need to move away from the current perpetual GDP growth economic model that depends on and is controlled by fossil fuels, banks, and corporations that are focused on satisfying financial returns to shareholders, and redesign our economic system using a degrowth, circular, doughnut economics model with a new goal of achieving an optimal "sustainable" return on investment rather than a maximum "financial" return on investment. This model must include living within the Earth's planetary boundaries, repairing and restoring unhealthy ecosystems, and creating a more just, equitable, and thriving society of farmers, farm workers, communities, and global societies.

Economic sustainability is one of the three pillars of the United Nations SDGs that is interconnected with environment and social sustainability, but it is not well defined. Most published scientific studies have defined or created their own measures to represent the perspective or specific context and components of the system being evaluated. For example, in the strategic management literature, economic sustainability is associated with different levels of economic competitiveness. Pitts and Lagnevik (1998) defined economic sustainability as a competitive industry that possesses the capability to profitably gain and maintain market share in domestic and foreign markets. The OECD (2003) defined economic sustainability as "agricultural production that is economically viable and does not degrade the environment over the long run." The FAO (2018c) emphasised that sustainability must include social, institutional, and economic components while not excluding environmental sustainability. The economic objectives of the European Commission Common Agricultural Policy 2023–27 are to (1) ensure a fair income for farmers, (2) increase competitiveness, and (3) improve the position of farmers in the food chain (https://agriculture.ec.europa.eu/index_en).

Similar to the differences in economic sustainability definitions, different metrics, and economic sustainability indicators have been proposed. Bathaei and Štreimikienė (2023) suggested that profitability, liquidity, stability, and productivity be used as primary indicators of economic sustainability. Competitive performance is determined by profitability, liquidity, and stability which are measured by margins and income, while competitive potential is measured by technical efficiency and is a function of productivity and efficiency (Latruffe et al. 2016). However, there are some inadequacies of using margins and income along with technical efficiency as economic sustainability metrics. Measures of margins and income are useful for assessing costs and returns and relative resilience to high-cost scenarios, but they do not account for opportunity costs of owned resources, viability, competitiveness with other sectors in the wider economy, and sources of competitiveness. Technical efficiency is an indicator of progress over time and distinguishes between actual performance and potential, but it is only a quantitative indicator that does not account for price or cost and requires a more detailed analysis to be useful. Therefore, additional data and measures are needed to establish benchmarks and provide a more comprehensive evaluation of progress and areas needing improvement in economic sustainability in food animal production systems. Ideally, data and measures associated with the economic sustainability of livestock and poultry production systems should include (1) more precise measurement of owned versus contracted capital and labor (time), (2) income disparities, (3) market concentration, (4) diversification and innovation, (5) externalities and public goods, (6) resilience to market shocks, (7) long-term impact assessment, and (8) external factors and market dynamics.

6.4 Resilient Livestock Production Systems and Supply Chains Must Prepare to Overcome Numerous Threats

The resiliency of livestock and poultry production systems and supply chains not only requires achieving economic sustainability but there are numerous threats to food animal production systems and supply chains that must be considered. Climate change is already having several direct effects on livestock production systems, including increased frequency, intensity, and duration of heat stress which reduces productivity, growth, and reproduction rates, and increases mortality and morbidity (Table 2.9). In addition, climate change is increasing the frequency of occurrence of extreme weather events and natural disasters (floods, drought, blizzards, earthquakes, wildfires; Table 2.9).

Climate change also has several indirect effects on animal production systems including altered weather patterns that reduce feed crop yields, pasture biomass, and ingredient supply; reduced nutritional quality of grains, oilseeds, and by-products by increasing the presence and concentrations of mycotoxins; reduced quality and quantity of freshwater; increased spread of vector-borne diseases; and increased likelihood of emerging pathogens (Table 2.9). Climate change is affecting the distribution and causing shifts in the production of feed and animals which is adding disproportional pressure to global supplies on different continents (Habib-ur-Rahman et al. 2022). Water scarcity has become a major global problem (Hoekstra et al. 2012; D'Odorico et al. 2019; Rosa et al. 2019) that must be addressed as a key part of creating livestock production resiliency plans and strategies.

Climate change is also a major contributor to emerging contaminants (Mukherjee 2023), emerging pandemic diseases (Morens and Fauci 2020), and zoonoses transmission from meat and milk (Pradhan and Karanth 2023). Increased globalization of agriculture and food supply chains combined with climate change have increased the risk of exposure to foodborne contaminants including several pathogens and parasites, toxic compounds from algae blooms, heavy metals, and mycotoxins (Mukherjee 2023). The disruptions caused by the COVID-19 pandemic have served as a stark reminder of the urgent need to understand and mitigate the underlying factors causing zoonotic spillover events that occur at the human-wildlife and domesticated animal interface (Shepon et al. 2023). Antimicrobial resistance is a serious One Health concern that has broad implications for human, animal, plant, and environmental health (Velazquez-Meza et al. 2022). In addition, the detrimental effects of emerging One Health concerns involving endocrine disruptors, per- and polyfluoroalkyl substances (PFAS), and micro- and nanoplastics have created another dimension of complex challenges than need to be urgently addressed to overcome their detrimental effects on human and animal health (Table 2.9). Emergence and re-emergence of existing diseases are determined by dynamic

Table 2.9 Summary of major threats to food livestock and poultry production systems

Threat	Main effects	References
Climate change—direct effects	Heat stress causes metabolic disorders (lameness, ketosis), oxidative stress, immune suppression, reduced health status, reduced growth rate, reduced reproductive performance, reduced meat, milk, and egg production, yield, and quality, increased morbidity, mortality, and economic losses	St-Pierre et al. (2002), Thornton (2010), Nardone et al. (2010), Henry et al. (2012), Lacetera (2019), Ali et al. (2020), and Thornton et al. (2022)
Climate change—indirect effects	Reduced nutritional quality and availability of feed resources; increased prevalence of mycotoxins in feed; reduced availability of clean water; increased vector-borne pests that can increase the spread of vector-borne diseases; increased parasites, helminths, and other pathogens; increased likelihood of emerging pathogens; loss of biodiversity; supply chain disruptions from natural disasters	Thornton (2010), Bett et al. (2017), Zhou et al. (2017), Carozzi et al. (2022), Caminade et al. (2019), Raza et al. (2019), Ali et al. (2020), and Jägermeyr et al. (2021)
Antimicrobial resistance	Antimicrobial resistance (AMR) is a leading global One Health problem; animal antimicrobial consumption is linked to animal AMR and human antimicrobial consumption is linked to human AMR; animal antibiotic consumption is positively linked to AMR of pathogens of critical priority to humans, and human antibiotic consumption was positively linked with animal AMR; soil is a reservoir of pathogens and antibiotic-resistant genes; reduced rates of antibiotic consumption are not sufficient to stop increasing global prevalence of AMR; changes in hygiene, biosecurity, and management practices are required to reduce the need for antimicrobials in livestock production; socioeconomics and governance play a significant role on AMR rates in humans and animals	Allel et al. (2023), LeJeune et al. (2023), Meier et al. (2023), Singh et al. (2023), and van den Honert and Hoffman (2023)

(continued)

Table 2.9 (continued)

Threat	Main effects	References
Regulated, co-occurring, emerging, and masked mycotoxins	Climate change is increasing the prevalence, concentrations, geographical distribution, and types of mycotoxins; mycotoxins are major dietary contributors to oxidative stress and reduce nutritional efficiency, productivity, and health of animals; deoxynivalenol, zearalenone, aflatoxins, ochratoxin A, patulin, fumonisins, and T-2 toxins are the most common mycotoxins measured in animal feeds; multiple mycotoxins are usually present in the same feed sample because some fungal species produce several mycotoxins and multiple fungal species can be found in the same ingredient, but the toxic effects of combinations of co-occurring mycotoxins are poorly understood; emerging mycotoxins are secondary metabolites produced by produced by *Fusarium* fungi species that are not routinely measured or regulated but can have detrimental effects on livestock and poultry; masked mycotoxins are derivatives of commonly measured and regulated mycotoxins that are undetectable due to their modified chemical structure when using conventional analytical methods but can be a feed and food safety concern	Escrivá et al. (2015), Vanhoutte et al. (2016), Medina et al. (2017), Peng et al. (2018), Liu et al. (2022), Xu et al. (2022), Borutova et al. (2012), Khoshal et al. (2021), Fraeyman et al. (2017), Streit et al. (2013), Berthiller et al. (2013), Rychlik et al. (2014), Zhang et al. (2019b), and Kos et al. (2023)
Endocrine disruptors	Endocrine disruptors are exogenous synthetic chemical substances or mixtures that interfere with natural hormones of livestock and humans by altering functions of the endocrine system that cause adverse health or reproductive effects on an organism, its progeny, or population which can have lifelong effects for next generation of offspring; new endocrine disruptor chemicals are emerging that have detrimental effects on growth, reproduction, and obesity; routes of entry to animal bodies include use of synthetic steroid hormones, grazing contaminated pasture, drinking water, concentrate feed, inhalation, placenta, milk exposure, skin contact	Gaddafi et al. (2021), Kolesnyk and Prodanchuk (2023), Magnusson and Persson (2015)
Various chemical contaminants	High doses of acute or low doses of chronic exposure to heavy metals (As, Pb, Cd, Hg), metalloids, herbicides, pesticides, insecticides, phthalates, and polychlorinated biphenyls can cause disruption of multiple organs and reproduction	Guvvala et al. (2020)

(continued)

Table 2.9 (continued)

Threat	Main effects	References
Per- and polyfluoroalkyl substances (PFAS)	Perfluorinated and polyfluorinated alkyl substances (PFAS) are a group of more than 4700 synthetic chemicals, also known as "forever chemicals," that resist environmental degradation, are present everywhere in the environment, and accumulate in all living things over time; PFAS have many adverse effects on human and animal health including liver damage, thyroid disease, obesity, infertility, and cancer; livestock are exposed to PFAS from air, water, soil, or feed	Brunn et al. (2023), Jha et al. (2021), Obsekov et al. (2023), Peritore et al. (2023), Death et al. (2021), and Fernandes et al. (2019)
Micro- and nanoplastics	Plastics slowly degrade in the environment to micro and nanoplastic particles (MNPs), move across trophic levels, adsorb harmful chemical pollutants, and migrate to various tissues in animals; MNPs are comprised of polymers (polyethylene, polypropylene, polystyrene) and can contain phthalates and bisphenol which are ecologically harmful compounds, serve as carriers for harmful chemical compounds (polychlorinated biphenyls, polyaromatic hydrocarbons), and biofilms comprised of pathogens; MNPs affect multiple systems from particle toxicity, release of leachates, sorption of hydrophobic compounds, changes in processes and population that affect ecosystem and public health; MNPs can be ingested by organisms, accumulate in the body, be transferred along food chains, adversely affect growth, reproduction, and metabolism; MNPs have been detected in all major livestock production systems, directly affect animal health, and indirectly affect human health; plastic mulching, fragmentation of plastic waste, and water runoff are main entry routes in grazing and mixed livestock production systems	Bouwmeester et al. (2015), Wu et al. (2019), Shen et al. (2019), Peng et al. (2020), Prata et al. (2021), Ramachandraiah et al. (2022), Kibria et al. (2023), and Pandey et al. (2023)

(continued)

Table 2.9 (continued)

Threat	Main effects	References
Water quality degradation and scarcity	Freshwater scarcity is a major global concern with substantial ecological consequences including complete desiccation of river basins during dry seasons, loss of aquatic biodiversity, and economic losses; 52% of global irrigation is unsustainable and driven by domestic and export demand for agricultural products; unsustainable irrigation practices deplete freshwater reserves, disrupt environmental flows, and impair aquatic ecosystems; determining the water footprint of local and regional domains can be used to compare, value, and negotiate quantities and sources of water for on-farm consumption as well as assess farming measures to improve precipitation and technical water productivity in livestock production; climate change is causing altered precipitation on pastures for grazing	Hoekstra et al. (2012), D'Odorico et al. (2019), Rosa et al. (2019), and Drastig et al. (2021)
Animal genetic diversity	Animal genetics are an essential part of agricultural and ecosystems comprised of more than 7000 distinct local breeds, and 1000 transboundary livestock breeds of about 40 species plus domesticated honeybees and other pollinators; breeds in less-intensive systems often develop specific adaptive features to those environments; sustainable management of livestock requires long-term government policies to conserve genetic diversity of food-producing animals; gene banking may be the most secure approach for long-term conservation of rare breed genetics; genome editing can accelerate genetic improvements for pest and disease resistance, animal welfare, adapting to climate change, and reducing environmental impacts of animal agriculture	Leroy et al. (2018), Blackburn et al. (2019), Martyniuk (2021), Wray-Cahen et al. (2022), and Perini et al. (2023)

(continued)

Table 2.9 (continued)

Threat	Main effects	References
Transboundary transmission of foreign animal diseases	Global trade is a major contributor to the spread of disease; disease surveillance systems provide necessary data for risk assessments of disease transmission, but capabilities vary among countries; global wildlife trade provides routes of disease transmission that cause human and animal disease outbreaks and threaten rural livelihoods, native wildlife populations, and health of ecosystems; pathogen introduction pathways include formal trade of animals and animal products, informal trade involves illegal smuggling and cross-border transfer of animals and animal products for personal use, and fomites; international travel, migration and local mobility, intensification of plant, animal, and animal product trade have contributed to increased disease spread	Zepeda et al. (2001), Karesh et al. (2005), Beltran-Alcrudo et al. (2019), and Baker et al. (2022)

balances or imbalances among humans, animals, pathogens, and the environment in complex, global ecosystems (Morens and Fauci 2020). Zoonotic pathogens (e.g., *Escherichia coli*, *Salmonella*, *Listeria*, Herpevirus, and *Toxoplasma gondii*) that can infect humans from their bovine, porcine, ovine, and avian reservoirs via consumption of animal-derived foods are increasing due to climate change, animal production intensification, global trade, and population expansion (Pradhan and Karanth 2023). Unfortunately, limited success has been achieved in integrating human, animal, and environmental health in decisions supporting sustainable agriculture and food systems. However, Zhang et al. (2022) developed and proposed a useful framework for a holistic global One Health index using human, animal, and environmental health indicators for developing government policies and implementing One Health practices.

Livestock and poultry producers must also manage risks associated with labor shortages; disease outbreaks; market price fluctuations; environmental degradation, invasive species, and pests; conflict and displacement; policy and regulatory issues; access to knowledge and training; meeting market and value chain requirements; and navigating social and cultural changes associated with rapid urbanisation and changing cultural norms. Therefore, practical contingency plans are needed to overcome potential threats to livestock production systems (Table 2.10). Securing alternative feed and water supplies, shifting from large to small ruminants based on available land and water resources, preventing and containing disease outbreaks, minimizing heat stress, ensuring adequate farm labor, reducing market risks, drought monitoring, and developing contingency plans for extreme weather events and supply chain disruptions are key examples necessary for improving the resilience of livestock and poultry production systems in an ever-changing uncertain world.

Resilient livestock systems provide stable and diverse sources of income and production with the capacity to recover from adverse events, have effective social mechanisms that reduce vulnerability, and can adapt to climate change and environmental disruptions (FAO/OECD 2012). The resilience of livestock systems is determined by three factors: exposure, sensitivity, and adaptive capacity, where exposure × sensitivity determines vulnerability. Exposure is determined by duration, frequency, severity, location of animals, and location of essential assets (feed, water, housing). Sensitivity is determined by breed, type of housing system, type of feeding system, animal health risk (vaccination rate, level of biosecurity), and value to the household (livelihoods, food security). Adaptive capacity is determined by species, breeds, availability and adaptability to alternate feed resources, production system, accessibility of animals for health services, outbreak response (surveillance, compensation), and household wealth status (FAO/OECD 2012). Therefore, resilient livestock systems can be achieved by reducing exposure (e.g., moving to another area), reducing sensitivity (e.g., diversification), and increasing adaptive capacity (e.g., disease surveillance training). Resilience can be built at all levels, from animal to social system, but it always requires locally specific options. Biodiversity (composition, structure, and function) of ecosystems with livestock production systems embedded within them is essential for climate change adaptation at the ecosystem, species, and gene levels. Only healthy ecosystems can provide services that are essential to livestock production (biomass production, water, disease regulation), especially due to climate change. Shifting species to small ruminants in arid climates and geography is necessary to match feed resources and availability by utilising plant species that are drought-resistant. Genetic diversity provided by animal genetic resources is essential for adapting to harsh conditions and providing resilience to extreme climatic events and diseases. Greater emphasis needs to be focused on sustainable management of genetic diversity of gene pools because 27% of livestock breeds are at risk of extinction. Although it may be possible to develop more heat-tolerant and disease-resistant genetic lines and breeds using animal breeding programs focused on building resilience in the future (Berghof et al. 2019; Poppe et al. 2021), most of the current breeds and genotypes of food-producing animals are not well adapted to these extremes.

7 Food Fraud and Inadequate Cybersecurity Are Significant Threats to Our Global Food System

Food fraud and inadequate cybersecurity are threats to the resiliency of our global food and agricultural system. Food fraud is a deliberate act intended to deceive customers by intentional adulteration of food for economic gain or the purpose of causing harm. Food fraud costs the food industry and consumers $10 to $15 billion per year (Grocery Manufacturers Association 2010). Everstine et al. (2018) reported that about 44% of 1670 food fraud incidents in the United States were considered to be potentially hazardous. Food fraud occurs by substituting one product for another; dilution with less valuable substances; removal of a valuable constituent; using

Table 2.10 Key strategies to overcome potential threats and improve resiliency of livestock and poultry production systems

Strategy	Strategy
Securing adequate quality and quantity of feed	*Ensure adequate farm labor*
Adapt sourcing to shifting crop production patterns and yields	Recruit and train farm workers in best practices
Develop alternative sourcing strategies to minimise and mitigate adverse effects of increased prevalence and concentrations of mycotoxins	Develop action plans to effectively respond to human health pandemics
Identify opportunities to source locally available agri-food by-products and upcycle appropriate thermally processed food waste sources into animal diets	Develop action plans to respond to adverse weather events and natural disasters
Shift production systems from large to small ruminants based on available local feed and water resources	Provide optimal health and well-being by requiring appropriate vaccinations and following worker health and safety protocols
Develop community reserves of key ingredients for use during supply chain disruptions	*Develop contingency plans for supply chain disruptions*
Protect land and water resources from foreign ownership	Establish alternative transportation logistics to minimize shipping disruptions caused by pandemics, severe weather events, and conflicts
Securing adequate quality and quantity of water	Evaluate the use of insurance policies for financial protection from disease outbreak risks
Identify potential alternative water sources and evaluate suitability for animal consumption	*Reduce market risks*
Match animal density with locally available water and feed resources	Seek professional advice on use of marketing Contracts and various pricing mechanisms to reduce market risks
Prevent and contain disease outbreaks	Identify opportunities to access different markets
Local biosecurity, surveillance, and quarantine protocols	Develop alternate transportation options
Only use antibiotics for therapeutic purposes	*Install on-farm renewable energy generators*
Consider using antiviral and nonantibiotic feed additives	Solar, wind, biogas
Use appropriate vaccines	*Use silviculture and agroforestry practices*
Minimize heat stress	Windbreaks
Ensure adequate access to water	Riparian forest buffers and alley cropping
Minimize sun exposure	Silvopasture
Implement cooling systems (e.g., fans, drip or spray cooling) in confinement facilities	*Develop animal breeding programs to increase resilience to disease and heat stress*

unapproved, illegal, or undeclared enhancements or additives; misrepresenting nutritional content, country of origin, expiration date, or production method; using counterfeit documentation; animal species substitution; stolen food shipments; or intentional contamination with chemicals, biological agents, or other substances harmful to human health (Gendel 2023). Common examples include substituting food products with unique characteristics (e.g., value-added, certified, organic, free-range, wild-caught, natural, and Protected Designation of Origin) with conventional or inferior products (Kelly 2023). Food fraud is a public health concern because of the intended and unintended food safety consequences that can lead to toxicity, food-borne illness, or death (Kelly 2023). Therefore, food fraud is a significant threat to sustainability and One Health that requires mitigation to interrupt the flow of fraudulent ingredients and information along supply chains (Gendel 2023). Preventing food fraud requires developing a vulnerability assessment and mitigation plan (Global Food Safety Initiative 2020).

During the COVID-19 pandemic, several organised cyberattacks on the food industry occurred using ransomware, which exposed cybersecurity vulnerability and emphasised the need for organisations throughout food supply chains to develop effective food defense plans that reduce the risk of a ransomware attack and facilitate targeted, quick responses if an event occurs (Manning and Kowalska 2023). Cybercrime has increased more than 50% since 2018, and costs the global economy nearly 1 trillion USD annually (Cremer et al. 2022). Cybersecurity is an essential component of agricultural and food system supply chains and is becoming critically important in a global economy increasingly driven by data and information flows. Cyberattacks are increasing and causing data breaches, and disruptions that severely impact production and entire supply chains, and create fear and mistrust among consumers (Alqudhaibi et al. 2024). Current cybersecurity frameworks are inadequate to protect against this increasing threat to the global food system and require a framework that includes continuous employee training and education (Alqudhaibi et al. 2024). There is also an increasing need for better cyber information sources, standardised databases, mandatory reporting, and public awareness (Cremer et al. 2022). At the farm level, cybersecurity on dairy farms in Finland was evaluated and found to be woefully inadequate due to inappropriate network topology, malware protection, system backups, nonfunctioning security cameras, and lack of farmer awareness of technologies (Nikander et al. 2020). It is highly probable that these problems are not unique to dairy farms in Finland and are widespread throughout global livestock production systems. Supply chain disruptions can be minimized by detecting fake news and disinformation using AI and machine learning applications (Akhtar et al. 2023). The Internet of Things, blockchain, and tiny machine learning (TinyML) are emerging technologies capable of assisting with food supply chain security and the detection of malicious behavior (Tsoukas et al. 2022). Therefore, although there are several useful cybersecurity technologies available to protect all stakeholders in global food supply chains and livestock production systems, they have been poorly implemented which has resulted in a substantial threat to global food safety, security, and sustainability.

8 How Do We Manage the Use of Novel Genetic Resources and Prevent Biopiracy of Natural Products from Diverse Ecosystems?

Historically, new seeds and new foods have been exchanged and traded among societies and countries without any legal regulation of intellectual property (IP) for many centuries. However, the emergence of biotechnology applications in agriculture, especially genetic modifications from plant breeding, has led governments and corporations in developed countries to realize the enormous potential for economic growth and increased international competitiveness. As a result, legal protection mechanisms were created. Intellectual property rights in agriculture and food systems have become an important component of sustainability because they affect food security, promote scientific and technical innovations, encourage multinational plant genetic company investment, reduce the cost of research and development, and improve agricultural productivity, but they may also lead to increased biodiversity loss, increase production costs and management for farmers, block food technology innovation and dispersion, and increase farmer litigation risks (Zhang and Ge 2022). China has received much attention regarding IP in agriculture because of enormous pressures to find expedient and inexpensive ways of feeding billions of people in the country. Acquiring IP illegally through agricultural espionage has become a convenient way for China to quickly improve productivity and become more competitive in global markets, and has led to stealing plant genetics IP in the United States rather than investing and conducting the necessary research in China (Hvistendahl 2020). There are also concerns that stolen genetically modified seeds from the United States could be used to reverse engineer and hack the code of DNA to create biowarfare applications (Greenwood 2022). Therefore, because of the potential risks of unknown consequences of genetic engineering and the potential misuse of these genetic sources for biowarfare, there is a compelling need to continue with current laws to protect plant and seed IP, but it should also involve coordinating the relationship between IP rights, food rights, and farmer rights, as well as establishing a mechanism to provide oversight from multiple participating institutions to prevent IP from controlling markets, restricting competition, limiting innovation, and being exploited for destructive activities (Zhang and Ge 2022). As digital technology applications, including sensors, robotics, software-defined networks, and the Internet of Things continue to develop in agriculture and food systems, they can also play an important role in IP theft protection (Hossain et al. 2021).

Biodiverse ecosystems can be a rich resource for discovering novel, natural chemical and biological compounds of benefit to society. The increasing need to find novel medicinal chemical compounds and biological agents to treat a variety of human and animal health ailments, along with the need to identify and develop alternatives to antibiotics to combat antimicrobial resistance, have led pharmaceutical and agricultural companies to bioprospecting. Bioprospecting involves exploring biodiverse ecosystems to discover novel chemicals, genes, metabolic pathways, structures, materials, and behaviors of wild species that have social and commercial value (Beattie et al. 2011). However, when bioprospecting is conducted in local ecosystems and using local medicinal knowledge without acknowledging and compensating Indigenous people for this IP, it becomes biopiracy. Biopiracy is biological theft that involves the unethical and illegal collection of Indigenous medicinal plants, seeds, genetic resources, and traditional medicines by corporations who patent them for commercial use without fair compensation to the people or government of the region or country where they are grown (Imran et al. 2021). Pharmaceutical companies are the most common violators, and biopiracy may also include illegal acquisition of Indigenous knowledge, farming practices, and extraction processes that exploit less affluent communities around the world. The World Intellectual Property Organization (WIPO) Intergovernmental Committee on Intellectual Property and Genetic Resources, Traditional Knowledge and Folklore (IGC) is developing legal guidelines that ensure protection of these unique biological resources and fair compensation for those who use them (https://www.wipo.int/tk/en/igc/).

Farmer's rights have been extensively debated in international public law regarding the "tug-of-war" conflict of legal protection between plant genetics corporations and smallholder farmer's rights to save, use, exchange, and sell seeds saved from farms (Muzaka 2020). During the past three decades, the internationally recognised and legally binding IP rights of plant breeders have become a dominant force restricting the use of plant genetics for crop production. This trend is contrary to the doughnut economics principles of privatising natural resources (plant genetics) for profits by large corporations while restricting freedoms of traditional farmers for using resources to support their livelihoods (Muzaka 2020).

9 Opportunities and Challenges of Emerging Artificial Intelligence, Robotics, Quantum Computing, and Synthetic Biology Technologies

Science and technology have provided numerous benefits to improve the well-being of billions of people for many centuries. Technology has impacted almost every aspect of people's lives and is a significant component of geopolitical strategies where it has created or supported global flows of people, information, ideas, knowledge, commodities, finished goods, capital, and wealth (Suleyman and Bhaskar 2023). The possibilities of the coming wave of emerging technologies involving artificial intelligence (AI), quantum computing, robotics, advanced biotechnology, nanotechnology, gene synthesis, and their combinations have the potential to transform the human experience and the world in which we live in unbelievable ways.

Artificial intelligence is based on computer programs that can rapidly analyze enormous amounts of data, finding connections, learning, conducting simulations, and forecasting to enable better decision-making. There are numerous applications and possibilities for using AI to help solve global climate change and ecosystem health problems such as optimising the supply and demand of natural resources for food production, identifying systems strategies capable of safe and efficient carbon sequestration, and modelling and forecasting floods and other natural disasters to improve the preparedness of local governments. The use of AI enables addressing complex issues in quantitative and predictive epidemiology by extracting information from massive amounts of data to monitor animal health, study host-pathogen interactions, and implement precision-based human and animal medicine through administration of more effective treatments (Ezanno et al. 2021).

Applications of AI technology in crop and livestock production are beginning to emerge as part of precision farming practices. In crop production, AI has been used to predict weather and crop yields; characterize crop quality; is often combined with sensors, computers, and robotics for early identification of plant diseases; and can improve soil management and optimise irrigation water, pesticide and herbicide use (Awasthi 2020; Ayed and Hanana 2021; Eli-Chukwu 2019; European Parliamentary Research Service 2023; Hossen et al. 2023; Patrício and Rieder 2018; Talaviya et al. 2020). Most of the AI applications in animal agriculture thus far involve assessment of various animal behaviors for health and welfare monitoring, early disease detection, individual animal identification, and live weight estimation (Bao and Xie 2022; Neethirajan 2021; Priya et al. 2022; Mamehgol Yousefi et al. 2022).

Ryan (2022) discussed the social and ethical impacts of AI use in agriculture and proposed 11 overarching principles that need to be addressed including transparency, justice and fairness, non-maleficence, responsibility, privacy, beneficence, freedom and autonomy, trust, dignity, sustainability, and solidarity. There are also numerous concerns regarding the use of AI in agriculture that require discussion, including (1) technology-mediated unemployment; (2) transformation of the nature of agricultural work; (3) concentration of ownership; (4) exploitation and surveillance; (5) vulnerability of hacking; (6) bias; (7) explainability, trust, and responsibility; (8) transformation of rural and urban relations; (9) economic vulnerability; (10) environmental risks; and (11) alienation from the natural world (Sparrow et al. 2021).

Simulating complex agricultural systems is challenging using current digital technologies because they do not have the computational capacity to deal with the multiple interacting variables. Fortunately, quantum computing is another rapidly emerging technology that uses quantum mechanics, computer science, physics, and mathematics to solve complex problems at unprecedented speeds and capacity compared with current computing technology. Although quantum computing is still in its developmental stages, it has tremendous potential to process enormous datasets and conduct simulations that could lead to improved sustainable farming practices and more efficient resource allocation, along with weather forecasting, crop modeling, genetic analysis, optimising supply chain logistics, and discovering new pharmaceutical compounds. However, there are many legal and political issues related to adoption, potential areas of market disruption, and numerous research questions that need to be addressed to strategically position quantum computing in various applications (Ukpabi et al. 2023).

Agricultural robots are an emerging technology that are being used in agriculture to reduce labor demands on farms and accelerate data-driven precision farming practices, but several challenges exist for ground robots in agricultural environments due to wide variations in environmental conditions (Vougioukas 2019). Most robots used in agriculture require intelligence solutions which limits their use to small-scale applications without integration with AI (Cheng et al. 2023). However, various AI techniques are being applied to some robotics applications used in agriculture and ecosystems (Wakchaure et al. 2023). Although robots used for harvesting fruits and vegetables are still less efficient than humans, research is rapidly advancing the capabilities of this application (Morar et al. 2020). Robotics and autonomous systems can also be used to decarbonise agricultural production by increasing nitrogen use efficiency, accelerating plant breeding, supporting regenerative agriculture, electrifying robotic vehicles, and reducing food waste (Pearson et al. 2022). Although there are several robotic milking technologies that have been developed for pasture-grazed dairy farming that could benefit farmers and society, they have not been readily adopted (Eastwood et al. 2022).

Similar to AI, synthetic biology is a rapidly emerging technology that will provide numerous new capabilities at a much lower cost. Synthetic biology is the ability to design and engineer new organisms or redesign existing biological systems by modifying and using DNA (Suleyman and Bhaskar 2023). Recent scientific discoveries and innovations enable the ability to alter DNA encoding and direct its effects in food production and medicine. Scientists now have the capability to sequence entire genomes of humans, plants, animals, and all other organisms, edit and write genetic code using CRISPR (clustered regularly interspaced short palindromic repeats) technology, and synthesize and print to create new strands of DNA (Suleyman and Bhaskar 2023). Companies are emerging that have the technological capabilities of reading, editing, and writing DNA to create completely new molecules in life processes (Suleyman and Bhaskar 2023). Many technological innovations are being developed that may benefit the environment such as algae that reduces carbon in the atmosphere, plants that consume toxic waste, and gene drives that move disease-spreading mosquitoes out of habitats (Suleyman and Bhaskar 2023). Various nanotechnology applications are also emerging that will have many beneficial effects, including reducing adverse environmental impacts, improving food security and productivity, and promoting social and commercial equity (Gupta et al. 2023). There are numerous medical applications, materials, manufacturing processes, and consumer goods that will likely be transformed using synthetic biology technologies when combined with AI (Suleyman and Bhaskar 2023). In agriculture, synthetic biology can be used to engineer biosensors, synthetic speciation, microbial metabolic engineering, and novel antimicrobials (Goold et al. 2018). Additional applications in food production can result in producing healthier food ingredients with specifically designed nutritional characteristics (Shi et al. 2022). Gene editing of farm animals can lead to improved health and productivity (Tait-Burkard et al. 2018) and enable the creation of genetic resistance of poultry to avian influenza (Idoko-Akoh et al. 2023). Synthetic biology can also be used to produce customized functional feed resources that improve nutrient utilisation efficiency and health of food-producing animals (Wang et al. 2022b).

However, Suleyman and Bhaskar (2023) warn us that although AI, quantum computing, and synthetic biology can be incredibly powerful technologies that can provide an extraordinary number of benefits to food systems and societies, their power can also lead to immense dangers and ethical dilemmas. In general, individuals, organizations, and corporations do not like regulations and being told what they can and cannot do, but we need organized and established rules and regulations to prevent bad actors from abusing these technologies and the potentially devastating consequences (e.g., bioterrorism, cybercrimes) that could occur. Suleyman and Bhaskar (2023) remind us that introducing these new technologies into our current world that is under enormous social stress (i.e., political instability, waves of populism, increased concentration of wealth, loss of trust, and deep-seated social anxiety and resentment caused by increasing inequality) will make the containment of these technologies incredibly challenging to ensure that they are used to provide a net benefit to society. We have already learned from experience and extensive debate, that there are different perspectives regarding the safety and appropriateness of producing genetically modified crops and organisms that have led to different government regulations and trade policies in different countries. Containment and control of bad actors who use these technologies will be difficult because they will have diverse and disproportionate impacts, are evolving at an exponential pace, can be used for many different purposes, and are independent from previous technologies (Suleyman and Bhaskar 2023). However, Suleyman and Bhaskar (2023) proposed ten collective strategies that may be effective if they are implemented. The fate of human society and the planet will depend on our ability to manage these powerful emerging technologies to an even greater extent than the enormous challenges we are trying to overcome in transforming our global economy, livestock and food production systems, ecosystems, and societies for a sustainable and resilient future.

10 The Path Forward

Human behaviors and actions have shaped the current state of the world where we live including the many good things we enjoy and the many bad things we need to overcome. A different type of economy is possible, but it requires all of us to work on behalf of the common good to create more equitable, just, and safer global societies while living within the Earth's carrying capacity to support 9+ billion people. If we all agree that this should be our new goal, then achieving it depends on whether we have the collective courage to alter our current trajectory from a growth and fossil fuel–dependent economic system toward one that reconnects people, agriculture and food production, and ecosystems in sustainable ways. To do this, there are many "tug-of-wars" that need to be concluded with outcomes that optimise the sustainable return on investment rather than maximize the financial return on investment. The global scientific community does not have all of the answers to achieving all of the United Nations SDGs nor achieving One Health, but restoring our faith in the truthfulness of science and using it to guide decisions to solve these complex problems can be a starting point for protecting the planet and human society, and reversing our current trajectory toward catastrophic devastation. We know the problems and their root causes. There are no "silver bullets" or quick fixes. Change is difficult, but courage is the root of change. A better human society and world is possible, and we have the capability to achieve it if we have the courage to change our behavior. Artificial intelligence, and eventually quantum computing, can be enormously helpful in finding systems solutions that account for the many complexities, tradeoffs, and unintended consequences of changes beyond our current capabilities. Solving global problems of the magnitude we are currently experiencing requires all communities to focus on what is good for "us" rather than what is good for "me." Nature is constantly changing and adapting. Can we?

Acknowledgments Special thanks to Derek M. Shurson (DCO, LLC, Inver Grove Heights, Minnesota, USA; dshurs20@gmail.com) for generously providing his graphic design talents and collaborating to develop Fig. 2.1 in this chapter.

References

Adcock F, Anderson D, Rosson P (2015) The economic impacts of immigrant labor on U.S. dairy farms. National Milk Federation, Texas A & M AgriLife Research, College Station, pp 1–21. https://www.nmpf.org/wp-content/uploads/immigration-survey-090915.pdf

Adesogan AT, Havelaar AH, McKune SL, Eilittä M, Dahl GE (2020) Animal source foods: sustainability problem or malnutrition and sustainability solution? Perspective matters. Global Food Security 25:100325. https://doi.org/10.1016/j.gfs.2019.100325

Akhtar F, Lodhi SA, Khan SS (2015) Permaculture approach: linking ecological sustainability to business strategies. Manag Environ Qual Inter J 26:795–809. https://doi.org/10.1108/MEQ-01-2015-0001

Akhtar P, Ghouri AM, Khan HUR, ul Haq MA, Awan U, Zahoor N, Khan Z, Ashraf A (2023) Detecting fake news and disinformation using artificial intelligence and machine learning to avoid supply chain disruptions. Ann Oper Res 327:633–657. https://doi.org/10.1007/s10479-022-05015-5

Alders RG, Campbell A, Costa R, Guèye EF, Hoque MA, Perezgrovas-Garza R, Rota A, Wingett K (2021) Livestock across the world: diverse animal species with complex roles in human societies and ecosystem services. Anim Front 11:20–29. https://doi.org/10.1093/af/vfab047

Ali MZ, Carlile G, Giasuddin M (2020) Impact of global climate change on livestock health: Bangladesh perspective. Open Veterinary J 10:178–188. https://doi.org/10.4314/ovj.v10i2.7

Allel K, Day L, Hamilton A, Lin L, Furuya-Kanamori L, Moore CE, van Boeckel T, Laxminarayan R, Yakob L (2023) Global antimicrobial-resistance drivers: an ecological country-level study at the human-animal interface. Lancet Planet Health 7:e291–e303. https://doi.org/10.1016/S2542-5196(23)00026-8

Alqudhaibi A, Krishna A, Jagtap S, Williams N, Afy-Shararah M, Salonitis K (2024) Cybersecurity 4.0: safeguarding trust and production in the digital food industry era. Discov Food 4:2. https://doi.org/10.1007/s44187-023-00071-7

Altizer S, Ostfeld RS, Johnson PTJ, Kutz S, Harvell CD (2013) Climate change and infectious diseases: From evidence to a predictive framework. Science 341:514. https://doi.org/10.1126/science.1239401

American Immigration Council (2022) The essential role of immigrants in America's meat and dairy industries. Labor Market Series Factsheet, Tending to America's Food Supply, p 11. https://www.americanimmigrationcouncil.org/sites/default/files/research/tending_to_americas_food_supply_meat_dairy_industries.pdf

Andretta I, Pomar C, Rivest J, Pomar J, Lovatto PA, Neto JR (2014) The impact of feeding growing-finishing pigs with daily tailored diets using precision feeding techniques on animal per-

formance, nutrient utilization, and body and carcass composition. J Anim Sci 92:3925–3936. https://doi.org/10.2527/jas2014-7643
Andretta I, Pomar C, Rivest J, Pomar J, Radünz J (2016) Precision feeding can significantly reduce lysine intake and nitrogen excretion without compromising the performance of growing pigs. Animal 10:1137–1147. https://doi.org/10.1017/S1751731115003067
Andretta I, Hauschild L, Kipper M, Pires PGS, Pomar C (2018) Environmental impacts of precision feeding programs applied in pig production. Animal 12:1990–1998. https://doi.org/10.1017/S1751731117003159
Asafu-Adjaye, J. Blomqvist, L., Brand, S., Brook, B., Defries, R., Ellis, E., Foreman, C., Keith, D., Lewis, M., Lynas, M., Nordhaus, T., Pielke, Jr., R., Pritzker, R., Roy, J., Sagoff, M., Shellenberger, M., Stone, R., & Teague, P. (2015): An Ecomodernist manifesto. https://static1.squarespace.com/static/5515d9f9e4b04d5c3198b7bb/t/552d37bbe4b07a7dd69fcdbb/1429026747046/An+Ecomodernist+Manifesto.pdf
Aslett K, Sanderson Z, Godel W, Persily N, Nagler J, Tucker JA (2023) Online searches to evaluate misinformation can increase its perceived veracity. Nature 625:548. https://doi.org/10.1038/s41586-023-06883-y
Auer D, Bonoli G, Fossati F, Liechti F (2018) The matching hierarchies model: evidence from a survey experiment on employers. Int Migr Rev 53:90–121. https://doi.org/10.1177/0197918318764872
Awasthi Y (2020) Press "A" for artificial intelligence in agriculture: a review. JOIV Int J Info Visual 4:112–116. https://doi.org/10.30630/joiv.4.3.387
Ayed RB, Hanana M (2021) Artificial intelligence to improve the food and agriculture sector. J Food Qual 5584754:1. https://doi.org/10.1155/2021/5584754
Ayoub M, Abdullah AZ (2012) Critical review on the current scenario and significance of crude glycerol resulting from biodiesel industry towards more sustainable renewable energy industry. Renew Sust Energ Rev 16:2671–2686. https://doi.org/10.1016/j.rser.2012.01.054
Badger PC (2002) Ethanol from cellulose: A general review. In: Janick J, Whipkey A (eds) Trends in new crops and new uses. ASHS Press, Alexandria, VA
Baker RE, Mahmud AS, Miller IF, Rajeev M, Rasambainarivo F, Rice BL, Takahashi S, Tatem AJ, Wagner CE, Wang J-F, Wesolowski A, Metcalf CJE (2022) Infectious disease in an era of global change. Nat Rev Microbiol 20:193–205. https://doi.org/10.1038/s41579-021-00639-z
Balogh JM, Jámbor A (2020) The environmental impacts of agricultural trade: A systematic literature review. Sustain For 12:1152. https://doi.org/10.3390/su12031152
Baltenweck I, Enahoro D, Frija A, Tarawali S (2020) Why is production of animal source foods important for economic development in Africa and Asia? Anim Front 10:22–29. https://doi.org/10.1093/af/vfaa036
Bao J, Xie Q (2022) Artificial intelligence in animal farming: A systematic literature review. J Clean Prod 331:129956. https://doi.org/10.1016/j.jclepro.2021.129956
Barnharst T, Sun X, Rajendran A, Urriola P, Shurson G, Hu B (2021) Enhanced protein and amino acids of corn ethanol co-product by *Mucor indicus* and *Rhizopus oryzae*. Bioprocess Biosyst Eng 44:1989–2000. https://doi.org/10.1007/s00449-021-02580-0
Barth H, Zalkat G (2023) Harder than you think—immigrant labor market integration in agricultural sector. Nord J Work Life Stud 13:27–48. https://tidsskrift.dk/njwls/index
Bathaei A, Štreimikienė D (2023) A systematic review of agricultural sustainability indicators. Agriculture 13:241. https://doi.org/10.3390/agriculture13020241
Baur S, Nauels A, Nicholls Z, Sanderson BM, Schleussner C-F (2023) The deployment length of solar radiation modification: an interplay of mitigation, net-negative emissions and climate uncertainty. Earth Syst Dynam 14:367–381. https://doi.org/10.5194/esd-14-367-2023
Beal T, Ortenzi F, Fanzo J (2023) Estimated micronutrient shortfalls of the EAT-lancet planetary health diet. Lancet Planet Health 7:e233–e237. https://doi.org/10.1016/S2542-5196(23)00006-2
Beattie AJ, Hay M, Magnusson B, de Nys R, Smeathers J, Vincent JV (2011) Ecology and bioprospecting. Austral Ecol 36:341–356. https://doi.org/10.1111/j.1442-9993.2010.02170.x

Beltran-Alcrudo D, Falco JR, Raizman E, Dietze K (2019) Transboundary spread of pig diseases: the role of international trade and travel. BMC Vet Res 15:64. https://doi.org/10.1186/s12917-019-1800-5

Berckmans D (2017) General introduction to precision livestock farming. Anim Front 7:6–11. https://doi.org/10.2527/af.2017.0102

Berghof TV, Poppe M, Mulder HA (2019) Opportunities to improve resilience in animal breeding programs. Front Genet 9:692. https://doi.org/10.3389/fgene.2018.00692

Berthiller F, Crews C, Dall'Asta C, De Saeger S, Haesaert G, Karlovsky P, Oswald IP, Seefelder W, Speijers G, Stroka J (2013) Masked mycotoxins: a review. Mol Nutr Food Res 57:165–186. https://doi.org/10.1002/mnfr.201100764

Bett B, Kiunga P, Gachohi J, Sindato C, Mbotha D, Robinson T, Lindahl J, Grace D (2017) Effects of climate change on the occurrence and distribution of livestock diseases. Prev Vet Med 137, Part B:119–129. https://doi.org/10.1016/j.prevetmed.2016.11.019

Biesbroek S, Kok FJ, Tufford AR, Bloem MW, Darmon N, Drewnowski A, Fan S, Fanzo J, Gordon LJ, Hu FB, Lähteenmäki L, Nnam N, Ridoutt BG, Rivera J, Swinburn B, van't Veer P (2023) Toward healthy and sustainable diets for the 21st century: importance of sociocultural and economic considerations. Proc Natl Acad Sci 120:e2219272120. https://doi.org/10.1073/pnas.2219272120

Blackburn HD, Wilson CS, Krehbiel B (2019) Conservation and utilization of livestock genetic diversity in The United States of America through gene banking. Diversity 11:244. https://doi.org/10.3390/d11120244

Blackstone NT, Rodriguez-Huerta E, Battaglia K, Jackson B, Jackson E, Norris CB, Sparks JLD (2023) Forced labour risk in pervasive in the US land-based food supply. Nature Food 4:596–606. https://doi.org/10.1038/s43016-023-00794-x

Bock AK, Krzysztofowicz M, Rudkin J, Winthagen V (2020) Farmers of the future. EUR 30464 EN, publications Office of the European Union, Luxembourg. https://doi.org/10.2760/680650

Borutova R, Aragon YA, Nährer K, Berthiller F (2012) Co-occurrence and statistical correlations between mycotoxins in feedstuffs collected in the Asia-Oceania in 2010. Anim Feed Sci Technol 178:190–197. https://doi.org/10.1016/j.anifeedsci.2012.09.015

Bouwman L, Goldewijk KK, van der Hoek KW, Beusen AHW, van Vuuren DP, Willems J, Rufino MC, Stehfest E (2013) Exploring global changes in nitrogen and phosphorus cycles in agriculture induced by livestock production over the 1900–2050 period. Proc Natl Acad Sci 110:20882–20887. https://doi.org/10.1073/pnas.1012878108

Bouwmeester H, Hollman PCH, Peters RJB (2015) Potential health impact of environmentally released micro- and nanoplastics in the human food production chain: experiences from nanotoxicology. Environ Sci Technol 49:8932–8947. https://doi.org/10.1021/acs.est.5b01090

Brauman KA, Goodkind AL, Kim T, Pelton REO, Schmitt J, Smith TM (2020) Unique water scarcity footprints and water risks in US meat and ethanol supply chains identified via subnational commodity flows. Environ Res Lett 15:105018. https://doi.org/10.1088/1748-9326/ab9a6a

Brennan M, Hennessy T, Dillon E, Meredith D (2023) Putting social into agricultural sustainability: integrating assessments of quality of life and wellbeing into farm sustainability indicators. Sociol Rural 63:629–660. https://doi.org/10.1111/soru.12417

Breslow SJ, Allen M, Holstein D, Sojka B, Barnea R, Basurto X, Carothers C, Charnley S, Coulthard S, Dolšak N, Donatuto J, Garcia-Quijano C, Hicks CC, Levine A, Mascia MB, Norman K, Poe M, Satterfield T, St. Martin, K., & Levin, P.S. (2018) Evaluating indicators of human Well-being for ecosystem-based management. Ecosyst Health Sustain 3:1–18. https://doi.org/10.1080/20964129.2017.1411767

Brunn H, Arnold G, Körner W, Rippen G, Steinhäuser KG, Valentin I (2023) PFAS: forever chemicals—persistent, bioaccumulative and mobile. Reviewing the status and the need for their phase out and remediation of contaminated sites. Environ Sci Europe 35:20. https://doi.org/10.1186/s12302-023-00721-8

Bryant C, Szejda K, Parekh N, Desphande V, Tse B (2019) A survey of consumer perceptions of plant-based and clean meat in the USA, India, and China. Front Sustain Food Sys 3:11. https://doi.org/10.3389/fsufs.2019.00011

Burton RJF (2004) Seeing through the 'good farmer's' eyes: towards developing an understanding of the social symbolic value of 'productivist' behaviour. Sociologia Ruralis 44:195–215. https://doi.org/10.1111/j.1467-9523.2004.00270.x

Caldwell JM, Mills ENC (2023) Safety, nutrition and sustainability of plant-based meat alternatives. Chapter 66. In: Present knowledge in food safety, pp 1016–1031. https://doi.org/10.1016/B978-0-12-819470-6.00063-9

Cambra-Lopez M, Aarinink AJA, Zhao Y, Calvet S, Torres AG (2010) Airborne particulate matter from livestock production systems: a review of an air pollution problem. Environ Pollut 158:1–17. https://doi.org/10.1016/j.envpol.2009.07.011

Caminade C, McIntyre KM, Jones AE (2019) Impact of recent and future climate change on vector-borne diseases. Ann N Y Acad Sci 1436:157–173. https://doi.org/10.1111/nyas.13950

Capper JL (2023) The impact of controlling diseases of significant global importance on greenhouse gas emissions from livestock production. One Health Outlook 5:17. https://doi.org/10.1186/s42522-023-00089-y

Carlisle L (2022) Healing grounds: climate, justice, and the deep roots of regenerative farming. Island Press, Washington

Carozzi M, Martin R, Klumpp K, Massad RS (2022) Effects of climate change in European croplands and grasslands: productivity, greenhouse gas balance and soil carbon storage. Biogeosciences 19:3021–3050. https://doi.org/10.5194/bg-19-3021-2022

Carr JA, D'Odorico P, Suweis S, Seekell DA (2016) What commodities and countries impact inequality in the global food system? Environ Res Lett 11:095013. https://doi.org/10.1088/1748-9326/11/9/095013

Castillo-Diaz FJ, Belmonte-Ureña LJ, López-Serrano MJ, Camacho-Ferre F (2023) Assessment of the sustainability of the European Agri-food sector in the context of the circular economy. Sustainable Prod Consumption 40:398–411. https://doi.org/10.1016/j.spc.2023.07.010

Challis JK, Sura S, Cantin J, Curtis AW, Shade KM, McAllister TA, Jones PD, Giesy JP, Larney FJ (2021) Ractopamine and other growth-promoting compounds in beef cattle operations: fate and transport in feedlot pens and adjacent environments. Environ Sci Technol 55:1730–1739. https://doi.org/10.1021/acs.est.0c06450

Champlin D, Hake E (2006) Immigration as industrial strategy in American meat packing. Rev Political Economy 18:49–70. https://doi.org/10.1080/09538250500354140

Chang J, Ciais P, Gasser T, Smith P, Herrero M, Havlík P, Obersteiner M, Guenet B, Goll DS, Li W, Naipal V, Peng S, Qiu C, Tian H, Viovy N, Yue C, Zhu D (2021) Climate warming from managed grasslands cancels the cooling effect of carbon sinks in sparsely grazed and natural grasslands. Nat Commun 12:118. https://doi.org/10.1038/s41467-020-20406-7

Charlier J, Morgan ER, Kyriazakis I (2021) Letter: quantifying the interrelationship between livestock infections and climate change: Response to Ezenwa et al. Trends Ecol Evol 36:576–577. https://doi.org/10.1016/j.tree.2021.02.003

Chaudhary A, Kastner T (2016) Land use biodiversity impacts embodied in international trade. Glob Environ Chang 38:195–204. https://doi.org/10.1016/j.gloenvcha.2016.03.013

Chen X, Hou Y, Kastner T, Liu L, Zhang Y, Yin T, Li M, Malik A, Li M, Thorp KR, Han S, Liu Y, Muhammad T, Liu J, Li Y (2023) Physical and virtual nutrient flows in global telecoupled agricultural trade networks. Nat Commun 14:2391. https://doi.org/10.1038/s41467-023-38094-4

Cheng C, Fu J, Su H, Ren L (2023) Recent advancements in agriculture robots: benefits and challenges. Mach Des 11:48. https://doi.org/10.3390/machines11010048

Cherry MA, Sneirson JF (2012) Chevron, greenwashing, and the myth of "Green oil companies". Washington Lee J Energy Climate Environ 3(1) https://scholarlycommons.law.wlu.edu/jece/vol3/iss1/5

Childress C, Farrell A, Bhimani S, Maass KL (2023) Disrupting labor trafficking in the agricultural sector: looking at opportunities beyond law enforcement interventions. Vict Offenders 18:473–511. https://doi.org/10.1080/15564886.2022.2133036

Chowdhury A, Kabir KH, Abdulai A-R, Alam MF (2023) Systematic review of misinformation in social and online media for the development of an analytical framework for agri-food sector. Sustain For 15:4753. https://doi.org/10.3390/su15064753

Chozhavendhan S, Praveen Jumar R, Elavazhagen S, Barathiraja B, Jayakumar M, Varjani SJ (2018) Utilization of crude glycerol from biodiesel industry for the production of value-added bioproducts. In: Singhania R, Agarwal R, Kumar R, Sukumaran R (eds) Waste to wealth. Energy, environment, and sustainability. Springer, Singapore. https://doi.org/10.1007/978-981-10-7431-8_4

Clark M, Tilman D (2017) Comparative analysis of environmental impacts of agricultural production systems, agricultural input efficiency, and food choice. Environ Res Lett 12:064016. https://doi.org/10.1088/1748-9326/aa6cd5

Clark M, Springman M, Rayner M, Scarborough P, Hill J, Tilman D, Macdiarmid JI, Fanzo J, Bandy L, Harrington RA (2022) Estimating the environmental impacts of 57,000 food products. Proc Natl Acad Sci 119:e2120584119. https://doi.org/10.1073/pnas.2120584119

Coffey D, Dawson K, Ferket P, Connolly A (2016) Review of the feed industry from a historical perspective and implications for its future. J Appl Animal Nutr 4:e3. https://doi.org/10.1017/jan.2015.11

Cremer F, Sheehan B, Fortmann M, Kia AN, Mullins M, Murphy F, Materne S (2022) Cyber risk and cybersecurity: a systematic review of data availability. Geneva Pap Risk Insur Issues Pract 47:698–736. https://doi.org/10.1057/s41288-022-00266-6

Crippa M, Guizzardi D, Pagani F, Banja M, Muntean M, Schaaf E, Becker W, Monforti-Ferrario F, Quadrelli R, Risquez Martin A, Taghavi-Moharamli P, Köykkä J, Grassi G, Rossi S, Brandao De Melo J, Oom D, Branco A, San-Miguel J, Vignati E (2023) GHG emissions of all world countries. Publications Office of the European Union, Luxembourg. https://doi.org/10.2760/953322. JRC134504

Cue R, Doornink M, George R, Griffiths B, Jorgensen MW, Rogers R, Saha A, Taysom K, Cabrera VE, Wangen SR, Fadul-Pacheco L (2021) Data governance in the dairy industry. Animals 11:2981. https://doi.org/10.3390/ani11102981

D'Odorico P, Rulli MC (2013) The fourth food revolution. Nat Geosci 6:417–418. https://doi.org/10.1038/ngeo1842

D'Odorico P, Rulli MC, Dell'Angelo J, Davis KF (2017) New frontiers of land and water commodification: socio-environmental controversies of large-scale acquisitions. Land Degrad Dev 28:2234–2244

D'Odorico P, Carr J, Dalin C, Dell'Angelo J, Konar M, Laio F, Ridolfi L, Rosa L, Suweis S, Tamea S, Tuninetti M (2019) Global virtual water trade and the hydrological cycle: patterns, drivers, and socio-economic impacts. Environ Res Lett 14:053001. https://doi.org/10.1088/1748-9326/ab05f4

Dalin C, Rodríguez-Iturbe I (2016) Environmental impacts of food trade via resource use and greenhouse gas emissions. Environ Res Lett 11:035012. https://doi.org/10.1088/1748-9326/11/3/035012

Dalla RL, Ellis A, Cramer SC (2005) Immigration and rural America—Latinos' perceptions of work and residence in three meatpacking communities. Community Work Fam 8:163–185. https://doi.org/10.1080/13668800500049639

Daum T (2023) Mechanization and sustainable agri-food system transformation in the global south. A review. Agron Sustain Develop 43:16. https://doi.org/10.1007/s13593-023-00868-x

de Quelen F, Brossard L, Wilfart A, Dourmad J-Y, Garcia-Launay F (2021) Eco-friendly feed formulation and on-farm feed production as ways to reduce the environmental impacts of pig production without consequences on animal performance. Front Veterinary Sci 8:689012. https://doi.org/10.3389/fvets.2021.689012

Death C, Bell C, Champness D, Milne C, Reichman S, Hagen T (2021) *Per*- and polyfluoroalkyl substances (PFAS) in livestock and game species: a review. Sci Total Environ 774:144795. https://doi.org/10.1016/j.scitotenv.2020.144795

Dedieu B, Contzen S, Nettle R, de Schiavi SA, Sraïri MT (2022) The multiple influences on the future of work in agriculture: global perspectives. Front Sustain Food Sys 6:889508. https://doi.org/10.3389/fsufs.2022.889508

DeFoliart GR, Paoletti MG (2005) Overview of role of edible insects in preserving biodiversity. In: Ecological implications of Minilivestock: potential of insects, rodents, frogs, and snails, pp 123–140

Dell'Angelo J, D'Odorico P, Rulli MC (2017) Threats to sustainable development posed by land and water grabbing. Curr Opin Environ Sustain 26–27:120–128. https://doi.org/10.1016/j.cosust.2017.07.007

Denver S, Christensen T, Lund TB, Olsen JV, Sandøe P (2023) Willingness-to-pay for reduced carbon footprint and other sustainability concerns relating to pork production—A comparison of consumers in China, Denmark, Germany and the UK. Livestock Sci 276:105337. https://doi.org/10.1016/j.livsci.2023.105337

Dernbach JC, Cheever F (2015) Sustainable development and its discontents. Trans Environ Law 4:247–287. https://doi.org/10.1017/S2047102515000163

Dogan B (2022) Does FDI in agriculture promote food security in developing countries? The role of land governance. Trans Corp J 29:47–74

Domingo NGG, Balasubramanian S, Thakrar SK, Clark MA, Adams PJ, Marshall JD, Muller NZ, Pandis SN, Polasky S, Robinson AL, Tessum CW, Tilman D, Tschofen P, Hill JD (2021) Air quality-related health damages of food. Proc Natl Acad Sci 118:e2013637118. https://doi.org/10.1073/pnas.2013637118

Dou Z, Galligan D, Shurson G (2021) Food supply chain and waste in climate mitigation. In: The role of agricultural science and Technology in Climate 21 project implementation. Council for Agricultural Science and Technology, Ames, IA, QTA2021-1. pp 14–17. https://www.cast-science.org/publication/the-role-of-agricultural-science-and-technology-in-climate-21-project-implementation/

Dou Z, Toth JD, Pitta DW, Bender JS, Hennessy ML, Vecchiarelli B, Indugu N, Chen T, Li Y, Sherman R, Deutsch J, Hu B, Shurson GC, Parsons B, Baker L (2022) Proof of concept for developing novel feeds for cattle from wasted food and crop biomass to enhance agri-food system efficiency. Sci Rep 12:13630. https://doi.org/10.1038/s41598-022-17812-w

Drastig K, Vellenga L, Qualitz G, Singh R, Pfister S, Boulay A-M, Wiedemann S, Prochnow A, Chapagain A, De Camillis C, Opio C, Mottet A (2021) Accounting for livestock water productivity—how and why? In: Land and Water Discussion Papers. FAO, Rome. https://doi.org/10.4060/ca7565en

Dyson E, Helbig R, Avermaete T, Halliwell K, Calder PC, Brown LR, Ingram J, Popping B, Verhagen H, Boobis AR, Guelinckx I, Dye L, Boyle N (2023) Impacts of the Ukraine-Russia conflict on the global food supply chain and building future resilience. EuroChoices 22:14–19. https://doi.org/10.1111/1746-692X.12380

Eastwood CR, Dela Rue B, Edwards JP, Jago J (2022) Responsible robotics design – A systems approach to developing design guides for robotics in pasture-grazed dairy farming. Front Robotics AI 9:914850. https://doi.org/10.3389/frobt.2022.914850

Egoh BN, Nyelele C, Holl KD, Bullock JM, Carver S, Sandom CJ (2021) Rewilding and restoring nature in a changing world. PLoS One 16:e0254249. https://doi.org/10.1371/journal.pone.0254249

Eli-Chukwu NC (2019) Applications of artificial intelligence in agriculture: a review. Eng Technol Appl Sci Res 9:4377–4383. https://doi.org/10.48084/etasr.2756

Elliott KC, Werkheiser I (2023) A framework for transparency in precision livestock farming. Animals 13:3358. https://doi.org/10.3390/ani13213358

Energy Institute (2023) Statistical review of world energy 2023, 72nd edn https://www.energyinst.org/statistical-review

Enthoven L, Van den Broeck G (2021) Local food systems: reviewing two decades of research. Agric Syst 193:103226. https://doi.org/10.1016/j.agsy.2021.103226

Escrivá L, Font G, Manyes L (2015) *In vivo* toxicity studies of fusarium mycotoxins in the last decade: a review. Food Chem Toxicol 78:185–206. https://doi.org/10.1016/j.fct.2015.02.005

European Parliamentary Research Service (2023) Artificial intelligence in the agri-food sector – application, risks and impacts study—panel for the future of science and technology, Scientific foresight Unit, PE 734711, March 2023

Everstine K, Abt E, McColl D, Popping B, Morrison-Rowe S, Lane RW, Scimeca J, Winter C, Ebert A, Moore JC, Chin HB (2018) Development of a hazard classification scheme for substances used in the fraudulent adulteration of foods. J Food Prot 81:31–36. https://doi.org/10.4315/0362-028X.JFP-17-173

Ezanno P, Picault S, Beaunée G, Bailly X, Muñoz F, Duboz R, Monod H, Guégan J-F (2021) Research perspectives on animal health in the era of artificial intelligence. Vet Res 52:40. https://doi.org/10.1186/s13567-021-00902-4

Ezenwa VO, Civitello DJ, Barton BT, Becker DJ, Brenn-White M, Classen AT, Deem SL, Johnson ZE, Kutz S, Malishev M, Penczykowski RM, Preston DL, Vannatta JT, Koltz AM (2020) Infectious diseases, livestock and climate: A vicious cycle? Trends Ecolog Evol 35:959–962. https://doi.org/10.1016/j.tree.2020.08.012

Fadul-Pacheco L, Wangen SR, da Silva TE, Cabrera VE (2022) Addressing data bottlenecks in the dairy farm industry. Animals 12:721. https://doi.org/10.3390/ani12060721

Fajack H (2014) Agenda 21: pathway to a better tomorrow or global conspiracy to subjugate individual rights? Critical Planning 21., https://escholarship.org/uc/item/043715n8. https://doi.org/10.5070/CP8211024787

Falsetti B, Vallino E, Ridolfi L, Laio F (2020) Is water consumption embedded in crop prices? A global data-driven analysis. Environ Res Lett 15:104016. https://doi.org/10.1088/1748-9326/aba782

FAO (2012) Energy-smart food at FAO: an overview. In: Environment and natural resource management working paper no. 53. Food and Agriculture Organization of the United Nations, Rome. https://www.fao.org/3/an913e/an913e.pdf

FAO (2018a) The 10 elements of Agroecology: guiding the transition to sustainable food and agricultural systems. Food and Agriculture Organization of the United Nations, Rome. http://www.fao.org/3/i9037en/i9037en.pdf

FAO (2018b) World livestock: transforming the livestock sector through the sustainable development goals. Food and Agriculture Organization of the United Nations, Rome. https://doi.org/10.4060/ca1201en

FAO (2018c) Sustainable food systems—concept and framework. Food and Agriculture Organization of the United Nations, Rome. http://www.fao.org/3/ca2079en/CA2079EN.pdf

FAO (2019) The state of food and agriculture 2019. In: Moving forward on food loss and waste reduction. Food and Agriculture Organization of the United Nations, Rome, p 182. https://www.fao.org/3/ca6030en/ca6030en.pdf

FAO (2020) In: Food and Agriculture Organization of the United Nations (ed) In brief. Five practical actions toward resilient, low-carbon livestock systems, Rome

FAO/OECD (2012) Building resilience for adaptation to climate change in the agriculture sector. In: Meybeck A, Lankoski J, Redfern S, Azzu N, Gitz V (eds) Proceedings of a Joint FAO/OECD Workshop. Food and Agriculture Organization of the United Nations Organisation for Economic Co-operation and Development, Rome

Fernandes AR, Lake IR, Dowding A, Rose M, Jones NR, Petch R, Smith F, Panton S (2019) The potential of recycled materials used in agriculture to contaminate food through uptake by livestock. Sci Total Environ 667:359–370. https://doi.org/10.1016/j.scitotenv.2019.02.211

Fernández-Llamazares Á, Virtanen PK (2020) Game masters and Amazonian indigenous views on sustainability. Curr Opin Environ Sustain 43:21–27. https://doi.org/10.1016/j.cosust.2020.01.004

Fierros-González I, López-Feldman A (2021) Farmers' perception of climate change: a review of the literature from Latin America. Front Environ Sci 9:672399. https://doi.org/10.3389/fenvs.2021.672399

Figueroa-Helland L, Thomas C, Aguilera AP (2018) Decolonizing food systems: food sovereignty, indigenous revitalization, and agroecology as counter-hegemonic movements. Perspect Global Develop Technol 17:173–201. https://doi.org/10.1163/15691497-12341473

Findlay A, McCollum D (2013) Recruitment and employment regimes: migrant labour channels in the UK'S rural agribusiness sector, from accession to recession. J Rural Stud 30:10–19. https://doi.org/10.1016/j.jrurstud.2012.11.006

Fletcher RJ Jr, Robertson BA, Evans J, Doran PJ, Alavalapati JRR (2010) Biodiversity conservation in the era of biofuels: risks and opportunities. Front Ecol Environ 9:161–168. https://doi.org/10.1890/090091

Fraeyman S, Croubels S, Devreese M, Antonissen G (2017) Emerging *Fusarium* and *Alternaria* mycotoxins: occurrence, toxicity and toxicokinetics. Toxins 9:228. https://doi.org/10.3390/toxins9070228

Fung L, Urriola PE, Shurson GC (2019a) Energy, amino acid, and phosphorus digestibility of thermally-treated food waste sources for swine. Trans Animal Sci 3:676–691. https://doi.org/10.1093/tas/txz028

Fung L, Urriola PE, Baker L, Shurson GC (2019b) Estimated energy and nutrient composition of different food waste sources and their potential for use in sustainable swine feeding programs. Transl Animal Sc 3:143–152. https://doi.org/10.1093/tas/txy099

Gaddafi S, Garba MG, Yahaya MA (2021) Understanding exposure routes of endocrine disruptors in livestock. Nigerian J Animal Sci 23:39–45

Garcia-Launay F, Dusart L, Espagnol S, Laisse-Redoux S, Gaudré D, Méda B, Wilfart A (2018) Multiobjective formulation is an effective method to reduce environmental impacts of livestock feeds. Br J Nutr 120:1298–1309. https://doi.org/10.1017/S0007114518002672

Gaudaré U, Pellerin S, Benoit M, Durand G, Dumont B, Barbieri P, Nesme T (2021) Comparing productivity and feed-use efficiency between organic and conventional livestock animals. Environ Res Lett 16:024012. https://doi.org/10.1088/1748-9326/abd65e

Gendel SM (2023) Preventing food fraud. Chapter 16. In: Present knowledge in food safety, p 260. https://doi.org/10.1016/B978-0-12-819470-6.00013-5

Gennari P, Navarro DK (2019) The challenge of measuring agricultural sustainability in all its dimensions. J Sustain Res 1:e190013. https://doi.org/10.20900/jsr20190013

Gerber PJ, Steinfeld H, Henderson B, Mottet A, Opio C, Dijkman J, Falcucci A, Tempio G (2013) Tackling climate change through livestock – A global assessment of emissions and mitigation opportunities. Food and agriculture Organization of the United Nations (FAO), Rome

Gerber PJ, Uwizeye A, Schulte RPO, Opio CI, de Boer IJM (2014) Nutrient use efficiency: a valuable approach to benchmark the sustainability of nutrient use in global livestock production? Curr Opin Environ Sustain 9-10:122–130. https://doi.org/10.1016/j.cosust.2014.09.007

Giampietro M (2023) Reflections on the popularity of the circular bioeconomy concept: the ontological crisis of sustainability science. Sustain Sci 18:749–754. https://doi.org/10.1007/s11625-022-01267-z

Gilio-Whitaker D (2019) As Long as grass grows: the indigenous fight for environmental justice, from colonization to standing rock. Beacon Press, Boston

Giller KE, Delaune T, Silva JV, van de Descheemaeker K, Ven G, Schut AGT, van Wijk M, Hammond J, Hochman Z, Taulya G, Chikowo R, Narayanan S, Kishore A, Bresciani F, Teixeira HM, Andersson JA, van Ittersum MK (2021a) The future of farming: who will produce our food? Food Secur 13:1073–1099. https://doi.org/10.1007/s12571-021-01184-6

Giller KE, Hijbeek R, Andersson JA, Sumberg J (2021b) Regenerative agriculture: An agronomic perspective. Outlook Agri 50:13–25. https://doi.org/10.1177/0030727021998063

Glatzle A (2014) Questioning key conclusions of FAO publications 'livestock's Long shadow' (2006) appearing again in 'tackling climate change through livestock' (2013). *Pastoralism 4*:1. http://www.pastoralismjournal.com/content/4/1/1

Global Food Safety Initiative (2020) Benchmarking requirements. https://mygfsi.com/news-and-resources/?type=publications

Goglio P, Knudsen MT, Van Mierlo K, Röhrig N, Fossey M, Maresca A, Hashemi F, Waqas MA, Yngvesson J, Nassy G, Broekema R, Moakes S, Pfeifer C, Borek R, Yanez-Ruiz D, Cascante MQ, Syp A, Zylowsky T, Romero-Huelva M, Smith LG (2023) Defining common criteria for harmonizing life cycle assessments of livestock systems. Cleaner Prod Lett 4:100035. https://doi.org/10.1016/j.clpl.2023.100035

Goold HD, Wright P, Hailstones D (2018) Emerging opportunities for synthetic biology in agriculture. Genes 9:341. https://doi.org/10.3390/genes9070341

Greenwood L (2022) China's interests in U.S. agriculture: augmenting food security through investment abroad. U.S.-China economic and security review commission, staff research report

Grenz J, Armstrong CG (2023) Pop-up restoration in colonial contexts: applying an indigenous food systems lens to ecological restoration. Front Sustain Food Sys 7:1244790. https://doi.org/10.3389/fsufs.2023.1244790

Grocery Manufacturers Association (2010) Consumer product fraud: deterrence and detection. Grocery Manufacturers Association, Washington, DC

Groom MJ, Gray EM, Townsend PA (2008) Biofuels and biodiversity: principles for creating better policies for biofuel production. Conserv Biol 22:602–609. https://doi.org/10.1111/j.1523-1739.2007.00879.x

Gupta A, Rayeen F, Mishra R, Tripathi M, Pathak N (2023) Nanotechnology applications in sustainable agriculture: an emerging eco-friendly approach. Plant Nano Biology 4:100033. https://doi.org/10.1016/j.plana.2023.100033

Guptill A (2021) Understanding and measuring social sustainability. SARE quality-of-life working group, sustainable agriculture research and education. University of Maryland, College Park. https://www.sare.org/resources/understanding-and-measuring-social-sustainability/

Guvvala PR, Ravindra JP, Selvaraju S (2020) Impact of environmental contaminants on reproductive health of male domestic ruminants: a review. Environ Sci Pollut Res 27:3819–3836. https://doi.org/10.1007/s11356-019-06980-4

Habib-ur-Rahman M, Ahmad A, Raza A, Hasnain MU, Alharby HF, Alzahrani YM, Bamagoos AA, Hakeem KR, Ahmad S, Nasim W, Ali S, Mansour F, Sabagh AEL (2022) Impact of climate change on agricultural production; issues, challenges, and opportunities in Asia. Front Plant Sci 13:925548. https://doi.org/10.3389/fpls.2022.925548

Haedicke MA (2013) From collective bargaining to social justice certification: workers' rights in the American meatpacking industry. Sociol Focus 46:119–137. https://doi.org/10.1080/00380237.2013.766877

Hartung J, Banhazi T, Vranken E, Guarino M (2017) European farmers' experiences with precision livestock farming systems. Anim Front 7:38–44. https://doi.org/10.2527/af.2017.0107

Hassen TB, Bilali HE (2022) Impacts of the Russia-Ukraine war on global food security: towards more sustainable and resilient food systems? Food Secur 11:2301. https://doi.org/10.3390/foods11152301

Hathaway MD (2016) Agroecology and permaculture: addressing key ecological problems by rethinking and redesigning agricultural systems. J Environ Stud Sci 6:239–250. https://doi.org/10.1007/s13412-015-0254-8

Havens SM, Hedman CJ, Hemming JDC, Mieritz MG, Shafer MM, Schauer JJ (2020) Occurrence of estrogens, androgens and progestogens and estrogenic activity in surface water runoff from beef and dairy manure amended crop fields. Sci Total Environ 710:136247. https://doi.org/10.1016/j.scitotenv.2019.136247

Hawes C, Iannetta PPM, Squire GR (2021) Agroecological practices for whole-system sustainability. CABI Rev 16:1–19. https://doi.org/10.1079/PAVSNNR202116005

Hayek MN, Miller SM (2021) Underestimates of methane from intensively raised animals could undermine goals of sustainable development. Environ Res Lett 16:063006. https://doi.org/10.1088/1748-9326/ac02ef

Heinke J, Lannerstad M, Gerten D, Havlik P, Herrero M, Notenbaert AMO, Hoff H, Müller C (2020) Water use in global livestock production—opportunities and constraints for increasing water productivity. Water Resour Res 56:e2019WR026995. https://doi.org/10.1029/2019WR026995

Hellegers P (2022) Food security vulnerability due to trade dependencies on Russia and Ukraine. Food Secur 14:1503–1510. https://doi.org/10.1007/s12571-022-01306-8

Henrekson M, Oner O, Sanandaji T (2019) The refugee crisis and the reinvigoration of the nation state: does the European Union have a common refugee policy? IFN working paper no. 1265. Research Institute of Industrial Economics (IFN), Stockholm. https://www.econstor.eu/bitstream/10419/210906/1/1067644768.pdf

Henry B, Charmley E, Eckard R, Gaughan JB, Hegarty R (2012) Livestock production in a changing climate: adaptation and mitigation research in Australia. Crop Pasture Sci 63:191–202. https://doi.org/10.1071/CP11169

Hernandez J (2022) Fresh Banana leaves: healing indigenous landscapes through indigenous science. North Atlantic Books

Hernandez J, Meisner J, Jacobs LA, Rabinowitz PM (2022) Re-centering indigenous knowledge in climate change discourse. PLOS Climate 1:e0000032. https://doi.org/10.1371/journal.pclm.0000032

Herrero M, Gerber P, Vellinga T, Garnett T, Leip A, Opio C, Westhoek HJ, Thornton PK, Olesen J, Hutchings N, Montgomery H, Soussana J-F, Steinfeld H, McAllister TA (2011) Livestock and greenhouse gas emissions: the importance of getting the numbers right. Anim Feed Sci Technol 166-167:779–782. https://doi.org/10.1016/j.anifeedsci.2011.04.083

Herrero M, Thornton PK, Power B, Bogard JR, Remans R, Fritz S, Gerber JS, Nelson G, See L, Waha K, Watson RA, West PC, Samberg LH, van de Steeg J, Stephenson E, van Mark W, Havlik P (2017) Farming and the geography of nutrient production for human use: a transdisciplinary analysis. Lancet Planet Health 1:e33–e42. https://doi.org/10.1016/S2542-5196(17)30007-4

Hickel J (2020) Quantifying national responsibility for climate breakdown: an equality-based attribution approach for carbon dioxide emissions in excess of the planetary boundary. Lancet Planet Health 4:e399–e404. https://doi.org/10.1016/S2542-5196(20)30196-0

Hickel J (2022) Less is more—how degrowth will save the world. Penguin Books

Hill R, Adem Ç, Alangui WV, Molnár Z, Aumeeruddy-Thomas Y, Bridgewater P, Tengö M, Thaman R, Adou Yao CY, Berkes F, Carino J, da Cunha MC, Diaw MC, Díaz S, Figueroa VE, Fisher J, Hardison P, Ichikawa K, Kariuki P, Karki M, Lyver POB, Malmer P, Masardule O, Oteng Yeboah AA, Pacheco D, Pataridze T, Perez E, Roué M-M, Roba H, Rubis J, Saito O, Xue D (2020) Working with indigenous, local and scientific knowledge in assessments of nature and nature's linkages with people. Curr Opin Environ Sustain 43:8–20. https://doi.org/10.1016/j.cosust.2019.12.006

Hoekstra AY, Mekonnen MM, Chapagain AK, Mathews RE, Richter BD (2012) Global monthly water scarcity: blue water footprints versus blue water availability. PLoS One 7:e32688. https://doi.org/10.1371/journal.pone.0032688

Holmgren D (2002) Permaculture—principles and pathways beyond sustainability. Holmgren Design Services, Victoria

Hormio S (2023) Collective responsibility for climate change. WIREs Climate Change 14:e830. https://doi.org/10.1002/wcc.830

Hossain MS, Rahman MH, Rahman MS, Hosen ASMS, Seo C, Cho GH (2021) Intellectual property theft protection in IoT based precision agriculture using SDN. Electronics 10:1987. https://doi.org/10.3390/electronics10161987

Hossen I, Fahad N, Sarkar R, Rabbi MR (2023) Artificial intelligence in agriculture: A systematic literature review. Turkish J Comput Mathematics Educ 14:137–146

Hvistendahl M (2020) The scientist and the spy: A trues story of China, the FBI, and industrial espionage. Riverhead Books

Iannacci-Manasia L (2023) Unprotected youth workers in US agriculture. Front Public Health 11:1064143. https://doi.org/10.3389/fpubh.2023.1064143

Ibidhi R, Ben Salem H (2020) Water footprint of livestock products and production systems: a review. Anim Prod Sci 60:1369–1380. https://doi.org/10.1071/AN17705

Idoko-Akoh A, Goldhill DH, Sheppard CM, Bialy D, Quantrill JL, Sukhova K, Brown JC, Richardson S, Campbell C, Taylor L, Sherman A, Nazki S, Long JS, Skinner MA, Shelton H, Sang HM, Barclay WS, McGrew MJ (2023) Creating resistance to avian influenza infection through genome editing of the ANP32 gene family. Nat Commun 14:6136. https://doi.org/10.1038/s41467-023-41476-3

Imoro ZA, Imoro AZ, Duwiejuah AB, Abukari A (2021) Harnessing indigenous technologies for sustainable management of land, water, and food resources amidst climate change. Front Sustain Food Sys 5:691603. https://doi.org/10.3389/fsufs.2021.691603

Imran Y, Wijekoon N, Gonawala L, Chiang Y-C, De Silva KRD (2021) Biopiracy: abolish corporate hijacking of indigenous medicinal entities. Sci World J 2021:8898842. https://doi.org/10.1155/2021/8898842

International Labour Organization (2014) The meanings of forced labour. https://www.ilo.org/global/topics/forced-labour/news/WCMS_237569/lang-en/index.htm

IPBES (2019) In: Brondizio ES, Settele J, Diaz S, Ngo HT (eds) Summary for policymakers of the global assessment report on biodiversity and ecosystem services of the intergovernmental science-policy platform on biodiversity and ecosystem services. IPBES Secretariat, Bonn, 1144 pp

IPCC (2023) Climate change 2023: synthesis report. In: Core Writing Team, Lee H, Romero J (eds) Contribution of working groups I, II, and III to the sixth assessment report of the intergovernmental panel on climate change. IPCC, Geneva, p 184. https://doi.org/10.59327/IPCC/AR6-9789291691647

Jägermeyr J, Müller C, Ruane AC, Elliott J, Balkovic J, Castillo O, Faye B, Folberth C, Franke JA, Fuchs K, Guarin JR, Heinke J, Hoogenboom G, Iizumi T, Jain AK, Kelly D, Khabarov N, Lange S, Lin T-S, Liu W, Mialyk O, Minoli S, Moyer EJ, Okada M, Phillips M, Porter C, Rabin SS, Scheer C, Schneider JM, Schyns JF, Skalsky R, Smerald A, Stella T, Stephens H, Webber H, Zabel F, Rosenzweig C (2021) Climate impacts on global agriculture emerge earlier in new generation of climate and crop models. Nature Food 2:873–885. https://doi.org/10.1038/s43016-021-00400-y

Jang J-C, Zeng Z, Urriola PE, Shurson GC (2021) Effects of feeding corn distillers dried grains with solubles diets with and without supplemental enzymes on growth performance of pigs: a meta-analysis. Trans Animal Sci 5:1–15. https://doi.org/10.1093/tas/txab029

Jepson P, Schepers F, Helmer W (2018) Governing with nature: a European perspective on putting rewilding principles into practice. Philos Trans R Soc B 373:20170434. https://doi.org/10.1098/rstb.2017.0434

Jessen TD, Ban NC, Claxton NX, Darimont CT (2021) Contributions of indigenous knowledge to ecological and evolutionary understanding. Front Ecol Environ 20:93–101. https://doi.org/10.1002/fee.2435

Jha G, Kankaria V, McLennon E, Pal S, Sihi D, Dari B, Diaz D, Nocco M (2021) Per- and polyfluoroalkyl substances (PFAS) in integrated crop-livestock systems: environmental exposure and human health risks. Int J Environ Res Public Health 18:12550. https://doi.org/10.3390/ijerph182312550

Jiang B, Tang W, Cui L, Deng X (2023) Precision livestock farming research: a global scientometric review. Animals 13:2096. https://doi.org/10.3390/ani13132096

Jones GD, Benchetler PV, Tate KW, Kolodziej EP (2014) Mass balance approaches to characterizing the leaching potential of trenbolone acetate metabolites in agro-ecosystems. Environ Sci Technol 48:3715–3723. https://doi.org/10.1021/es405701f

Jørgensen D (2015) Rethinking rewilding. Geoforum 65:482–488. https://doi.org/10.1016/j.geoforum.2014.11.016

Jouanjean M-A, Casalini F, Wiseman L, Gray E (2020) Issues around data governance in the digital transformation of agriculture: the farmers' perspective. In: OECD Food, agriculture and fisheries papers, no. 146. OECD Publishing, Paris. https://doi.org/10.1787/53ecf2ab-en

Karesh WB, Cook RA, Bennett EL, Newcomb J (2005) Wildlife trade and global disease emergence. Emerg Infect Dis 11:1000–1002. https://doi.org/10.3201/eid1107.050194

Kastner T, Chaudhary A, Gingrich S, Marques A, Persson UM, Bidoglio G, Le Provost G, Schwarzmüller F (2021) Global agricultural trade and land system sustainability: implications for ecosystem carbon storage, biodiversity, and human nutrition. One Earth 4:1425–1443. https://doi.org/10.1016/j.oneear.2021.09.006

Kaur J, Fard SMH, Amiri-Zarandi M, Dara R (2022) Protecting farmer's data privacy and confidentiality: recommendations and considerations. Front Sustain Food Sys 6:903230. https://doi.org/10.3389/fsufs.2022.903230

Kealiikanakaoleohaililan K, Giardina CP (2016) Embracing the sacred: an indigenous framework for tomorrow's sustainability science. Sustain Sci 11:57–67. https://doi.org/10.1007/s11625-015-0343-3

Keith DW (2003) Geoengineering the climate: history and prospect. Annual Rev Energy Environ 25:245–284. https://doi.org/10.1146/annurev.energy.25.1.245

Kelly SD (2023) Emerging contaminants in the context of food fraud. Chapter 23. In: Present knowledge in food safety, pp 315–319. https://doi.org/10.1016/B978-0-12-819470-6.00052-4

Kerr BJ, Shurson GC, Johnston LJ, Dozier WA III (2011) Utilization of crude glycerin in nonruminants. In: Montero G, Stoytcheva M (eds) Biodiesel – Quality, Emissions and By-Products. InTech. http://www.intechopen.com/books/show/title/biodiesel-quality-emissions-and-by-products

Kerr BJ, Urriola PE, Jha R, Thomson J, Curry SM, Shurson GC (2019) Amino acid composition and digestible amino acid content in animal protein by-product meals fed to growing pigs. J Anim Sci 97:4540–4547. https://doi.org/10.1093/jas/skz294

Khoshal AK, Novak B, Martin PGP, Jenkins T, Neves M, Schatzmayr G, Oswald IP, Pinton P (2021) Co-occurrence of DON and emerging mycotoxins in worldwide finished pig feed and their combined toxicity in intestinal cells. Toxins 11:727. https://doi.org/10.3390/toxins11120727

Kibria MG, Masuk NI, Safayet R, Nguyen HQ, Mourshed M (2023) Plastic waste: challenges and opportunities to mitigate pollution and effective management. Int J Environ Res 17:20. https://doi.org/10.1007/s41742-023-00507-z

Kim SW, Less JF, Wang L, Yan T, Kiron V, Kaushik SJ, Lei XG (2019) Meeting global feed protein demand: challenge, opportunity, and strategy. Annual Rev Animal Biosci 7:221–243. https://doi.org/10.1146/annurev-animal-030117-014838

Kipling RP, Bannink A, Bartley DJ, Blanco-Penedo I, Faverdin P, Graux A-I, Hutchings NJ, Kyriazakis I, Macleod M, Østergaard S, Robinson TP, Vitali A, Vosough Ahmadi B, Özkan Ş (2021) Short communication: identifying key parameters for modeling the impacts of livestock health conditions on greenhouse gas emissions. Animal 15:1000023. https://doi.org/10.1016/j.animal.2020.100023

Kleen JL, Guatteo R (2023) Precision livestock farming: what does it contain and what are the perspectives? Animals 13:779. https://doi.org/10.3390/ani13050779

Klein N (2014) This changes everything: capitalisms vs. the climate. Simon & Schuster, New York

Klinenberg E, Araos M, Koslov L (2020) Sociology and the climate crisis. Annu Rev Sociol 46:649–669. https://doi.org/10.1146/annurev-soc-121919-054750

Klink-Lehmann J, Tatic M, Weingarten N, Hartmann M (2023) Sustainability trade-offs in animal husbandry: consumers' choice when they can't have it all. Q Open 3:1–25. https://doi.org/10.1093/qopen/qoad025

Kolesnyk S, Prodanchuk M (2023) Endocrine disruptors. Chapter 19. In: Present knowledge in food safety, p 281. https://doi.org/10.1016/B978-0-12-819470-6.00051-2

Kos J, Anić M, Radić B, Zadravec M, Hajnal EJ, Pleadin J (2023) Climate change – A global threat resulting in increasing mycotoxin occurrence. Food Secur 12:2704. https://doi.org/10.3390/foods12142704

Krampe C, Serratosa J, Niemi JK, Ingenbleek PTM (2021) Consumer perceptions of precision livestock farming – A qualitative study in three European countries. Animals 11:1221. https://doi.org/10.3390/ani11051221

Krebs J, Bach S (2018) Permaculture—scientific evidence of principles for agroecological design of farming systems. Sustain For 10:3218. https://doi.org/10.3390/su10093218

Kunz N, Chesney T, Trautrims A, Gold S (2023) Adoption and transferability of joint interventions to fight modern slavery in food supply chains. Int J Prod Econ 258:108809. https://doi.org/10.1016/j.ijpe.2023.108809

Lacetera N (2019) Impact of climate change on animal health and welfare. Anim Front 9:26–31. https://doi.org/10.1093/af/vfy030

Lähdesmäki M, Suutari T (2020) Good workers, good firms? Rural SMEs legitimizing immigrant workforce. J Rural Stud 77:1–10. https://doi.org/10.1016/j.jrurstud.2020.04.035

Lal R (2020) Regenerative agriculture for food and climate. J Soil Water Conserv 75:123A–124A. https://doi.org/10.2489/jswc.2020.0620A

Latruffe L, Diazabakana A, Bockstaller C, Desjeux Y, Finn J, Kelly E, Ryan M, Uthes S (2016) Measurement of sustainability in agriculture: a review of indicators. Stud Agri Econ 118:123–130. https://doi.org/10.7896/j.1642

LEAP (2015) Environmental performance of animal feed supply chains: guidelines for assessment. In: Livestock environmental assessment performance partnership. Food and Agriculture Organization, Rome

Lee U, Hawkins TR, Yoo E, Wang M (2021a) Using waste CO_2 from corn ethanol biorefineries for additional ethanol production: life-cycle analysis. Biofuels Bioprod Biorefin 15:468–480. https://doi.org/10.1002/bbb.2175

Lee U, Kwon H, Wu M, Wang M (2021b) Retrospective analysis of the U.S. corn ethanol industry for 2005–2019: implications for greenhouse gas emission reductions. Biofuels Bioprod Biorefin 15:1318–1331. https://doi.org/10.1002/bbb.2225

Lejano RP, Kan WS, Chau CC (2020) The hidden disequities of carbon trading: carbon emissions, air toxics, and environmental justice. Front Environ Sci 8:593014. https://doi.org/10.3389/fenvs.2020.593014

LeJeune JT, Garcia AD, Latronico F (2023) Antimicrobial resistance and antimicrobial residues in the food chain. Chapter 20. In: Present knowledge in food safety, pp 297–302. https://doi.org/10.1016/B978-0-12-819470-6.00045-7

Lenton TM, Xu C, Abrams JF, Ghadiali A, Loriani S, Sakschewski B, Zimm C, Ebi KL, Dunn RR, Svenning J-C, Scheffer M (2023) Quantifying the human cost of global warming. Nature Sustain 6:1237–1247. https://doi.org/10.1038/s41893-023-01132-6

Leoneti AB, Aragão-Leoneti V, de Oliveira SVWB (2012) Glycerol as a by-product of biodiesel production in Brazil: alternatives for the use of unrefined glycerol. Renew Energy 45:138–145. https://doi.org/10.1016/j.renene.2012.02.032

Lerner JS, Li Y, Valdesolo P, Kassam KS (2015) Emotion and decision making. Annu Rev Psychol 66:799–823. https://doi.org/10.1146/annurev-psych-010213-115043

Leroy G, Baumung R, Boettcher P, Besbes B, From T, Hoffman I (2018) Animal genetic resources diversity and ecosystem services. Glob Food Sec 17:84–91. https://doi.org/10.1016/j.gfs.2018.04.003

Leroy F, Abraini F, Beal T, Dominguez-Salas P, Gregorini P, Manzano P, Rowntree J, van Vliet S (2022) Animal board invited review: animal source foods in healthy, sustainable, and ethical diets—an argument against drastic limitation of livestock in the food system. Animal 16:100457. https://doi.org/10.1016/j.animal.2022.100457

Lewis H (2009) From Mexico to Iowa: new immigrant farmers' pathways and potentials. Community Dev 40:139–153. https://doi.org/10.1080/15575330903011058

Lim T, Massey R, McCann L, Canter T, Omura S, Willett C, Roach A, Key N, Dodson L (2023) Increasing the value of animal manure for farmers. AP-109, U.S. Department of Agriculture, Economic Research Service. https://www.ers.usda.gov/webdocs/publications/106089/ap-109.pdf?v=7367.1

Lin F, Li X, Jia N, Feng F, Huang H, Huang J, Fan S, Ciais P, Song X-P (2023) The impact of Russia-Ukraine conflict on global food security. Glob Food Sec 36:100661. https://doi.org/10.1016/j.gfs.2022.100661

Liu M, Zhao L, Gong G, Zhang L, Shi L, Dai J, Han Y, Wu Y, Khalil MM, Sun L (2022) Invited review: remediation strategies for mycotoxin control in feed. J Animal Sci Biotechnol 13:19. https://doi.org/10.1186/s40104-021-00661-4

Lun, F., Liu, J., Ciais, P., Nesme, T., Chang, J., Wang, R., Goll, D., Sardans, J., Peñuelas, J., Obersteiner, M. 2018. Global and regional phosphorus budgets in agricultural systems and their implications for phosphorus-use efficiency. Earth Syst. Sci. Data 10, 1-18. https://doi.org/10.5194/essd-10-1-2018

Loring PA (2020) Threshold concepts and sustainability: features of a contested paradigm. FACETS 5:182–199. https://doi.org/10.1139/facets-2019-0037

Loring PA (2023) A vernacular for living systems: alternative framings for the future of food. Futures 154:103276. https://doi.org/10.1016/j.futures.2023.103276

Lu Q, Zhou W, Min M, Ma X, Chandra C, Doan YTT, Ma Y, Zheng H, Cheng S, Griffith R, Chen P, Chen C, Urriola PE, Shurson GC, Gislerød HR, Ruan R (2015) Growing *chlorella* sp. on meat processing wastewater for nutrient removal and biomass production. Bioresour Technol 198:189–197. https://doi.org/10.1016/j.biotech.2015.08.133

Lu Q, Zhou W, Min M, Ma Y, Chen P, Zheng H, Doan YT, Liu H, Chen C, Urriola PE, Shurson GC, Ruan R (2016) Mitigating ammonia nitrogen deficiency in dairy wastewaters for algae cultivation. Bioresour Technol 201:33–40. https://doi.org/10.1016/j.biortech.2015.11.029

Lyon TP, Montgomery AW (2015) The means and end of greenwash. Organ Environ 28:223–249. https://doi.org/10.1177/1086026615575332

Mackenzie SG, Leinonen I, Ferguson N, Kyriazakis I (2016) Towards a methodology to formulate sustainable diets for livestock: accounting for environmental impact in diet formulation. Br J Nutr 115:1860–1874. https://doi.org/10.1017/S0007114516000763

Magnusson U, Persson S (2015) Endocrine disruptors in domestic animal reproduction: A clinical issue? Reprod Domest Anim 50(Suppl. 3):15–19. https://doi.org/10.1111/rda.12563

Makkar HPS, Tran G, Heuzé V, Ankers P (2014) State-of-the-art on use of insects as animal feed. Anim Feed Sci Technol 197:1–33. https://doi.org/10.1016/j.anifeedsci.2014.07.008

Mamehgol Youscfi DB, Mohd Rafie AS, Al-Haddad SAR, Azrad S (2022) A systematic literature review on the use of deep learning in precision livestock detection and localization using unmanned aerial vehicles. IEEE Access 10:80071–80091. https://doi.org/10.1109/ACCESS.2022.3194507

Manning L, Kowalska A (2023) The threat of ransomware in the food supply chain: a challenge for food defense. In: Trends in organized crime. Springer. https://doi.org/10.1007/s12117-023-09516-y

Manyi-Loh C, Mamphweli S, Meyer E, Okoh A (2018) Antibiotic use in agriculture and its consequential resistance in environmental sources: potential public health implications. Molecules 23:795. https://doi.org/10.3390/molecules23040795

Mapes BR, Prager SD, Béné C, Gonzalez CE (2022) Healthy and sustainable diets from today to 2050—the role of international trade. PLoS One 17:e0264729. https://doi.org/10.1371/journal.pone.0264729

Martin P (2015) Immigration and farm labor: challenges and opportunities. AgBioforum 18:252–258

Martin PJ, Midgley E (2006) Immigration: shaping and reshaping America. Popul Bull 61:1–29

Martyniuk E (2021) Policy effects on the sustainability of animal breeding. Sustain For 13:7787. https://doi.org/10.3390/su13147787

Massicotte M-J, Kelly-Bisson C (2019) What's wrong with permculture design courses? Brazilian lessons for agroecological movement-building in Canada. Agric Hum Values 36:581–594. https://doi.org/10.1007/s10460-018-9870-8

McGregor D, Whitaker S, Sritharan M (2020) Indigenous environmental justice and sustainability. Curr Opin Environ Sustain 43:35–40. https://doi.org/10.1016/j.cosust.2020.01.007

McLennon E, Dari B, Jha G, Sihi D, Kankarla V (2021) Regenerative agriculture and integrative permaculture for sustainable and technology driven global food production and security. Agron J 113:4541–4559. https://doi.org/10.1002/agj2.20814

Meadows DH (2008) In: Wright D (ed) Thinking in systems: a primer. Chelsea Green Publishing, VT

Méda B, Garcia-Launay F, Dusart L, Ponchant P, Espagnol S, Wilfart A (2021) Reducing environmental impacts of feed using multiobjective formulation: what benefits at the farm gate for pig and broiler production? Animal 15:100024. https://doi.org/10.1016/j.animal.2020.100024

Medina Á, González-Jartín JM, Sainz MJ (2017) Impact of global warming on mycotoxins. Curr Opin Food Sci 18:76–81. https://doi.org/10.1016/j.cofs.2017.11.009

Meemken E-M, Qaim M (2018) Organic agriculture, food security, and the environment. Ann Rev Resour Econ 10:39–63. https://doi.org/10.1146/annurev-resource-100517-023252

Meier H, Spinner K, Crump L, Kuenzli E, Schuepbach G, Zinsstag J (2023) State of knowledge on the acquisition, diversity, interspecies attribution and spread of antimicrobial resistance between humans, animals and the environment: A systematic review. Antibiotics 12:73. https://doi.org/10.3390/antibiotics12010073

Menger LM, Pezzutti F, Tellechea T, Stallones L, Rosecrance J, Roman-Muniz IN (2016) Perceptions of health and safety among immigrant Latino/a dairy workers in the U.S. Front Public Health 4:106. https://doi.org/10.3389/fpubh.2016.00106

Mercier S (2014) Employing agriculture: how Midwest farm and food sector relies on immigrant labor. In: The Chicago council on global affairs, Chicago, p 16. https://globalaffairs.org/sites/default/files/2021-07/Midwest_Ag_final.pdf
Meyer JH, Land R (2005) Threshold concepts and troublesome knowledge (2): epistemological considerations and a conceptual framework for teaching and learning. High Educ 49:373–388. https://doi.org/10.1007/s10734-004-6779-5
Meyer JH, Land R (2006) Overcoming barriers to student understanding: threshold concepts and troublesome knowledge. Routledge, New York
Miltenberger O, Jospe C, Pittman J (2021) The good is never perfect: why the current flaws of voluntary carbon markets are services, not barriers to successful climate action. Front Climate 3:686516. https://doi.org/10.3389/fclim.2021.686516
Monasterolo I (2020) Climate change and the financial system. Ann Rev Resour Econ 12:299–320. https://doi.org/10.1146/annurev-resource-110119-031134
Monteiro A, Santos S, Gonçalves P (2021) Precision agriculture for crop and livestock farming—brief review. Animals 11:2345. https://doi.org/10.3390/ani11082345
Morar CA, Doroftei IA, Doroftei I, Hagan MG (2020) Robotic applications on agricultural industry. A review. IOP Conf Ser Mater Sci Eng 997:012081. https://doi.org/10.1088/1757-899X/997/1/012081
Morens DM, Fauci AS (2020) Emerging pandemic diseases: how we got COVID-19. Cell 182:1077–1092. https://doi.org/10.1016/j.cell.2020.08.021
Moyce SC, Schenker M (2018) Migrant workers and their occupational health and safety. Annu Rev Public Health 39:351–365. https://doi.org/10.1146/annurev-publhealth-040617-013714
Mukherjee K (2023) Climate change as a driving factor for emerging contaminants. Chapter 21. In: Present knowledge in food safety, pp 303–308. https://doi.org/10.1016/B978-0-12-819470-6.00048-2
Müller MF, Penny G, Niles MT, Ricciardi V, Chiarelli DD, Davis KF, Dell'Angelo J, D'Odorico P, Rosa L, Rulli MC, Mueller ND (2021) Impact of transnational land acquisitions on local food security and dietary diversity. Proc Nat Acad Sci 118:e2020535118. https://doi.org/10.1073/pnas.2020535118
Muzaka V (2020) Stealing the common from the goose: the emergence of farmers' rights and their implementation in India and Brazil. J Agrar Chang 21:356–376. https://doi.org/10.1111/joac.12398
Nardone A, Ronchi B, Lacetera N, Ranieri MS, Bernabucci U (2010) Effects of climate changes on animal production and sustainability of livestock systems. Livest Sci 130:57–69. https://doi.org/10.1016/j.livsci.2010.02.011
Navarre C, Daniels A, Johnston MO, Mathis C, Perrett T, Posey D (2020) Impact of recruitment and retention of food animal veterinarians on the U.S. In: Food Supply Council for Agricultural Science and Technology Issue Paper, vol 67, p 16
Neethirajan S (2021) The use of artificial intelligence in assessing affective states in livestock. Front Veterinary Sci 8:715261. https://doi.org/10.3389/fvets.2021.715261
Nesme T, Metson GS, Bennett EM (2018) Global phosphorus flows through agricultural trade. Glob Environ Chang 50:133–141. https://doi.org/10.1016/j.gloenvcha.2018.04.004
Neuenschwander M, Stadelmaier J, Eble J, Grummich K, Szczerba E, Kiesswetter E, Schlesinger S, Schwingshacki L (2023) Substitution of animal-based with plant-based foods on cardiometabolic health and all-cause mortality: a systematic review and meta-analysis of prospective studies. BMC Med 21:404. https://doi.org/10.1186/s12916-023-03093-1
Neumayer E (2003) Weak versus strong sustainability—exploring the limits of two opposing paradigms, 2nd edn. Edward Elgar Publishing, Cheltenham/Northampton
Newman L, Fraser E, Newell R, Bowness E, Newman K, Glaros A (2023) Chapter 1 – Cellular agriculture and the sustainable development goals. In: Genomics and the Global Bioeconomy, pp 3–23. https://doi.org/10.1016/B978-0-323-91601-1.00010-9
Newton P, Civita N, Frankel-Goldwater L, Bartel K, Johns C (2020) What is regenerative agriculture? A review of scholar and practitioner definitions based on processes and outcomes. Front Sustain Food Sys 4:577723. https://doi.org/10.3389/fsufs.2020.577723

Nibert D (2014) Animals, immigrants, and profits: slaughterhouses and the political economy. In: Sorenson J (ed) Critical animal studies: thinking the unthinkable. Canadian Scholars' Press Inc., Toronto

Nikander J, Manninen O, Laajalahti M (2020) Requirement for cybersecurity in agricultural communication networks. Comput Electron Agric 179:105776. https://doi.org/10.1016/j.compag.2020.105776

Nitah S (2021) Indigenous peoples proven to sustain biodiversity and address climate change: now it's to recognize and support this leadership. One Earth 4:907–909. https://doi.org/10.1016/j.oneear.2021.06.015

Nori M (2017) The shades of green: migrants' contribution to EU agriculture. In: Context, trends, opportunities, challenges. European University Institute in Florence: Migration Policy Centre

Núñez-Regueiro MM, Siddiqui SF, Fletcher RJ Jr (2021) Effects of bioenergy on biodiversity arising from land-use change and crop type. Conserv Biol 35:77–87. https://doi.org/10.1111/cobi.13452

Obsekov V, Kahn LG, Trasande L (2023) Leveraging systematic reviews to explore disease burden and costs of per- and polyfluoroalkyl substance exposures in the United States. Expo Health 15:373–394. https://doi.org/10.1007/s12403-022-00496-y

OECD (2003) Glossary of statistical terms—sustainable agriculture. The Organisation for Economic C-operation and Development (OECD), Paris. https://stats.oecd.org/glossary/detail.asp?ID=2624

OECD/FAO (2023) OECD-FAO Agricultural Outlook 2023–2032. OECD Publishing, Paris. https://doi.org/10.1787/08801ab7-en

Oonincx DGAB, de Boer IJ (2012) Environmental impact of the production of mealworms as a protein source for humans—a life cycle assessment. PLoS One 7:e51145. https://doi.org/10.1371/journal.pone.0051145

Padeyanda Y, Jang Y-C, Ko Y, Yi S (2016) Evaluation of environmental impacts of food waste management by material flow analysis (MFA) and life cycle assessment. J Material Cycle Waste Manag 18:493–508. https://doi.org/10.1007/s10163-016-0510-3

Palumbo L, Corrado A, Triandafyllidou A (2022) Migrant labour in the Agri-food system in Europe: unpacking the social and legal factors of exploitation. European Journal of Migration Law 24:179–192. https://doi.org/10.1163/15718166-12340125

Pandey P, Dhiman M, Kansal A, Subudhi SP (2023) Plastic waste management for sustainable environment: techniques and approaches. Waste Disposal Sustain Ener 5:205–222. https://doi.org/10.1007/s42768-023-00134-6

Pardey PG, Beintema NM, Dehmer S, Wood S (2006) Agricultural research: A growing Global Divide? Agricultural Science and Technology Indicators Initiative, International Food Policy Research Institute, Washington, DC

Patrício DI, Rieder R (2018) Computer vision and artificial intelligence in precision agriculture for grain crops: A systematic review. Comput Electron Agric 153:69–81. https://doi.org/10.1016/j.compag.2018.08.001

Pearse R, Böhm S (2014) Ten reasons why carbon markets will not bring about radical emissions reduction. Carbon Manag 5:325–337. https://doi.org/10.1080/17583004.2014.990679

Pearson S, Camacho-Villa TC, Valluru R, Gaju O, Rai MC, Gould I, Brewer S, Sklar E (2022) Robotics and autonomous systems for net zero agriculture. Curr Robot Rep 3:57–64. https://doi.org/10.1007/s43154-022-00077-6

Pelton REO, Spawn-Lee SA, Lark TJ, Kim T, Springer N, Hawthorne P, Ray DK, Schmitt J (2021) Land use leverage points to reduce GHG emissions in U.S. agricultural supply chains. Environ Res Lett 16:115002. https://doi.org/10.1088/1748-9326/ac2775

Peng W-X, Marchal JLM, van der Poel AFB (2018) Strategies to prevent and reduce mycotoxins for compound feed manufacturing. Anim Feed Sci Technol 237:129–153. https://doi.org/10.1016/j.anifeedsci.2018.01.017

Peng L, Fu D, Qi H, Lan CQ, Yu H, Ge C (2020) Micro- and nano-plastics in marine environment: source, distribution and threats – A review. Sci Total Environ 698:134254. https://doi.org/10.1016/j.scitotenv.2019.134254

Perini F, Ceccobelli S, Crooijmans RPMA, Tiambo CK, Lasagna E (2023) Editorial: global green strategies and capacities to manage a sustainable animal biodiversity. Front Genet 14:1213080. https://doi.org/10.3389/fgene.2023.1213080

Peritore AF, Gugliandolo E, Cuzzocrea S, Crupi R, Britti D (2023) Current review of increasing animal health threat of per- and polyfluoroalkyl substances (PFAS): harms, limitations, and alternatives to manage their toxicity. Int J Mol Sci 24:11707. https://doi.org/10.3390/ijms241411707

Perry BD, Robinson TP, Grace DC (2018) Review: animal health and sustainable global livestock systems. Animal 12:1699–1708. https://doi.org/10.1017/S1751731118000630

Phelps LN, Kaplan JO (2017) Land use for animal production in global change studies: defining and characterizing a framework. Glob Chang Biol 23:4457–4471. https://doi.org/10.1111/gcb.13732

Piñeiro G, Jobbágy EG, Baker J, Murray BC, Jackson RB (2009) Set-asides can be better climate investment than corn ethanol. Ecol Appl 19:277–282. https://doi.org/10.1890/08-0645.1

Pinotti L, Ottoboni M (2021) Substrate as insect feed for bio-mass production. J Insects Food Feed 7:585–596. https://doi.org/10.3920/JIFF2020.0110

Pitts E, Lagnevik M (1998) Chapter 1: what determines food industry competitiveness? In: Traill B, Pitts E (eds) Competitiveness in the food industry. Blackie Academic & Professonal, London, pp 1–34

Pomar C, van Milgen J, Remus A (2019) Precision livestock feeding, principle and practice. In: Poultry and pig nutrition. Wageningen Academic Publishers. ISBN: 978-90-8686-333-4. https://hal.archives-ouvertes.fr/hal-02263872

Poore J, Nemecek T (2018) Reducing food's environmental impacts through producers and consumers. Science 360:987–992. https://doi.org/10.1126/science.aaq0216

Poppe M, Bonekamp G, van Pelt ML, Mulder HA (2021) Genetic analysis of resilience indicators based on milk yield records in different lactations and at different lactation stages. J Dairy Sci 104:P1967–P1981. https://doi.org/10.3168/jds.2020-19245

Pradhan AK, Karanth S (2023) Zoonoses from animal meat and milk. Chapter 28. In: Present knowledge in food safety, pp 394–411. https://doi.org/10.1016/B978-0-12-819470-6.00029-9

Prata JC, da Costa JP, Lopes I, Andrady AL, Duarte AC, Rocha-Santos T (2021) A one health perspective of the impacts of microplastics on animal, human and environmental health. Sci Total Environ 777:146094. https://doi.org/10.1016/j.scitotenv.2021.146094

Priya DT, Singh M, Kumar Gaur G, Sahu V (2022) 'Artificial intelligence' as an intelligent tool for the future animal production and better management: a review. Agric Rev. https://doi.org/10.18805/ag.R-2297

Ramachandraiah K, Ameer K, Jiang G, Hong G-P (2022) Micro- and nanoplastic contamination in livestock production: entry pathways, potential effects and analytical challenges. Sci Total Environ 844:157234. https://doi.org/10.1016/j.scitotenv.2022.157234

Ramos AK, Carlo G, Grant KM, Bendixsen C, Fuentes A, Gamboa R (2018) A preliminary analysis of immigrant cattle feedyard worker perspectives on job-related safety training. Safety 4:37. https://doi.org/10.3390/safety4030037

Ramsey S, Williams B, Jarrell P, Hubbs T (2023) Global demand for fuel ethanol through 2030, BIO-05, U.S. Department of Agriculture, Economic Research Service

Ramus CA, Montiel I (2005) When are corporate environmental policies a form of greenwashing? Bus Soc 44:377–414. https://doi.org/10.1177/0007650305278120

Raworth K (2017) Doughnut economics: seven ways to think like a 21st century economist. Chelsea Green Publishing

Raza A, Razzaq A, Mehmood SS, Zou X, Zhang X, Lv Y, Xu J (2019) Impact of climate change on crops adaptation and strategies to tackle its outcome: a review. Plan Theory 8:34. https://doi.org/10.3390/plants8020034

Renewable Fuels Association (2023) Ready. Set. Go! Ethanol Industry Outlook. https://d35t1syewk4d42.cloudfront.net/file/2432/2023%20RFA%20Outlook%20FINAL.pdf

Reynolds JL (2019) Solar geoengineering to reduce climate change: a review of governance proposal. Proc Royal Soc A 475:20190255. https://doi.org/10.1098/rspa.2019.0255

Ritchie H (2020) You want to reduce the carbon footprint of your food? Focus on what you eat, not whether your food is local. Published online at OurWorldInData.org. https://ourworldindata.org/food-choice-vs-eating-local

Robinson GM (2018) Globalization of agriculture. Ann Rev Resour Econ 10:133–160. https://doi.org/10.1146/annurev-resource-100517-023303

Robling H, Hatab AA, Säll S, Hansson H (2023) Measuring sustainability at farm level—A critical view on data and indicators. Environ Sustain Indicator 18:100258. https://doi.org/10.1016/j.indic.2023.100258

Rockström J, Gupta J, Qin D, Lade SJ, Abrams JF, Andersen LS, McKay DIA, Bai X, Bala G, Bunn SE, Ciobanu D, DeClerck F, Ebi K, Gifford L, Gordon C, Hasan S, Kanie N, Lenton TM, Loriani S, Liverman DM, Mohamed A, Nakicenovic N, Obura D, Ospina D, Prodani K, Rammelt C, Sakschewski B, Scholtens J, Stewart-Koster B, van Tharammal VD, Verburg PH, Winkelmann R, Zimm C, Bennett EM, Bringezu S, Broadgate W, Green PA, Huang L, Jacobson L, Ndehedehe C, Pedde S, Rocha J, Scheffer M, Schulte-Uebbing L, de Vries W, Xiao C, Xu C, Xu X, Zafra-Calvo N, Zhang X (2023) Safe and just earth system boundaries. Nature 619:102–117. https://doi.org/10.1038/s41586-023-06083-8

Rosa L, Chiarelli DD, Tu C, Rulli MC, D'Odorico P (2019) Global unsustainable virtual water flows in agricultural trade. Environ Res Lett 14:114001. https://doi.org/10.1088/1748-9326/ab4bfc

Rosenbloom D, Markard J, Geels FW, Fuenfschilling L (2020) Why carbon pricing is not sufficient to mitigate climate change—and how "sustainability transition policy" can help. Proc Natl Acad Sci 117:8664–8668. https://doi.org/10.1073/pnas.2004093117

Ruffatto K, Emaminejad SA, Juneja A, Kurambhatti C, Margenot A, Singh V, Cusick RD (2022) Mapping the national phosphorus recovery potential from centralized wastewater and corn ethanol infrastructure. Environ Sci Technol 56:8691–8701. https://doi.org/10.1021/acs.est.1c07881

Ruffatto K, Shurson GC, Muenich RL, Cusick RD (2023) Modeling national embedded phosphorus flows of corn ethanol distillers' grains to elucidate nutrient reduction opportunities. Environ Sci Technol 57:14429–14441. https://doi.org/10.1021/acs.est.3c02228

Ryan M (2022) The social and ethical impacts of artificial intelligence in agriculture: mapping the agricultural AI literature. AI & Soc 38:2473–2485. https://doi.org/10.1007/s00146-021-01377-9

Rychlik M, Humpf H-U, Marko D, Dänicke S, Mally A, Berthiller F, Klaffke H, Lorenz N (2014) Proposal of a comprehensive definition of modified and other forms of mycotoxins including "masked" mycotoxins. Mycotoxin Res 30:197–205. https://doi.org/10.1007/s12550-014-0203-5

Sabillón BH, Gerster-Bentaya M, Knierim A (2022) Measuring farmers' Well-being: influence of farm-level factors on satisfaction with work and quality of life. J Agric Econ 73:452–471. https://doi.org/10.1111/1477-9552.12457

Salemdeeb R, Vivanco DF, Al-Tabbaa A, zu Ermgassen, E.K.H.J. (2017) A holistic approach to the environmental evaluation of food waste prevention. Waste Manag 59:442–450. https://doi.org/10.1016/j.wasman.2016.09.042

Salomone R, Saija G, Mondello G, Giannetto A, Fasulo S, Savastano D (2017) Environmental impact of food waste bioconversion by insects: application of life cycle assessment to process using *Hermetia illucens*. J Clean Prod 140:890–905. https://doi.org/10.1016/j.jclepro.2016.06.154

Santangelo GD (2018) The impact of FDI in land in agriculture in developing countries on host country food security. J World Bus 53:75–84. https://doi.org/10.1016/j.jwb.2017.07.006

Scanes CG (2018) Chapter 18: impact of agricultural animals on the environment. In: Animals and human society, pp 427–449. https://doi.org/10.1016/B978-0-12-805247-1.00025-3

Scheufele DA, Krause NM (2019) Science audiences, misinformation, and fake news. Proc Natl Acad Sci 116:7662–7669. https://doi.org/10.1073/pnas.1805871115

Schmitz OJ, Sylvén M, Atwood TB, Bakker ES, Berzaghi F, Brodie JF, Cromsigt JPGM, Davies AB, Leroux SJ, Schepers FJ, Smith FA, Stark S, Svenning J-C, Tilker A, Ylänne H (2023) Trophic rewilding can expand natural climate solutions. Nat Clim Chang 13:324–333. https://doi.org/10.1038/s41558-023-01631-6

Schneider KR, Fanzo J, Haddad L, Herrero M, Moncayo JR, Herforth A, Remans R, Guarin A, Resnick D, Covic N, Béné C, Cattaneo A, Aburto N, Ambikapathi R, Aytekin D, Barquera S, Battersby J, Beal T, Molina PB, Cafiero C, Campeau C, Caron P, Conforti P, Damerau K, Di Girolamo M, DeClerk F, Dewi D, Elouafi I, Fabi C, Foley P, Frazier TJ, Gephart J, Golden C, Fischer CG, Hendriks S, Honorati M, Huang J, Kennedy G, Laar A, Lal R, Lidder P, Loken B, Marshall Q, Masuda YJ, McLaren R, Miachon L, Muñoz H, Nordhagen S, Qayyum N, Saisana M, Suhardiman D, Sumaila UR, Cullen MT, Tubiello FN, Vivero-Pol J-L, Webb P, Wiebe K (2023) The state of food systems worldwide in the countdown to 2030. Nat Food 4:1090–1110. https://doi.org/10.1038/s43016-023-00885-9

Scialabba NE-H (2022) Chapter 1: introduction to livestock systems. In: Managing healthy livestock and consumption. https://doi.org/10.1016/B978-0-12-823019-0.00010-6

Senay SD, Shurson G, Cardona C, Silverstein K (2022) Big data, data privacy, and plant and animal disease research using GEMS. Agron J 114:2644–2652. https://doi.org/10.1002/agj2.20933

Shen M, Zhang Y, Zhu Y, Song B, Zeng G, Hu D, Wen X, Ren X (2019) Recent advances in toxicological research of nanoplastics in the environment: a review. Environ Pollut 252(Part A):511–521. https://doi.org/10.1016/j.envpol.2019.05.102

Shepon A, Wu T, Kremen C, Dayan T, Perfecto I, Fanzo J, Eshel G, Golden CD (2023) Exploring scenarios for the food system-zoonotic risk interface. Lancet Planet Health 7:e329–e335. https://doi.org/10.1016/S2542-5196(23)00007-4

Shi S, Wang Z, Shen L, Xiao H (2022) Synthetic biology: a new frontier in food production. Trends Biotechnol 40:781–803. https://doi.org/10.1016/j.tibtech.2022.01.002

Shurson GC (2017) The role of biofuels co-products in feeding the world sustainably. Annual Rev Animal Biosci 5:229–254. https://doi.org/10.1146/annurev-animal-022516-022907

Shurson GC (2020) "What a waste" – Can we improve sustainability of food animal production systems by recycling food waste streams into animal feed in an era of health, climate, and economic crises? Sustain For 12:7071. https://doi.org/10.3390/su12177071

Shurson GC, Kerr BJ (2023) Challenges and opportunities for improving nitrogen utilization efficiency for more sustainable pork production. Front Animal Sci 4:1204863. https://doi.org/10.3389/fanim.2023.1204863

Shurson GC, Urriola PE (2022) Invited review: sustainable swine feeding programs require the convergence of circular agriculture and food systems with one health. Anim Front 12:30–40. https://doi.org/10.1093/af/vfac077

Shurson GC, Urriola PE, van de Ligt JLG (2022) Can we effectively manage parasites, prions, and pathogens in the global feed industry to achieve one health? Transbound Emerg Dis 69:4–30. https://doi.org/10.1111/tbed.14205

Shurson GC, Dierenfeld ES, Dou Z (2023) Rules are meant to be broken—rethinking the regulations restricting use of food waste as animal feed. Resour Conserv Recycl 199:107273. https://doi.org/10.1016/j.resconrec.2023.107273

Singh P, Kaur A (2022) A systematic review of artificial intelligence in agriculture. In: Deep Learning for Sustainable Agriculture. https://doi.org/10.1016/B978-0-323-85214-2.00011-2

Singh BK, Yan Z-Z, Whittaker M, Vargas R, Abdelfattah A (2023) Soil microbiomes must be explicitly included in one health policy. Nat Microbiol 8:1367–1372. https://doi.org/10.1038/s41564-023-01386-y

Sinke P, Swartz E, Sanctorum H, van der Giesen C, Odegard I (2023) Ex-ante life cycle assessment of commercial-scale cultivated meat production in 2030. Int J Life Cycle Assess 28:234–254. https://doi.org/10.1007/s11367-022-02128-8

Siragusa L, Arzyutov D (2020) Nothing goes to waste: sustainable practices of re-use among indigenous groups in the Russian north. Curr Opin Environ Sustain 43:41–48. https://doi.org/10.1016/j.cosust.2020.02.001

Smith TM, Goodkind AL, Kim T, Pelton REO, Suh K, Schmitt J (2017) Subnational mobility and consumption-based environmental accounting of US corn in animal protein and ethanol supply chains. Proc Natl Acad Sci 114:E7891–E7899. https://doi.org/10.1073/pnas.1703793114

Sogari G, Amato M, Biasato I, Chiesa S, Gasco L (2019) The potential role of insects as feed: A multi-perspective review. Animals 9:119. https://doi.org/10.3390/ani9040119

Soleimani T, Gilbert H (2021) An approach to achieve overall farm feed efficiency in pig production: environmental evaluation through individual life cycle assessment. Int J Life Cycle Assess 26:455–469. https://doi.org/10.1007/s11367-020-01860-3

Santhanam-Martin, M., Wilkinson, R., and Cowan, L. 2021. What is important for attracting and retaining farm employees? Evidence from the Australian orchard industry. In: 2nd International Symposium on Work in Agriculture, Clermont-Ferrand, France. Available online at: https://www.researchgate.net/profile/Michael-Santhanam-Martin/publication/355179770_What_is_important_for_attracting_and_retaining_farm_employees_Evidence_from_the_Australian_orchard_industry/links/6164bbcce7993f536cc82b99/What-is-important-for-attracting-and-retaining-farm-employees-Evidence-from-the-Australian-orchard-industry.pdf

Sovacool BK (2021) Reckless or righteous? Reviewing the sociotechnical benefits and risks of climate change geoengineering. Energy Strategy Rev 35:100656. https://doi.org/10.1016/j.esr.2021.100656

Sparrow R, Howard M, Degeling C (2021) Managing the risks of artificial intelligence in agriculture. NJAS: Impact Agri Life Sci 93:172–196. https://doi.org/10.1080/27685241.2021.2008777

Spawn-Lee SA, Lark TJ, Gibbs HK, Houghton RA, Kucharik CJ, Malins C, Pelton REO, Robertson GP (2021) Comment on 'carbon intensity of corn ethanol in the United States: state of the science'. Environ Res Lett 16:118001. https://doi.org/10.1088/1748-9326/ac2e35

Stein AJ, Santini F (2021) The sustainability of "local" food: a review for policy-makers. Rev Agri Food Environ Stud 103:77–89. https://doi.org/10.1007/s41130-021-00148-w

Steinfeld H, Gerber P, Wassenaar T, Castel V, Rosales M, de Haan C (2006) Livestock's Long shadow. In: The livestock, environment and development initiative (LEAD). Food and Agricultural Organization of the United Nations, Rome. http://www.fao.org/docrep/010/a0701e/a0701e00.HTM

Stoitsis G, Papakonstantinou M, Karvounis M, Manouselis N (2023) The role of big data and artificial intelligence in food risk assessment and prediction. In: Present knowledge in food safety, pp 1032–1044

St-Pierre NR, Cobanov B, Schnitkey G (2002) Economic losses from heat stress by US livestock industries. J Dairy Sci 86:E52–E77. https://doi.org/10.3168/jds.S0022-0302(03)74040-5

Streit E, Schwab C, Sulyok M, Naehrer K, Krska R, Schatzmayr G (2013) Multi-mycotoxin screening reveals the occurrence of 139 different secondary metabolites in feed and feed ingredients. Toxins 5:504–523. https://doi.org/10.3390/toxins5030504

Suleyman M, Bhaskar M (2023) The coming wave—technology, Power, and the 21st century's greatest dilemma. Crown Publishing Group, Penguin Random House, LLC, New York

Sun X, Devi D, Urriola P, Tiffany D, Jang J-C, Shurson G, Hu B (2021a) Feeding value improvement of corn ethanol co-product and soybean hulls by fungal fermentation: fiber degradation and digestibility improvement. Food Bioprod Process 130:143–153. https://doi.org/10.1016/j.fbp.2021.09.013

Sun X, Chen Y, Luo L, Heidari F, Tiffany D, Urriola P, Shurson G, Hu B (2021b) Feeding value improvement of corn ethanol co-product and agro-industrial residues by *Rhizopus oryzae*. Process Biochem 111:144–153. https://doi.org/10.1016/j.procbio.2021.10.029

Sun X, Tiffany D, Urriola P, Shurson G, Hu B (2021c) Nutrition upgrading of corn ethanol co-product by fungal fermentation: amino acid enrichment and anti-nutritional factor degradation. Food Bioprod Process 130:1–13. https://doi.org/10.1016/j.fbp.2021.09.004

Sun X, Dou Z, Shurson G, Hu B (2023a) Fungal bioprocessing of wheat straw with fresh fruit and vegetable discards to produce cattle feeds for enhanced sustainability. Resour Conserv Recycl 199:107251. https://doi.org/10.1016/j.resconrec.2023.107251

Sun X, Urriola PE, Shurson G, Tiffany D, Hu B (2023b) Enhancing feeding value of corn distiller's grains with solubles via fungal co-cultured solid-state fermentation for monogastric animal nutrition. Anim Feed Sci Technol 303:115673. https://doi.org/10.1016/j.anifeedsci.2023.115673

Sun X, Dou Z, Shurson G, Hu B (2024) Bioprocessing to upcycle agro-industrial and food wastes into high-nutritional animal feed for sustainable food and agriculture systems. Resour Conserv Recycl 201:107325. https://doi.org/10.1016/j.resconrec.2023.107325

Swanson BE, Davis K (2014) Status of agricultural extension and rural advisory services Worldwide, Summary Report. Global Forum for Rural Advisory Services, Lindau. https://www.g-fras.org/en/knowledge/gfras-publications.html?download=391:status-of-agricultural-extension-and-rural-advisory-services-worldwide

Sykuta ME (2016) Big data in agriculture: property rights, privacy and competition in ag data services. Int Food Agribusiness Manag Rev Special Issue 19, Issue A:57–74

Tait-Burkard C, Doeschl-Wilson A, McGrew MJ, Archibald AL, Sang HM, Houston RD, Whitelaw CB, Watson M (2018) Livestock 2.0 – genome editing for fitter, healthier, and more productive farmed animals. Genome Biol 19:204. https://doi.org/10.1186/s13059-018-1583-1

Talaviya T, Shah D, Patel N, Yagnik H, Shah M (2020) Implementation of artificial intelligence in agriculture for optimization of irrigation and application of pesticides and herbicides. Artif Intell Agric 4:58–73. https://doi.org/10.1016/j.aiia.2020.04.002

Tang KL, Caffrey NP, Nóbrega DB, Cork SC, Ronksley PE, Barkema HW, Polachek AJ, Ganshorn H, Sharma N, Kellner JD, Ghali WA (2017) Restricting the use of antibiotics in food-producing animals and its associations with antibiotic resistance in food-producing animals and human beings: a systematic review and meta-analysis. Lancet Planet Health 1:E316–E327. https://10.1016/S2542-5196(17)30141-9

Thompson PB (2017) Spirit of the soil: agriculture and environmental ethics, 2nd edn. Routledge, New York. https://doi.org/10.4324/9781315559971

Thornton PK (2010) Livestock production: recent trends, future prospects. Philos Trans R Soc B 365:2853–2867. https://doi.org/10.1098/rstb.2010.0134

Thornton P, Nelson G, Mayberry D, Herrero M (2022) Impacts of heat stress on global cattle production during the 21st century: a modelling study. Lancet Planet Health 6:e192–e201

Tittonell P, El Mujtar V, Felix G, Kebede Y, Laborda L, Luján Soto R, de Vente J (2022) Regenerative agriculture—agroecology without politics? Front Sustain Food Sys 6:844261. https://doi.org/10.3389/fsufs.2022.844261

Tsoukas V, Gkogkidis A, Kampa A, Spathoulas G, Kakarountas A (2022) Enhancing food supply chain security through the use of blockchain and TinyML. Information 13:213. https://doi.org/10.3390/info13050213

Tuomisto HL, Hodge ID, Riordan P, Macdonald DW (2012) Does organic farming reduce environmental impacts?—A meta-analysis of European research. J Environ Manag 112:309–320. https://doi.org/10.1016/j.jenvman.2012.08.018

Tuyttens FAM, Molento CFM, Benaissa S (2022) Twelve threats of precision livestock farming (PLF) for animal welfare. Front Veterinary Sci 9:889623. https://doi.org/10.3389/fvets.2022.889623

Twine R (2021) Emissions from animal agriculture—16.5% is the new minimum figure. Sustain For 13:6276. https://doi.org/10.3390/su13116276

Tzanidakis C, Simitzis P, Panagakis P (2023) Precision livestock farming (PLF) systems: improving sustainability and efficiency of animal production. In: Márquez FPG, Lev B (eds) Sustainability; cases and studies in using operations research and management science methods, vol 33, pp 285–339. https://doi.org/10.1007/978-3-031-16620-4

Ukpabi D, Karjaluoto H, Bötticher A, Nikiforova A, Petrescu D, Schindler P, Valterbergs V, Lehmann L (2023) Framework for understanding quantum computing use cases from a multidisciplinary perspective and future research directions. Futures 154:103277. https://doi.org/10.1016/j.futures.2023.103277

Upton M (2004) The role of livestock in economic development and poverty reduction. Pro-poor livestock policy initiative, Working Paper No. 10. Food and Agriculture Organization, Rome, 66 pp

Urriola PE, Mielke JA, Mao Q, Hung YT, Kurtz JF, Johnston LJ, Shurson GC, Chen C, Saqui-Salces M (2018) Evaluation of partially de-oiled microalgae product as a feed source for nursery pigs. Trans Animal Sci 2:169–183. https://doi.org/10.1093/tas/txy013

Uwizeye A, Gerber PJ, Schulte RPO, de Boer IJM (2016) A comprehensive framework to assess the sustainability of nutrient use in global livestock supply chains. J Clean Prod 129:647–658. https://doi.org/10.1016/j.jclepro.2016.03.108

Uwizeye A, Gerber PJ, Opio CI, Tempio G, Mottet A, Makkar HPS, Falcucci A, Steinfeld H, de Boer IJM (2019) Nitrogen flows in global pork supply chains and potential improvement from feeding swill to pigs. Res Conserv Recycl 146:168–179. https://doi.org/10.1016/j.resconrec.2019.03.032

Uwizeye A, de Boer IJM, Opio CI, Schulte RPO, Falcucci A, Tempio G, Teillard F, Casu F, Rulli M, Galloway JN, Leip A, Erisman JW, Robinson TP, Steinfeld H, Gerber PJ (2020) Nitrogen emissions along the global livestock supply chains. Nat Food 1:437–446. https://doi.org/10.1038/s43016-020-0113-y

Van Berkum S (2021) How trade can drive inclusive and sustainable food system outcomes in food deficit low-income countries. Food Secur 13:1541–1554. https://doi.org/10.1007/s12571-021-01218-z

Van den Honert M, Hoffman L (2023) Drug-resistant bacteria from "farm to fork": impact of antibiotic use in animal production. Chapter 58. In: Present knowledge in food safety, pp 871–892. https://doi.org/10.1016/B978-0-12-819470-6.00011-1

Van Middelaar CE, van Zanten HHE, de Boer IJM (2019) Future of animal nutrition: the role of life cycle assessment. In: Hendriks WH, Verstegen MWA, Babinsky L (eds) Poultry and pig nutrition—challenges of the 21st century. Wageningen Academic Publishers, pp 307–314. https://doi.org/10.3920/978-90-8686-884-1_14

Van Wagenberg CPA, de Haas Y, Hogeveen H, van Krimpen MM, Meuwissen MPM, van Middelaar CE, Rodenburg TB (2017) Animal board invited review: comparing conventional and organic livestock production systems on different aspects of sustainability. Animal 11:1839–1851. https://doi.org/10.1017/S175173111700115X

Van Zanten HHE, Mollenhorst H, Oonincx DGAB, Bikker P, Meerburg BG, de Boer IJM (2015) From environmental nuisance to environmental opportunity: housefly larvae convert waste to livestock feed. J Clean Prod 102:362–369. https://doi.org/10.1016/j.jclepro.2015.04.106

Vanhoutte I, Audenaert K, De Gelder L (2016) Biodegradation of mycotoxins: Tales from known and unexplored worlds. Front Microbiol 7:561. https://doi.org/10.3389/fmicb.2016.00561

Velazquez-Meza ME, Galarde-López M, Carrillo-Quiróz B, Alpuche-Aranda CM (2022) Antimicrobial resistance: one health approach. Veterinary World 15:743–749. https://doi.org/10.14202/vetworld.2022.743-749

Vijayan D, Ludwig D, Rybak C, Kaechele H, Hoffman H, Schönfeldt HC, Mbwana HA, Rivero CV, Löhr K (2022) Indigenous knowledge in food systems transformations. Commun Earth Environ 3:213. https://doi.org/10.1038/s43247-022-00543-1

Vinyeta K, Lynn K (2013) Exploring the role of traditional ecological knowledge in climate change initiatives. In: General Technical Report PNW-GTR-879. U.S. Department of Agriculture, Forest Service, Pacific Northwest Research Station, Portland, p 37

Virtanen PK, Siragusa L, Guttorm H (2020) Introduction: toward more inclusive definitions of sustainability. Curr Opin Environ Sustain 43:77–82. https://doi.org/10.1016/j.cosust.2020.04.003

Vougioukas SG (2019) Agricultural robotics. Annual Rev Control Robotics Autonomous Sys 2:365–392. https://doi.org/10.1146/annurev-control-053018-023617

Vranken E, Berckmans D (2017) Precision livestock farming for pigs. Anim Front 7:33–37. https://doi.org/10.2527/af.2017.0106

Wakchaure M, Patle BK, Mahindrakar AK (2023) Application of AI techniques and robotics in agriculture: a review. Artif Intell Life Sci 3:100057. https://doi.org/10.1016/j.ailsci.2023.100057

Wang JM, Liu Q, Hou Y, Qin W, Bai ZH, Zhang FS, Oenema O (2022a) Impacts of international food and feed trade on nitrogen balances and nitrogen use efficiencies of food systems. Sci Total Environ 838(Part 3):156151. https://doi.org/10.1016/j.scitotenv.2022.156151

Wang G, Wu X, Yin Y (2022b) Synthetic biology-driven customization of functional feed resources. Trends Biotechnol 40:777–780. https://doi.org/10.1016/j.tibtech.2022.03.002

Wang H, Qi J-F, Qin R, Ding K, Graham DW, Zhu Y-G (2023a) Intensified livestock farming increases antibiotic resistance genotypes and phenotypes in animal feces. Commun Earth Environ 4:123. https://doi.org/10.1038/s43247-023-00790

Wang X, Dou Z, Feng S, Zhang Y, Ma L, Zou C, Bai Z, Lakshmanan P, Shi X, Liu D, Zang W, Deng Y, Zhang W, Chen X, Zhang F, Chen X (2023b) Global food nutrients analysis reveals

alarming gaps and daunting challenges. Nat Food 4:1007–1017. https://doi.org/10.1038/s43016-023-00851-5
Weersink A, Fraser E, Pannell D, Rotz S (2018) Opportunities and challenges for big data in agricultural and environmental analysis. Annual Rev Res Econ 10:19–37. https://doi.org/10.1146/annurev-resource-100516-053654
Weindl I, Bodirsky BL, Rolinski S, Biewald A, Lotze-Campen H, Müller C, Dietrich JP, Humpenöder F, Stevanović M, Schaphoff S, Popp A (2017a) Livestock production and the water challenge of future food supply: implications of agricultural management and dietary choices. Glob Environ Chang 47:121–132. https://doi.org/10.1016/j.gloenvcha.2017.09.010
Weindl I, Popp A, Bodirsky BL, Rolinski S, Lotze-Campen H, Biewald A, Humpenöder F, Dietrich JP, Stevanović M (2017b) Livestock and human use of land: productivity trends and dietary choices as drivers of future land and carbon dynamics. Glob Planet Chang 159:1–10. https://doi.org/10.1016/j.gloplacha.2017.10.002
Weinrich R, Strack M, Neugebauer F (2020) Consumer acceptance of cultured meat in Germany. Meat Sci 162:107924. https://doi.org/10.1016/j.meatsci.2019.107924
Weins J, Fargione J, Hill J (2011) Biofuels and biodiversity. Ecol Appl 21:1085–1095. https://doi.org/10.1890/09-0673.1
Werkheiser I (2018) Precision livestock farming and farmers' duties to livestock. J Agri Environ Ethics 31:181–195. https://doi.org/10.1007/s10806-018-9720-0
Wezel A, Herren BG, Kerr RB, Barrios E, Gonçalves ALR, Sinclair F (2020) Agroecological principles and elements and their implications for transitioning to sustainable food systems. A Rev Agron Sustain Develop 40:40. https://doi.org/10.1007/s13593-020-00646-z
Wilgenbusch JC, Pardey PG, Hospodarsky N, Lynch BJ (2022) Addressing new data privacy realities affecting agricultural research and development: a tiered-risk, standards-based approach. Agron J 114:2653–2668. https://doi.org/10.1002/agj2.20968
Willett W, Rockström J, Loken B, Springmann M, Lang T, Vermeulen S, Garnett T, Tilman D, DeClerck F, Wood A, Jonell M, Clark M, Gordon LJ, Fanzo J, Hawkes C, Zurayk R, Rivera JA, De Vries W, Sibanda LM, Afshin A, Chaudhary A, Herrero M, Agustina R, Branca F, Lartey A, Fan S, Crona B, Fox E, Bignet V, Troell M, Lindahl T, Singh S, Cornell SE, Reddy KS, Narain S, Nishtar S, Murray CJL (2019) Food in the Anthropocene: the EAT-*lancet* commission on healthy diets from sustainable food systems. Lancet Comm 393:447–492. https://doi.org/10.1016/S0140-6736(18)31788-4
Wiseman L, Sanderson J, Zhang A, Jakku E (2019) Farmers and their data: an examination of farmers' reluctance to share their data through the lens of the laws impacting smart farming. NJAS: Wageningen J Life Sci:90, 100301–91. https://doi.org/10.1016/j.njas.2019.04.007
Wiśniewska J (2011) Economic sustainability of agriculture – conceptions and indicators. Acta Scientiarum Polonorum, Oeconomia 10:119–137. https://js.wne.sggw.pl/index.php/aspe/article/view/4002
World Meteorological Organization (2023) Provision State of the Global Climate. https://wmo.int/sites/default/files/2023-11/WMO%20Provisional%20State%20of%20the%20Global%20Climate%202023.pdf
World Wildlife Fund. (2021). Driven to waste: the global impact of food loss and waste on farms. https://files.worldwildlife.org/wwfcmsprod/files/Publication/file/5p58sxloyr_technical_report_wwf_farm_stage_food_loss_and_waste.pdf?_ga=2.136657523.25368299.1705772770-1654357082.1705772769
Wray-Cahen D, Bodnar A, Rexroad C, Siewerdt F, Kovich D (2022) Advancing genome editing to improve the sustainability and resiliency of animal agriculture. CABI Agri Biosci 3:21. https://doi.org/10.1186/s43170-022-00091-w
Wu P, Huang J, Zheng Y, Yang Y, Zhang Y, He F, Chen H, Quan G, Yan J, Li T, Gao B (2019) Environment occurrences, fate, and impacts of microplastics. Ecotoxicol Environ Saf 184:109612. https://doi.org/10.1016/j.ecoenv.2019.109612
Xu X, Sharma P, Shu S, Lin T-S, Ciais P, Tubiello FN, Smith P, Campbell N, Jain AK (2021) Global greenhouse gas emissions from animal-based food are twice those of plant-based foods. Nat Food 2:724–732. https://doi.org/10.1038/s43016-021-00358-x

Xu R, Kiarie EG, Yiannikouris A, Sun L, Karrow NA (2022) Nutritional impact of mycotoxins in food animal production and strategies for mitigation. J Animal Sci Biotechnol 13:69. https://doi.org/10.1186/s40104-022-00714-2

Yang M, Chen L, Wang J, Msigwa G, Osman AI, Fawzy S, Rooney DW, Yap P-S (2023a) Circular economy strategies for combatting climate change and other environmental issues. Environ Chem Lett 21:55–80. https://doi.org/10.1007/s10311-022-01499-6

Yang Z, Urriola PE, Johnston LJ, Shurson GC (2023b) A systems approach to evaluate nitrogen utilization efficiency and environmental impacts of growing-finishing feeding programs in U.S. pork production systems. J Anim Sci 101:skad188. https://doi.org/10.1093/jas/skad188

Zapeda C, Salman M, Ruppanner R (2001) International trade, animal health and veterinary epidemiology: challenges and opportunities. Prev Vet Med 48:261–271. https://doi.org/10.1016/S0167-5877(00)00200-2

Zhang Z, Ge Z (2022) Protection and restriction of grain-related intellectual property rights in the context of food security based on industrial internet of things. Mob Inf Syst 2022:4554447. https://doi.org/10.1155/2022/4554447

Zhang Z, Moore JC, Huisingh D, Zhao Y (2015) Review of geoengineering approaches to mitigating climate change. J Clean Prod 103:898–907. https://doi.org/10.1016/j.jclepro.2014.09.076

Zhang Y, Li Y, Hubacek K, Tian X, Lu Z (2019a) Analysis of CO_2 transfer processes involved in global trade based on ecological network analysis. Appl Energy 233–234:576–583. https://doi.org/10.1016/j.apenergy.2018.10.051

Zhang Z, Nie D, Fan K, Yang J, Guo W, Meng J (2019b) A systematic review of plant-conjugated masked mycotoxins: occurrence, toxicology, and metabolism. Crit Rev Food Sci Nutr 60:1523–1537. https://doi.org/10.1080/10408398.2019.1578944

Zhang X-X, Liu J-S, Han L-F, Xia S, Li S-Z, Li OY, Kassegne K, Li M, Yin K, Hu Q-Q, Xiu L-S, Zhu Y-Z, Huang L-Y, Wang X-C, Zhang Y, Zhao H-Q, Yin J-X, Jiang T-G, Li Q, Fei S-W, Gu S-Y, Chen F-M, Zhou N, Cheng Z-L, Xie Y, Li H-M, Chen J, Guo Z-Y, Feng J-X, Ai L, Xue J-B, Ye Q, Grant L, Song J-X, Simm G, Utzinger J, Guo X-K, Zhou X-N (2022) Towards a global one health index: a potential assessment tool for one health performance. Infect Dis Poverty 11:57. https://doi.org/10.1186/s40249-022-00979-9

Zepeda, C., Salman, M., Ruppanner, R. (2001) International trade, animal health and veterinary epidemiology: challenges and opportunities. Preventive Veterinary Medicine 48, 261-271. https://doi.org/10.1016/S0167-5877(00)00200-2

Zhou C, Liu B, Piao S, Wang X, Lobell DB, Huang Y, Huang M, Yao Y, Bassu S, Ciais P, Durand J-L, Elliott J, Ewert F, Janssens IA, Li T, Lin E, Liu Q, Martre P, Müller C, Peng S, Peñuelas J, Ruane AC, Wallach D, Wang T, Wu D, Liu Z, Zhu Y, Zhu Z, Asseng S (2017) Temperature increase reduces global yields of major crops in four independent estimates. Proc Natl Acad Sci 114:9326–9331. https://doi.org/10.1073/pnas.1701762114

Zu Ermgassen EKHJ, Balmford A, Salemdeeb R (2016) Reduce, relegalize, and recycle food waste. Science 352:1526. https://doi.org/10.1126/science.aaf9630

3 Intensive Livestock Production: Mechanisms and Structures

Stephan Brosig, Franziska Hauff, and Alfons Balmann

The production and consumption of livestock products has increased significantly in recent decades. As the world's population and income levels continue to grow, so does the demand for protein-rich foods, including meat and dairy products. Globally, livestock production has responded to increasing demand primarily by shifting to specialised production units that are typically large (to take advantage of economies of scale), with high stocking densities (to maximise production per unit area), and geographically concentrated (to benefit from network externalities). These intensive forms of production tend to use advanced technologies (such as automated feeding systems, climate control systems, genetic engineering, and data analysis), intensive health management, and specialised breeding and genetics to optimise production efficiency.

Monogastric species such as pigs and poultry are particularly well suited for rapid intensification of production due to their high feed conversion ratios and short generation intervals (Robinson et al. 2011). However, intensification of livestock production has raised serious concerns about sustainability, environmental impact, animal welfare, and public health, leading to calls for more responsible and efficient practices within the industry.

The first two sections outline major structural trends in global and EU livestock production and typical forms of value chain coordination for selected animal products. The third section focuses on the structural development of intensive livestock production in Germany.

S. Brosig (✉) · F. Hauff · A. Balmann
Leibniz Institute of Agricultural Development in Transition Economies (IAMO), Halle (Saale), Germany
e-mail: brosig@iamo.de; hauff@iamo.de; balmann@iamo.de

B. Grabkowsky, T. Blaha (eds.), *Intensive Livestock Production in Transition*,
https://doi.org/10.1007/978-3-031-97872-2_3

1 Structural Developments in Global and EU-Wide Livestock Production

On a global scale, meat production has risen substantially in recent decades. While beef and veal production has increased relatively moderately over the past 20 years, pig production has increased by more than a third, and milk production by 52% (Table 3.1). Poultry meat (Table 3.1 refers exclusively to chicken) has shown the highest growth rate, almost doubling over the same period.

Brazil, China, the EU (despite declining beef and pork production in the last decade), and the United States are the "big four" in global meat production (Fig. 3.1). Together, they account for more than 60% of global poultry and beef/veal production in 2023, and even more than 80% of global pork production. China has been by far the world's largest producer of pork for years. After the drastic declines in Chinese pig stocks due to the impact of African Swine Fever, particularly in 2019 and 2020, production has now almost fully recovered. India has become the world's largest producer of milk in recent years. More than 31% of the liquid milk produced globally in 2023 came from India, while the EU, formerly the world's top producer, is now in second place behind India with a considerable gap.

As global production has increased, so has global trade in meat and milk (Table 3.2). Comparing the absolute volumes of the world's top four meat and milk producers and exporters (Fig. 3.2), it is obvious that the global expansion in production has not all gone to exports, but that most of the additional production has remained in the producing countries and larger regions (e.g., the EU). Moreover, not all leading exporters necessarily have a high export share. For example, trade in

Table 3.1 Development of global meat* and milk production 2000–2023

1000 MT	Beef and veal	Pig	Chicken	Milk, fluid
2000	52,912	84,732	54,547	439,565
2005	54,063	92,934	65,298	483,528
2010	57,110	102,831	79,058	513,791
2015	57,467	112,034	91,840	585,047
2020	57,634	96,084	99,659	649,540
2021	58,361	107,954	101,060	656,881
2022	59,281	114,533	101,840	659,127
2023	59,374	115,215	102,389	666,595
Index (2000 = 100)				
2000	100.0	100.0	100.0	100.0
2005	102.2	109.7	119.7	110.0
2010	107.9	121.4	144.9	116.9
2015	108.6	132.2	168.4	133.1
2020	108.9	113.4	182.7	147.8
2021	110.3	127.4	185.3	149.4
2022	112.0	135.2	186.7	149.9
2023	112.2	136.0	187.7	151.6

PSD Online (2024); * in CWE: carcass weight equivalents

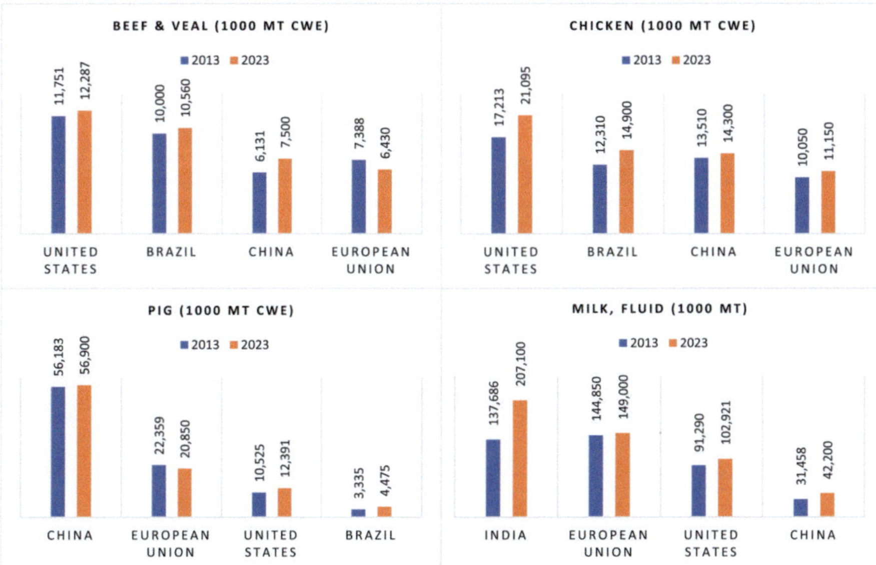

PSD Online *in CWE: Carcass weight equivalent

Fig. 3.1 Global top four meat* and milk producers, 2013 and 2023

Table 3.2 Development of global meat* and milk exports 2000–2023

1000 MT	Beef and veal	Pig	Chicken	Milk, fluid
2000	5,943	3,084	4,856	441
2005	7,036	4,819	6,779	402
2010	7,439	5,884	9,102	717
2015	9,123	7,077	10,564	1,531
2020	11,229	12,569	13,102	3,297
2021	11,440	12,220	13,308	3,504
2022	12,030	10,943	13,574	3,237
2023	11,932	10,007	13,559	2,970
Index (2000 = 100)				
2000	100.0	100.0	100.0	100.0
2005	118.4	156.3	139.6	91.2
2010	125.2	190.8	187.4	162.6
2015	153.5	229.5	217.5	347.2
2020	188.9	407.6	269.8	747.6
2021	192.5	396.2	274.1	794.6
2022	202.4	354.8	279.5	734.0
2023	200.8	324.5	279.2	673.5

PSD Online (2024); * in CWE: carcass weight equivalents

Fig. 3.2 Global top four meat and milk exporters, 2013 and 2023

fluid milk outside the EU remains minimal, despite increases over the last decade. Exports, almost exclusively of pasteurised milk in small end-consumer packages, have historically been directed to the United Kingdom (UK), North Africa, and Middle Eastern countries. China has become the dominant export destination for EU milk in recent years, accounting for about 0.5% of EU production (Polet 2022). However, Chinese imports of EU-packaged milk have recently declined, in part due to increased domestic production.

Livestock production has also adjusted and shifted in a tendency toward larger and more intensive production structures. Intensity can be measured for each type of production by the annual volume of production in relation to the annual average of the stock size, i.e., to the annual average of the number of animals kept at any given time (Robinson et al. 2011). It is often related (though not necessarily proportional) to the livestock size per farm and livestock density in animals per hectare. Stock size per farm can hence be used as an indicator of structural change in livestock production. Significant heterogeneity persists across countries and species. In aggregate terms, the EU's total livestock population has declined considerably over the past two decades. Between 2002 and 2022, the total number of pigs, bovine animals, sheep, and goats in the EU-27 (2020) dropped by an estimated 14.4%, from 326 million to 280 million (Eurostat 2023e). However, this decline has been particularly pronounced among smaller-scale pig holdings (Fig. 3.3 (right-hand bar charts). Between 2010 and 2020, the share of pigs kept in stocks of fewer than 500 livestock units (LSU)[1] declined markedly, while the proportion held in larger units (>500 LSU) continued to increase. A comparable trend is observed in the dairy

[1] According to Eurostat, the reference unit used for the calculation of livestock units (=1 LSU) is the grazing equivalent of one adult dairy cow producing 3000 kg of milk annually, without additional concentrated foodstuffs.

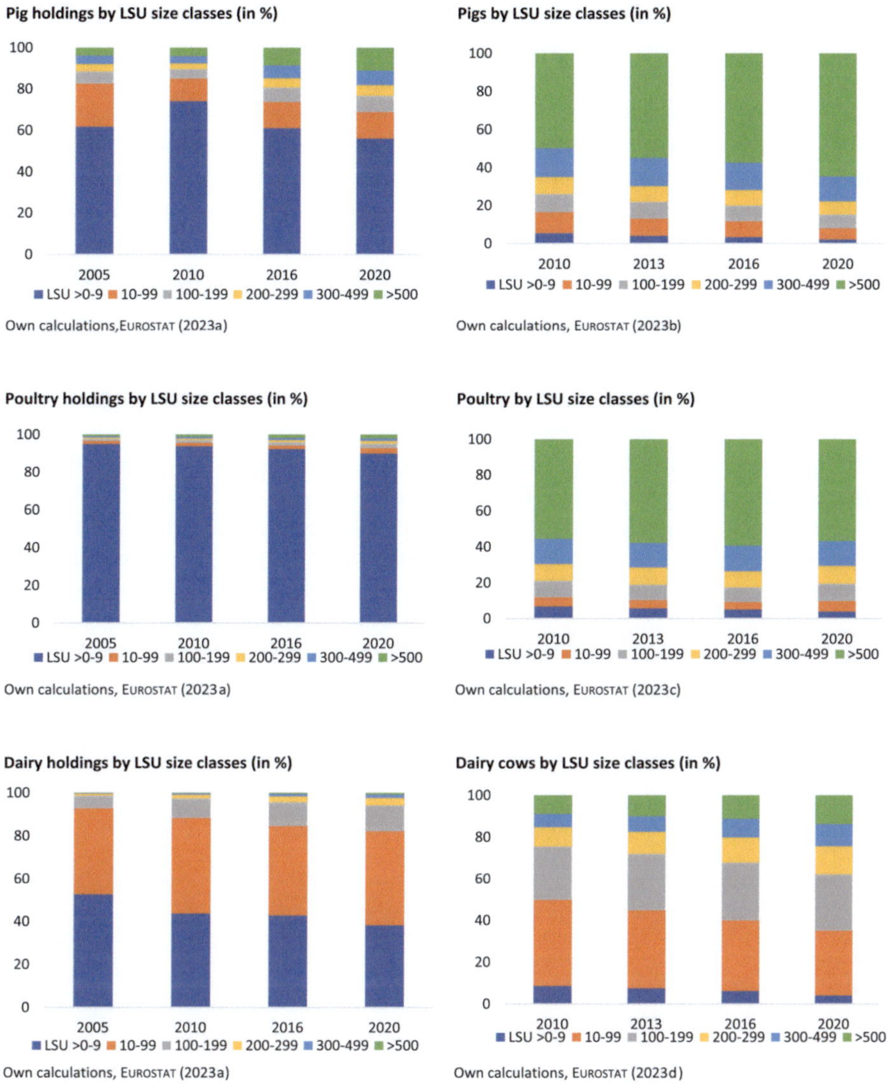

Fig. 3.3 Change in livestock production structures over time, EU 27 (2020)

sector, where the share of animals in holdings exceeding 200 LSU rose steadily over the same period. These developments reflect a broader tendency toward structural consolidation and increasing specialization.

Despite the trend towars concentration, farms with small and medium stock sizes still represent the quantitative majority of holdings (Fig. 3.3 (left-hand bar charts)). Particularly in many Eastern European countries, small-scale production systems – often referred to as 'backyard farms' – numerically dominate and are frequently operated on a part-time basis. At the other end of the spectrum, however, there are very large, intensively managed farms that account for a significant share of total production. This dual farm structure is a characteristic feature of agricultural production systems in Eastern and Central Europe.

In general, and across the EU, it can be seen that the majority of the EU's livestock (and thus the majority of production) is concentrated in larger types of operations, the number of which has in turn tended to increase across various sectors over recent decades (Fig. 3.3). This has tended to shift production toward intensive indoor confinement systems, which have the potential to raise large numbers of animals in a very cost-effective manner with very high productivity.

Particularly in emerging and developing countries, where economies have developed and incomes have risen, consumers have increasingly been able to afford animal products as part of their diets. Rapid urbanisation, particularly in developing regions, has also led to dietary shifts toward higher meat, dairy, and egg consumption (Delgado et al. 1999; OECD/FAO 2023). According to projections by the OECD/FAO, this trend is expected to continue and will continue to significantly influence livestock production (Fig. 3.4). In particular, countries with middle-income levels—characterised by economic dynamism and urbanisation—are expected to experience a marked increase in meat consumption. In contrast, in low-income countries, strong population growth is seen as the main driver. Nevertheless, per capita meat consumption in developing countries will remain considerably lower and is projected to reach only 40% of the average consumption level in developed countries by 2032.

In developed countries, demand for beef and pork is expected to rise only slightly, as per capita consumption is already high. The strongest increase in total

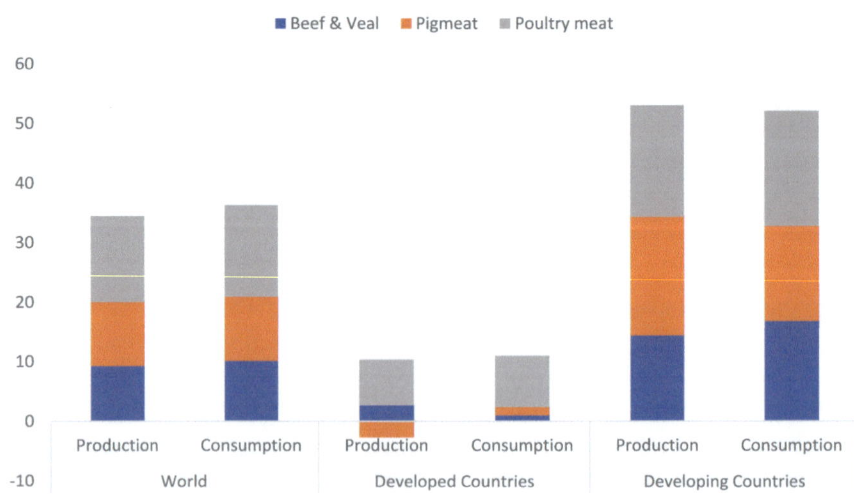

Developed and developing countries: See summary table for country grouping in the Agricultural Outlook 2023-2032
Own calculations, OECD/FAO (2023)

Fig. 3.4 Total meat production and consumption, change in 2032 compared to 2020/22 av. (in %)

consumption—more than 8% by 2032—is projected for poultry meat (Fig. 3.4). While disposable income used to play a central role in meat consumption, its importance has diminished. Instead, factors such as health concerns, environmental awareness, and animal welfare have become increasingly important. These factors are leading consumers to reconsider their diets—for example, by favoring certain types of meat (e.g., more poultry, less red meat) or by generally reducing meat consumption (OECD/FAO 2023). Furthermore, global livestock and poultry populations are expected to continue growing. Combined with advances in breeding, management, infrastructure, and technology, this is likely to enhance production and productivity, especially in pig and poultry farming, particularly in developing countries. By contrast, several EU countries are projected to see a decline in pork production. Reasons for this include rising feed and energy costs, disease risks, stricter environmental regulations, and increasing demands for animal welfare (OECD/FAO 2023).

2 Typical Forms of Value Chain Coordination

Livestock farming is integrated into complex value chains of suppliers and buyers organised in different types of coordination. These types of coordination occur in different constellations, from the spot market to long-term supply relationships, from simple marketing contracts, production contracts, and contract fattening to vertical integration. Vertical integration means that companies are organised at different levels of the value chain within one company. This can be driven, for example, by the backward integration of meatpackers or processors (typical development in the United States) or by the backward integration of actors further downstream, such as processors and food retailers (typical for Brazil) (WBAE 2015). It should be noted that there is a wide variation in coordination mechanisms and concentration levels. The following paragraphs therefore focus on selected development trends in value chain coordination.

In many countries, including Germany, the poultry market is today dominated by a few large companies that have strong global networks and trade their animals and meat products worldwide. They often have key stages of the value chain in one agri-food holding: Breeding, hatcheries, feed mills, slaughtering, and processing. Other parts, particularly contract fattening, are outsourced to contract farmers who receive chicks and feed from the holding (Martinez 1999; MacDonald et al. 2023). The result is the emergence of integrated value chains often coordinated by a "chain captain" with contractually secured supply and procurement obligations (Hanf and Kühl 2005). The degree of freedom of the farmers involved, for example with regard to the procurement of inputs, the organisation of production, or price negotiations with retailers, is severely restricted. Contract farming is the dominant form of organisation in the poultry industry. An example of this business model is the PHW Group, the largest poultry producer in Germany, which markets its products under the "Wiesenhof" brand. This close vertical integration is neither new nor specific to Germany. On a global scale, the Brazilian holding company JBS S.A. is in 2023 the largest poultry producer—a highly vertically integrated meatpacker with an annual processing capacity of 4.4 billion head of poultry (WATTPoultry 2023).

Consolidation and concentration processes are also taking place in pig production, again with regional variations. In the United States, for example, meat packers and processors have built much larger facilities in recent decades, while at the same time developing closer links with livestock producers through direct ownership or contracts. As a result, by the early 2000s, the majority of pigs were raised by contract farmers. Today, contract farmers have become even more specialized in selected stages of production such as farrowing, farrowing to weaning, or weaning to finishing (MacDonald et al. 2023). While taking advantage of significant economies of scale and realizing productivity gains, this development has promoted further concentration and growth processes at the production level. According to the US Department of Agriculture (USDA), less than 10% of pigs are currently traded on spot markets, where prices are determined on the day of delivery and transactions occur without prior contracts. This is an indication of the high degree of integration in the industry (MacDonald et al. 2023).

The pig meat chain in the EU is very heterogeneous, as there are large differences in the organisation of farming systems and types of farms in and between Member States. Production has recently declined across Europe and Spain has become the EU's leading pig producer and exporter in 2023 (ESS-Feed 2023). There, the production stage is characterised by a high degree of vertical integration, where farmers are contracted to breed and fatten the animals (Augère-Granier 2020). Denmark also has highly developed integrated production systems, while in Germany, pig production is still largely carried out by vertically independent companies and production contracts are relatively less pronounced. There, the marketisation of slaughter pigs is mostly based on long-term business relationships with livestock traders, livestock marketing cooperatives, or slaughterhouses. Such arrangements usually include sales and purchase obligations but do not interfere in the production processes of the contracting parties (WBAE 2015).

Another recent development is the integration of the previously separate stages of cutting, packaging, and processing. There have also been some inter-industry mergers: Tyson Foods, the largest broiler processor in the United States, acquired IPB, the largest cattle and hog packer; and Brazil's JBS acquired beef packer Swift, Cargill's pork business, and chicken processor Pilgrim's Pride (MacDonald et al. 2023).

Large retail chains can play a significant role in shaping meat production through the standards and specifications they set for their supply chains. While not necessarily directly involved in food processing and agricultural production, they contract with processors and packers of fresh produce, influencing various aspects of meat production and processing, including quality, safety, sustainability, and packaging (MacDonald et al. 2023). In recent years, societal expectations have often been directed at the food retailer due to the lack of high-profile companies at the production stage. Producers in the dairy and especially the meat sectors often act as anonymous private label suppliers. The retailers' position as intermediaries between producers and consumers gives them a significant degree of control over product specifications and requirements.

Private label initiatives, coordinating cooperation with partners in the production chain, play an important role in responding to growing concerns about animal welfare. For example, the Dutch non-governmental organisation "Dierenbescherming" (Dutch Society for the Protection of Animals) has introduced the popular "Beter leven" (Better Life) label, which aims to improve the welfare of on-farm livestock via the market. A three-star system reflects the different levels of animal welfare that can be found on the packaging of meat, dairy products, and eggs. The Better Life Label Foundation and certification bodies audit all parties involved in the chain, such as suppliers and buyers.

3 Structure and Structural Change of German Livestock Production

Nearly two-thirds of German farms keep livestock. The most common type of livestock is cattle (41% of farms), especially dairy cows (21%), laying hens (18%), and fattening pigs (11%) (BMEL 2023a). A large share of agricultural land is used for feedstuff production, but the perception of such share hinges largely on how using arable land for the production of coupled outputs (e.g., rapeseed oil and extraction meal) is accounted for and how land cultivated for food production is counted when inferior quality degrades the respective crop to feed use (Windisch 2021). Official German agricultural statistics report that 59% of agricultural land is used for livestock (grassland and 44% of arable land) (BMEL 2024).

The contribution of livestock production to agricultural income and the economy in Germany's rural areas is considerable. Its share of the total production value in German agriculture fluctuated between 44% and 50% in the years 2016–22 (BMEL 2024). In the 2021/22 financial year, 63% of the agricultural workforce (222,800 full-time equivalents) was accounted for by livestock farms, 29% solely by specialised dairy farms, and 9% by farms specialising in monogastric species, i.e., pigs and poultry (BMEL 2023b).

The importance of German livestock farming for the income (and development) of rural areas goes far beyond the level of farms as such. Feed producers and traders, manufacturers of stable buildings and equipment, and slaughtering and meat and dairy processors are among the largest economic sectors in German rural regions. The two largest branches of the downstream sector alone accounted in 2022 for 150,000 (slaughter and meat processing) and 42,000 (dairy processing) jobs (BMEL 2024), i.e., comparable figures to the workforce on farms themselves. In regions of intensive livestock production in north-western Germany, food processing accounts for up to 50% of jobs in manufacturing and total regional income (Nier and Tamásy 2015).

The annual income per person working in livestock production during recent financial years is illustrated in Fig. 3.5. It varied considerably over the years due to price fluctuations and differed according to the type of livestock farming (with dairy

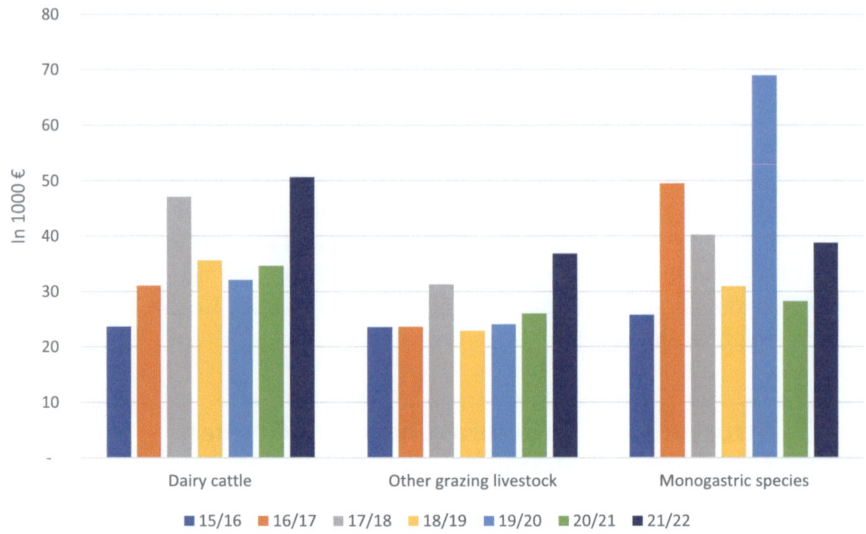

* Full-time farms and corporate farms, in current prices.
BMEL (var. issues-a), BMEL (2024, p. 43)

Fig. 3.5 Average farm income per annual work unit (AWU), Germany 2016–2022 [€/year]*

cattle and monogastric species generating higher income than other grazing livestock). A slight upward trend could be observed on average. German livestock production is subject to dynamic structural change, and a considerable heterogeneity of patterns can be observed. Figure 3.6 shows the development of production in the most important livestock sectors since the turn of the millennium. Total meat production increased by almost 40% until 2015 but has declined continuously since then. The patterns differ considerably between individual types of meat. Poultry meat production, which had almost doubled over two decades, has slightly declined from 2020 to 2022. After a moderate increase, pork production almost returned to the level of 2000, while beef production has continuously declined by a third. Egg production declined by 25% between 2003 and 2010, which is linked to stricter animal welfare regulations that have phased out cage farming and the delisting of cage eggs by German retailers during the years 2003 to 2009 (WBAE 2015). However, egg producers have switched to other housing systems for laying hens and have continuously increased production by 50% since 2011. Milk production was almost constant for many years due to production quotas, but phasing out of the quota system and increased procurement prices have led to a 20% increase in deliveries to dairy processors between 2008 and 2021 (Fig. 3.7).

The range of intensity levels in German livestock production is considerable. As mentioned above, developments in stock sizes per farm reflect trends of production intensity in livestock sectors. The stock size developments in German cattle and pig sectors (illustrated for dairy cows in Fig. 3.8 and for fattening pigs in Fig. 3.9) follow similar patterns. In the period from 2001 to 2023, the number of dairy cows

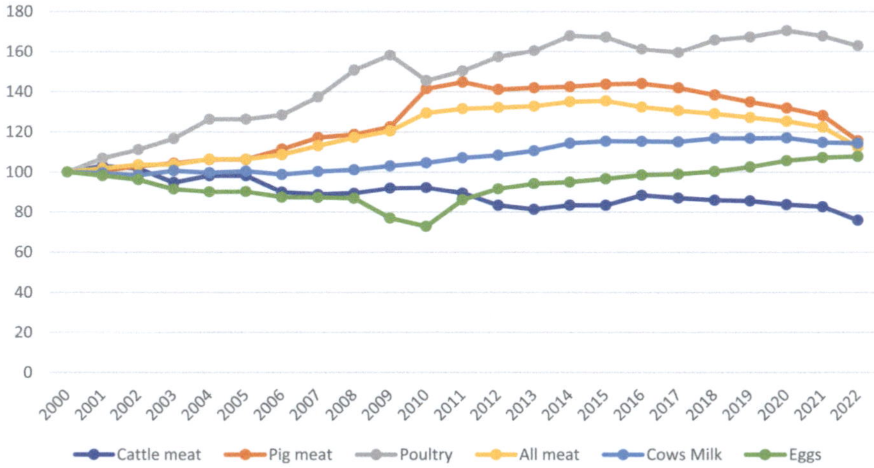

1) meat: carcass weight, milk: deliveries to milk processors, eggs: weight of usable production

BMEL (var. issues-b)

Fig. 3.6 Development of livestock production quantities[1] 2000–2022, Germany

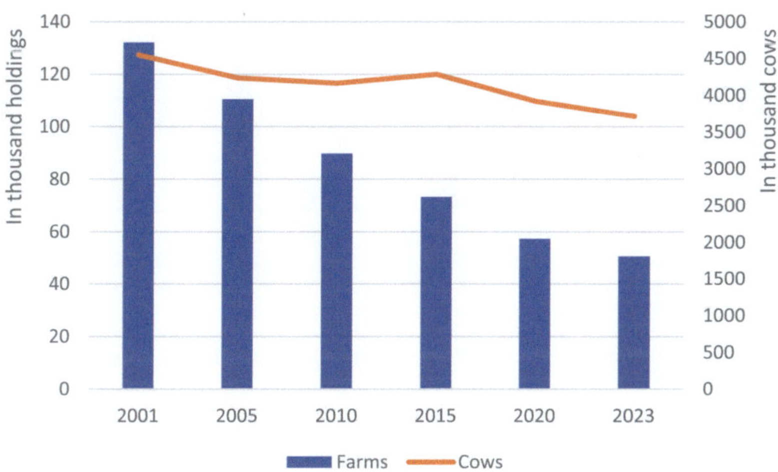

DESTATIS (2024b, var. issues-a)

Fig. 3.7 Number of dairy holdings and dairy cows, 2000–2022, Germany

declined by 18% (red line in Fig. 3.7) while the volume of milk production has increased by almost the same percentage (blue line in Fig. 3.6). This corresponds to a considerable increase in the average milk yield per cow by 1.5% per year (BMEL 2024, p. 113). These productivity gains are mainly caused by progress in breeding (genetic factors) and larger shares of dairy cattle kept under modern management methods (housing and milking systems, feeding, health, and fertility management).

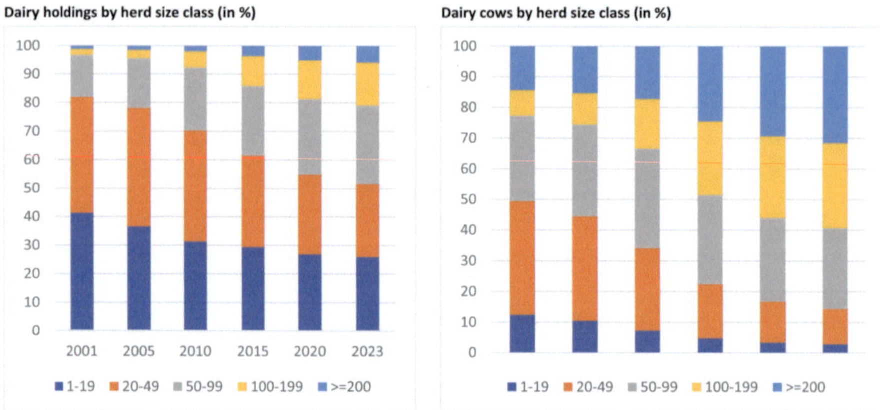

DESTATIS (2024b, var. issues-a)

Fig. 3.8 Dynamics of stock size categories in dairy farming, 2000–2022, Germany

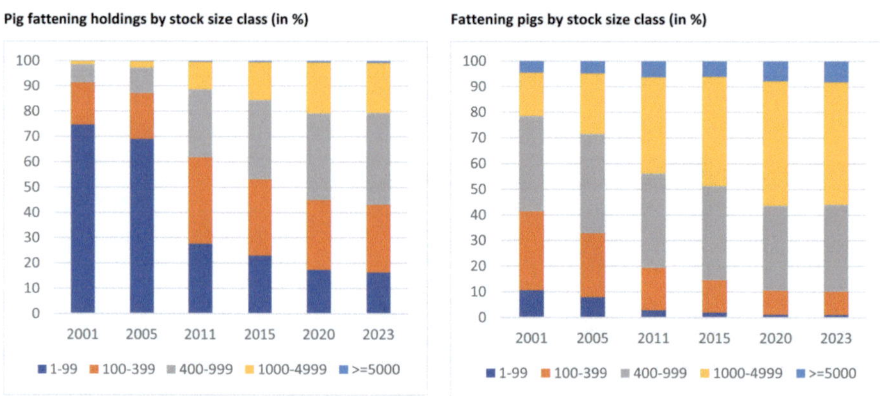

DESTATIS (2024b, var. issues-a)

Fig. 3.9 Dynamics of stock size categories in pig fattening, 2000–2022, Germany

While the number of dairy farms has dropped by 62%, the number of pig fattening farms dropped even by 84% over the last two decades. This decline has mainly affected smaller farms that have either expanded their stock sizes or given up the respective activity (or farming entirely). The right-hand panels of Figs. 3.8 and 3.9 show the proportions of dairy cows and fattening pigs in various stock size classes. The share of dairy cows in stocks of up to 50 cows has decreased, while that of cows in stock size classes above 100 cows has increased. A similar pattern can be observed for pig fattening, where the number of pigs fattened in facilities with up to 400 pigs has declined, while the number of pigs in facilities above 1000 pigs has increased.[2]

[2] The term "facilities" refers to one single production location of a pig fattening enterprise potentially comprising several buildings. Pig fattening enterprises can be comprised of several production locations.

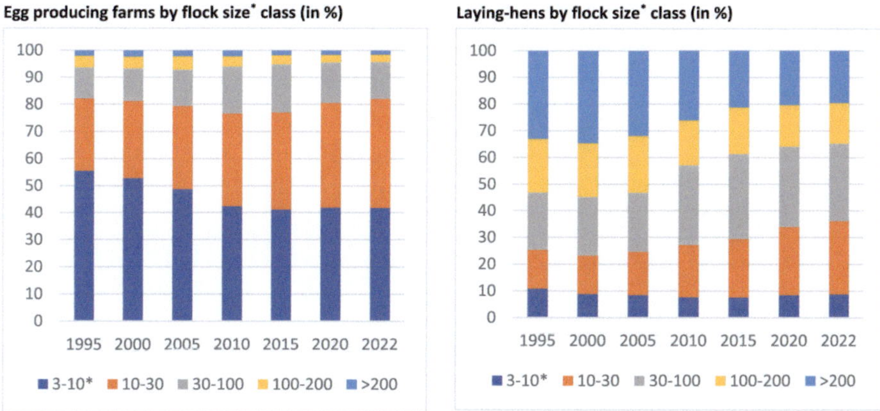

* Flock size in Thousands
DESTATIS (2024a, var. issues-b)

Fig. 3.10 Dynamics of flock size categories in egg production, 2000–2022, Germany

A particularly pronounced pattern of structural change toward larger production units can be observed in the piglet production sub-sector, where the number of farms with less than 50 sows fell by 82% between 2010 and 2023 alone (DESTATIS 2024c). Smaller piglet producers have on average a lower productivity in terms of piglets per sow per year; they achieve lower prices per piglet, as fattening farms prefer to buy larger quantities. In addition, higher animal welfare standards since 2013 have necessitated investments into housing systems that were economically not viable for smaller piglet producers and led to many farm exits.

The structural change in egg production as reflected in laying hens' flock size structures is shown in Fig. 3.10. The decline in flock sizes above 200,000 laying hens between 2005 and 2015 followed the phasing out of cage farming, which was widespread and typically large scale. While the share of laying hens kept in flock sizes between 10,000 and 30,000 birds increased by 130% between 2001 and 2023, the shares kept in smaller and larger flocks declined. Figure 3.11 shows that after a production decline between 2000 and 2010, largely influenced by the dismantling of cage housing, the establishment of floor and free-range systems has led egg production in Germany back to 2000 levels already in 2012. Fears that banning cage systems could supplant egg production in countries with lower animal welfare standards were not justified. A fast shift toward floor systems (barn and free litter) more than compensated for the dismantling of cages, and the continuous growth of free-range and organic (mostly free-range) systems of annually 8% and 12%, respectively, between 2007 and 2022 further contributed.

The structural change toward larger farms in the pig and poultry sector is likely to be even more pronounced than described here, as a considerable number of enterprises are registered as legally independent entities even though they are owned and managed as part of a single holding. Moreover, the trend toward larger units is likely to continue in Germany. As mentioned above, in other countries—e.g., in the United

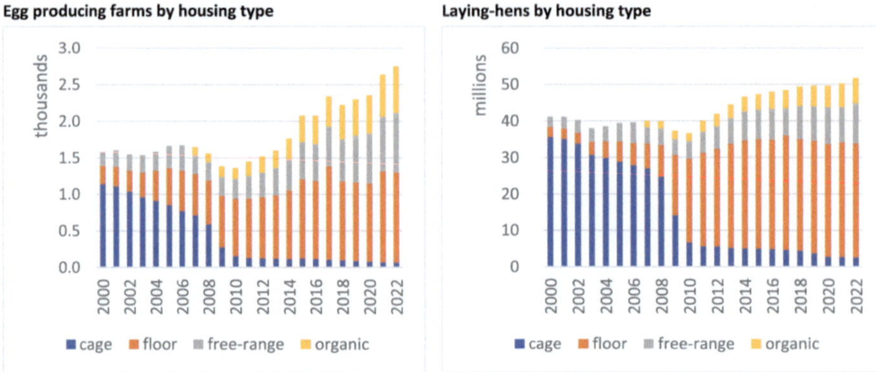

BMEL (2024)

Fig. 3.11 Dynamics of housing types for laying hens, 2000–2022

States—large proportions of pork, poultry, and eggs are produced in facilities under the umbrella of a large corporation, and such structures with standardised processes and high productivity may become more prevalent in Germany, too.

The regional distribution of livestock production is highly uneven, with high densities in the regions in North Rhine-Westphalia and Lower Saxony for pork and poultry as well as in the Alps and Alpine foothills for cattle. In many arable farming regions, livestock production has largely disappeared (Fig. 3.12). The density has declined in most regions between 2000 and 2022 (Fig. 3.13). Increases can be observed in the north-western regions of Germany, where high densities already prevailed, and in some eastern German regions, where the livestock industry revived locally after the slump following reunification. In monogastric livestock hot spots in the northwest, there were increases of 20% for pigs and 25% for poultry. Production in these regions benefits from the vicinity of seaports (comparatively low costs for imported feed components, especially soybean meal), proximity to large centers of demand (Ruhr-region, Hamburg, Bremen), relatively poor conditions for arable farming (i.e., low competition from cash-crop production), and positive agglomeration and cluster effects. Cluster effects result from the proximity to highly specialized and innovative suppliers of animal husbandry technology and processors of meat and milk. This facilitates cost advantages through horizontal and vertical cooperation and technological spillover effects. Milk production has been rather evenly distributed due to region-specific milk quota allocation. As the quotas could be traded between regions and after the expiry of the quota after 2015, production has shifted from southern to northern regions. By that, concentration and specialisation increased.

On the one hand, agglomeration and livestock concentration promote economic development in the regions concerned, which is reflected in employment and income throughout the value chain. On the other hand, negative agglomeration effects affect these regions, in particular nutrient surpluses, as well as odor, dust, and noise nuisances. In addition, the risk of spreading animal diseases and human health risks

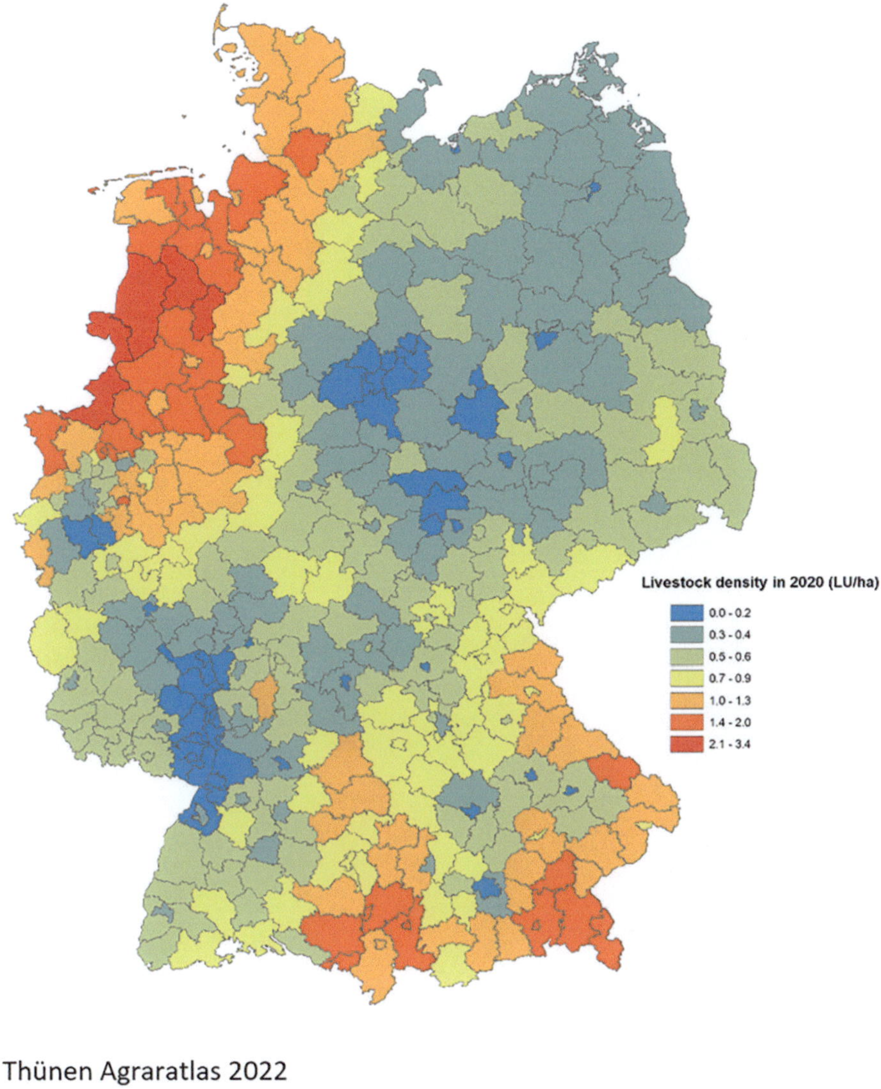

Fig. 3.12 Livestock density in Germany 2020, district level

through bi-aerosol emissions and antibiotic resistance phenomena increases in regions of dense livestock production. Furthermore, residential, nature conservation, and recreation areas are increasingly competing with the spatial expansion of livestock production. As a result, measures have been introduced to reduce livestock density: One type of measure at the national level concerns the management of large quantities of manure typically produced by intensive livestock farming. Since 1980, the amount of manure allowed to be applied per hectare has been limited in Germany, and regulations have been established on how it is applied. These have been amended

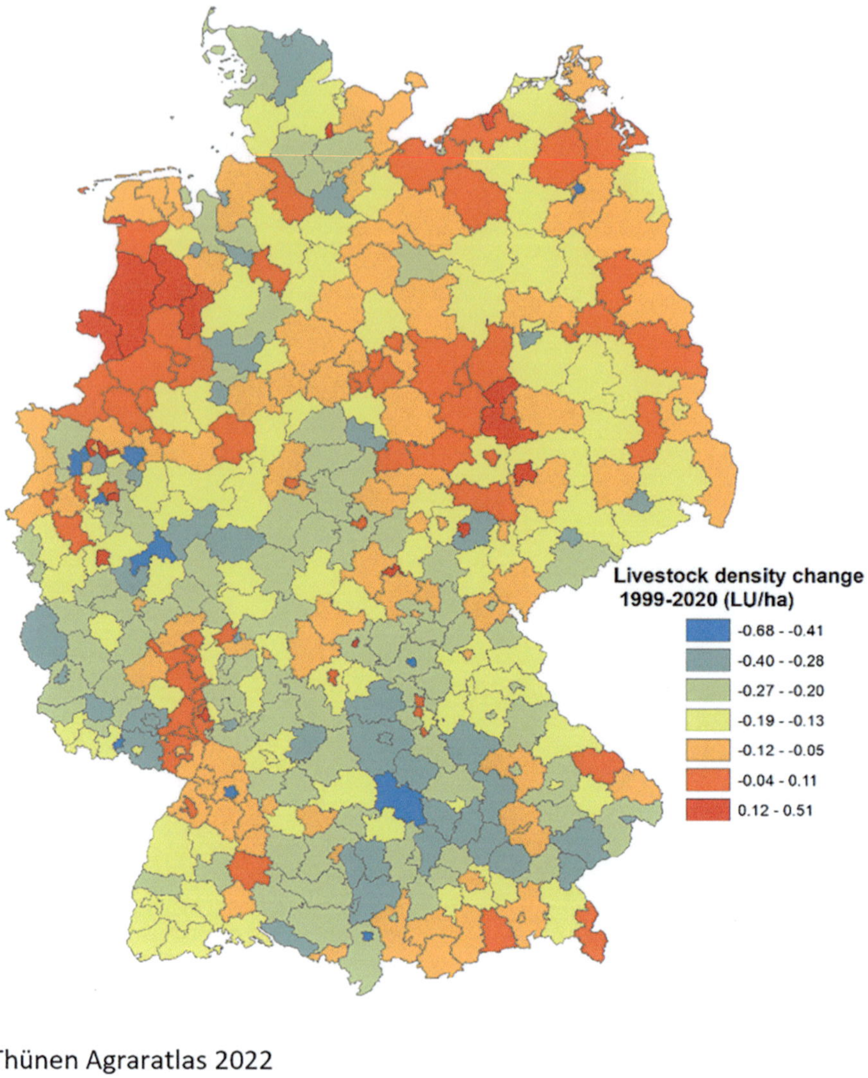

Fig. 3.13 Changes of livestock density 2000–2022, district level. (Thünen Agraratlas 2022)

by further regulations to reduce emissions and to regulate the transport of manure across regions. Since 2000, several federal and state laws have required the filtering of exhaust air for livestock facilities above a certain size. Finally, revised building laws have tightened the requirements for obtaining building permits for livestock facilities outside commercial and residential areas. These changes in the regulatory framework for livestock production have increased production costs and contributed to the decline in production in recent years. They have also stimulated research and development into technical options for concentrating manure and converting it to products with better storability and transportability.

4 Conclusions

Over the past decades, production of and demand for animal-based products have substantially increased. This process went along with enormous intensification (particularly in terms of farm sizes and capital intensity) and productivity increases (particularly per animal and per unit of labor input), though these differ substantially across as well as within countries. Moreover, substantial changes occurred within the supply chains driven by vertical integration and globalisation (Boehlje 1999; Balmann et al. 2006). Typical characteristics were increasing farm size and structural change and increasing knowledge intensity of agricultural production as well as standardisation, particularly to address the needs of the so-called supermarket revolution which started in the United States already in the 1930s, in Germany and other Western industrialised countries in the 1960s, in the former socialist countries in the 1990s, and the emerging economies in the early 2000s (Reardon et al. 2012).

These developments allowed for a resource-efficient production of livestock products as well as high sanitary standards. However, also substantial rebound effects emerged. While the availability of safe and cheap livestock products addressed the ever-increasing demand for these products, substantial negative side effects exist (e.g., WBAE 2020): Negative environmental effects include for example greenhouse gas emissions, nitrogen and phosphorous emissions to the surface and groundwater, as well as a loss in biodiversity due to the production of large amounts of required feed; negative impacts on animal welfare resulted from a strong focus of breeding, husbandry systems, and economic incentives on productivity; negative health effects result from too much fat and protein in animal-based products; and negative social effects result from the exploitation of workers in the meat industry. In a globalised food sector, it will be challenging to address these issues. However, ignoring them will impose substantial uncertainties and risks for the sector, including policy responses such as more restrictive legislation and consumer reactions such as a sudden and strong decline in demand for animal-based products. Emerging limitations and tradeoffs in intensive livestock production, as mentioned in this overview, comprise core topics of the subsequent chapters in this volume, and the way how food systems will deal with them will be of major importance for societal development.

References

Augère-Granier M-L (2020) The EU pig meat sector, European Parliamentary Research Service (EPRS), PE 652.044, European Union. https://www.europarl.europa.eu/RegData/etudes/BRIE/2020/652044/EPRS_BRI(2020)652044_EN.pdf

Balmann A, Dautzenberg K, Happe K, Kellermann K (2006) On the dynamics of structural change in agriculture: internal frictions, policy threats and vertical integration. Outlook Agric 35(2):115–121. https://doi.org/10.5367/000000006777641543

BMEL (2023a) Agrarpolitischer Bericht der Bundesregierung 2023. https://www.bmel.de/SharedDocs/Downloads/DE/Broschueren/agrarbericht-2023.html

BMEL (2023b) Die wirtschaftliche Lage der Landwirtschaftlichen Betriebe 2022. https://www.bmel-statistik.de/fileadmin/daten/0111001-2022.pdf
BMEL (2024) Statistisches Jahrbuch über Ernährung, Landwirtschaft und Forsten 2023. Bundesministerium für Ernährung und Landwirtschaft. https://www.bmel-statistik.de/fileadmin/SITE_MASTER/content/Jahrbuch/Agrarstatistisches_Jahrbuch_2023.pdf
BMEL (n.d.-a) Die wirtschaftliche Lage der Landwirtschaftlichen Betriebe – Buchführungsergebnisse der Testbetriebe [Economic situation of agricultural enterprises] issues 2015/16–2021/22. https://www.bmel-statistik.de/landwirtschaft/testbetriebsnetz/testbetriebsnetz-landwirtschaft-buchfuehrungsergebnisse
BMEL (n.d.-b) Statistisches Jahrbuch über Ernährung, Landwirtschaft und Forsten [Statistical Yearbook on Nutrition, Agriculture, and Forestry] Item 3110100, (issues 2006, 2011, 2014, 2019, 2023). https://www.bmel-statistik.de/archiv/statistisches-jahrbuch
Boehlje M (1999) Structural changes in the agricultural industries: how do we measure, analyze and understand them? Am J Agric Econ 81(5):1028–1041. https://doi.org/10.2307/1244080
Delgado CL et al (1999) Livestock to 2020: the Next Food Revolution. Food, Agriculture and the Environment Discussion Paper No. 28, FAO, Rome
DESTATIS (2024a) Geflügelstatistik: Unternehmen mit Legehennenhaltung, Tabelle 41323–0001. https://www-genesis.destatis.de/genesis/online?operation=statistic&levelindex=0&levelid=1711013304249&code=41323
DESTATIS (2024b) Viehbestandserhebung Rinder, Tabelle 41312–004. https://www-genesis.destatis.de/genesis/online?operation=statistic&levelindex=0&levelid=1711013304249&code=41312
DESTATIS (2024c) Viehbestandserhebung Schweine, Betriebe mit Zuchtsauen, Tabelle 41313–0003. https://www-genesis.destatis.de/genesis/online?operation=statistic&levelindex=0&levelid=1711447819841&code=41313
DESTATIS (n.d.-a) Fachserie 3.2.1.3 (Viehhaltung der Betriebe), issue 2001, 2003, 2005, 2007, 2010. https://www.statistischebibliothek.de/mir/receive/DESerie_mods_00000027
DESTATIS (n.d.-b) Fachserie 3.4.2.3 (Geflügel), issue 1995, 2000, 2005, 2010. https://www.statistischebibliothek.de/mir/receive/DESerie_mods_00000051
ESS-Feed (2023) European Pork industry 2023 review: challenging times. https://essfeed.com/european-pork-industry-2023-review-challenging-times/. Accessed 13 Mar 2024
EUROSTAT (2023a) Main livestock indicators by NUTS 2 regions, Online data code: ef_lsk_main, https://doi.org/10.2908/ef_lsk_main
EUROSTAT (2023b) Pig types by utilised agricultural area, size classes of livestock and NUTS 2 regions, Online data code: ef_lsk_pigs, https://doi.org/10.2908/ef_lsk_pigs
EUROSTAT (2023c) Poultry types by utilised agricultural area, size classes of livestock and NUTS 2 regions, Online data code: ef_lsk_poultry, https://doi.org/10.2908/ef_lsk_poultry
EUROSTAT (2023d) Bovine types by utilised agricultural area and share of fodder area, size classes of livestock and NUTS 2 regions, Online data code: ef_lsk_bovine, https://doi.org/10.2908/ef_lsk_bovine
EUROSTAT (2023e) Key Figures on the European Food Chain – 2023 edition. https://ec.europa.eu/eurostat/documents/15216629/18054337/KS-FK-23-001-EN-N.pdf/048e130f-79fa-e870-6c46-d80c9408620b?version=7.0&t=1707290893751
Hanf JH, Kühl R (2005) Branding and its consequences for German agribusiness. Agribusiness 21:177–189. https://doi.org/10.1002/agr.20042
MacDonald JM, Dong X, Fuglie K (2023) Concentration and competition in U.S. agribusiness, Report No. EIB-256, U.S. Department of Agriculture, Economic Research Service. https://doi.org/10.32747/2023.8054022.ers
Martinez SW (1999) Vertical coordination in the Pork and Broiler industries: implications for Pork and chicken products, report no. 77, U.S. Department of Agriculture, Economic Research Service. https://www.ers.usda.gov/webdocs/publications/40999/17957_aer777b_1_.pdf?v=1309.2

Nier S, Tamásy C (2015) Die Agrar- und Ernährungswirtschaft in Niedersachsen. https://www.uni-vechta.de/fileadmin/user_upload/ISPA/Publikationen/Weisse_Reihe/Weisse_Reihe_Band_41.pdf

OECD/FAO (2023) OECD-FAO Agricultural Outlook 2023–2032, OECD Publishing, Paris. https://doi.org/10.1787/08801ab7-en. Accessed 29 Feb 2024

Polet Y (2022) Dairy and Products Annual, GAIN Report E 42022–0066, Global Agricultural Information Network (GAIN), USDA Foreign Agricultural Service. https://apps.fas.usda.gov/newgainapi/api/Report/DownloadReportByFileName?fileName=Dairy%20and%20Products%20Annual_Brussels%20USEU_European%20Union_E42022-0066

PSD Online (2024) Production, supply & distribution online (PSD Online), USDA Foreign Agricultural Service: Custom Query. https://apps.fas.usda.gov/psdonline/app/index.html#/app/advQuery. Accessed 29 Feb 2024

Reardon T, Timmer CP, Minten B (2012) Supermarket revolution in Asia and emerging development strategies to include small farmers. Proc Natl Acad Sci USA 109(31):12332–12337. https://doi.org/10.1073/pnas.1003160108

Robinson TP, Thornton PK, Franceschini G, Kruska RL, Chiozza F, Notenbaert A, Cecchi G, Herrero M, Epprecht M, Fritz S, You L, Conchedda G, See L (2011) Global livestock production systems, Food and Agriculture Organization of the United Nations (FAO) and International Livestock Research Institute (ILRI), p 152, Rome. https://www.fao.org/4/i2414e/i2414e.pdf

Thünen Agraratlas (2022) Landwirtschaftliche Nutzung Version 2022. Method: Gocht A, Röder N (2014) Using a Bayesian estimator to combine information from cluster analysis and remote sensing data to estimate high-resolution data for agricultural production in Germany. Int J Geogr Inform Sci 28(9):1744–1764, https://doi.org/10.1080/13658816.2014.897348. Data: THÜNEN AGRARATLAS (2022) Landwirtschaftliche Nutzung Version 2022. Daten: Stat. Ämter der Länder, Kreisdaten der Landwirtschaftszählung 2020 (eigene Berechnungen); FDZ der Stat. Ämter des Bundes und der Länder, Landwirtschaftszählung 2010/2020 (10.21242/41141.2010.00.00.1.1.0; 10.21242/41141.2020.00.00.1.1.0) und AFiD-Panel Agrarstruktur 1999, 2003, 2007, 2016 (10.21242/41121.2007.00.01.1.1.0; 10.21242/41121.2020.00.01.1.1.0) (eigenen Berechnung: Kreisdaten 1999–2020. Clusterschätzer); © GeoBasis-DE/BKG (2020) Retrieved on March 24, 2024. https://atlas.thuenen.de/atlanten/agraratlas

WATTPoultry (2023) The world's leading poultry producers. https://www.wattagnet.com/top-poultry-companies. Accessed 06 May 2024

WBAE (2015) Wege zu einer gesellschaftlich akzeptierten Nutztierhaltung. Gutachten. Berlin. https://www.bmel.de/SharedDocs/Downloads/DE/_Ministerium/Beiraete/agrarpolitik/GutachtenNutztierhaltung.pdf?__blob=publicationFile&v=2

WBAE – Scientific Advisory Board on Agricultural Policy, Food and Consumer Health Protection of the BMEL (2020): Promoting sustainability in food consumption – developing an integrated food policy and creating fair food environments. Expertise, Berlin. https://www.bmel.de/SharedDocs/Downloads/EN/_Ministry/Scientific_Advisory_Board-Promoting-sustainability-food-consumption.pdf

Windisch W (2021) Can we do without livestock? J Consum Prot Food Saf 16(3):193–194. https://doi.org/10.1007/s00003-021-01338-w

Ecological Footprint of Livestock Production

4

Lina Sofie von Fricken, Caroline Timme,
and Irene Rosique-Conesa

1 Introduction

Livestock production faces complex challenges spanning environmental, ethical, health and economic dimensions. The environmental impacts of livestock production are the most prominent ones and have been coined *Livestock's Long Shadow* by the FAO (FAO 2006). Examples of environmental impacts include greenhouse gas emissions, deforestation and land use, water usage and pollution, as well as climate change and necessary adaptation. Other impacts range from animal welfare, disease management and usage of antibiotics, health and safety concerns for workers, and the ethics of industrial farming practices to consumer demands for more transparent, humane and environmentally friendly livestock farming. Overall, industrial livestock systems face increasing pressure to operate more efficiently and sustainably while meeting global demands for food.

Addressing these issues will require a *transformation of livestock systems*—that is, innovation in farming practices, shifts in consumer behaviour and changes in policy to create more sustainable and resilient livestock production systems. The goal is to move towards systems that respect animal welfare, protect the environment and ensure food security while addressing growing global demands.

L. S. von Fricken (✉)
Big Dutchman AG, Vechta, Germany
e-mail: lvonfricken@bigdutchman.de

C. Timme
FutterAllianz Nordwest GmbH & Co. KG, Schneiderkrug, Germany
e-mail: caroline.timme@futterallianz.de

I. Rosique-Conesa
Opteinics™, Chemovator GmbH, Mannheim, Germany
e-mail: irene@opteinics.com

B. Grabkowsky, T. Blaha (eds.), *Intensive Livestock Production in Transition*,
https://doi.org/10.1007/978-3-031-97872-2_4

Amongst the much-discussed solutions to transform livestock systems are a shift towards regenerative and agroecological practices as well as the promotion of localised, small-scale and diversified livestock production systems. On the other hand, many solutions focus on enhancing the efficiency of livestock systems to be more sustainable. Examples include the optimisation of feed and resource efficiency, a reduction of the greenhouse gas emissions of livestock farming, adopting circular economy models such as waste-to-resources or closed-loop systems, improving livestock health and welfare and adopting cutting-edge technology and innovations. Overall, transforming livestock systems is not a one-size-fits-all solution, but a multifaceted approach that involves shifting farming practices, changing consumer behaviour, encouraging policy reforms and incorporating new technologies.

In this chapter, the ecological footprint of pig husbandry in the region of North-West Germany is analysed using the life cycle analysis or assessment (LCA) methodology. Therefore, the concept of ecological footprinting is shortly introduced (Chap. 2), as is the online tool Nuvio Planet that helped to perform ecological footprinting on pig farms (Chap. 3). Afterwards, the results of 165 cycles at 20 pig farms between December 2022 and December 2024 are displayed and elaborated upon (Chap. 4), before providing a short summary and discussion (Chap. 5). The study shows that ecological footprinting at the farm level is possible and a viable tool to measure the sustainability of pig husbandry.

2 Ecological Footprinting (of Livestock Production)

There are different approaches for calculating the ecological footprint of livestock production systems, ranging from generic ones to sector-specific ones. Yet one similarity is that the approaches use life the LCA methodology. An LCA is a systematic analysis of the environmental impacts of a product over the entire period of its life. Part of every LCA is determining its system boundaries, since different boundaries might make sense for different products and analyses (such as "Cradle-to-gate" or "Cradle-to-grave").

Among the *generic LCA guidelines* are the following:

- The *Greenhouse Gas (GHG) Protocol* is a multistakeholder partnership of businesses, non-governmental organisations (NGOs), governments and others convened by the World Resources Institute (WRI) and the World Business Council for Sustainable Development (WBCSD). Launched in 1998, the mission of the GHG Protocol is to develop internationally accepted greenhouse gas (GHG) accounting and reporting standards and tools and to promote their adoption so as to achieve a low-emissions economy worldwide (World Resources Institute & WBCSD 2011).
- *ISO 14044:2006* covers life cycle assessment (LCA) studies and life cycle inventory (LCI) studies. It specifies requirements and provides guidelines for life cycle assessment (LCA) including: definition of the goal and scope of the LCA, the life cycle inventory analysis (LCI) phase, the life cycle impact assessment (LCIA) phase, the life cycle interpretation phase, reporting and critical review of

the LCA, limitations of the LCA, relationship between the LCA phases, and conditions for use of value choices and optional elements (ISO 2006a).

- *ISO 14040:2006* describes the principles and framework for life cycle assessment (LCA) including: definition of the goal and scope of the LCA, the life cycle inventory analysis (LCI) phase, the life cycle impact assessment (LCIA) phase, the life cycle interpretation phase, reporting and critical review of the LCA, limitations of the LCA, the relationship between the LCA phases, and conditions for use of value choices and optional elements. It does not describe the LCA technique in detail, nor does it specify methodologies for the individual phases of the LCA. The intended application of LCA or LCI results is considered during the definition of the goal and scope, but the application itself is outside the scope of this standard (ISO 2006b).
- *ISO 14067:2018* specifies principles, requirements and guidelines for the quantification and reporting of the carbon footprint of a product (CFP), in a manner consistent with international standards on life cycle assessment (LCA) (ISO 14040 and ISO 14044). Requirements and guidelines for the quantification of a partial CFP are also specified. ISO 14067 addresses only a single impact category: climate change. Carbon offsetting and communication of CFP or partial CFP information are outside the scope of the ISO norm. Also, social or economic aspects and impacts, or any other environmental aspects and related impacts potentially arising from the life cycle of a product, are not addressed in ISO 14067 (ISO 2018).

Farmers, consumers and other livestock stakeholders are increasingly in need of more information about the environmental impact and the sustainability of livestock supply chains. Hence, there are also *livestock-sector guidelines* for environmental footprinting:

- The FAO's *LEAP* (Livestock Environmental Assessment and Performance) partnership is a multistakeholder initiative that is committed to improving the environmental performance of livestock supply chains whilst ensuring its economic and social viability. Although a wide range of environmental assessment methods have been developed, there is a need for comparative and standardised indicators to change the dialogue with stakeholders from methodological issues to improvement measures. Also, there is a need for better metrics and data. LEAP develops comprehensive guidance and methodology for understanding the environmental impact of livestock supply chains so as to shape evidence-based policy measures and business strategies (FAO LEAP 2025). Among its publications are guidelines for different species:
  - The *guidelines for poultry supply chains* aim at introducing a harmonised international approach to the assessment of the environmental impact of poultry supply chains in a manner that takes account of the specificity of the various production systems involved. They aim to increase the understanding of poultry supply chains and help improve their environmental impact. The guidelines are a product of the LEAP Partnership (FAO 2016a).

  - There are similar guidelines for *pig supply chains* (FAO 2018a), *large ruminant supply chains* (FAO 2016b) and many more.
- The *Product Environmental Footprint (PEF)* and *Organisation Environmental Footprint (OEF)* methods, collectively known as Environmental Footprint methods, provide knowledge and tools to quantify and convey the environmental impact of companies in a reliable way. The Environmental Footprint methods provide meticulous guidelines for modelling, calculating and reporting life cycle environmental impacts of products and organisations. The Environmental Footprint methods build upon established international practices and standards, such as ISO 14000/44. These methods are instrumental in advancing towards a more sustainable economy by providing producers and buyers with the tools they need to make informed and more sustainable choices (European Commission 2025a). Among its publications are guidelines for different products:
  - The *Dairy PEF*—also known as Product Environmental Footprint Category Rules or PEFCR—is a methodology for assessing the environmental footprint of different dairy products, considering a broad array of environmental criteria and covering the full life cycle of the product, from feeding the animal to washing your yoghurt spoon. It attests to the European dairy sector's continuous effort to improve its long-term environmental sustainability (European Dairy Association 2025).
  - There are other examples of PEFCR category rules, for example, for *feed* (FEFAC 2025) or the *space sector* (European Commission 2025b).

The above-mentioned methodologies are just a variety of many others. And on top of the different approaches, there are also a variety of software solutions and tools for performing LCA analysis of livestock production systems. For this research, Nuvio Planet is used and will be introduced in the next chapter.

3 Nuvio Planet

Nuvio Planet is a proprietary Software-as-a-Service (SaaS) platform of BASF (Opteinics 2024), used to calculate the life cycle assessments (LCA) of animal feed and pork production. It evaluates the complete life cycle of pork production systems, considering the life cycle stages of feed production, farming, harvesting (slaughterhouse) and the associated inbound and outbound transportation processes.

Nuvio Planet complies with ISO 14040 and ISO 14044 (ISO 2006c) standards, as well as the European Commission's Product Environmental Footprint (EU PEF) (European Commission 2021a) models for impact assessment, assessing 11 environmental factors. The feed and pork LCA models adhere to sector-specific guidelines based on PEF (PEFCR 2018) and the Livestock Environmental Assessment and Performance Partnership (LEAP) (FAO 2018b) for the livestock sector. Nuvio Planet relies on background data from various external databases, including the Global Feed LCA Institute (GFLI) (GFLI 2024) for feed production, emissions factors from the IPCC (2019) for greenhouse gas calculations and the EMEP/EEA guidebook (EMEP/EEA 2019) for estimating other agricultural emissions.

The Nuvio Planet pork and feed model has been critically reviewed by NSF Sustainability (NSF 2025), ensuring compliance with ISO 14040, ISO 14044 (ISO 2006c) and the LEAP (FAO 2018b) partnership guidelines and methodology.

The goal of Nuvio Planet is to improve the environmental sustainability of animal-based food production by providing a transparent LCA software for analysing and optimising the environmental impacts of pork production systems.

Nuvio Planet covers the scope from cradle to cold carcass for pork products. In line with LEAP (FAO 2018b) guidelines, the functional unit (FU) at the final analysis stage is defined as the weight of the product intended for human consumption, specifically one ton of cold carcass weight (FAO 2018b).

Allocation, as defined in ISO 14044:2006 (ISO 2006c), allows for the partitioning of input or output flows between the product system under study and one or more other product systems. Co-product allocation in Nuvio Planet aligns with the LEAP (FAO 2018b) guidelines for the production of pork, feed and piglets. Nuvio Planet uses economic allocation by default to distribute environmental impacts among various outputs. Key aspects of co-product allocation in Nuvio Planet include:

- Feed production: utilises GFLI (GFLI 2024) datasets that include predefined economic allocations.
- Farm stage: manure is classified as waste or residual.
- Slaughterhouse stage: environmental impacts are allocated to both edible and non-edible by-products using economic allocation.

The LCA conducted in Nuvio Planet follows a “Cradle-to-Gate” approach, covering all stages from raw material extraction to the production of the functional unit (FU) of one ton of cold carcass weight. This system boundary includes three main stages: feed production, farm stage and slaughterhouse stage.

- Feed production: This stage involves the cultivation, milling and transportation of feed ingredients, primarily sourced from crop cultivation. It requires inputs such as manure, fertilisers, energy and water, with the full lifecycle of these inputs considered, including transport and capital goods depreciation, primarily sourced from the GFLI (GFLI 2024).
- Farm stage: This phase involves rearing piglets into fattening pigs, requiring feed, water and utilities. The production process is optimised through efficient management and infrastructure with ongoing manure management.
- Slaughterhouse stage: This final stage includes the slaughtering, evisceration and processing of pork into cold carcass weight, with the associated environmental impacts calculated and included in the overall assessment.

The diagram in Fig. 4.1 illustrates the material flows within this boundary.

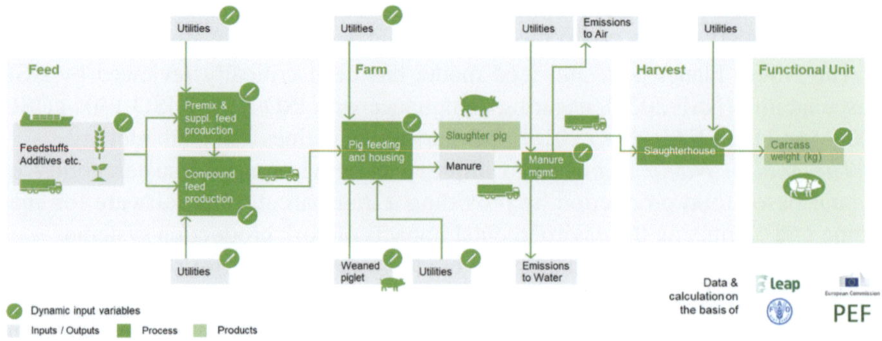

Fig. 4.1 System boundaries of Nuvio Planet for pork production

The second step of the LCA is the *life cycle inventory (LCI)*. Therefore, relevant input data are entered by the user in the user interface of Nuvio Planet to conduct the LCA. Key input data include:

- Pork production:
  - Number of piglets weaned per sow per year.
  - Carcass weight percentage.
- Characteristics of piglet and fattening stages, such as mortality rates, starting and ending weights, feed conversion ratio (FCR) and phase length.
- Origin and type of feed ingredients.
- Electricity and fuel consumption at the feed mill, farm and slaughterhouse, including the proportion from fossil and renewable sources.
- Total drinking and non-drinking water usage on the farm (including water for cleaning and cooling).
- Type of manure management system (IPCC 2019, see Chapter 3.2.2.) and whether manure is treated as waste or residual.

Generic secondary data from reliable sources is used in place of primary data for various aspects, including crop cultivation (GFLI 2024), purchased piglets, slaughterhouse data (Reckman et al. 2013) and the transportation of feed ingredients based on PEF guidelines (PEFCR 2018).

Ultimately, a typical *life cycle impact assessment (LCIA)* is performed by Nuvio Planet. The 11 environmental impact categories, methods and indicators align with the latest guidance from EU PEF (European Commission 2021a). The methods, indicators and reference versions used comply with the standards outlined in the EF reference package 3.1 (European Commission 2022).

4 Farm-Level Study Results

The described methodologies, i.e. life cycle assessment according to FAO LEAP and the EU's PEF, were employed on pig farms in North-West Germany between December 2022 and December 2024. Using Nuvio Planet as an LCA platform, 165 pig husbandry cycles at 20 farms were analysed regarding their environmental impact. Involved companies are the cooperative *GS—Die Genossenschaft* as well as the slaughterhouse *Boeseler Goldschmaus*.

Necessary data were acquired from the so-called ANW-Stallprofi, a digital and real-time data solution for pig farmers, including start and end dates of pig husbandry, livestock sales, energy and water consumption, feed deliveries, mortality rates and slaughterhouse parameters—just to name a few. Since data availability is one of the most important success factors of environmental footprinting, it was essential to have access to Stallprofi data. That way, farmers did not have to contribute significantly to the calculations except for granting access to the data. And as a result, a high number of pig husbandry cycles were analysed.

Necessary data inputs are:

- Feed ingredients and recipes.
- Energy (and water) usage at the feed mill, farm and slaughterhouse level.
- Characteristics of piglet and fattening stages, such as mortality rates, start and end weights, feed conversion ratio (FCR) and phase length.
- Biological parameters such as weaned piglets per sow and year.
- Type of the manure management system and whether manure is treated as waste or residual.
- The carcass weight percentage as well as slaughterhouse indicators if available.

The participating pig farms are very diverse, ranging from so-called closed systems including sows, piglets and fattening to simple pig fattening farms that buy piglets from breeders. Also, farms differ regarding their energy sources (mostly grid, but sometimes also solar, wind or biogas energy were monitored), feeding concepts (from two to four piglet feeding phases and from two to seven pig fattening phases) and manure management systems (reaching from usage of untreated manure at own acres to slurry discharge to external off-takers or processing of slurry in biogas plants).

The 165 husbandry cycles are analysed as there are, i.e. without optimisation factors such as advanced feeding concepts, innovative manure management systems, improved energy usage or similar aspects. Consequently, the study's results need to be interpreted as status quo and not as environmentally optimised pig husbandry scenarios.

In Table 4.1, the average environmental impact of 165 pig husbandry cycles at 20 farms is displayed. The total environmental impact is displayed in person years, which is a normalisation factor used in the PEF Method and based on global emissions from, for example, agriculture, industrial activities, transportation, production of consumer goods and so on (European Commission 2021b). In addition, the

Table 4.1 Average environmental impact of 165 pig husbandry cycles at 20 farms

	Total(person years)	Acidification (mol H+ eq)[a]	Climate change (kg CO_2 eq)[a]	Freshwater eutrophication (kg P eq)[a]	Marine eutrophication (kg N eq)[a]	Terrestrial eutrophication (mol N eq)[a]	Land use (soil quality index)[a]	Ozone depletion (kg CFC-11 eq)[a]	Particulate matter (disease incidences)[a]	Fossil resource use (MJ)[a]	Water use (m3 water eq. deprived)[a]
Farm 1	0.60	74.86	3,522.96	1.61	25.46	307.13	378,737.45	0.00010	0.00058	18,859.35	1,436.71
Farm 2	0.66	81.91	3,933.01	1.82	27.59	336.11	440,264.43	0.00011	0.00064	19,537.23	1,817.06
Farm 3	0.62	73.03	3,931.36	1.77	25.59	296.17	422,585.00	0.00011	0.00058	20,409.61	1,626.93
Farm 4	0.61	71.88	3,878.26	1.58	24.96	294.14	361,469.86	0.00009	0.00058	19,833.66	1,769.10
Farm 5	0.70	90.01	3,963.96	1.71	30.10	371.48	393,115.89	0.00010	0.00069	19,871.60	1,729.86
Farm 6	0.71	89.99	4,105.93	1.84	29.59	370.44	440,071.54	0.00012	0.00070	21,182.40	1,677.58
Farm 7	0.61	71.15	3,853.97	1.73	25.02	288.55	412,981.96	0.00011	0.00056	20,007.21	1,594.56
Farm 8	0.62	76.62	3,720.91	1.64	26.03	314.45	386,946.63	0.00010	0.00060	19,279.78	1,582.65
Farm 9	0.66	83.68	3,763.16	1.72	26.84	349.64	392,745.42	0.00010	0.00066	18,034.86	1,680.19
Farm 10	0.69	90.35	3,773.98	1.78	28.70	372.96	392,339.16	0.00010	0.00069	19,361.04	1,597.46
Farm 11	0.57	66.99	3,712.85	1.60	24.24	271.28	384,759.60	0.00010	0.00053	19,214.44	1,506.97
Farm 12	0.67	83.62	3,920.31	1.74	27.80	343.99	413,056.71	0.00011	0.00065	20,279.88	1,627.73
Farm 13	0.61	71.84	3,936.91	1.75	25.20	291.63	418,050.86	0.00011	0.00057	20,213.42	1,643.22
Farm 14	0.61	71.48	3,788.96	1.72	24.76	294.25	395,261.63	0.00011	0.00057	19,609.22	1,658.72
Farm 15	0.68	86.03	3,910.01	1.76	28.10	355.61	418,279.00	0.00011	0.00067	19,632.85	1,620.25
Farm 16	0.70	89.79	3,938.81	1.80	29.13	369.60	423,541.50	0.00011	0.00069	20,613.86	1,605.97
Farm 17	0.67	85.24	3,906.99	1.75	27.75	352.80	412,750.97	0.00011	0.00066	19,860.81	1,619.18
Farm 18	0.66	82.56	3,890.24	1.75	27.49	339.17	414,087.69	0.00011	0.00064	19,860.52	1,636.80
Farm 19	0.66	82.86	3,833.59	1.71	27.44	340.74	403,494.93	0.00011	0.00064	19,916.23	1,589.08
Farm 20	0.62	77.41	3,642.09	1.66	26.22	318.10	392,841.21	0.00010	0.00060	19,264.18	1,495.90
Min	**0.57**	**66.99**	**3,522.96**	**1.58**	**24.24**	**271.28**	**361,469.86**	**0.00009**	**0.00053**	**18,034.86**	**1,436.71**
Max	**0.71**	**90.35**	**4,105.93**	**1.84**	**30.10**	**372.96**	**440,264.43**	**0.00012**	**0.00070**	**21,182.40**	**1,817.06**
Mean	**0.66**	**82.24**	**3,884.25**	**1.73**	**27.14**	**337.64**	**408,122.95**	**0.00011**	**0.00064**	**19,847.09**	**1,623.59**

[a]Per 1,000 kg slaughter weight

following ten environmental indicators were analysed for each of the cycles (each of them per 1,000 kg of the pigs' slaughter weight):

- Acidification.
- Climate change.
- Freshwater eutrophication (i.e. phosphorus emissions).
- Marine eutrophication (i.e. nitrogen emissions to marine or coastal ecosystems).
- Terrestrial eutrophication (i.e. nitrogen emissions to soils and inland waters).
- Land use.
- Ozone depletion.
- Particulate matter.
- Fossil resource use.
- Water use.

For each of the indicators, minimum, maximum and mean values are displayed. For each of the environmental indicators, there is a significant spread between minimum and maximum values. This clearly shows that there is great potential to improve the environmental impact of pig husbandry, for example, via optimising feed recipes, manure management or housing of the pigs. Taking climate change as a prominent example, the values range from 4.1 kg CO_2e (maximum) to 3.5 kg CO_2e (minimum) per kilogram of a pig's slaughter weight, with an average of 3.9 kg CO_2e per kilogram of slaughter weight. This means that the pigs that were analysed for the purpose of this study come with a carbon footprint of circa 3.9 kg CO_2e per kg slaughter weight. As a comparison, the average global emission intensity of pig meat is indicated with 6.1 kg CO_2e per kg carcass weight by the FAO (FAO 2013). Hence, pig husbandry in North-West Germany seems to be more sustainable than average pig husbandry. Again, this figure is without any environmentally friendly interventions but a simple status quo.

When aiming at reducing the environmental footprint of pig husbandry, feeding concepts and manure management are the two most important interventions. Feeding concepts come with circa 50% of input parameter influence, and nutrient management with another 20–30% input parameter influence (depending on whether the farm is a closed system including sows and piglets or a pig fattening farm). Hence, using multi-phased feeding concepts with pig-specific nutrient supply and crude protein drawdown at the end of the fattening cycle can significantly improve the environmental footprint of pig husbandry and at the same time improve animal welfare. The same applies to innovative slurry storage concepts and up-to-date spreading techniques. The other influential aspects relate to herd and housing management, such as energy consumption, biological performance of the sows and mortality rates.

In Fig. 4.2, the relevance of the 10 environmental indicators for the 165 pig husbandry cycles is displayed (ozone depletion does not show in the pie chart due to its small impact). It shows that the indicators climate change, acidification, particulate matter and terrestrial eutrophication (i.e. nitrogen runoff) are the most significant

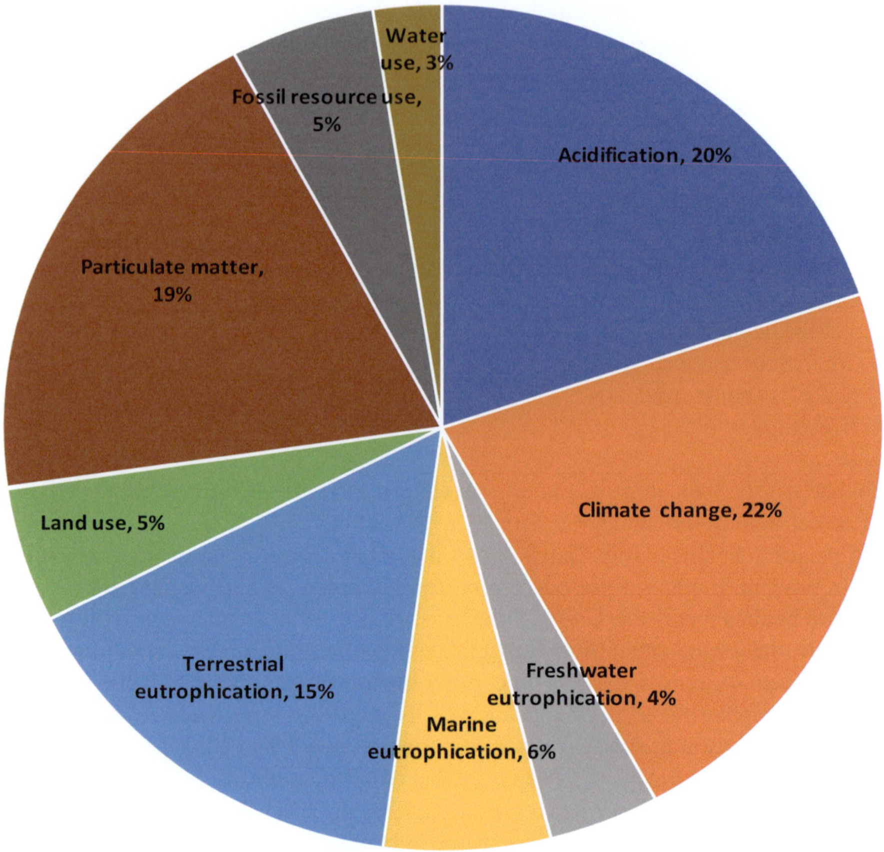

Fig. 4.2 Relevance of environmental indicators for the 165 pig husbandry cycles at 20 farms

ones. This finding is very relevant and important because it clearly shows that it is insufficient to only discuss the carbon footprint (i.e. CO_2 emissions) of pig husbandry. In addition to carbon emissions, livestock production (and pig husbandry in particular) does have an impact on aspects such as acidification, eutrophication and particulate matter emissions (i.e. fine dust). Hence, scientists and practitioners should also focus on these aspects of pig husbandry.

5 Summary and Discussion

The study's findings are both clear and significant. The central message is that pig husbandry in North-West Germany, as it currently exists, operates at a high level of efficiency and, as a result, is more sustainable than the average pig production system. However, an even more compelling insight emerges from the substantial variation in the environmental impact of the 165 different pig production cycles examined across ten key environmental indicators. This variation underscores both the potential for improvement and the necessity of enhancing sustainability in the future.

The study identifies feeding concepts and manure management as the two most influential factors in improving the environmental impact of pig husbandry in the region. Consequently, local farmers and value chain actors should collaborate more closely to further reduce the environmental footprint of pig farming. Also, it shows that farmers do not need to fundamentally rethink their way of pig husbandry but can achieve great improvements in environmental impact by adopting good agricultural practices.

Another crucial finding is that livestock production should not be assessed solely based on its carbon footprint. While CO_2 emissions are widely recognised and easy to communicate, relying on a single indicator provides an incomplete and potentially misleading assessment of sustainability. Therefore, a more holistic approach that incorporates multiple sustainability dimensions is essential, even if it adds complexity. Farmers and agricultural organisations should advocate for a broader and more comprehensive perspective on sustainability in livestock production.

Lastly, it is important to recognise that the study's conclusions are based on an analytical definition of sustainability. While the results indicate that intensive and modern pig farming is sustainable due to its efficiency, they also acknowledge the existence of alternative husbandry systems, such as organic, outdoor or straw-based pig farming. These systems may be less efficient but are not necessarily less sustainable. Instead, all forms of pig husbandry should continuously strive to enhance their sustainability over time. This finding should also be addressed in political and societal discussions, where efficient livestock farming is often considered less sustainable because of its impact on animals and nature. However, a modern and efficient livestock industry that truly deals with issues of animal welfare can be more sustainable than its ecological and regenerative alternatives which lack efficiency.

In summary, environmental footprinting serves as a valuable tool for evaluating and improving sustainability in livestock production. This study demonstrates that environmental footprinting of pig husbandry is both feasible and effective in guiding future sustainability efforts.

References

EMEP/EEA (2019) EMEP/EEA air pollutant emission inventory guidebook. European Environment Agency. Available at: https://www.eea.europa.eu/en/analysis/publications/emep-eea-guidebook-2019?activeTab=22266594-97f5-4524-946f-095a50759ae7

European Commission (2021a) Commission Recommendation (EU) 2021/2279 of 15 December 2021 on the use of the Environmental Footprint methods to measure and communicate the life cycle environmental performance of products and organisations. Available at: https://eur-lex.europa.eu/legal-content/EN/TXT/?uri=CELEX%3A32021H2279

European Commission (2021b) Annexes 1 to 2: environmental footprint methods. Retrieved from https://environment.ec.europa.eu/system/files/2021-12/Annexes%201%20to%202.pdf

European Commission (2022) The EF reference package includes all information to perform the LCIA phase. Available at: https://eplca.jrc.ec.europa.eu/LCDN/developerEF.xhtml

European Commission (2025a) Environmental footprint methods. Available online: https://green-business.ec.europa.eu/environmental-footprint-methods_en
European Commission (2025b) Product Environmental Footprint Category Rules (PEFCR) for the space sector. Available online: https://defence%2D%2Dindustry%2D%2Dspace-ec-europa-eu.translate.goog/product-environmental-footprint-category-rules-pefcr-space-sector_en?_x_tr_sl=en&_x_tr_tl=de&_x_tr_hl=de&_x_tr_pto=rq
European Dairy Association (2025) Dairy product environmental footprint. Available online: https://eda.euromilk.org/food-environment-health/sustainability-environment/dairy-product-environmental-foot-print/
FAO (2006) Executive summary – climate change and Livestock. Available at: https://www.europarl.europa.eu/climatechange/doc/FAO%20report%20executive%20summary.pdf
FAO (2013) Tackling climate change through livestock: a global assessment of emissions and mitigation opportunities (p 35). Food and Agriculture Organization of the United Nations. Retrieved from https://www.fao.org/4/i3437e/i3437e.pdf
FAO (2016a) Greenhouse gas emissions and fossil energy use from poultry supply chains: guidelines for assessment (Version 1). Available online: https://openknowledge.fao.org/server/api/core/bitstreams/0bed0058-9507-4791-b127-758e722064d7/content
FAO (2016b) Environmental performance of large ruminant supply chains: guidelines for assessment (Version 1). Available online: https://openknowledge.fao.org/server/api/core/bitstreams/a9dc20d8-d6ed-4c99-8fa5-75fc21e2d804/content
FAO (2018a) Environmental performance of pig supply chains: guidelines for assessment (Version 1). Available online: https://openknowledge.fao.org/server/api/core/bitstreams/c60ff115-96b0-458c-9d5b-a3f9092c636a/content
FAO (2018b) Environmental performance of pig supply chains: guidelines for assessment (Version 1). Livestock Environmental Assessment and Performance Partnership. FAO: Rome, Italy. Available at: https://openknowledge.fao.org/items/4fe037fa-9eb1-4b83-9e05-d5b9930e39ca
FAO LEAP (2025) Livestock Environmental Assessment and Performance Partnership (LEAP). Available online: https://www.fao.org/partnerships/leap/en/
FEFAC (2025) The PEFC Feed for Food-Producing Animals. Available online: https://fefac.eu/priorities/sustainability/pefcr-feed/
GFLI (2024) LCIA Feed Database, Environmental Footprint 3.1. Available at: https://globalfeedlca.org/
IPCC (2019) 2019 refinement to the 2006 IPCC guidelines for national greenhouse gas inventories. Available at: https://www.ipcc.ch/report/2019-refinement-to-the-2006-ipcc-guidelines-for-national-greenhouse-gas-inventories/
ISO (2006a) Environmental management – life cycle assessment – requirements and guidelines. ISO 14044:2006. Available online: https://www.iso.org/standard/38498.html
ISO (2006b) Environmental management – life cycle assessment – principles and framework. ISO 14040:2006. Available online: https://www.iso.org/standard/37456.html
ISO (2006c) International Organization for Standardization: ISO 14040/14044: environmental management – life cycle assessment – principles, framework, requirements, and guidelines. Geneva: International Organization for Standardization. Available online: https://www.iso.org/standard/37456.html
ISO (2018) Greenhouse gases – carbon footprint of products – requirements and guidelines for quantification. ISO 14067:2018. Available online: https://www.iso.org/standard/71206.html
NSF (2025) National Sanitation Foundation. Available at: https://www.nsf.org/
Opteinics (2024). Available at: https://opteinics.com/
PEFCR (2018) PEFCR Feed for food producing animals, version 4.1. Available at: https://fefacfeedpefcr.eu/#p=4
Reckman K, Traulsen I, Krieter J (2013) Life cycle assessment of pork production: a data inventory for the case of Germany. Livest Sci 157(2–3):586–596. https://doi.org/10.1016/j.livsci.2013.09.001
World Resources Institute; World Business Council for Sustainable Development (2011) Product life cycle accounting and reporting standard. Available online: https://ghgprotocol.org/sites/default/files/standards/Product-Life-Cycle-Accounting-Reporting-Standard_041613.pdf

The Transformation of Livestock Production in a Globalised World: Barriers and Opportunities

5

Stephan von Cramon-Taubadel and Karen Arcia

1 Introduction

Livestock products—meat and milk—are an especially controversial subject of debates over the sustainable transformation of the global agri-food system. Livestock production and the consumption of livestock products affect various dimensions of the environment as well as human health and animal welfare. The objective of this chapter is to discuss how attempts to manage these effects using local and national policy initiatives can be compromised by telecoupling that leads to possible leakage effects. We begin by briefly reviewing the negative and positive effects generated by the production and consumption of animal-based products. Despite a broad consensus that reducing the production and consumption of livestock products would generate crucial benefits, most projections indicate that global demand for these products will continue to grow in the coming decades. We then introduce the concepts of telecoupling and leakage effects and discuss how trade in livestock products between the EU and the rest of the world, and within the EU, leads to complex interactions between local policy initiatives designed to transform the livestock sector and regional or global outcomes. We conclude that the transformation of the livestock system is necessary but cannot focus exclusively on production. Instead, policymakers in high-income countries such as Germany must also address the sensitive issue of how to reduce consumption. Based on a comparison of greenhouse gas emissions (GHG) from beef production in Brazil and Germany, we illustrate the importance of international coordination to establish methods for defining and implementing border adjustments that reflect

S. von Cramon-Taubadel (✉) · K. Arcia
Department of Agricultural Economics and Rural Development, University of Göttingen, Göttingen, Germany
e-mail: scramon@gwdg.de; karcia@uni-goettingen.de

B. Grabkowsky, T. Blaha (eds.), *Intensive Livestock Production in Transition*,
https://doi.org/10.1007/978-3-031-97872-2_5

GHG footprints. Ensuring that the prices of livestock products reflect their total production costs (economic and environmental) is necessary to ensure that the production required to meet global demand for livestock products takes place where it is most efficient.

2 Externalities Associated with Livestock Production and the Consumption of Livestock Products

2.1 Environmental and Climate Effects

Animal-based foods have much larger negative effects on the environment and on climate change than plant-based foods, especially relative to their contributions to global human dietary energy (18%) and protein (25%) consumption (Mottet et al. 2017). Livestock production uses land directly, and indirectly for feed production. The resulting land-use change often entails the conversion of forests and natural grasslands, causing biodiversity loss and degradation of ecosystem services. The disruption of nitrogen (N) and phosphorus (P) cycles leads to fertility depletion and nutrient pollution such as acidification and eutrophication of terrestrial and aquatic ecosystems. Additionally, GHG emissions from the animal food chain contribute to climate change.

Globally, animal-based foods account for one-quarter of humanity's freshwater footprint, 97% of which is due to feed production (Hoekstra 2012). Livestock production uses 78% of the world's agricultural land, causing 40% of tropical deforestation and 30% of human-induced biodiversity loss (Westhoek et al. 2011). Moreover, 49% of global grasslands have been degraded, mainly by livestock activities such as overgrazing, feed crop cultivation, and intensive feedlot operations, although other factors also play a role (Bardgett et al. 2021). Livestock production is also linked to 33% of N pollution (Uwizeye et al. 2020), uses 88% of phosphate rock deposits (Lun et al. 2018), and accounts for 44% of eutrophication and 13–14.5% of human-induced GHG emissions (Gerber et al. 2013; Poore and Nemecek 2018).

In the EU, about 65% of the available agricultural land (55 million ha of cropland and 51 million ha of grassland) is used to produce animal-based food (European Commission 2020b). In addition, animal-based production in the EU uses imported feed which is produced on agricultural land in other countries: Fischer et al. (2017) estimate that this indirect land use amounts to 66.5 million ha. Animal-based production is also responsible for 18% of the EU's total GHG emissions (861 million tons of CO_2 eq., Leip et al. 2015). 81% and 73% of the agricultural reactive N and P that reach aquatic systems in the EU can be attributed to animal-based production, as can 82% of agriculture's ammonia emissions (Leip et al. 2015). Crenna et al. (2019) estimate that 71% of the biodiversity loss that results from food consumption in the EU can be attributed to livestock production.

In Germany, animal-based food production uses 61.1 billion m^3 of water annually, which is equivalent to 32% of national freshwater availability (Brindha 2020). Domestic livestock production uses over 60% of Germany's agricultural land (BMEL 2021) and is responsible for 59% of Germany's N surplus (Klement et al. 2021) and 77% of its P surplus (Mayer and Kalschmith 2022). Various studies (e.g., Socher et al. 2013; Gossner et al. 2016) show that intensive use and conversion of grassland in parts of Germany has led to a loss of biodiversity and the release of carbon stocks. Not all of these effects can be attributed to animal-based food production (for example, some grassland is farmed intensively for biogas production, and not all converted grassland is used to produce animal feed), but animal-based production does contribute substantially.

Valuing these environmental effects in monetary terms is fraught with conceptual and technical difficulties. While some forms of pollution generate costs that can, in principle, be measured, other effects such as biodiversity loss involve elusive questions of intrinsic value. According to Funke et al. (2021), GHG emissions from livestock production in high-income countries generate costs of 5.8–19 $/kg of beef, 0.6 $/kg of pork, and 0.5 $/kg of poultry meat. Corresponding estimated costs for N pollution are 3.5–3.8 $/kg of beef, 1.3 $/kg of pork, and 1 $/kg of poultry meat (Funke et al. 2021).

2.2 Consumption and Health

The levels of animal-based food consumption observed in many middle- and high-income countries are associated with a variety of health risks. Increased consumption of red and processed meats is associated with cardiovascular diseases (Micha et al. 2010), colorectal cancer (Lippi et al. 2016), type 2 diabetes (Pan et al. 2013), and dementia (Zhang et al. 2021). Parlasca and Qaim (2022) review the literature on these associations and stress that establishing causal links between meat consumption and health is difficult. Health is affected by many factors that interact with meat consumption and are challenging to untangle. For example, health-conscious individuals might consume less meat on average than others but also exercise more and smoke less. Nevertheless, even if the magnitude of the associated health costs is difficult to pin down, the average individual in many middle- and most high-income countries consumes meat at levels that exceed nutritional requirements and increase health risks.

Animal-based food production can also generate indirect health effects. For example, excessive and improper use of antibiotics in livestock production systems can contribute to antimicrobial resistance (Schrijver et al. 2018). Furthermore, health risks associated with livestock production include exposure to agrochemicals and reactive nutrient pollution (e.g. nitrates in groundwater) in areas where livestock are raised, where feed is produced, or where minerals to fertilise feed crops are extracted.

2.3 Animal Welfare

The use of animals for food production is speciesism, meaning that it is rooted in the conviction that humans are more important than other species (Espinosa and Treich 2021). Some individuals take a fundamentally anti-speciesist position that emphasises (nonhuman) animals' intrinsic value and denies the right of humans to use animals. Others take an animal-welfare position that accepts the use of animals provided it is subject to restrictions that reduce animal suffering. In many high-income countries, anti-speciesist and animal welfare concerns are driving an increasing prevalence of vegan and vegetarian diets. In many low-income countries, dietary choices are dictated by the non-affordability or non-availability of animal-based foods, or by religious beliefs, rather than by a recognition of animal rights or welfare (Legwegoh and Riley 2014; Parlasca et al. 2023). According to Alonso et al. (2020), less than 1% of the world's population views meat consumption as inherently unethical. Although definitive data are lacking, likely less than 10% of the global population is vegan, ranging from 13% in India to almost 0% in Japan (Statista Consumer Insights 2023). In a recent survey, 91% of EU citizens and 92% of Germans responded that protecting the welfare of farm animals is very or somewhat important (Eurobarometer 2023). However, only 3.2% of the EU population and 1.3% of Germans are vegan.

Animal welfare standards vary widely across countries and regions. The Treaty on the Functioning of the European Union includes an article that explicitly recognises that animals are sentient beings (TFEU, Title II, Article 13). In addition, the EU has adopted rules for the protection of animal welfare based on the so-called Five Freedoms framework that, in international comparison, are stringent and well-enforced (World Animal Protection 2024). Of course, even the strictest animal welfare standards will not satisfy fundamentally anti-speciesist individuals, and societal consensus over what constitutes acceptable standards can evolve over time. This creates difficulties for farmers who wish to modernise or expand their livestock holding systems. While the associated investments, for example in hog barns, typically have useful lives that extend over two decades and more, there is no guarantee that animal welfare standards in effect when an investment is made will remain unchanged over this useful life. Hence, farmers often oppose stricter animal welfare standards. They are joined by agribusiness stakeholders who worry that stricter standards will lead to reduced production and, therefore, reduced demand for inputs such as feed and for processing (i.e. slaughterhouses and meat packers). In addition, there are concerns that stricter animal welfare standards will increase costs and disrupt traditional farming practices (Bonnet et al. 2020).

In high-income countries such as Germany, concern about animal welfare generally focuses on intensive livestock farming systems, which generally involve confinement, (over)crowding, restricted natural behaviour, and physical mutilations such as castration and docking tails. However, animal welfare can also suffer in extensive livestock systems, for example, due to malnutrition and dehydration if access to food and water is inconsistent, as well as increased vulnerability to disease, predators, and environmental stressors (Temple and Manteca 2020).

3 Livestock's Role in a Sustainable Agri-Food System

Based on the concerns outlined in the previous section, there is a wide and growing consensus that a shift to more plant-based diets would generate threefold benefits—to the environment, to animal welfare, and to human health. However, this shift to more plant-based diets does not mean a complete cessation of livestock production, which has an important role to play in a sustainable global agri-food system.

First, global demand for livestock products is projected to continue to increase in the coming decades. Global demand for meat almost doubled from 23 kg per capita and year in 1961 to 43 kg per capita and year in 2021 (Fig. 5.1). Pork and especially poultry consumption have accounted for most of this increase. The demand for milk has also increased, albeit not as rapidly, from 75 to 87 kg per capita and year over the same period. The highest levels of meat demand are observed in the Americas and Europe, but the highest growth rates have been observed in Asia, where consumption per capita and year increased more than sixfold between 1961 and 2021 (Fig. 5.2).

Predictions of future meat demand vary. The OECD-FAO Agricultural Outlook (OECD-FAO 2023) projects a 2% increase in global average per capita meat demand by 2032 compared with the base period 2020–2022. Komarek et al. (2021) predict that global demand for protein from meat, eggs, and milk will increase by 38% between 2020 and 2050 if current trends continue. Differences in time horizon and product coverage notwithstanding, all projections agree that global demand for meat and other livestock products will continue to increase in the foreseeable future. Demand growth will take place primarily in populous middle-income countries,

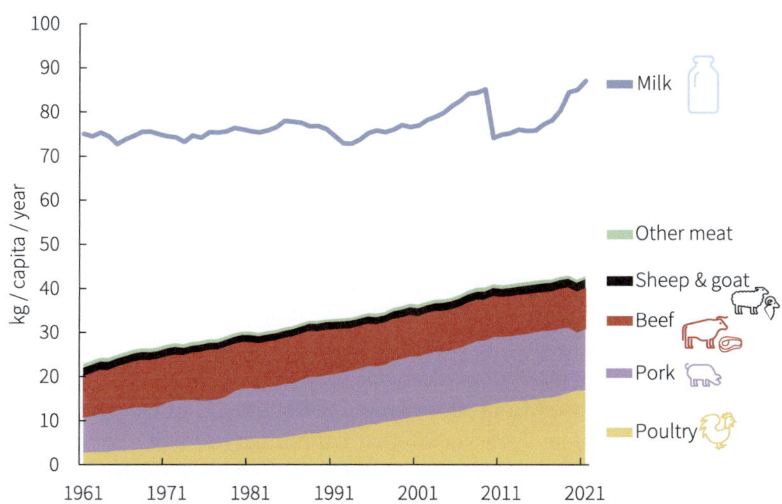

Fig. 5.1 Global demand for livestock products (kg per capita and year, 1961–2021). (Source: FAOSTAT 2024 and own calculations)

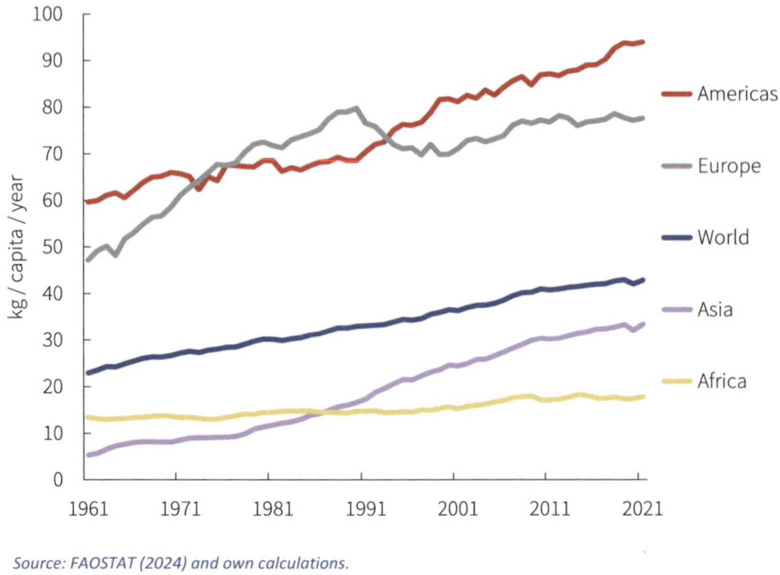

Fig. 5.2 Per capita meat demand in regions of the world (kg per capita and year, 1961–2021). (Source: FAOSTAT 2024 and own calculations)

while demand in high-income countries will tend to stagnate and shift from red to white meat.

Second, while many individuals especially in middle- and high-income countries consume more meat than is healthy, undernutrition remains a pressing global problem. Moderate consumption of meat, milk, and eggs, which are rich sources of amino acids, fats, and micronutrients such as iron, can make important contributions to reducing undernutrition. Studies such as Headey et al. (2018) show that including meat in diets can reduce nutritional deficiencies among children and other groups. The relationship between the consumption of animal-based foods and nutrition is complex (Parlasca and Qaim (2022) provide a review). Purely plant-based healthy nutrition is possible, and often more affordable because meat and other livestock products are often a comparatively expensive source of nutrients. However, purely plant-based nutrition is not always feasible, for example in colder or drier regions where crop production is difficult, and availability is highly seasonal. In addition, purely plant-based diets often require more nutritional knowledge and planning than diets that include livestock products.

Third, while livestock production does use feed (primarily grains) and, therefore, cropland that could be used to produce food for humans, it also makes use of by-products, crop residues, and grasslands that cannot serve as or produce food for humans. According to Mottet et al. (2017), 14% of the global livestock feed ration is composed of grains (13%) and other materials (1%) that could be consumed by humans. Especially ruminants make it possible to produce food for humans on roughly 1.3 billion ha of grassland which Mottet et al. (2017, p. 5) consider to be "non-convertible", i.e. which could not be used as cropland. If these grasslands are managed appropriately, they can provide a variety of important ecosystem services such as carbon sequestration and maintenance of biodiversity (Bengtsson et al. 2019; Zhao et al. 2020). According to Liu et al. (2022), the value of the regulating services provided by grasslands is eight times the value of their food supply services.

Finally, some livestock production systems—e.g. the transhumant pastoralism that characterises many mountain regions in Europe—generate important benefits in addition to food production (Liechti and Biber 2016). These include the maintenance of so-called high-nature-value farmland which supports biodiversity, as well as socio-cultural services that range from the preservation of scenic landscapes that support tourism to the observance of traditional holidays and cultural practices, the production of food specialties, and regional identities.

In summary, the global demand for livestock products is expected to continue to grow in the coming decades. The production of animal-based foods is associated with a variety of negative effects on the environment, human health, and animal welfare. However, animal-based food production can also contribute to the environment and help feed the world. The challenge is to transform the livestock sector in a way that addresses its negative effects without foregoing its potential benefits. To this end, it is important to account for complex interactions between local changes and global outcomes, as we explain in the following section.

4 Transforming the Livestock Sector: Local Changes and Global Outcomes

4.1 Telecoupling and Leakage

The global agri-food system is characterised by complex socioeconomic and environmental interactions in space that are often referred to as telecoupling (Liu et al. 2019). As a result, production and consumption decisions at one location can have effects on socioeconomic and environmental outcomes at other locations. A key driver of telecoupling is the fact that most agricultural products are tradeable. Since the spatial distribution of the world's population, consumption preferences, and agricultural production potential do not overlap perfectly, all countries are more or less dependent on trade (Kinnunen et al. 2020). In consequence, changes in one country's production will not only affect domestic prices in that country, but they will lead to changes in trade flows that affect prices and production in other countries. The environmental effects that result from a change in production in one country can therefore not be considered in isolation. First, some of these environmental effects might be local in nature, but others—such as pollution in transnational watersheds, or GHG emissions—might be regional or even global. Second, the production changes triggered in other countries will generate additional environmental effects in those countries.

An important consequence of this telecoupling is the danger of so-called leakage. Leakage takes place when changes in policy designed to reduce negative effects at one location lead to shifts in economic activity between locations that cause the spatial displacement of negative effects (Marcu et al. 2013). Leakage is most commonly considered in connection with GHG emissions and climate change: the IPPC (2007) for example defines carbon leakage as: “The increase in CO_2 emissions outside the countries taking domestic mitigation action divided by the reduction in the emissions of these countries”. However, leakage can also refer to other environmental effects as well as socioeconomic effects or animal welfare, for example, if a reduction in intensive pork production in one country leads to increased production in other countries with lower labour standards in the meat processing industry, or lower animal welfare standards.

The danger of leakage can be illustrated using the EU’s so-called Farm to Fork (F2F) proposals. Published in May 2020 as part of the Green Deal, the “Farm to Fork Strategy for a fair, healthy and environmentally-friendly food system” (EU Commission 2020a) calls for a number of measures to reduce the intensity of farming in the EU by 2030. These include a 20% reduction in mineral fertiliser use, a 50% reduction in the use of chemical pesticides, and ensuring that at least 25% of all farmland in the EU is farmed organically. Several studies have simulated the effects of these measures on agricultural production in the EU, and Table 5.1 summarises some of their results. Note that the studies covered in Table 5.1 were produced by a wide variety of researchers: Barreiro-Hurle and co-authors work for the EU’s Joint Research Centre, while the Beckmann and co-authors produced their study for the American USDA. The study by Henning and co-authors was commissioned by the Grain Club (an alliance of German agricultural associations representing the grain, oilseed, and feed industry), while Kempen and Johannes prepared their study for NABU, an environmental NGO in Germany.

Despite these varied interests and the different modelling approaches and assumptions that underlie their studies, all come to the basic conclusion that implementation of the F2F measures would lead to substantial reductions in agricultural production in the EU. Kempen and Johannes (2023, p. 12) conclude that “[t]he … simulations show that the farm areas outside Germany that are needed to satisfy German food demand would remain unchanged despite production reductions due to the F2F-Strategy if German meat consumption were reduced by 50%”. In other words, if the F2F measures were implemented but food demand in Germany

Table 5.1 Reductions in the production of agricultural products due to the implementation of the Farm to Fork Strategy (results of selected simulation studies, in %)

	Production change in %				
Study	Cereals	Oilseeds	Fruits and vegetables	Beef	Milk
Barreiro-Hurle et al. (2021)	−15.0	−15.0	−12.0	−13.0	−10.0
Beckman et al. (2020)	−48.5	−60.7	−5.2	−13.5	−11.6
Bremmer et al. (2021)	−18.0				
Henning et al. (2021)	−23.6	−7.3	−13.0	−17.0	−6.0
Noleppa and Cartsburg (2021)	−26.0	−24.0			
Kempen and Johannes (2023)	−29.0	−19.0	−7.0	−42.0	−22.0

Source: Wesseler (2022) and Kempen and Johannes (2023)

remained unchanged, production would increase elsewhere, leading to leakage effects. Reducing German meat consumption by 50% would suffice to compensate for the production reductions due to F2F and thus prevent leakage.

Telecoupling and leakage are usually considered with reference to national borders because statistics on production, consumption, and trade are collected at the national level. However, agricultural products are also traded within countries, and telecoupling exists between jurisdictions at all scales from local to global, not only across national borders. The potential for intra-national leakage effects is obvious in geographically large countries such as Brazil, Canada, and China. However, this potential is also relevant within the EU. Movements of agricultural products between, for example, Germany and Luxembourg are reported as trade, despite the fact that the EU is a single market with free trade between member states. At the same time, movements over greater distances within individual member states, for example, movements from Bavaria in the south to Schleswig-Holstein in the north of Germany, are not reported, even though they might be associated with stronger telecoupling and potential for leakage.

4.2 Implications for the Transformation of the Livestock Sector

Telecoupling and the associated danger of leakage have important implications for the outcomes of local, national, and EU-wide policy efforts to transform the livestock sector. In a nutshell, good intentions do not guarantee good outcomes because telecoupling implies that initiatives that generate positive effects at one location can trigger negative effects elsewhere. If such leakages are not accounted for, policies that improve outcomes in the EU as a whole or in a part of the EU, such as livestock production in northwest Germany, can end up worsening aggregate outcomes at higher levels such as the EU or the planet as a whole.

Figure 5.3 shows that for the EU as a whole, imports of livestock products from third countries play a comparatively small role. In 2023, the EU imported 288 thousand tons of beef, which is equivalent to 4.5% of the total supply of beef in the EU. For poultry meat, imports accounted for 3% of supply; for the other products depicted in Fig. 5.3, the share of imports in total supply was 1% or less. Farm groups in the EU sometimes argue that the EU risks being 'flooded' with cheap imports of livestock products. However, livestock products from other countries are subject to the high and often prohibitive import tariffs that the EU maintains for these products, coupled with strict controls and inspection requirements.[1] Most of the livestock-based imports that enter the EU take place under special trade agreements with individual partners that define so-called tariff rate quotas (TRQs), which permit a limited quantity of imports to enter at reduced or zero tariff rates.

Since production costs are often lower in meat-exporting countries such as Brazil than they are in the EU, the imports that enter the EU under TRQs are highly

[1] See, for example, Government of Canada (2024) and AHDB (2024).

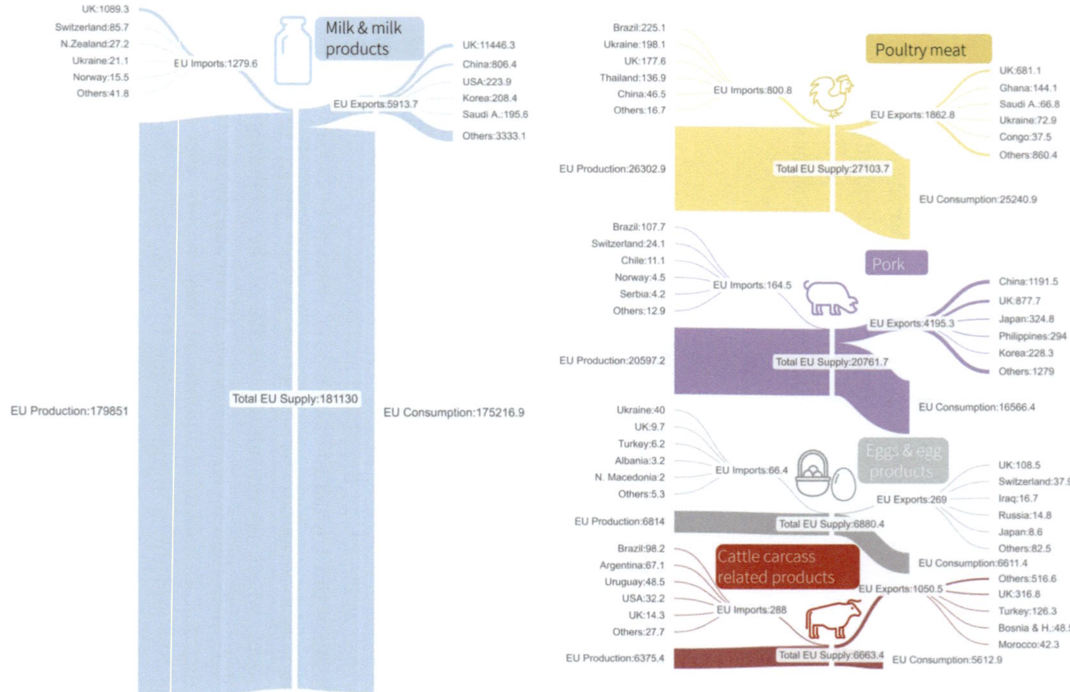

Fig. 5.3 The EU's trade in livestock products (2023, in thousand tons). (Source: European Commission 2024 and own calculations)

competitive on the EU market and displace EU production. This explains why farmers in the EU oppose trade agreements, such as the agreement that is being negotiated between the EU and Mercosur in Latin America, which would likely lead to an expansion of the EU's TRQs for beef and other livestock products, for example, from Argentina, Brazil and Uruguay. However, as we explain below for the example of beef from Brazil, if environmental costs were considered in addition to production costs, then many imported livestock products would no longer be competitive on the EU market. This underlines the importance of developing internationally accepted rules that ensure that the prices of traded products reflect not only their production costs but also the environmental costs associated with their production.

While EU imports of livestock products are small and restricted, exports are larger, making the EU a net exporter of each of the five products depicted in Fig. 5.3. The EU exported 20.4% of its pork supply, 16.5% of its beef supply, and 7.1% of its poultry meat supply in 2023. The implication is that if the EU were to reduce its production of livestock products while its demand remained unchanged, it would export less. As a result, production would increase elsewhere outside the EU to meet the continued and growing demand of importing countries such as China. The result would be leakage that erodes at least some of the benefits to the environment or animal welfare that are felt in the EU.

As we noted above, leakage can also take place within the EU. Table 5.2 provides an overview of German production, trade, and use of pork, beef, and poultry meat in 2023. In Table 5.2, imports and exports are broken down by source/destination into EU-27 and the rest of the world (ROW). We see that Germany is a net exporter of pork and beef, but a net importer of poultry meat. We also see that for all products, intra-EU trade between Germany and other EU-27 members is considerably larger than trade between Germany and countries outside the EU.

While the EU can use trade measures such as tariffs and TRQs to restrict imports from third countries, Germany cannot use such measures to restrict imports from other EU member states. In the EU single market, member states are not permitted to restrict imports from other member states, provided the products in question are produced in accordance with EU standards. A member state can impose standards

Table 5.2 German production, trade, and use of meat (2023, in thousand tons carcass weight)

Product	Production	Imports (from EU-27) [from ROW]	Exports (to EU-27) [to ROW]	Net exports (to EU-27) [to ROW]	Use	Self-sufficiency (%)
Pork	3921	962 (919) [43]	2222 (1889) [333]	1260 (970) [290]	2924	134
Beef	1032	503 (430) [73]	556 (504) [51]	53 (74) [−21]	947	109
Poultry	1713	1135 (1080) [55]	982 (865) [117]	−153 (−215) [62]	1678	102

Source: BMEL (2024), own calculations. *ROW* rest of the world (non-EU-27)

that are more stringent than the EU standards. However, it must nevertheless continue to allow imports from other EU member states to enter without restriction. Of course, if a member state imposes standards that exceed the EU minimum standards, then production costs in that member state will increase relative to production costs elsewhere in the EU, and production will be displaced to other member states.

For example, if Germany imposes animal welfare standards in pork production—such as minimum space requirements for fattening pigs—that exceed the EU standards, then all other things being equal, German pork production will become more expensive relative to production in other member states. As a result, production will shift from Germany to those other member states. The pigs that continue to be held in Germany will benefit from higher German standards. But some of the pigs that used to be held under lower EU standards in Germany will now be held under the same lower EU standards in other member states. In addition, if production in those other member states is, on average, costlier in environmental terms (e.g. if it generates more GHG emissions per kg of pork), the increased animal welfare in Germany might be achieved at the cost of higher environmental damage overall.

4.3 Policy Implications

How can policymakers ensure that the transformation of the livestock sector in the EU or in Germany alone does not trigger leakage effects? Two policy implications emerge from the above discussion:

- First, policymakers in Germany and the EU should focus on reducing consumption rather than production. If consumption falls (or grows less rapidly), production will fall (or grow less rapidly) as a result. Each kilogram of meat or litre of milk that is not consumed is a kilogram or litre that does not need to be produced, somewhere on the planet, with attendant benefits for the environment and animal welfare.
- Second, if production is reduced, it should be reduced in those systems/locations in which it generates the highest total (economic plus environmental) costs. Hence, policy should ensure that the prices of livestock products reflect their total production costs, including the environmental costs that result from production. This will ensure that production takes place where it is most competitive in a comprehensive sense that reflects not only economic but also environmental costs. Ensuring that production and trade are based on a comprehensive accounting of economic and environmental costs is a crucial and difficult task that will require international coordination, for example, to establish methods for measuring and implementing border adjustments that reflect products' GHG footprints.

A consequence of these considerations is that reductions in production and consumption need not match at the local or national level. For example, if the German government takes steps to reduce meat consumption in Germany by a certain amount, it does not follow that production in Germany should fall by the same amount. The benefits of reductions in German meat consumption[2] are maximised if the economically and environmentally most expensive production on the planet is reduced correspondingly, wherever it has taken place in the past. Global consumption (which is comprised of the remaining consumption in Germany, and consumption in the rest of the world) should be satisfied by the production that generates the lowest total (economic and environmental) cost. This will likely include continued, and possibly even increased, German production. To illustrate this, in the following section, we compare the GHG emissions of beef production in Brazil and Germany.

4.4 The Example of Beef Production in Germany and Brazil

As we saw in Fig. 5.3, Brazil is the largest exporter of beef to the EU, supplying just under 100 thousand tons in 2023. Brazil is also a source of Germany's beef imports, supplying 6.1 of the 73 thousand tons of beef that Germany imported from non-EU countries in 2023 (BMEL 2024). Figure 5.4 presents estimated GHG emissions in kg CO_2 equivalents per kg of carcass weight (CO_2eq/CW) in Brazil and Germany. These estimates are compiled from a range of sources as outlined in the Annex. The estimates cover the entire chain from production (in Brazil or Germany) to retail in Germany, including land-use change, feed production and animal farming, trade, and post-farm stages such as slaughtering, packaging, and transport to retail including, in the case of Brazilian beef, transport to Germany. We present three estimates: the average GHG footprint of beef produced in Brazil and consumed in Germany; the average GHG footprint of beef produced and consumed in Germany (accounting for the imported feed used in German production); and the average GHG footprint of the beef actually consumed in Germany (which includes both domestic production and imported beef).

The estimates show that if beef consumption in Germany were met exclusively by imports from Brazil, GHG emissions would be much higher per kg of beef consumed than is currently the case. Beef imported from Brazil has a GHG footprint of 140.5 CO_2eq/CW, 63% of which is due to land-use change (88 CO_2eq/CW) and 32% of which is emitted in production on the farm (44.4 CO_2eq/CW). Note that the GHG emissions generated post-farm in Brazil to process the beef and transport it to Germany are comparatively small (7.9 CO_2eq/CW). Hence, it is the production of beef in Brazil, not its subsequent processing and transport, that accounts for the bulk of its GHG footprint.

In contrast, Germany's domestic production, even accounting for the land-use change associated with imported feed, has a lower footprint of 39.5 CO_2eq/CW. This is mainly because Germany's coupled dual-purpose cattle system produces dairy products in addition to meat. In addition, in Brazil's pure beef production system, cattle are fattened over longer periods of time, leading to higher emissions per kg of beef. Note that GHG emissions due to trade are higher for beef produced in Germany

[2] These benefits include health benefits of reduced meat consumption for consumers in Germany.

• Both from **dairy & suckler herd** • CO_2eq /kg carcass weight (CW)	Stage in the beef supply chain /source of GHG emissions Land use changes	Farm gate (i.e., enteric fermentation, N_2O soil, fuel and electricity use, fertilizer production, and organic soil and liming)	Trade (i.e., beef/ feed from barn/slaughterhouse to origin harbor, then to EU harbor, via maritime	Post-farm stages (e.g., slaughtering, packaging, retail transport)	Σ Total
Brazil domestic production Beef produced & imported from Brazil to the EU/Germany	88.0	44.4	0.2	7.9	140.5
Germany domestic production	13.1 =(7.9 abroad + 5.1 domestic)	21.9	0.4	4.1	39.5
Germany domestic consumption	24.0	25.2	0.4	4.6	54.1

Fig. 5.4 Greenhouse gas (GHG) emissions of beef production in Brazil and Germany
Notes: GHG emissions of beef production in Brazil and Germany in kg CO_2eq/kg carcass weight (CW). The estimates cover the entire chain from production (in Brazil or Germany) to retail in Germany, including land-use change, feed production, and animal farming, trade, and post-farm stages such as slaughtering, packaging, and transport to retail including, in the case of Brazilian beef, export to Germany. Own calculations based on sources and assumptions are outlined in the Appendix

than for beef produced in Brazil and exported to Germany. This is because the GHG footprint associated with shipping Germany's feed imports per kg of beef produced in Germany is larger than the GHG footprint of moving a kg of beef from Brazil to Germany. Figure 5.4 shows that on average, one kg of beef produced in Brazil generates roughly 100 kg of CO_2 equivalent more than one kg of beef produced in Germany. Using a current carbon permit price of roughly 70 €/ton CO_2eq or 75 $/ton CO_2eq (Statista 2024), this is equivalent to roughly US$ 7.50 of additional environmental costs per kg of beef produced in Brazil. If the price of Brazilian beef reflected these costs in addition to its economic production costs, Brazilian beef would presumably be considerably less competitive on EU markets.

The estimated GHG footprint of current beef consumption in Germany is 54.1 CO_2eq/CW. This estimate is based on Germany's import share of 85.5% from within the EU and 14.5% outside the EU in 2023. We assume that all beef imported from within the EU has the same GHG footprint as beef produced in Germany and that all beef imported from outside the EU has the same GHG footprint as Brazilian beef. These are rough assumptions. Besides Brazil, Germany also imports beef from Argentina, Uruguay and the USA, where GHG footprints for beef are likely similar to those in Brazil. The GHG footprint of beef imported from within the EU likely varies considerably depending on the country and production system. Furthermore, the averages reported in Fig. 5.4 hide considerable within-country variation. On average, producing one kg of beef in Brazil generates more GHG than producing one kg of beef in Germany, but the underlying distributions will likely overlap, and some Brazilian production will have a smaller footprint than some German production. Despite these uncertainties, it is safe to assume that the GHG footprint of beef consumed in Germany lies somewhat above the estimated GHG footprint of beef produced in Germany.

5 Conclusions

The production and consumption of animal-based foods are associated with negative effects on the environment and animal welfare. In addition, over-consumption especially of meat, which is increasingly prevalent not only in high-income countries, is associated with negative health outcomes. At the same time, the production and consumption of animal-based foods can also generate important ecosystem services and contribute to feeding the world. The challenge is to transform the livestock sector in a way that reduces its negative effects without foregoing its potential benefits.

In addressing this challenge, it is important to account for complex interactions between local changes and global outcomes. Since most livestock products are tradeable, production and consumption decisions at one location on the planet can have effects on socioeconomic and environmental outcomes at other locations. An important consequence of this telecoupling is the danger of so-called leakage. Leakage takes place when changes in policy designed to reduce negative effects at one location lead to shifts in economic activity between locations that cause the displacement of negative effects between locations. As a result, policies that improve outcomes in the EU as a whole or in a part of the EU, such as livestock production in northwest Germany, can end up worsening aggregate outcomes at higher levels, such as the EU or the planet as a whole.

Studies that simulate the effects of implementing the EU's so-called Farm to Fork proposals demonstrate that leakage effects can be substantial. Policies that reduce agricultural production in the EU will, if not accompanied by matching reductions in food consumption, lead to increases in production elsewhere on the planet. The result will be a shift in the locations where negative environmental effects are generated, but not necessarily a reduction in the global volume of these negative effects. The danger of shifts in the location of production and associated leakage effects is especially present within the EU, where all goods that meet EU standards can move freely between member states. Hence, proposals that lead to higher production or animal welfare standards in Germany alone risk generating substantial leakage effects and, for the EU as a whole, few benefits.

This leads us to two policy recommendations. First, policymakers in Germany and the EU should focus on reducing consumption rather than production. Each kilogram of meat or litre of milk that is not consumed is a kilogram or litre that does not need to be produced, somewhere on the planet, with attendant benefits for the environment and animal welfare. Second, if production is reduced, it should be reduced in those systems/locations in which it generates the highest total (economic plus environmental) costs. Hence, policy should ensure that the prices of livestock products reflect their total production costs, including the environmental costs that result from production. This will ensure that production takes place where it is most competitive in a comprehensive sense that reflects not only economic but also environmental costs. Using the example of beef in Brazil and Germany, we demonstrate that taking GHG emissions into account considerably improves the competitiveness of German beef production.

These recommendations are easy to write, but much more difficult to implement. Transforming the livestock sector is not a win-win proposition. Less production and shifts to less environmentally-costly production systems imply difficult decisions, risky investments, and, in some cases, the cessation of production for farmers. Ensuring that the prices of livestock products also reflect their environmental costs entails food price increases. If food becomes more expensive, consumers will have less disposable income to spend on other goods and services. Moreover, food price increases are a regressive measure that disproportionately burden households with lower incomes.

Finally, ensuring that production and trade are based on a comprehensive accounting of economic and environmental costs requires international coordination, for example, to establish methods for measuring and implementing border adjustments that reflect products' GHG footprints. International cooperation is always challenging and appears to be especially elusive at the moment. Moreover, we acknowledge that there is a fundamental tension between, on the one hand, harnessing international trade based on comprehensive cost accounting to minimise the environmental footprint of global food production and, on the other hand, worries in many countries about food security and becoming overly dependent on trade. International trade for many agricultural products is highly concentrated on the export side. In 2022, the five largest exporters of animal products (the EU, the USA, Brazil, Australia, and Canada) accounted for 68% of global trade (WTO 2024). In 2023/24, the six largest exporters of wheat (Russia, the EU, Canada, Australia, Ukraine, and the USA) accounted for 80% of global trade (USDA 2024b). The second example in particular highlights the underlying tension: Russia's military attack on Ukraine in February 2022 sent international grain and other food prices skyrocketing and sparked great anxiety in many importing countries (von Cramon-Taubadel 2022). Russia and the Black Sea region in general are agro-climatically predestined to make a major contribution to sustainable global food production. However, if importers—with good reason—question Russia's reliability as a source of grain and other food products, they will take steps to diversify their sources of imports and increase domestic self-sufficiency, at the expense of the sustainability of the global agri-food system. Without international cooperation to ensure a reliable, rules-based global trading system, transforming the global agri-food system becomes an even greater challenge. In this regard, Germany can achieve much more in concert with its fellow EU member states than by pursuing a narrow domestic agenda.

Acknowledgements Karen Arcia thanks the Research Training Group 2654 "Sustainable Food Systems" funded by the German Research Foundation (DFG), the Chair of Agricultural Policy at the University of Gottingen, Germany, as well as the Colombian Ministry of Science, Technology, and Innovation under the scholarship "Doctorados en el Exterior, call 885-2020", for the funding provided.

Competing Interests The authors have no conflicts of interest to declare that are relevant to the content of this chapter.

Ethics Approval Not applicable.

Appendix

How We Calculate GHG Footprints for Brazilian and German Beef

Beef Produced in Brazil

Land-use change: 6% of beef production in Brazil is associated with deforestation. The resulting emissions vary from 22 to 370 CO_2eq/kg live weight (LW) depending on farming system, location, year, management practices, allocation, and study boundaries (Cederberg 2011). We use the average effect (44 CO_2eq/LW), amortised over 20 years and allocated to all beef production in Brazil. Assuming 50% of LW is carcass weight (CW), this is equivalent to 88 CO_2eq/CW.

Farm-gate stage: We calculate emissions at the farm-gate stage as the sum of the emissions from different production systems, each weighted by its share of the total cattle herd in Brazil, which we estimate using data from de Oliveira Silva (2017), Dick et al. (2015), Cardoso et al. (2016), and Ruviaro et al. (2015) as follows:

- 94% of Brazilian beef is from suckler-beef herds, of which:
  - 16.2% are held in confined systems (2.6% in feedlots plus 14.5% in other confined systems).
  - 77.8% are held on pasture (33.1% extensive pasture, 49.7% in intensive pasture).
- 6% of Brazilian beef is from dairy herds, of which:
  - 0.14% are held in confined systems.
  - 5.86% are held on pasture (extensive or Intensive).

We calculate GHG emissions from the suckler-beef systems based on Dick et al. (2015), Cardoso et al. (2016), Ruviaro et al. (2015), and Morais et al. (2023). For beef from dairy cows, we allocate the GHG footprint estimated by Léis (2015) to beef and milk according to their shares in production value, based on average prices from 2015 to 2023 for beef (13.2 Real/kg) and milk (2.5 Real/kg) (Table 5.3).[3]

Trade stage: We use the average emissions of 8.8 tons of CO_2 for a 26-ton container on a common transport route between Brazil and the EU, as reported by Velazco-Bedoya et al. (2011). We adjust this value based on the assumption that 60% of the carcass is boneless meat.

Post-farm stage: We draw on Poore and Nemecek (2018), who estimate that on global average, the post-farm stage accounts for 3.4% of the total GHG footprint of beef from suckler-herds.

[3] Since 2020, the value of one Brazilian Real has fluctuated between 0.15 and 0.20 Euro.

Table 5.3 Greenhouse gas (GHG) emissions from different beef production systems in Brazil (in kg CO_2eq/kg CW)

Herd	Production system	Average GHG emissions (CO_2eq/CW)	Range of emissions (CO_2eq/CW)
Suckler	Extensive pasture	63.0	40.9–85.1
	Intensive pasture	23.9	18.3–29.6
	Confined	34.7	29.4–40.0
	Feedlot	64.7	44.1–85.4
Dairy	Feedlot	0.002	–
	Pasture	0.027	–

Note: *CW* Carcass weight

Beef Produced in Germany

Land use stage, feed production in Germany: We assume that rewetting peatland in Germany that is used to produce cattle feed will reduce CO_2 emissions by 24.9 million tons (Arcia and von Cramon-Taubadel 2024). Beef accounts for 22.6% of the economic output of the cattle sector in Germany, and milk accounts for the remaining 77.4%. Hence, beef production accounts for 5.6% of the 24.9 million tons of CO_2. Assuming 1.1 million tons of beef production in Germany (carcass weight), this is equivalent to 5.1 CO_2eq/CW.

Land use stage, feed production abroad: Hentschl et al. (2023) estimate that land-use change-related GHG emissions for livestock feed imported by Germany are 135.4 million tons CO_2eq. Of the 5.6 million tons of commercial feed imported into Germany in 2020, we assume that 1.6 million tons (28.6%) were fed to cattle. We therefore conclude that imported feed for cattle accounted for 38.7 of the 135.4 million tons of GHG emissions. As above, we assume that beef accounts for 22.6% of the economic output of cattle in Germany, and that Germany produces 1.1 million tons of beef, leading to an estimate of 7.95 CO_2eq/CW.

Farm-gate stage: We sum the emissions from dairy-based and suckler-based beef production systems, weighted by their proportion in the total cattle herd. While there are no representative estimates for German systems specifically, a review for the EU by Pishgar-Komleh and Beldman (2022) suggests that dairy-based emissions range from 7.5 to 20 CO_2eq/CW (average 13.5), and that emissions from suckler-based systems range from 11 to 46 CO_2eq/CW (average 25.5). We estimate a culling rate of around 20% for dairy cows in Germany. This implies that roughly 30% of all cattle meat produced in Germany comes from dairy cows, and the other 70% is based on excess calves from dairy production and specialised beef production. Thus, beef production in Germany generates average GHG emissions of 21.9 CO_2eq/CW.

Trade stage: Drawing on Fliehr (2013) and USDA (2024a), we estimate the average GHG emissions for soy bulk carriers from Brazil and the USA to the EU to be approximately 0.25 tons CO_2/ton soy. Assuming 1.6 million tons of high-protein feed imported and 1.1 million tons of beef production in Germany, we estimate trade-related GHG emissions of approximately 0.4 CO_2eq/CW.

Post-farm stages: We use Poore and Nemecek's (2018) estimate that the post-farm stage accounts for 10.5% of the total GHG footprint of beef from dairy herds.

References

AHDB (Agriculture and Horticulture Development Board) (2024) EU and UK import tariff rates for selected beef products. https://ahdb.org.uk/eu-and-uk-import-tariff-rates-for-selected-beef-products

Alonso ME, González-Montaña JR, Lomillos JM (2020) Consumers' concerns and perceptions of farm animal welfare. Animals 10(3):385. https://doi.org/10.3390/ani10030385

Arcia K, von Cramon-Taubadel S (2024) Greenhouse gas emissions mitigation potential from Germany's cattle supply chain. Unpublished manuscript

Bardgett RD, Bullock JM, Lavorel S, Manning P, Schaffner U, Ostle N et al (2021) Combating global grassland degradation. Nat Rev Earth Environ 2(10):720–735. https://doi.org/10.1038/s43017-021-00207-2

Barreiro-Hurle J, Bogonos M, Himics M, Hristov J, Perez Dominguez I, Sahoo A, Salputra G, Weiss F, Baldoni E, Elleby C (2021) Modelling Environmental and Climate Ambition in the Agricultural Sector with the CAPRI Model. Exploring the Potential Effects of Selected Farm to Fork and Biodiversity Strategies Targets in the Framework of the 2030 Climate Targets and the Post 2020 Common Agricultural Policy. Luxembourg: Publications Office of the European Union

Beckman J, Ivanic M, Jelliffe JL, Baquedano FG, Scott SG (2020) Economic and Food Security Impacts of Agricultural Input Reduction Under the European Union Green Deal's Farm to Fork and Biodiversity Strategies." EB-30, U.S. Department of Agriculture, Economic Research Service

Bremmer J, Gonzalez-Martinez A, Jongeneel R, Huiting H, Stokkers R (2021) Impact Assessment Study on EC 2030 Green Deal Targets for Sustainable Food Production. Wageningen, Economic Research, Report 2021–150

Bengtsson J, Bullock JM, Egoh B, Everson C, Everson T, O'Connor T, O'Farrell PJ, Smith HG, Lindborg R (2019) Grasslands – more important for ecosystem services than you might think. Ecosphere 10(2):e02582. https://doi.org/10.1002/ecs2.2582

BMEL (2024) Versorgungsbilanz Fleisch. https://www.bmel-statistik.de/ernaehrung/versorgungsbilanzen/fleisch

BMEL (Bundesministerium für Ernährung und Landwirtschaft) (2021) Understanding Farming. Facts and figures about German farming. https://www.bmel.de/SharedDocs/Downloads/EN/Publications/UnderstandingFarming.pdf

Bonnet C, Bouamra-Mechemache Z, Réquillart V, Treich N (2020) Regulating meat consumption to improve health, the environment and animal welfare. Food Policy 97:101847. https://doi.org/10.1016/j.foodpol.2020.101847

Brindha K (2020) Virtual water flows, water footprint and water savings from the trade of crop and livestock products of Germany. Water Environ J 34(4):656–668. https://doi.org/10.1111/wej.12601

Cardoso AS, Berndt A, Leytem A, Alves BJ, de Carvalho IDN, de Barros Soares LH et al (2016) Impact of the intensification of beef production in Brazil on greenhouse gas emissions and land use. Agric Syst 143:86–96. https://doi.org/10.1016/j.agsy.2015.12.007

Cederberg C, Persson UM, Neovius K, Molander S, Clift R (2011) Including carbon emissions from deforestation in the carbon footprint of Brazilian beef. https://doi.org/10.1021/es103240z

Crenna E, Sinkko T, Sala S (2019) Biodiversity impacts due to food consumption in Europe. J Clean Prod 227:378–391. https://doi.org/10.1016/j.jclepro.2019.04.054

de Léis CM, Cherubini E, Ruviaro CF et al (2015) Carbon footprint of milk production in Brazil: a comparative case study. Int J Life Cycle Assess 20:46–60. https://doi.org/10.1007/s11367-014-0813-3

de Oliveira Silva R, Barioni LG, Hall JJ, Moretti AC, Veloso RF, Alexander P et al (2017) Sustainable intensification of Brazilian livestock production through optimized pasture restoration. Agric Syst 153:201–211. https://doi.org/10.1016/j.agsy.2017.02.001

Dick M, Da Silva MA, Dewes H (2015) Life cycle assessment of beef cattle production in two typical grassland systems of southern Brazil. J Clean Prod 96:426–434. https://doi.org/10.1016/j.jclepro.2014.01.080

Espinosa R, Treich N (2021) Animal welfare: antispeciesism, veganism and a "life worth living". Soc Choice Welf 56:531–548. https://doi.org/10.1007/s00355-020-01287-7

Eurobarometer (2023) EUROBAROMETER wave 99.1 Technical report. European Commission. https://europa.eu/eurobarometer/surveys/detail/2996

European Commission (2020a) Communication from the Commission to the European Parliament, the Council, the European Economic and Social Committee and the Committee of the Regions: a Farm to Fork Strategy for a fair, healthy and environmentally-friendly food system (COM/2020/381 final). https://eur-lex.europa.eu/legal-content/EN/TXT/?uri=CELEX:52020DC0381

European Commission (2020b) EU agricultural census 2020. https://ec.europa.eu/eurostat/web/agriculture/information-data/agricultural-census

European Commission (2024) Agri-Food markets: trade volume and value data for Eggs and Poultry, Pork, Cattle, Milk and Dairy Products (2000–2023). https://ec.europa.eu/agriculture/agri-food-markets_en

FAOSTAT (2024) Food and agriculture data. https://www.fao.org/faostat/en/#home

Fischer G, Tramberend S, Bruckner M, Lieber M (2017) Quantifying the land footprint of Germany and the EU using a hybrid accounting model (TEXTE 78/2017). Umweltbundesamt. https://www.umweltbundesamt.de/publikationen

Fliehr O (2013) Analysis of transportation and logistics processes for soybeans in Brazil (No. 4). Thünen Working Paper. https://doi.org/10.3220/WP_4_2013

Funke F, Mattauch L, van den Bijgaart I, Godfray C, Hepburn CJ, Klenert D et al (2021) Is meat too cheap? Towards optimal meat taxation. Towards Optimal Meat Taxation (March 9, 2021). https://doi.org/10.2139/ssrn.3801702

Gerber PJ, Steinfeld H, Henderson B, Mottet A, Opio C, Dijkman J et al (2013) Tackling climate change through livestock: a global assessment of emissions and mitigation opportunities. Food and agriculture Organization of the United Nations (FAO). http://www.fao.org/3/a-i3437e.pdf

Gossner MM, Lewinsohn TM, Kahl T, Grassein F, Boch S, Prati D et al (2016) Land-use intensification causes multitrophic homogenization of grassland communities. Nature 540(7632):266–269. https://doi.org/10.1038/nature20575

Government of Canada (2024) European Union (EU) – export requirements for meat and poultry products. https://inspection.canada.ca/en/exporting-food-plants-animals/food-exports/requirements/eu-meat-and-poultry-products

Headey D, Hirvonen K, Hoddinott J (2018) Animal sourced foods and child stunting. Am J Agric Econ 100:1302–1319

Hentschl M, Michalke A, Pieper M, Gaugler T, Stoll-Kleemann S (2023) Dietary change and land use change: assessing preventable climate and biodiversity damage due to meat consumption in Germany. Sustain Sci 1-17. https://doi.org/10.1007/s11625-023-01326-z

Henning C, Witzke P, Panknin L, Grunenberg M (2021) Ökonomische und Ökologische Auswirkungen des Green Deals in der Agrarwirtschaft. Kiel: Institut für Agrarökonomie, Abteilung Agrarpolitik, Christian-Albrechts-Universität https://www.bio-pop.agrarpol.uni-kiel.de/de/f2f-studie

Hoekstra AY (2012) The hidden water resource use behind meat and dairy. Anim Front 2(2):3–8. https://doi.org/10.2527/af.2012-0038

IPPC (2007) Climate change 2007: mitigation of climate change. Contribution of Working Group III to the Fourth Assessment Report of the Intergovernmental Panel on Climate Change

Kempen M, Johannes P (2023) Es geht: Wie wir unsere Ernährung sichern und gleichzeitig die Natur und das Klima schützen können. NABU-Statement basierend auf einer CAPRI-

Modellierungsstudie (1. Auflage). NABU-Bundesverband. https://www.nabu.de/imperia/md/content/nabude/landwirtschaft/230113-nabu_flaechennutzungsstudie.pdf
Kinnunen P, Guillaume MT, Taka M et al (2020) Local food crop production can fulfil demand for less than one-third of the population. Nat Food 1:229–237. https://doi.org/10.1038/s43016-020-0060-7
Klement L, Bach M, Geupel M, Breuer L (2021) Calculation of a food consumption nitrogen footprint for Germany. Environ Res Lett 16(7):075005. https://doi.org/10.1016/j.jenvman.2023.119023
Komarek AM, Dunston S, Enahoro D, Godfray HCJ, Herrero M, Mason-D'Croz D, Rich KM, Scarborough P, Springmann M, Sulser TB, Wiebe K, Willenbockel D (2021) Income, consumer preferences, and the future oflivestock-derived food demand. Glob Environ Chang 70:102343. https://doi.org/10.1016/j.gloenvcha.2021.102343
Legwegoh AF, Riley L (2014) Food, place, and culture in urban Africa: comparative consumption in Gaborone and Blantyre. J Hunger Environ Nutr 9(2):256–279. https://doi.org/10.1080/19320248.2013.845868
Leip A, Billen G, Garnier J, Grizzetti B, Lassaletta L, Reis S et al (2015) Impacts of European livestock production: nitrogen, sulphur, phosphorus and greenhouse gas emissions, land-use, water eutrophication and biodiversity. Environ Res Lett 10(11):115004
Liechti K, Biber J-P (2016) Pastoralisim in Europe: characteristics and challenges of highland-lowland transhumance. Rev Sci Tech 35(2):561–575
Lippi G, Mattiuzzi C, Cervellin G (2016) Meat consumption and cancer risk: a critical review of published meta-analyses. Crit Rev Oncol Hematol 97:1–14. https://doi.org/10.1016/j.critrevonc.2015.11.008
Liu J, Herzberger A, Kapsar K et al (2019) What is telecoupling? Palgrave studies in natural resource management. https://www.canr.msu.edu/csis/uploads/files/Chapter_2_2019_Book_Telecoupling.pdf
Liu H, Hou L, Kang N, Nan Z, Huang J (2022) The economic value of grassland ecosystem services: a global meta-analysis. Grassland Res 1(1):63–74. https://doi.org/10.1002/glr2.12012
Lun F, Liu J, Ciais P, Nesme T, Chang J, Wang R et al (2018) Global and regional phosphorus budgets in agricultural systems and their implications for phosphorus-use efficiency. Earth Syst Sci Data 10(1):1–18. https://doi.org/10.5194/essd-10-1-2018
Marcu A, Egenhofer C, Roth S, Stoefs W (2013) Carbon leakage: an overview. CEPS Special Report No. 79. Centre for European Policy Studies, Brussels
Mayer N, Kaltschmitt M (2022) Closing the phosphorus cycle: current P balance and future prospects in Germany. J Clean Prod 347:131272. https://doi.org/10.1016/j.jclepro.2022.131272
Micha R, Wallace SK, Mozaffarian D (2010) Red and processed meat consumption and risk of incident coronary heart disease, stroke, and diabetes mellitus: a systematic review and meta-analysis. Circulation 121(21):2271–2283. https://doi.org/10.1161/CIRCULATIONAHA.109.924977
Morais HB, Chardulo LAL, Baldassini WA, Lippi ICDC, Orsi GB, Ruviaro CF (2023) Environmental impacts of high-quality Brazilian beef production: a comparative life cycle assessment of premium and super-premium beef. Animals 13(22):3578. https://doi.org/10.3390/ani13223578
Mottet A, de Haan C, Falcucci A, Tempio OC, Gerber P (2017) Livestock: on our plates or eating at our table? A new analysis of the feed/food debate. Glob Food Sec 14:1. https://doi.org/10.1016/j.gfs.2017.01.001
Noleppa S, Cartsburg M (2021) The Socio-Economic and Environmental Values of Plant Breeding in the EU and for Selected EU Member States. Berlin: HFFA Research GmbH
OECD-FAO (2023) Agricultural Outlook 2023–2032: Chapter 6 "Meat". https://www.oecd-ilibrary.org/sites/f01f6101-en/index.html?itemId=/content/component/f01f6101-en
Pan A, Sun Q, Bernstein AM, Manson JE, Willett WC, Hu FB (2013) Changes in red meat consumption and subsequent risk of type 2 diabetes mellitus: three cohorts of US men and women. JAMA Intern Med 173(14):1328–1335. https://doi.org/10.1001/jamainternmed.2013.6633
Parlasca MC, Qaim M (2022) Meat consumption and sustainability. Ann Rev Resour Econ 14:17–41. https://doi.org/10.1146/annurev-resource-111820-032340

Parlasca M, Knößlsdorfer I, Alemayehu G, Doyle R (2023) How and why animal welfare concerns evolve in developing countries. Anim Front 13(1):26–33. https://doi.org/10.1093/af/vfac082
Pishgar-Komleh SH, Beldman A (2022) Literature review of beef production systems in Europe. Report 1364. Wageningen Livestock Research, Wageningen University and Research and the European Roundtable for Beef Sustainability (ERBS). https://edepot.wur.nl/567148
Poore J, Nemecek T (2018) Reducing food's environmental impacts through producers and consumers. Science 360(6392):987–992
Ruviaro CF, de Léis CM, Lampert VDN, Barcellos JOJ, Dewes H (2015) Carbon footprint in different beef production systems on a southern Brazilian farm: a case study. J Clean Prod 96:435–443. https://doi.org/10.1016/j.jclepro.2014.01.037
Schrijver R, Stijntjes M, Rodríguez-Baño J, Tacconelli E, Rajendran NB, Voss A (2018) Review of antimicrobial resistance surveillance programmes in livestock and meat in EU with focus on humans. Clin Microbiol Infect 24(6):577–590. https://doi.org/10.1016/j.cmi.2017.09.013
Socher SA, Prati D, Boch S et al (2013) Interacting effects of fertilization, mowing and grazing on plant species diversity of 1500 grasslands in Germany differ between regions. Basic Appl Ecol 14(2):126–136
Statista (2024) Current carbon permit price. Retrieved on 05.20.2024 from https://www.statista.com/markets/
Statista Consumer Insights (2023) Share of vegans in select countries worldwide in 2023. Retrieved on May 20, 2024, from https://www.statista.com/statistics/1280066/global-country-ranking-vegan-share/
Temple D, Manteca X (2020) Animal welfare in extensive production systems is still an area of concern. Front Sustain Food Syst 4:545902. https://doi.org/10.3389/fsufs.2020.545902
USDA (United States Department of Agriculture) (2024a) Bulk carriers GHG emission footprint from the US to the EU. https://agtransport.usda.gov/
USDA (United States Department of Agriculture) (2024b) Grain: World Markets and Trade. Retrieved on July 25, 2024, from https://apps.fas.usda.gov/psdonline/circulars/grain.pdf
Uwizeye A, de Boer IJ, Opio CI, Schulte RP, Falcucci A, Tempio G et al (2020) Nitrogen emissions along global livestock supply chains. Nat Food 1(7):437–446. https://doi.org/10.1038/s43016-020-0113-y
Velazco-Bedoya DM et al (2011) Characterization of "in natura" bovine meat transportation from Brazil to Europe, its carbon dioxide emission and its alternatives reduction Center for Advanced Studies on Applied Economics – CEPEA, ESALQ/USP Johann Heinrich von Thünen Institut – vTI. https://www.cepea.esalq.usp.br/en/documentos/texto/cattle-caracterization-of-in-natura-bovine-meat-transportation-from-brazil-to-europe-its-carbon-dioxide-emission-and-its-alternatives-reduction-1.aspx
von Cramon-Taubadel S (2022) Russia's invasion of Ukraine–implications for grain markets and food security. German J Agric Econ 71:1–13
Wesseler J (2022) The EU's farm-to-fork strategy: an assessment from the perspective of agricultural economics. Appl Econ Perspect Policy 44(4):1826–1843. https://doi.org/10.1002/aepp.13239
Westhoek H et al (2011) The protein puzzle: the consumption and production of meat, dairy and fish in the European Union (No. 500166001). PBL Netherlands Environmental Assessment Agency. https://library.wur.nl/WebQuery/wurpubs/406619
World Animal Protection (2024) Animal protection index. Retrieved on May 20, 2024, from https://api.worldanimalprotection.org/
WTO (World Trade Organisation) (2024) Charts – world trade in agricultural products. Retrieved on June 25, 2024, from https://www.wto.org/english/tratop_e/agric_e/ag_imp_exp_charts_e.htm
Zhang H, Greenwood DC, Risch HA, Bunce D, Hardie LJ, Cade JE (2021) Meat consumption and risk of incident dementia: cohort study of 493,888 UK biobank participants. Am J Clin Nutr 114(1):175–184. https://doi.org/10.1093/ajcn/nqab028
Zhao Y, Liu Z, Wu J (2020) Grassland ecosystem services: a systematic review of research advances and future directions. Landsc Ecol 35:793–814. https://doi.org/10.1007/s10980-020-00980-3

Public Ethical Discourse About Livestock Production

6

Andrea Nelke, Peter Kunzmann, and Tabea Weber

1 Characteristics of the Public Discourse on Animals

The discourse on animals is interspersed with ethically relevant topics in many places. The central driving force appears to be the clearly noticeable change in human-animal relationships in recent years and the associated questioning of many practices in intensive animal husbandry. In an intensive one-year multidisciplinary discourse project at the University of Veterinary Medicine Hannover, we have filtered elements of discourse that can significantly influence its failure or success and designed a structuring aid for discourses that can detoxify them (Weber et al. 2021). The findings presented below are therefore based on conditions in Germany, although certain patterns can be observed in other countries where intensive animal husbandry is practiced as well.

Asking people who have often been involved in such debates invariably express a desire for more objectivity. To create a more objective discourse environment, it makes sense to first take a closer look at the content-related and communicative components of such discussions.

A. Nelke (✉) · P. Kunzmann
University of Veterinary Medicine Hannover, Hannover, Germany
e-mail: andrea.nelke@tiho-hannover.de; peter.kunzmann@tiho-hannover.de

T. Weber
Bielefeld University, Bielefeld, Germany
e-mail: tabea_sabrina.weber@uni-bielefeld.de

B. Grabkowsky, T. Blaha (eds.), *Intensive Livestock Production in Transition*,
https://doi.org/10.1007/978-3-031-97872-2_6

2 Personal History and Communication Level

Due to the complexity, it is not possible to illuminate and systematically present all aspects at the communicative level, but the aim is to create awareness of which parts are likely to heavily influence how discourse will develop, both positively and negatively. Within the project, the team approached three levels of the livestock discourse: the analysis of and creation of consciousness with regard to "invisible bundles," the analysis and prevention of verbal communication barriers, and the orientation toward the development of a content-related focus model. By these means, sensitive points of discussion were localized as chances and structural transparency and empathy in moderation as potential game changers.

Personal components that each participant brings to the discourse and that unconsciously accompany every utterance are not visible to others in the discourse but are nevertheless highly relevant.

The culturally and socially formed, partly consolidated images of farm animals and their living conditions,[1] but also stereotypes of other participants in the discourse are part of this. In our project these images were called invisible bundles, referring to several packages of individual influences. Prejudices about certain stakeholder groups are difficult to ignore or even completely discard during the exchange. Farmers, for example, are often defensive and feel stigmatized, both in their immediate social environment and in social media, as several studies in Germany have impressively demonstrated (Wittmann 2021). The massive rejection of this occupational group is expressed in harsh phrases such as "You should be slaughtered" (Dürnberger 2019).

Additionally, the autostereotypes are crucial: What do I think how the others in this discussion think about myself? Will they listen to me? Will I have any space to describe my point of view, or will I only be confronted with attack, discrimination, or be made fun of? Do the others take me seriously? These may be some of the questions on the minds of participants of real-world live discussions about this topic and they put incredible stress on the individuals, which is firstly completely split from the actual content level of the discussion; it simply stems from the negative discourse history about human-animal relationships in many media and the own experience (Weber et al. 2021, pp. 15–17).

Open replies to a questionnaire answered by participants of the different discourse parties "farmers, consumers, animal welfare/animal rights persons, veterinarians" clearly showed that almost everybody had the impression not to be respected within the discourse by one or more of the other parties. Most persons also felt that not all the other parties had the right to engage in the discussion on animals in agricultural use (Weber et al. 2021, pp. 15–17).

Only a few exceptions were made where people thought that "everybody's perspective should be shared to develop a better future." The perception of being ridiculed, marginalized, or attacked by the confronting parties, rather than listened to,

[1] C.t. Christoph-Schulz (2018).

was shared by many of the participating farmers, animal welfare and rights members, and veterinarians.

Such previous experiences or views are very emotional and stand in the way of an objective dialogue. In the worst case, a discursive situation can be blocked from productivity a priori. It is very important to notice that such aggressive verbal behavior can be noticed from all different parties within the animal discourse and is often strongly supported not only by the shortened and triggering messages shared via social media but also in public discussion activities like talk shows on television where the "clash sells" as well. When arguments are medialized primarily to have a huge reach out and provoke further intensive reactions, rhetoric of confrontation is often dominating without showing real arguments sticking to the content facts which the participants, often desperately, try to communicate. This waste of possibilities is not merely disappointing, but alarming when looking at the pressing ethical questions to deal with the role of livestock production from diverse angles such as sustainability, animal ethics, human social factors, and economics.

While all the emotions involved in this discourse are important and have to be acknowledged during discussions—individual economic threat or stigmatization as experienced by many livestock-holding farmers or individual threat and despair about the treatment of animals that is perceived to be unethical from animal rights or animal welfare perspectives, to name just a few—they must be taken into account but not allowed to lead to discrimination of individuals, blocking any chance for constructive communicative exchange and just repeating the previously known prejudices and public statements everybody knew already before entering the specific discussions. The following aspects reveal value judgements about animals and the actions of others in an agricultural context: the invisible bundles, certain terms (see examples in Fig. 6.1 and 6.2) of expert language or buzz words, and the aggressive direct and indirect attacks on individual stakeholder parties (or worse, persons).

As one example of a buzzword in German-speaking countries, the term *Massentierhaltung*[2] can serve. It is naturally used to describe large-scale animal husbandry with many animals, not to provide neutral information about the number of animals, but to strongly criticize and devalue this type of animal housing (Christoph-Schulz 2018) from the perspectives of consumers, politicians, animal welfare, or animal rights. It has also been established in common everyday speech but is perceived as very pejorative by many farmers, while often being seen as descriptive and neutral by others. A counterpart in the English-speaking world with a similarly clear negative connotation, which is primarily used by opponents of intensive livestock farming, is *factory farming*. Although it emphasizes a different nuance, it fulfils an analogous function in the discourse. There is also a technical, sterile variant of the term for husbandry systems in which large numbers of animals are kept, so-called *concentrated animal feeding operations* (CAFO), which was defined by the United States Department of Agriculture. Whichever variant one chooses, no one describes livestock farming in a value-neutral way but reveals the fundamental attitude of the speaker.

[2] Translated literally: mass animal husbandry.

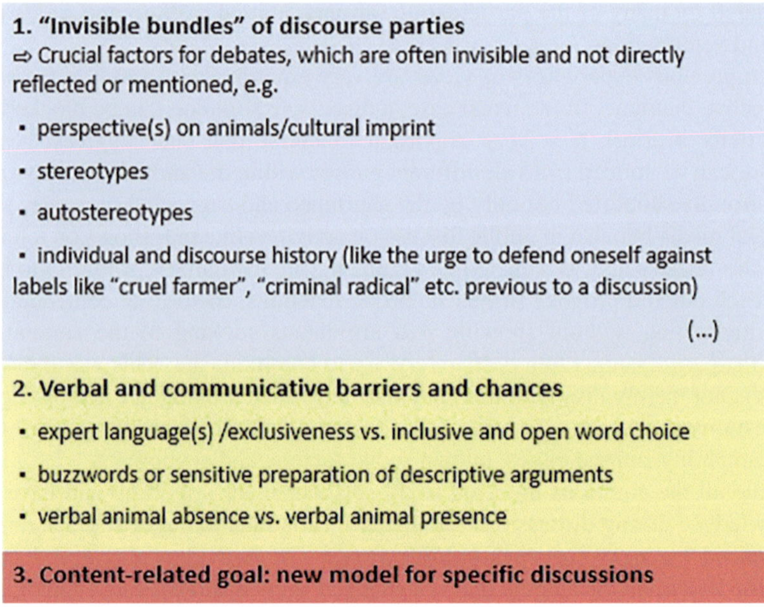

Fig. 6.1 3 Levels. (Weber et al. 2021 CC BY 4.0)

An example from the other side would be the term "weaning" used in agriculture and often perceived as sugarcoating anti-suckling devices and the separation from animal welfare or animal rights perspective. In both cases, it is evident that such terms can unnecessarily and unintentionally (or intentionally in some cases) heat or even block constructive individual discussion scenarios which could originally have been established to support arguments for improving situations for all involved parties, most importantly, the animals kept in livestock themselves.

In the rare occasions when people engage productively with the different points of view on animals, a real interest in the other positions is often recognizable, aiming at a better understanding of the other and the inherent goal to really be able to improve a situation for each, animals and humans, on the long run. Language thus unfolds its power in the dialogue itself and has a lasting impact on discourse.

Apart of avoiding buzzwords, throughout the project, the researchers thus tried to stick to Marshall Rosenberg's model of Nonviolent Communication (NVC), focusing on a language without threatening or judging and with emphasis on the mutual interest to have new insights into arguments or interaction. Rosenberg (2001) teaches four steps: (1) observations, (2) feelings, (3) needs, and (4) request. The steps should serve in all communicative situations, increasing reflected and respectful communication in all contexts, especially in arguments. Following this chronology, one's own feelings, e.g., fear or anger, are acknowledged and productively used to reach a goal in the end, developing a concrete request. Another principal factor was the previous reflection of stereotypes and autostereotypes before concrete

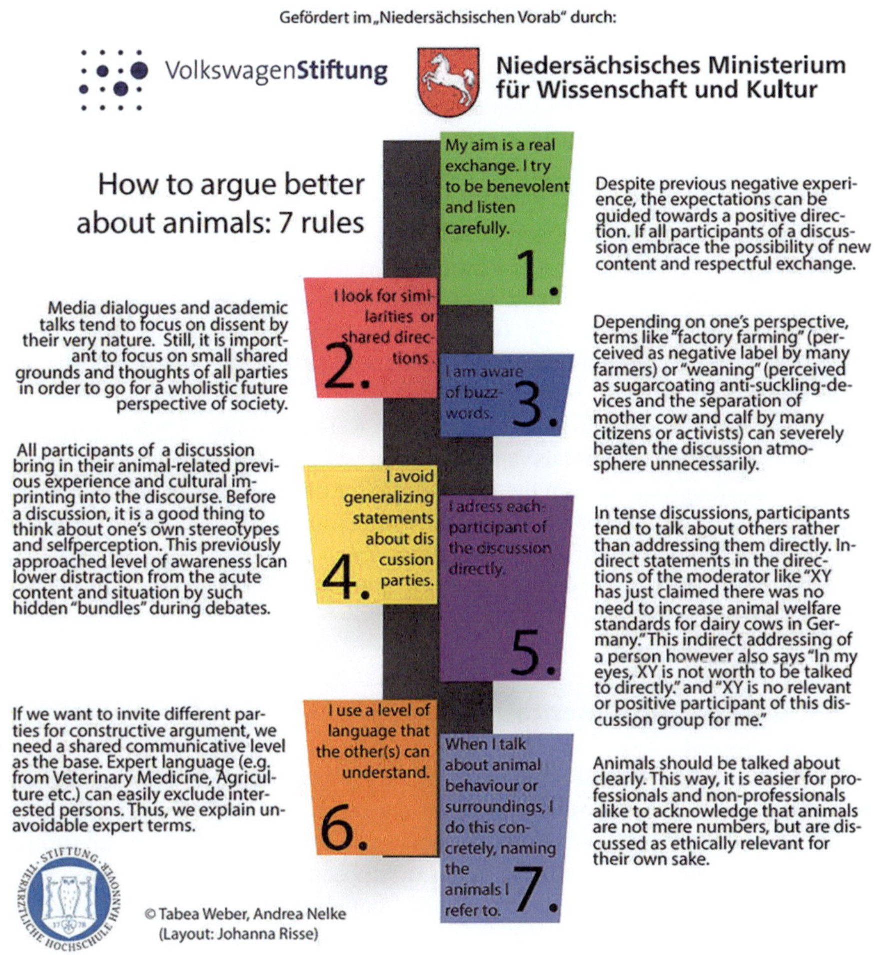

Fig. 6.2 Rules of dialogue in the context of animal farming. (Reproduced from Weber and Nelke 2021a: *Arguing better—7 rules to go*. CC BY 4.0)

situations of organized discussions such as stakeholder workshops, seminar discussions, or a public stakeholder discussion with all participants to create structure, transparency, and a guarantee to feel safer and more respected for everybody.

Also, to analyze specific patterns of escalation, the four-ear model by Friedemann Schulz von Thun (2019) was used. This model helped to sometimes localize seemingly trivial but key moments of discussions, which led to escalations. Schulz von Thun thinks that communication is an act of exchange between sender and receiver and that every message can be perceived quite differently, depending on which angle or focus ear it is perceived: It can be linked to either content, self-revelation, relationship, or appeal (Franck 2021, p. 124; Schulz von Thun 2019, pp. 27–35, 40–42). Like the "invisible bundles," this model shows how nobody is neutral in communication and that a will to concentrate on the other and listen carefully and with (previous) self-reflection is crucial.

It has not to be mentioned that this did not and cannot change the antagonistic perspectives about how animals should be treated by humans in general; contradicting perspectives must be allowed, and they will remain. But with these tools, it was possible to engage more productively and content-oriented with specific issues of ethical problems in livestock keeping.

During the project, several aspects of communication emerged that had a negative impact on the course of the dialogue and hindered a genuine, substantive exchange. To make future dialogues more conciliatory, we have formulated the following discussion rules (see Fig. 6.2), some of which are generally relevant for successful communication and some of which are specifically relevant for the discourse on farm animals:

Before many of the discussions, participants were given the "7 rules" (Fig. 6.2) for a better argument on animals; a small startup set to deal better with this important yet difficult discourse from the perspectives of the diverse protagonists. These rules of dialogue should not be mistaken as something final, but should be developed further and used in different arguments about animal use or human-animal relationships to increase a more respectful and constructive way of discussing animals.

3 Content-Argumentative Level

Livestock farming as an existential basis directly affects human interests on the one hand and animals, i.e., sentient beings, on the other. Roughly speaking, the pure content around which discourses on livestock revolve can be divided into an animal-centered and a human-centered part. It has often been observed that disputes focus primarily on human concerns, such as limited options for action for economic reasons. From an ethical point of view, it seems sensible to choose the situation of the animal in a specific issue as the starting point.

Step 1: Present State

A factual, ideally nonjudgmental description of the situation of the animal in question is a suitable starting point. In principle, a distinction should be made at the beginning as to whether deficits in certain forms of husbandry that violate current animal welfare legislation should be the subject of further discussion or whether currently valid legal minimum standards or sub-legal norms should be reflected upon. In the first case, there is already a consensus in advance that such treatment of animals is neither socially acceptable nor justifiable. There is no ethical conflict here, so the offence itself cannot be declared a morally justifiable act.

The second possible starting point focuses on clearly naming the currently legitimized livestock farming, which can certainly be questioned in several places. The aim of starting the discourse is to bring to light facts and circumstances that need to be examined in the further course, not least because their change or improvement is required.

Step 2: Evaluation of the Present State

People with hugely diverse levels of knowledge about livestock farming take part in discourses. Depending on the discipline, experts have various specialized tools at their disposal that they can use to assess the permissibility of issues. But even within their own profession, they can arrive at different judgments using the same tools.

In the context of livestock farming, voices from scientific, ethical, and legal contexts are particularly necessary, which must also be expanded to include facets of practical and experiential knowledge from those involved in the discourse. It is not possible to derive a universal assessment standard from this diversity of assessment grids that could be used to form an overall judgment. The specific challenge at this stage therefore lies in recognizing and naming those aspects that should carry weight in the balance, depending on the case. This requires the participants to be willing and able to recognize the assessment tools of other disciplines and filter out the added value.

Step 3: Defining the Ideal State

By balancing out all the value judgments that have been discussed, there is now an opportunity to combine them into an overall judgment. This in turn is to be translated into a common goal regarding the animal, but at this point without taking human interests into account. Admittedly, the question of the weighting of the respective value judgments in the overall structure could not be answered in our project. However, we see the explicitly named goal, which was jointly developed, as a key point of orientation for the overall debate.

Step 4: Realizability

At this stage of the discourse path, human-centered topics come to the fore. Both stakeholder-specific interests and the framework conditions that influence or limit the actions of those involved should be addressed explicitly. At present, economic circumstances and constraints often form the dominant counterweight in weighing up processes. In view of the growing importance of sustainability requirements, ecological factors will also come more into focus in the future. On the one hand, these framework conditions determine the current options for action of the stakeholders involved. On the other hand, however, many of them can certainly be changed, provided society is in favor of them and supports this. The specific objective at this stage is, in addition to reviewing the current possibilities for realization, to look for levers that are suitable for expanding the range of actions of people in favor of farm animals.

In general, such or similarly structured discussions as shown in Fig. 6.3—ideally guided by a moderator—offer several advantages. The fixed sequence of content makes it possible to cover the broad spectrum of topics associated with a specific question on current livestock farming. Each participant, usually bringing in their own area of specialization, can also be sure that their issue will have a place in the discourse at a certain point in time. This promise, this commitment, noticeably relieves the process, as it frees those involved from the urge to feed in their concerns now in fear that they might otherwise fall by the wayside.

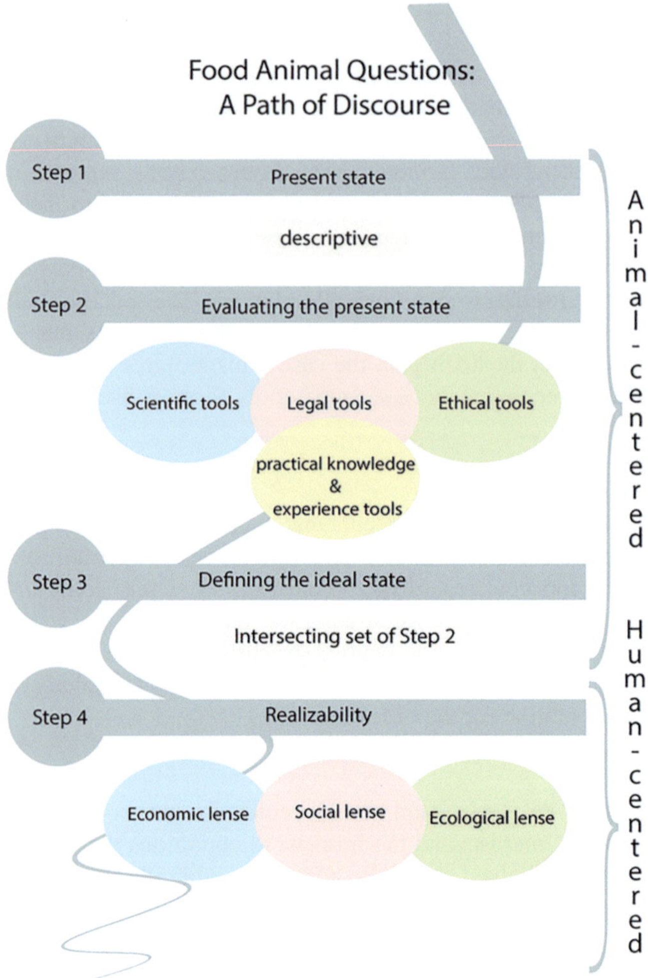

Fig. 6.3 Model of a structured conversation in which both animal and human concerns have a place. (Reproduced from Weber and Nelke 2021b: *Path of Discourse*. (The model was first published in: Kunzmann et al. (2021)) CC BY 4.0)

On the one hand, the step-by-step development of the topics uncovers problems; on the other hand, it inevitably leads to the central question of what animal husbandry should look like, to finally compare the currently possible framework for action and the many limitations. But primarily, it must be guaranteed that the voiceless living beings involved in these human evaluations are not forgotten. Ideally, the strict use of the *Path of Discourse* guarantees: not to lose sight of the animal.

References

Christoph-Schulz I (2018) SocialLab – Nutztierhaltung im Spiegel der Gesellschaft. J Consum Prot Food Saf 13:145–236. https://doi.org/10.1007/s00003-017-1144-7

Dürnberger C (2019) 'You should be slaughtered!' Experiences of criticism/hate speech, motives and strategies among German-speaking livestock farmers using social media. Int J Livest Prod 10(5):151–165. https://doi.org/10.5897/IJLP2018.0567

Franck N (2021) Handbuch Kommunikation. UTB, Schöningh, p 124

Kothe C, Wolfram J, Mergenthaler M (2020) Landwirtschaftliche Tierhaltung in den Medien – Eine qualitative Analyse von Experteninterviews mit Journalisten regionaler und überregionaler Allgemein- und Fachmedien. Berichte über Landwirtschaft 98. https://doi.org/10.12767/buel.v98i1.261

Kunzmann P, Nelke A, Weber T (2021) Das geht besser! In: DLG-Mitteilungen 5/2021, 53

Rosenberg MB (2001) Gewaltfreie Kommunikation. Eine Sprache des Lebens. Junfermann, Paderborn, pp 9–12, 21

Schulz von Thun F (2019) Miteinander Reden 1. Rowohlt, Hamburg, pp 27–35, 40–42

Sonntag WI, Ermann M, Spiller A, von Meyer-Höfer M (2021) Im Streit um die Nutztierhaltung: Gesellschaftsorientierte Kommunikationsstrategien für die Agrar- und Ernährungswirtschaft. Ger J Agric Econ 70(1):1–16. https://doi.org/10.30430/70.2021.1.1-16

Weber TS, Nelke A (2021a) Arguing better – 7 rules to go. Available at: https://wiemenschenuebertierestreiten.jimdofree.com/project-overview-english-version/arguing-better-7-rules-to-go-english-version/. Accessed 4 June 2024

Weber TS, Nelke A (2021b) Path of discourse. Available at: https://wiemenschenuebertierestreiten.jimdofree.com/project-overview-english-version/path-of-discourse-english-version/. Accessed 4 June 2024

Weber TS, Nelke A, Kunzmann, P (2021) Zukunftsdiskurse – Wie Menschen über Tiere streiten: Projektbericht. Stiftung Tierärztliche Hochschule Hannover, Institut für Tierhygiene, Tierschutz und Nutztierethologie, pp 15–17, 24–25, 35–36 and 38. https://doi.org/10.4119/unibi/2957406

Wittmann B (2021) Intensivtierhaltung – Landwirtschaftliche Positionierungen im Spannungsfeld von Ökologie, Ökonomie und Gesellschaft. Vandenhoeck & Ruprecht, Göttingen, pp 93–119

Systemic Complexity Makes the Transformation a "Wicked" Problem

7

Barbara Grabkowsky and Thomas Blaha

1 Introduction

The long-term viability of current livestock production is increasingly being questioned due to its severe environmental, social, and animal welfare challenges. Sustainability is the overarching goal, yet achieving it is highly complex. The transformation of livestock production toward sustainability is shaped by intricate interdependencies between economic, political, social, and cultural factors. While sustainability goals are widely acknowledged, their implementation across the agri-food system is hindered by structural barriers, competing interests, and conflicting policy objectives (Pingali 2012; Tilman et al. 2017). The systemic interdependencies of numerous endogenous and exogenous challenges, combined with the diversity of perspectives, interests, and demands of the various stakeholder groups, make this transformation a "wicked" problem (Rittel and Webber 1973). One of the major characteristics of "wicked" is that there is no one-dimensional solution but only constantly negotiated compromises that aim at balancing the many various values. This means that these compromises can only be achieved if negotiation processes are moderated by multi-disciplinary commissions that consider the needs and interests of the main stakeholder groups affected by the proposed changes (please refer to Chapter 30).

B. Grabkowsky (✉)
Center for Sustainability Transformation Research in AgriFood Systems, Vechta Institute of Sustainability Transformation in Rural Areas, University of Vechta,
Vechta, Germany
e-mail: barbara.grabkowsky@uni-vechta.de

T. Blaha
Former Head of the Field Station for Epidemiology (Bakum), University of Veterinary Medicine Hannover, Hannover, Germany
e-mail: thomas.blaha.ir@tiho-hannover.de

B. Grabkowsky, T. Blaha (eds.), *Intensive Livestock Production in Transition*,
https://doi.org/10.1007/978-3-031-97872-2_7

In Germany, Ireland, the Netherlands, and France, for example, mounting multidimensional pressures on the agricultural sector have led to social tensions and protests of the farming community in recent years. In 2019–2022, farmers' protests against new regulatory measures escalated to such an extent that the then-former German chancellor Angela Merkel initiated the Commission on the Future of Agriculture (Münch and Schulze Pals 2025) in Germany (pl. Compare article on ZKL, page XY, Chapter 30). The intensity of these protests reflects the deep frustration, existential concerns, and perceived powerlessness of farmers, which have accumulated over years of dynamic and often unpredictable pressures and burdens (Springmann et al. 2018). These tensions have been further exacerbated by unmet expectations toward policymakers, society, and the retail sector, as well as by persistent and difficult-to-resolve goal conflicts within the transformation process of restructuring livestock farming. This chapter explores the key endogenous and exogenous factors and their interdependencies that contribute to the complexity and wickedness of this transformation.

1.1 From Quantity-Driven Demand to Quality-Driven Demand Patterns

Following World War II, Germany and many other European nations faced severe food shortages, prompting a fundamental shift toward intensified agricultural production aimed at ensuring food security and economic stability (FAO 2022). The Green Revolution of the 1950s and 1960s led to major advancements in animal husbandry, genetics, feed efficiency, veterinary medicine, and mechanization, fostering highly specialized and intensive livestock systems (Pingali 2012).

In several regions, such as in northwestern Lower Saxony, agricultural innovation clusters emerged, where livestock production became highly integrated with upstream and downstream industries. This consolidated value chain drove employment and economic growth while simultaneously increasing production efficiency (Fink et al. 2022). Germany, Poland, Spain, the Netherlands, Italy, and Denmark subsequently became leading global meat exporters, supported by high demand in international markets (FAOSTAT 2025). Additionally, efficiency improvements in intensive livestock systems led to significant reductions in production costs, making meat and animal products more affordable and thus increasing consumer demand (Springmann et al. 2018).

As livestock farming has expanded in scale and intensity, its environmental impact has become increasingly evident. Intensive livestock production is now recognized as a major driver of greenhouse gas emissions, biodiversity loss, and resource depletion on a global scale (Steinfeld et al. 2006; Willett et al. 2019). The concept of planetary boundaries (Rockström et al. 2009; Richardson et al. 2023) underscores this impact, identifying livestock farming as a key contributor to ecosystem degradation, particularly through biodiversity loss, nitrogen cycle disruption, and climate change (Steinfeld et al. 2006; Willett et al. 2019).

As a result, both societal and scientific perspectives on agriculture and food production have shifted significantly. The intensive livestock model—once hailed as a success story of productivity and efficiency—is increasingly scrutinized for its ecological footprint and ethical implications (Poore and Nemecek 2018). This paradigm shift has led to mounting public and regulatory pressure to reduce emissions, improve animal welfare standards, and transition toward more sustainable food systems.

Moreover, this transition is not solely a technical or economic matter but reflects a deeper ethical transformation. The shift from an anthropocentric perspective (focusing on human needs) to pathocentric and biocentric approaches (acknowledging the intrinsic value of animals and ecosystems) (Selter and Salloch 2023) has profound implications for livestock production. This evolving ethical framework necessitates:

- Recognizing and respecting the intrinsic needs of animals, ensuring their welfare throughout housing, transport, and slaughtering processes (EFSA 2023).
- Adopting regenerative agricultural practices, which maintain ecosystem functions and ensure that resource extraction remains within the limits of natural replenishment rates (FAO (2021b)).—compare article Windisch and Roguet.

The shift toward a sustainable and resource-efficient agricultural model is now widely recognized as a prerequisite for future food security and climate resilience (EEA 2024). To align this shift with current regulatory, societal, and environmental expectations, future agricultural and environmental systems (AES) must integrate the requirements listed in Fig. 7.1.

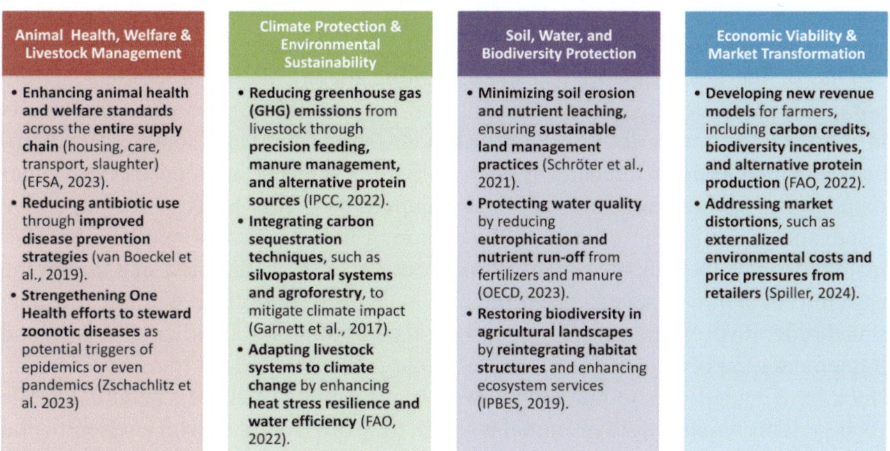

Fig. 7.1 Key requirements for future AES © Grabkowsky

2 Challenges in Achieving Sustainability Transformation

Meeting these fundamental sustainability requirements requires not only technical innovations and financial incentives but also systemic changes in governance structures, economic frameworks, and societal engagement (El Bilali et al. 2019). Sustainability transitions involve multiple interdependent factors that dynamically influence one another. Grasping the complexity of transformation requires examining both external drivers of change and internal, often overlooked, endogenous factors (El Bilali et al. 2019).

2.1 Exogenous Factors: Policy, Market, and Environmental Drivers

Exogenous factors shape the structural, regulatory, economic, and environmental context in which transformation processes unfold. Table 7.1 identifies key external drivers that affect livestock farming at multiple levels — ranging from overarching agrifood policy and global trade conditions to societal expectations, ecosystem dynamics, and market pressures. Selected developments and inherent trade-offs are included to illustrate the complexity and interdependencies involved in the sustainability transformation.

Among these external drivers, political and legal frameworks play a particularly central role. The sustainability requirements outlined in Table. 7.1 — especially those concerning animal welfare, climate protection, and biodiversity, are translated into agricultural policy goals and subsequently embedded in binding legal regulations. At the European level, the Common Agricultural Policy (CAP) and the Green Deal, including its Farm-to-Fork Strategy, provide the overarching regulatory framework. These instruments define EU-wide sustainability targets and help steer future investments via mechanisms such as the EU taxonomy.

One of the key challenges in implementing these policies in Germany is the politically motivated tightening of regulations beyond EU requirements. This results in shorter transition periods or more ambitious targets compared to other EU member states. In Germany's federal system, individual states often impose even stricter regulations, further exceeding national requirements. This leads to a competitive disadvantage for businesses and farms operating in international markets. Another major barrier is the lack of regulatory coherence in political decision-making for the agricultural and food sector. Many policies have been formulated without adequate stakeholder involvement, leading to technical contradictions, conflict of targets, and implementation obstacles. A striking example is the demand for transitioning livestock housing systems to open barns with higher animal welfare standards. While such systems would improve animal welfare, they often conflict with environmental regulations, particularly with the German Federal Immission Control Act (BImSchG) and the Technical Instructions on Air Quality Control (TA Luft). As a result, many municipal authorities deny building permits, causing uncertainty for farmers and leading to an investment vacuum (BMEL 2024).

Table 7.1 Exogenous drivers of transformation in agriculture and the food system

Exogenous factor	Examples of legal/policy frameworks or trends	Implications for transformation
Political and legal frameworks	EU Green Deal, Farm to Fork Strategy, CAP reform, Sustainable Use Regulation (SUR), CSRD	Shapes compliance obligations and long-term planning
Public funding and subsidies	CAP subsidies, agri-environmental schemes, investment support programs	Influences financial viability and adoption of practices
Market mechanisms	Price volatility, demand fluctuations, cost–price squeeze	Drives economic pressure or incentives to change systems
Global trade conditions	WTO rules, EU trade agreements, market liberalization	Affects competitiveness and exposure to global shocks
Geopolitical instability and macroeconomic trends	War, inflation, global supply chain disruptions	Creates uncertainty and need for robust strategies
Energy prices and availability	Volatile energy markets, fossil fuel phase-out policies	Modifies input costs and investment decisions
Technological innovation	Smart farming technologies, genetic innovation, alternative proteins	Enables efficiency but requires adaptation and capital
Digital transformation and data governance	EU Data Act, Data Governance Act, broadband expansion strategies, digital platform regulations	Determines access to digital tools and data ecosystems; raises challenges of interoperability, ownership, and infrastructure gaps
Broadband access and digital infrastructure	National broadband strategies, EU Digital Decade targets, regional infrastructure programs	Enables or constrains digitalization, access to advisory services, precision farming and smart technologies
Climate change and extreme weather events	Increased droughts, floods, heat stress affecting animals and feed	Requires adaptation and resilience strategies
Ecosystem integrity and resource availability	Water scarcity, soil degradation, biodiversity loss	Limits carrying capacity and long-term system viability
Livestock disease dynamics and zoonoses	ASF, BVD, avian influenza, antimicrobial resistance (AMR)	Increases risk and need for health-focused management
Public health regulations and biosecurity policies	EU Animal Health Law (AHL), national contingency plans	Necessitates coordinated prevention and response systems
Labor availability and rural demographic change	Aging rural populations, shortage of skilled labor in agriculture	Limits operational capacity, affects business continuity
Consumer behavior and preferences	Shift toward plant-based diets, animal welfare concerns	Pushes demand-driven shifts in production systems
Public opinion and societal values	Debates on sustainability, transparency, animal ethics	Shapes legitimacy and social acceptance of practices
NGO influence (e.g., environmental, animal welfare)	Campaigns and certification pressure (e.g., organic, animal welfare labels)	Drives voluntary standards and reputational pressure

Another fundamental challenge is the misalignment between political decision-making cycles and long-term agricultural planning. Political commitments often remain unfulfilled because policy frameworks operate within election cycles, whereas agricultural transformations require decades-long strategic planning. One example is the delayed implementation of the recommendations of the Borchert Commission (Münch & Schulze Pals 2025) and the Future Commission on Agriculture (Zukunftskommission Landwirtschaft (ZKL) 2021). Despite broad expert consensus, many recommendations have not been translated into binding policies. The dissolution of the Borchert Commission in 2023, without implementing key reforms, underscores the fragility of long-term agricultural transformation efforts (Hemme et al. 2023) (compare to article Grabkowsky & Zubek, Chapter 30).

Beyond regulatory and governance-related challenges, market-driven factors significantly hinder the sustainability transition in livestock farming. A critical issue is the Consumer-Citizen Gap: Consumers express strong preferences for sustainable, ethically produced food with high animal welfare standards. However, actual purchasing behavior does not reflect these values, as many consumers prioritize low-cost options due to inflation, rising energy costs, and economic insecurity (Strässner & Wirth 2024). This creates a dilemma for livestock producers: They must comply with increasingly strict sustain-ability and welfare standards. At the same time, they must remain economically viable, ensuring affordability for consumers while covering their rising costs.

The food retail sector plays a dominant role in shaping price structures and supply chain dynamics in the livestock industry: Supermarkets and discount chains wield significant influence over pricing strategies and procurement policies. Private-label products allow retailers to switch suppliers easily, increasing competition among producers and exerting downward pressure on farm-gate prices. In many EU countries, particularly Germany, aggressive price competition leads to low meat prices being used as marketing tools, resulting in farm-gate prices below production costs (FAO 2018; FAO 2021a, b). Despite repeated discussions, no mandatory and transparent labelling system for both fresh and processed meat products has been implemented at the EU level. Such a system would provide consumers with informed choices, potentially supporting domestic and EU-compliant production.

The discrepancy between societal expectations and economic realities has led to increasing dissatisfaction among both producers and consumers: Farmers face shrinking subsidies, increasing production costs, and growing administrative burdens. Consumers remain frustrated with rising food prices, even though they demand higher sustainability and welfare standards. Although livestock farming plays a system-critical role in food security, farmers often receive public recognition only during crises. The COVID-19 pandemic and the Ukraine war-induced energy crisis highlighted Europe's dependence on agricultural imports and the importance of resilient food systems (compare to article von Cramon, Chapter 5). However, ongoing challenges such as inflation, rising feed and energy costs, and labor shortages continue to strain the sector.

The combination of regulatory pressures, economic constraints, and uncertainty has triggered major structural shifts in the livestock sector. One of the most striking changes has occurred in pig farming in Germany. Between 2013 and 2023, the number of pig farms declined by 44%, while the total swine population dropped by 22% (Destatis 2023). Many small- and medium-sized farms have exited the industry, with larger farms consolidating to maintain economies of scale. Despite ambitious sustainability policies, the transition process in livestock production faces serious systemic barriers and has led to unintended consequences of current policies. The combination of fragmented regulatory frameworks, economic constraints, and the misalignment of political and agricultural planning cycles has triggered disruptive effects, disproportionately affecting small- and medium-sized farms. This underlines the urgent need for more coherent, participatory, and future-oriented governance structures that align regulatory instruments with economic realities and long-term transformation goals.

2.2 Endogenous Factors: Intra-Systemic Constraints

Endogenous factors originate within the agricultural and food system and reflect internal dynamics at the farm level. They shape the capacity, direction, and responsiveness of sustainability transformations and influence how individual actors adapt to change. They span managerial, economic, and socio-cultural dimensions, and their interplay with external drivers ultimately determines the pathway and pace of transformation. Table 7.2 provides an overview of the central endogenous factors relevant to livestock systems.

A key component of these aspects is the **qualification, skills and strategic mindset of farmers**. A high level of education and continuous training is necessary for farmers to adopt sustainable practices, manage changing regulations, and integrate innovative technologies (FAO 2023a, b). Lack of advisory services and training hinders the transformation process, especially among small- and medium-sized farms (OECD 2023). Additionally, **farm culture and business philosophy** play a central role, as they form the foundation for attitudes toward sustainability, ethics, and responsibility toward the environment and animals. The willingness of farm operators to transition toward sustainable systems depends on values, risk perceptions, and long-term business strategies (El Bilali et al. 2019). Moreover, individual decision-making styles and attitudes toward uncertainty can significantly influence whether and when transformation efforts are initiated. Farm managers with a more risk-averse profile may postpone investments, even in the face of external incentives or regulatory requirements. Traditional agricultural practices and entrenched cultural norms may reduce openness to change and make it harder for farmers to embrace new values or strategies (Lang et al. 2012). For example, business models in industrialized livestock farming are built on efficiency and cost minimization, making sustainability investments financially challenging (Grethe and Zöllmer 2024). **Farm size, structure, and organizational form** define the economic conditions that determine the feasibility of transformation efforts. Secure land tenure is

Table 7.2 Endogenous factors shaping farm-level transformation in agriculture

Endogenous factor	Description/implications
Farmer qualification and knowledge base	Education level, practical experience, and training shape decision-making and openness to transformation
Utilization of education and advisory services	Effective use of training, extension, and advisory services influences knowledge transfer, adoption of innovation, and strategic farm development
Farmer innovativeness and risk perception	Individual willingness to innovate and risk tolerance influence the adoption of transformative practices and technologies
Farm culture and sustainability philosophy	Values, traditions, and attitudes toward sustainability influence acceptance of new practices
Farm management and biosecurity practices	Planning, hygiene, and response strategies affect animal health and operational security
Housing systems and animal welfare standards	Design and condition of housing systems impact animal welfare, emissions, and system resilience
Farm size, structure, and organizational form	Farm structure affects risk tolerance, capacity for investment, and innovation potential
Land tenure security	Secure land ownership or long-term tenancy increases the likelihood of investing in sustainable and long-term transformations
Economic aspects: input costs, profitability, and market pressure	Financial health determines the ability to invest in sustainable practices under market pressure
Sustainable finance and EU taxonomy alignment	Eligibility for green finance and reporting obligations shape investment and strategy decisions
Diversification of farm income (e.g., alternative proteins, renewable energy)	Additional income sources increase resilience and support system transitions
Access to technology and on-farm digitalization	Digital readiness influences efficiency, monitoring capacity, and precision resource use
Environmental and climate-oriented concepts (e.g., soil, water, biodiversity)	Implementation of targeted environmental practices supports adaptation and sustainability
Circular economy approaches (e.g., nutrient recycling, carbon sequestration, agroecology)	Integration of circular practices fosters resource efficiency and long-term ecological viability
Management of externalities (e.g., nutrient surplus, emissions, social equity)	Ability to mitigate and internalize environmental and social impacts is key to legitimacy and sustainability

another important enabling condition for long-term investment decisions, particularly when structural changes such as barn conversions or land-based circular systems are required. Larger, vertically integrated farms have greater financial resilience and technological access to comply with sustainability regulations (FAO 2023a, b). Small- and medium-sized farms face higher transition costs and administrative burdens (BMEL 2024). Volatility in feed, energy, and input prices directly impacts the farms' viability and investment capacity in sustainability measures (OECD 2023;

FAO 2023a, b)—compare to article von Cramon-Taubadel, Chapter 5. Sustainability certification provides incentives but requires high compliance costs (FAO 2022). Access to financing and subsidies affects the rate of adoption of sustainable practices (BMEL 2024). The **sustainable finance concept** of the EU taxonomy determines the access to loans, and the degree of sustainability of a business determines the terms and conditions of the loan. At the same time, **effective farm management and high biosecurity standards** are essential to minimize disease risks, to enhance the farms' resilience (EFSA 2023) and to provide food safety from farm to fork (compare to article Meemken and Blaha, Chapter 10, and to article Conraths, Chapter 11). Moreover, **farms must balance manure management, emission reductions, and local environmental limits** (Wang-Erlandsson et al. 2022). In the context of the disruptive effects described in Chap. 2.2, diversification of agricultural income sources, such as integrating alternative protein sources and renewable energy, contributes to the economic stability of the agri-food system and individual farms. Here, agroforestry, biogas, or carbon sequestration projects offer additional revenue streams (PBL 2022). These practices contribute not only to climate mitigation but also to soil health and the overall resilience of farming systems. Farmers investing in alternative proteins (e.g., insect farming, plant-based substitutes) currently face high technological and market-entry barriers (Poore and Nemecek 2018, Chapters 19 and 27). Reliable broadband infrastructure and access to technological innovations play a crucial role in enabling farms to use precision livestock farming, **animal-centered, automated monitoring** to reduce resource use and improve welfare and to allow **multidirectional data analysis across the entire value chain** while improving resource efficiency (compare to article Müller & Hertzberg and Grethe and Zöllmer 2024). Ultimately, a farm's capacity to adopt new technologies and practices depends not only on financial resources but also on available labor, technical competence, and its integration in social and advisory networks that facilitate learning and experimentation.

Soil, water, and biodiversity-conserving practices are the foundation of **sustainable farm concepts and** are crucial for long-term productivity (Rockström et al. 2009). A **circular nutrient management system** aims to eliminate the need for mineral fertilizers while optimizing the use of organic fertilizers as efficiently and demand-driven as possible. By returning nutrients to natural cycles, the need for external fertilizers is reduced, minimizing external effects, improving cost efficiency, and thereby enhancing the long-term profitability of livestock farming (compare to Roguet, Chapter 17). Ultimately, this meets the demand for more environmentally friendly and ethically acceptable agricultural practices.

Additionally, promoting biodiversity can enhance climate resilience, particularly in grassland ecosystems. Carbon sequestration on farms presents several opportunities within the context of climate protection and sustainable agriculture. Agroforestry, cover cropping, and reduced tillage can contribute to storing carbon in the soil and reducing CO_2 accumulation in the atmosphere, aligning with global efforts to mitigate climate change and promote sustainable farming systems (Table 7.2). These endogenous dynamics highlight the importance of aligning internal farm capacities with external policy and market drivers to accelerate transformation.

2.3 Mechanisms of Exogenous and Endogenous Factors in the Sustainability Transformation of Livestock Farming

Figure 7.2 illustrates the systemic interplay between the exogenous and endogenous factors introduced in Sects. 2.1 and 2.2, highlighting dependencies and interrelationships through a cause-and-effect structure. Designed as an Ishikawa diagram, it assigns these factors to categories and shows how they jointly shape the transformation context. The figure visualises that the sustainability transformation of livestock production constitutes a typical "wicked" problem - characterized by complexity, feedback loops, and goal conflicts - and thus cannot be addressed through isolated or linear interventions (Dentoni et al. 2018). A systemic perspective, as described by Meadows 2008, offers an analytical lens to understand such challenges as interconnected systems rather than a sum of individual components. In this context, the interplay between endogenous and exogenous drivers gives rise to emergent properties that can only be adequately understood and managed through transdisciplinary research and collaborative governance.

Applying systems thinking to the transformation of livestock farming implies recognizing the interconnected components of the agrifood system. This includes not only farm-level factors but also broader influences, such as:

- Political and regulatory frameworks (e.g., EU agricultural policies, national legislation, subsidy schemes).
- Societal expectations and shifts in consumer behavior (BMEL 2024).
- Global and national market mechanisms, including supply chains, trade agreements, and commodity price fluctuations.
- Technological availability and infrastructure constraints.
- Ecological shifts driven by climate change and resource depletion.

A systemic approach helps tackle the "wicked" problem by capturing the complexity and dynamics of transformation processes in livestock farming, identifying path dependencies, and recognizing the potential consequences of policy decisions across the entire system. This perspective enables a more holistic understanding and facilitates the development of targeted, evidence-based strategies and holistic concepts for sustainability transformation. In the following chapters of the book, these several strategies will be analyzed and discussed.

3 Conclusions

The sustainability transformation of livestock systems cannot be achieved through isolated technical fixes or sector-specific regulations. It requires systemic, cross-sectoral approaches that address structural inconsistencies, empower stakeholders, and reconcile ethical imperatives with ecological and economic realities. Recognising the transformation as a wicked problem entails embracing complexity, fostering inclusive governance, and ensuring that long-term policy commitments

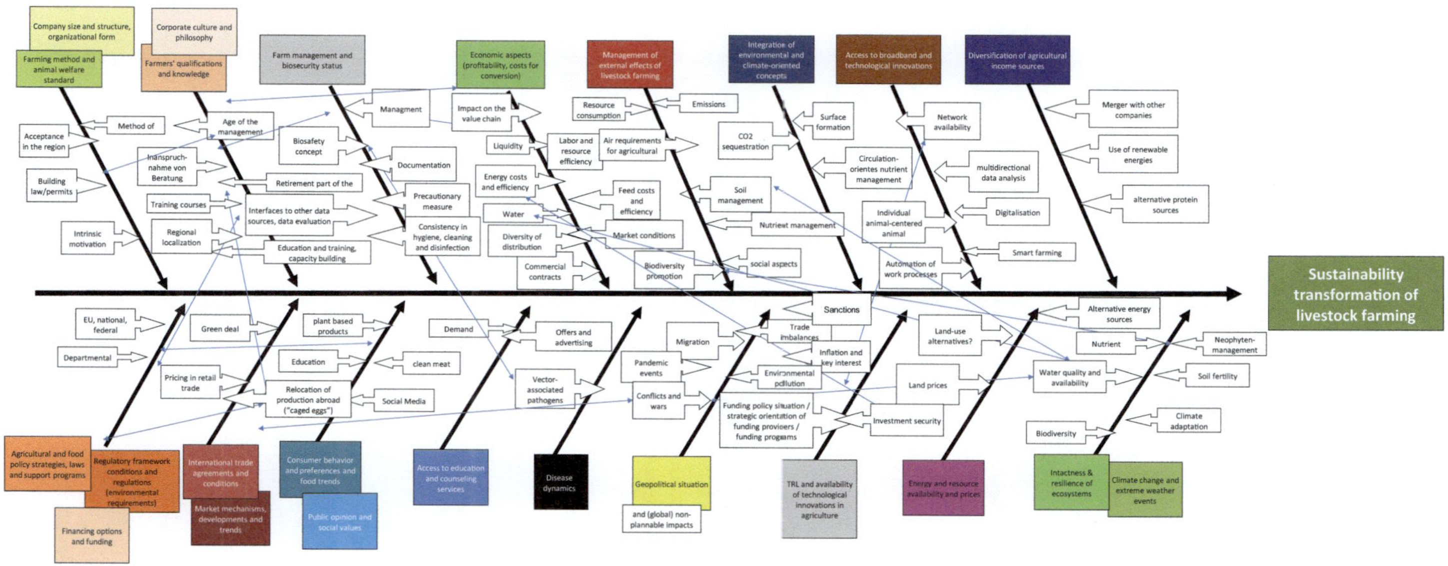

Fig. 7.2 Cause-and-effect diagram (Ishikawa diagram) illustrating the systemic interdependencies of exogenous and endogenous factors in the sustainability transformation of livestock farming (© Grabkowsky)

transcend short-term political cycles. To navigate this transformation, science, policy, and practice must collaborate in developing adaptive frameworks that align societal values with feasible implementation paths—towards a livestock sector that is climateresilient, ethically grounded, and economically viable.

References

BMEL (2024) Gutes Essen für Deutschland—Ernährungsstrategie der Bundesregierung.. Available at: https://www.bmel.de/DE/themen/ernaehrung/ernaehrungsstrategie/ernaehrungsstrategie.html. Accessed 27 Feb 2025

De Rosa M et al (2024) Circular economy practices in livestock farming. J Agric Sustain 29(2):101–119. ISBN: 978-953-7813-93-2

Dentoni D, Bitzer V, Schouten G (2018) Harnessing wicked problems in multi-stakeholder partnerships. J Bus Ethics 150(2):333–356. https://doi.org/10.1007/s10551-018-3858-6

Destatis (2023) Schweinebestand von Mai bis November 2023 leicht gestiegen. Available at: https://www.destatis.de/DE/Presse/Pressemitteilungen/2023/12/PD23_491_413.html. Accessed 8 Mar 2024

European Environment Agency (EEA) (2024). Greenhouse gas emissions from agriculture – EEA indicator assessment 2024. EEA, Copenhagen. Online verfügbar unter: https://www.eea.europa.eu/en/analysis/indicators/greenhouse-gas-emissions-from-agriculture

EFSA (2023) Animal welfare in the EU: current state and future challenges. https://doi.org/10.2903/sp.efsa.2023.EN-8475

El Bilali H, Callenius C, Strassner C, Probst L (2019) Food and nutrition security and sustainability transitions in food systems. Food Energy Secur 8:e00154. https://doi.org/10.1002/fes3.154

FAO (2018). Shaping the Future of Livestock. Available at: https://openknowledge.fao.org/items/050a596e-c861-4405-ba8f-bc5246b6a2ac

FAO (2021a). Nature-Based Solutions in agriculture. The case and pathway for adoption. Available at: https://openknowledge.fao.org/server/api/core/bitstreams/9c6d587e-1532-4252-852f-d2657634a66a/content

FAO (2021b). Nature-Based Solutions in agriculture. Sustainable management and conversation of land, water and biodiversity. Available at: https://openknowledge.fao.org/server/api/core/bitstreams/c9aa3745-ab2b-41bd-a3bac8ea3ef4582b/content

FAO (2022) The future of livestock in a changing climate. Available at: https://www.fao.org/documents/card/en/c/CB9543EN

FAO (2023a) The future of livestock production and climate mitigation. Available at: https://www.fao.org/documents/card/en/c/CC1981EN

FAO (2023b) The state of food and agriculture: nature-based solutions for Agri-food systems. Available at: https://www.fao.org/documents/card/en/c/CB9479EN

FAOSTAT (2025) FAOSTAT database. https://www.fao.org/faostat/en/#data-Data/Crops and livestock products. Data access: March 30, 2025

Fink A, Grabkowsky B, Hortmann-Scholten A, Lagemann A, Ohse S, Wedemeier J & Wolf A (2022) Transformationsszenarien der Agrar- und Ernährungswirtschaft in Nord-West-Niedersachsen (TRAIN), HWWI Policy Papers 136, Hamburg Institute of International Economics (HWWI). Available at: https://www.univechta.de/fileadmin/user_upload/Transformationsforschung_agrar/Projekte/HWWI_Policy_Paper_136.pdf

Grethe H, Zöllmer U (2024) Nachhaltige Agrarpolitik in der Transformation. In: Livestock production in transition. Springer

Hemme T, Deeken E, Eckstein J, Oltmer T (2023) Economic dynamics in the transformation of livestock farming: structural shifts in the European pork industry. J Agric Econ 74(3):510–533

Lang DJ, Wiek A, Bergmann M, Stauffacher M, Martens P, Moll P, Swilling M, Thomas CJ (2012) Transdisciplinary research in sustainability science: practice, principles, and challenges. Sustain Sci 7(S1):25–43. https://doi.org/10.1007/s11625-011-0149-x

Meadows DH (2008) Thinking in systems: a primer. Chelsea Green Publishing. ISBN: 978-1-84407-726-7 (pb), ISBN: 978-1-84407-725-0 (hb)
Méndez VE, Bacon CM, Cohen R, Gliessman SR (2017) Agroecology: a transdisciplinary, participatory and action-oriented approach. Agroecol Sustain Food Syst 41(9–10):1191–1193. https://doi.org/10.1080/21683565.2017.1358962
Midgley G (2000) Systemic intervention: philosophy, methodology, and practice. Springer, New York. ISBN: 0306464888
Münch, Rainer and Ludger Schulze Pals (2025): Brücken bauen: Für ein besseres Miteinander von Bürgern und Bauern. Der Erfolg der Zukunftskommission Landwirtschaft – und deren Scheitern? Analysen und Perspektiven für eine bessere Agrar-Politik. Landwirtschaftsvlg Münster. ISBN-13978-3784357997
OECD (2023) Policies for the future of farming and food in the Netherlands. OECD Agriculture and Food Policy Reviews, OECD Publishing, Paris. https://doi.org/10.1787/bb16dea4-e
Pingali P (2012) Green revolution: impacts, limits, and the path ahead. Proc Natl Acad Sci 109(31):12,302–12,308. https://doi.org/10.1073/pnas.0912953109
Poore J, Nemecek T (2018) Reducing food's environmental impacts through producers and consumers. Science 360(6392):987–992. https://doi.org/10.1126/science.aaq0216
Raworth K (2017) Doughnut economics: seven ways to think like a twenty-first-century economist. Chelsea Green Publishing. https://doi.org/10.33568/rbs.2409
Richardson K et al (2023) Earth beyond six of nine planetary boundaries. Sci Adv 9:eadh2458. https://doi.org/10.1126/sciadv.adh2458
Rittel HWJ, Webber MM (1973) Dilemmas in a general theory of planning. Policy Sci 4(2):155–169. Available at: https://www.andreasaltelli.eu/file/repository/Rittel_OCR.pdf
Rockström J, Steffen W, Noone K et al (2009) A safe operating space for humanity. Nature 461(7263):472–475. https://doi.org/10.1038/461472a
Selter F, Salloch S (2023) Whose health and which health? Two theoretical flaws in the one health paradigm. Bioethics 37(7):674–682. https://doi.org/10.1111/bioe.13192
Springmann M, Mason-D'Croz D, Robinson S, Wiebe K, Bodirsky BL, Herrero M (2018) Options for keeping the food system within environmental limits. Nature 562(7728):519–525. https://doi.org/10.1038/s41586-018-0594-0
Steinfeld H, Gerber P, Wassenaar T, Castel V, Rosales M, de Haan C (2006) Livestock's long shadow: environmental issues and options. FAO, Rome. ISBN: 978-92-5-105571-7
Strässner, Anna-Maria und Werner Wirth (2024): Shades and shifts in flexitarian and meat-oriented consumer profiles in a German panel study. In: Appetite, Volume 197. https://doi.org/10.1016/j.appet.2024.107298
Tilman D, Clark M, Williams DR, Kimmel K, Polasky S, Packer C (2017) Future threats to biodiversity and pathways to their prevention. Nature 546(7656):73–81. https://doi.org/10.1038/nature22900
Wang-Erlandsson L, Tobian A, van der Ent RJ et al (2022) A planetary boundary for green water. Nat Rev Earth Environ 3:380–392. https://doi.org/10.1038/s43017-022-00287-8
WEF (2025) Global risks report 2025. World Economic Forum, Geneva. Available at: https://reports.weforum.org/docs/WEF_Global_Risks_Report_2025.pdf
Willett W, Rockström J, Loken B, Springmann M, Lang T, Vermeulen S et al (2019) Food in the Anthropocene: the EAT–lancet commission on healthy diets from sustainable food systems. Lancet 393(10170):447–492. https://doi.org/10.1016/S0140-6736(18)31788-4
Zschachlitz T, Kümpfel R, Niemann H, Straff W (2023) Die Bedeutung der Konzepte One Health und Planetary Health für die Umweltmedizin im 21. Jahrhundert [The implications of the concepts One Health and Planetary Health for the environmental medicine of the twenty-first century]. Bundesgesundheitsblatt Gesundheitsforschung Gesundheitsschutz. 66(6):669–676. German. https://doi.org/10.1007/s00103-023-03711-6. Epub 2023 Jun 2. PMID: 37266666; PMCID: PMC10235829
Zukunftskommission Landwirtschaft (ZKL) (2021) Zukunft Landwirtschaft. Eine gesamtgesellschaftliche Aufgabe. *Empfehlungen der Zukunftskommission Landwirtschaft.* Bundesministerium für Ernährung und Landwirtschaft. PDF-Download www.bmel.de/goto?id=89464. Accessed 27 Feb 2025

Improving Animal Welfare and Animal Health

8

Thomas Blaha

1 Animal Welfare

1.1 The Need for Animal Welfare Improvements

The relationship between humans and animals has undergone significant changes in recent decades. Animals are no longer seen merely as objects to be used at the discretion of their owners; instead, they are increasingly recognized as sentient beings of intrinsic value. Consequently, individuals who own or care for animals bear a responsibility to guarantee them a decent quality of life (Kunzmann 2013).

Various frameworks have been developed to define and assess animal welfare. Fraser (2008) identifies three key welfare criteria: "health", "natural living", and "affective state". The EU's Welfare Quality® project (FOOD-CT-2004-506508) outlines four welfare domains: good feeding, good housing, good health, and appropriate behaviour (Temple et al. 2011). Mellor (2017) subsequently added a fifth domain, addressing animals' mental states. Globally, the World Organisation for Animal Health (WOAH = OIE) has defined animal welfare standards in its Terrestrial Animal Health Code, which are widely accepted.

A foundational concept for describing decent animal care remains the "Five Freedoms", originally proposed in the 1960s and formalized in the United Kingdom (Brambell 1965). These freedoms, reflecting the European perspective on animal welfare, include:

- Freedom from hunger and thirst: Ensuring access to fresh water and a diet promoting health and vigour

T. Blaha (✉)
Former Head of the Field Station for Epidemiology (Bakum), University of Veterinary Medicine Hannover, Hannover, Germany
e-mail: thomas.blaha.ir@tiho-hannover.de

B. Grabkowsky, T. Blaha (eds.), *Intensive Livestock Production in Transition*,
https://doi.org/10.1007/978-3-031-97872-2_8

- Freedom from discomfort: Providing a suitable environment including shelter and a comfortable resting area
- Freedom from pain, injury, and disease: Through prevention or rapid diagnosis and treatment
- Freedom to express normal behaviour: Offering adequate space, proper facilities, and the companionship of the animal's kind
- Freedom from fear and distress: By maintaining conditions and treatment that prevent mental suffering

While these freedoms form a valuable benchmark, the use and management of animals for human purposes often involve compromises. Animals may experience stress, restricted natural behaviours, and occasional pain or discomfort. As our understanding of human responsibility towards animals grows, a moral imperative emerges while striving to uphold the "Five Freedoms", any infringements should be minor, justified, and carefully managed to minimize avoidable suffering (Fraser 2008). This raises the question: What changes are necessary to ensure a decent life for animals used in food production?

In recent decades, the basic physical needs of farm animals—such as nutritious and safe feeding, access to clean water, and protective shelter—have largely been met. After all, only well-nourished and healthy animals produce meat, milk, eggs, and other products efficiently. However, a narrow focus on productivity and labour reduction has long overshadowed the recognition of animals' psychological, social, and emotional needs. These encompass the ability to exhibit natural behaviours, maintain stable social structures, and seek pleasant experiences while avoiding negative ones. Today, the concept of animal welfare has expanded beyond merely protecting animals from pain, injury, and suffering to encompass a broader understanding of their overall well-being.

In the 1970s, 1980s, and 1990s, many husbandry systems were designed primarily to maximize production and minimize human labour. This led to severe overcrowding, confinement, and barren environments for animals. Practices such as housing pigs and cattle on fully slatted floors, docking the tails of pigs and sheep, and trimming the beaks of poultry restricted natural behaviours and exacerbated stress. Animals in these systems often had no opportunity to express species-specific behaviours or escape aggression from pen-mates.

Today, the public discourse increasingly focuses on animal welfare. Criticism of the animals' living conditions in intensive husbandry systems—characterized by chronic suffering and severe behavioural constraints—has become a major factor in the growing societal rejection of current food production practices.

1.2 Ways Leading to More Animal Welfare for Food Animals

It is now more or less a societal consensus in high-income countries that all animals for food production should be kept *in circumstances that meet not only their physical but also their mental, societal, and emotional needs*, and should be *cared for so that they are provided with the utmost possible health and comfort*. To achieve this ambitious goal, the following two areas of room for improvement must be taken into consideration and addressed with a different set of measures:

The exclusively animal-performance-oriented housing systems that were developed in the 1970s and 1980s must be reconstructed into housing systems that provide the animals with conditions that improve the quality of their life by enabling them to meet not only their physical but also their mental, societal, and emotional needs. Thus, the predominantly current intensive husbandry systems (total confinement, fully slatted floors, no litter, no access to fresh air, lack of enrichment, unstructured pens, tethering, etc.) must be systematically reconstructed to be able to provide the animals at least the following features:

1. Animal housings for fattening must be designed and enriched to allow animals to move freely between different climate zones (resting areas with sufficient bedding, activity zones with separate feeding and defecating areas, and zones with access to fresh air—ideally with an outdoor area).
2. Gilts and empty and pregnant sows must be kept in groups, and farrowing sows must be kept in so-called free-farrowing pens without confinement during and after the birth of the piglets.
3. Cows and bulls must not be tethered. Cows must get daily access to a pasture or at least to an area where the animals can walk.
4. Poultry and turkeys must be kept in flock sizes that allow the individual animal to express its natural behaviour and its social needs. The animals should have access to fresh air—ideally to outdoor areas. Enrichment elements must allow for species-specific activities such as scratching in dry litter and climbing to higher sleeping places in the evening.
5. Any non-medically justified amputations like tail docking in pigs and beak trimming in poultry must be prohibited, for which the provision of animal-friendly living conditions, which offer the animals opportunities to meet their species-specific needs, is a precondition.
6. Reducing significantly the occurrence of animals suffering due to diseases by increasing the intensity of the daily care of every single animal and of preventive veterinary measures such as external and internal biosecurity, vaccinations, and early disease detection.

1.3 The Example of Pigtail Docking

The transformation of small-scale backyard pig farming on family farms—where single sows lived with their offspring and often had free access to outdoor areas—into intensive high-throughput pork production in the late 1960s and 1970s brought significant changes. Pigs were confined to barren pens with fully slatted floors and no litter, leading to tail biting among pen-mates. This behaviour is triggered by boredom due to the lack of environmental enrichment and by frustration caused by overcrowding, the absence of rooting material, inappropriate temperatures, and poor air quality. In some cases, these stressors triggered aggressive behaviour towards weaker animals within the group.

Tail biting remains a frequent issue in conventional pig farming systems with unstructured pens lacking litter and enrichment. To address the resulting pain and suffering, caretakers often resort to time-consuming interventions, such as separating injured animals, distracting aggressive individuals with chains or balls, or adding animal protein to the diet. However, these measures are labour-intensive and costly. To minimize tail biting, farmers in the early days of intensive confinement farming adopted tail docking—the removal of a portion of the pigs' tails. Shorter tails were bitten less frequently, leading to the belief that tail docking improved animal welfare by reducing pain and the incidence of spinal abscesses.

With society's evolving perception of animals as sentient beings deserving of a decent life, the practice of tail docking has faced growing criticism. It is now understood that tail docking not only causes pain but also compromises the animal's physical integrity and removes an important tool for pig communication. The pig's tail, often referred to as the "talking tail" (Zonderland et al. 2009), plays a vital role in social interactions among conspecifics and should be preserved.

Attempts to discontinue tail docking in conventional systems have often failed, as pigs in barren pens without sufficient enrichment inevitably revert to tail biting (D'Eath et al. 2016). While EU animal welfare laws prohibit amputations without medical justification, exemptions are granted when farmers can demonstrate that alternative measures—such as environmental enrichment or better management—are insufficient to prevent tail biting (De Briyne et al. 2018).

Evidence increasingly shows that transforming pig housing systems can effectively prevent tail biting without the need for tail docking. Modern systems with structured pens divided into three climate zones offer significant improvements:

1. *A resting area* with litter and adjustable coverings to regulate temperature
2. *A feeding area* with a solid concrete floor where small portions of feed are scattered throughout the day, mimicking the natural foraging behaviour of wild boars
3. *A slatted floor area* with access to fresh air or outdoor spaces

In Germany, a growing number of pig herds are raised in housing systems based on this model. These pigs are part of closed-value chains that produce "animal-welfare meat products" sold at premium prices to compensate farmers for their higher costs. In such a system, no tail biting has been observed, and all pigs retain intact, healthy tails. This demonstrates that housing systems designed to meet pigs' physical, social, and behavioural needs can prevent abnormal behaviours and associated suffering.

Despite a growing societal consensus over the last 20 years that animals deserve a decent life, and despite the knowledge about what the animals need, little change in the quality of life of our food animals has occurred until today: most animals are still kept in confinement without sufficient enrichment (many of them on fully slatted floors without any littering), they still lack choices of climatic zones, they still have no access to open areas, and little opportunities to express their behavioural repertoires; due to the missing optimization of the housing conditions, chicken beaks are still trimmed; pigtails are still docked; and by far, not all herds and flocks are healthy and get sufficient and competent care by the humans that have the responsibility for them.

The reasons for the discrepancies between the existing knowledge about what animals need and the societal expectations about improving the animals' quality of life on the one hand and the still-existing animal welfare flaws on the other are very complex:

1. The lowest-price policy for food and the global open agri-food markets prevent farmers from investing in upgrading their husbandry systems, from reducing the number of animals per herd or flock, and the current small consumer group that is willing to accept higher prices will not compensate for the higher running costs in case of reducing the number of animals per farm, and increased labour costs. This means that without funding only minor transformative activities can be expected. Scientific analyses have shown that the transition will not take place without funding from public and private capital (Grethe 2007; Grethe 2017; Spiller and Kühn 2022).
2. For many animal welfare activists, most animal welfare and animal health shortcomings seem to be easily solvable but solving them in the real world is not easy, since this is a genuine "wicked" problem (Rittel and Webber 1973): many interdependencies with other social issues, many stakeholders with completely different interests, and many people that must change their minds and habits are involved. The more radical and one-dimensional animal welfare demands are, the more they provoke resistance and counteractivities in those that are expected to change their attitudes and way of production, and the more other important societal values such as affordable food for all, biosecurity, and food safety are at risk of becoming compromised.

The consequence is that it is necessary to prevent any polarization, to orchestrate a consensual and civilized public debate, to include all stakeholders, and to increase international collaboration.

2 Animal Health

2.1 The Need for Animal Health Improvements

The impact of animal health impairments on livestock production is often vastly underestimated. Historically, animal epidemics such as rinderpest and foot-and-mouth disease have caused devastating famines. While such catastrophic scenarios have largely been mitigated, epidemic diseases remain a significant threat to national economies and international trade in animals and animal products (McElwain and Thumbi 2017).

Beyond the economic consequences, zoonotic diseases—those transmissible from animals to humans—pose an ongoing risk to human health. In cases where animal pathogens fully adapt to human hosts, as seen with the recent coronavirus pandemic, these diseases can escalate into global health crises. Highly pathogenic avian influenza (HPAI), originating in wild bird populations, is a prime example of this potential threat.

Furthermore, the growing global issue of antimicrobial resistance underscores the need for stringent disease prevention measures. Endemic bacterial diseases must be minimized through optimized biosecurity and hygienic practices across livestock populations. Reducing the need for veterinary antibiotic treatments is essential to curbing the broader antimicrobial resistance crisis.

The most effective strategies include:

- Strengthening *external biosecurity measures* to prevent disease introduction into farms
- Enhancing *internal biosecurity measures* to limit disease spread within farms
- Implementing comprehensive vaccination and surveillance programmes
- Ensuring optimal animal care to promote overall health and resilience

By focusing on prevention rather than treatment, these measures not only protect animal health but also contribute to safeguarding public health and ensuring sustainable livestock production.

The reduction of the animal caretaking intensity per animal that came automatically with the growth of the herds and flocks (the cost reduction for labour per animal was one of the intentions of the intensifications of the livestock industry in the 1970s) has partly led to remarkable deficiencies in the daily care for the individual animal. Animal disease due to deficiencies in prevention, biosecurity, diagnostics, and hygiene programmes is one of the major determinants (sick animals suffer, even under animal-friendly housing conditions). Thus, to improve the quality of life of the animals, a considerable systemic enhancement of animal care and disease prevention (biosecurity, hygiene programmes, vaccinations, early disease detection, and early treatments) is indispensable.

Weaknesses in the prevention of animal diseases have always consequences for the occurrence of various disease complexes:

1. Epidemic livestock diseases: The occurrence of epidemic diseases has again and again led to huge disruptions in livestock production regionally, nationally, and sometimes globally. In history, devastating famines were due to epidemic livestock diseases such as Rinderpest and Contagious Bovine Pleuropneumonia. Today, globally or nationally spreading diseases like Avian Influenza (AI) and African Swine Fever (ASF) are not only threats to the productivity of livestock production, but they are also trade barriers (WOAH 2008).
2. Zoonotic livestock diseases and infections: They pose enormous risks to human health. Zoonotic pathogens cause, e.g., food-borne disease in humans (*Salmonella*, *Campylobacter*, EHEC, and other *E. coli*-associated food infec-

tions, *Trichinella*, *Toxoplasma*, etc.), and they pose a risk for animal caretakers and people in close contact to animals (Q-fever, rickettsiosis, livestock-associated MRSA, etc.). Since COVID-19, with its assumed animal origin, the awareness about the potential risk of animal viruses becoming adapted to humans and causing epidemics, in worst cases pandemics, has grown. There is a common consensus in the medical and veterinary community that there will be another pandemic that will have its origin in animals in the years and decades to come. According to international and national Animal Health Authorities, one of the possible "candidates" for evolving an animal disease into a pandemic in humans is without question Avian Influenza due to the virus variant H5N1 (WHO 2024; EFSA 2024). This has become especially topical since this virus variant has proven that it can overwhelm the species barrier to mammals, identified by infections in cattle detected in the USA (Cohen 2024). And recently, even the infection of pigs from cows in the USA was reported.

3. Emerging and re-emerging livestock diseases: The increasing global tourism, trade with animals and animal products, but also climatic changes (Wang et al. 2021), are the drivers of the occurrence of diseases in regions that either had always been free from these diseases (emerging) or that had undergone successful eradication programmes and are now again affected (reemerging). Out of all emerging infectious diseases, approximately 75% are of zoonotic origin, with their source often traced back to animals (Sharan et al. 2023).
4. Endemic livestock diseases: The productiveness (and the societal acceptance) of livestock is the more at risk the more endemic diseases occur, since mortality and morbidity in livestock herds and flocks are an avoidable cause of animal suffering, and they contribute to the worldwide too high pre-harvest food loss, which is also one of the avoidable waste of resources (feed that was fed in vain, labour that was invested in vain). All bacterial endemic diseases are the reason for the need to use antibiotics in livestock. The higher the pressure of endemic diseases is, the higher the use of antimicrobial substances, and the higher the risk that the veterinary use of antibiotics contributes to the globally increasing bacterial resistance (AACTING-network 2018; Jensen et al. 2004). WHO regards antimicrobial resistance as one of the top 10 global public health threats facing humanity (Walsh et al. 2023). A significant *reduction in the use of antimicrobials in food animals is not just reducing the amount of antimicrobials but reducing the need for antimicrobial use by reducing bacterial endemic diseases* (Van Boeckel et al. 2017).

Apart from all the adverse impacts of animal disease on the animals' well-being, loss of productiveness, waste of resources, pre-harvest food losses, an increase of microbiological resistance, and threats to human health listed above, there is also evidence that the occurrence of animal diseases such as African Swine Fever (ASF), porcine reproductive and respiratory syndrome (PRRS) in pigs, foot and mouth disease (FMD) in cattle, and Avian Influenza (AI) in poultry and fowl remarkably increase greenhouse gas emissions (Perry et al. 2018; Capper 2023; Soliman et al. 2023).

All this is reason enough to *collectively intensify all efforts to improve the overall animal health status of our livestock populations as a major element of the transition, which is necessary to reduce the risks to the health of animals and humans and to restore the shrinking societal acceptance for producing food from and with animals.* Multisectoral collaboration is crucial for effective surveillance programmes. The following global and national activities are inevitable to considerably improve mankind's efforts to control *epidemic, zoonotic, and emerging and reemerging* diseases:

- Prioritizing zoonoses within the "One Health" approach is essential
- Shifting from reactive to proactive disease risk mitigation
- Early warning and sound interconnected surveillance systems for timely detection and response to zoonotic threats
- Global health security requires transparent data sharing and global commitment

As for *endemic* livestock diseases, the reality is that the health status in quite a lot of livestock herds and flocks all over the globe is currently far from optimal (Blaha and Koefer 2009; Huntington et al. 2021). The high prevalence of endemic animal diseases leads to avoidable pre-harvest food losses, including the death of animals and the waste of valuable resources. It also results in the excessive use of antimicrobials in livestock, contributing significantly to the global rise in antimicrobial-resistant bacteria (Aarestrup 2005).

This is a statement, which refers not only to intensive animal holdings but also to medium-sized and small holdings (and even to organic animal farming, since "organic" is not an automatic guarantee for the very best health status). The reasons for this high disease burden worldwide are manifold: economic constraints, lack of knowledge about the needs of the animals, misunderstood reduction of costs for vaccinations and hygiene programmes as supposedly improving the economic outcome, and missing empathy for the animals. Another reason is that the intensification of the livestock industry has steadily increased the number of animals that need to be cared for by one person, although there are huge differences in the caretakers' ability and engagement with the animals.

Improving the overall health status of livestock requires a significant increase in the number of food animal veterinarians—both practicing and official—who can support farmers in preventing disease and mitigating pre-harvest food safety risks (Navarre et al. 2020). Targeted programmes for regional or national benchmarking of animal health and welfare could further enhance these efforts. In Germany there are two initiatives: one is a national informative monitoring about health and animal welfare indicators; the other initiative is a programme (ITW = Initiative Animal Wellbeing) by the national quality assurance system "QS" that provides incentives for farmers that improve their husbandry system towards more animal friendliness. However, these programmes are only capable of providing general incentives to farmers to upgrade their husbandry system towards more animal-friendliness. They are not capable of identifying individual herds and flocks with suboptimal animal health and insufficient animal care. Developing and implementing herd-health benchmarking systems with meaningful animal health and welfare indicators could identify herds and flocks with specific health and welfare deficits, enabling more focused and effective interventions.

2.2 Animal Welfare and Health Databases for Benchmarking Purposes

It is well known that most of the dramatic animal welfare violations that are broadcasted by the media have little to do with the husbandry system or with the size of the herd or flock but with drastic deficiencies in the quality of animal care (Grandin 2009). The pictures of sick, injured, neglected, and suffering animals that the media present are clear proof of the fact that the veterinary authorities responsible for the control of the animal owners' compliance with Animal Welfare Acts and Directives for the Protection of Food Animals have no early warning system, which can identify herds or flocks with alarming suboptimal care for the animals early enough to make targeted counteractions possible. Before farm animals get into so poor a status of health that they may become the subject of a scandal, the ratio of dead animals and sick animals that reach the slaughter plant has been increasing for quite a long time, which the suggested benchmarking would indicate. The European regulations on controls of food-producing operations, including farmers, as "risk-oriented" controls (EU Reg. 178/2002; EU Reg. 217/625) are reasonable and are considerable theoretical progress towards efficient state interventions compared to the traditional mandatory randomly selected farm visits by official veterinarians. However, this progress is as long only theoretical as there is no monitoring of measurable health and animal welfare indicators at the farm level, which were comparable to quality assurance systems of successful food companies (More et al. 2017). The EU regulations on registering food animal operations and on the traceability of animal movements are mainly aimed at preventing emerging and re-emerging animal diseases at the national level. Since these data are neither standardized nor at the national level centralized, they do not provide databases for continuous monitoring for identifying individual food animal farms with very poor or suboptimal animal health and animal welfare statuses.

If there were national or regional databases that continuously record the mortality rates, the slaughter check data and other health and welfare indicators (both from life animals at unloading and in the lairage and from the carcasses at the slaughter line) cumulated for each herd and flock (Elbers 1991; Blaha 2005; Van Staaveren et al. 2017), early interventions by consultants ("yellow" and "red" herds), and, in case of non-responders to consulting, by veterinary authorities ("red" herds) would be possible. Such a benchmarking system would provide a simple and doable tool for the necessary "risk-oriented" state controls (see Fig. 8.1).

The databases should be anonymized in a way that except for the official veterinary authority, responsible for controlling the farms, only each farmer will know where his or her herd or flock is on the scale from a low to a high frequency of findings that indicate serious health and animal care shortcomings. Most farmers would

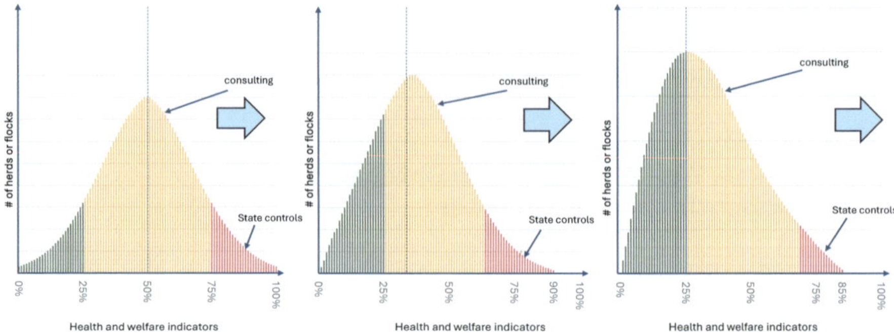

Fig. 8.1 A simplified benchmarking scheme for how to identify herds and flocks that need improvements in animal care and veterinary advice (© copyright Thomas Blaha)

use the data bank for their corrective measures since they do not know their herd's or flock's health status compared to those of other farmers. Most of them will look to being consulted if they know their shortcomings. Only those farmers who do not react to this early warning system will become objects of being controlled by the state veterinary authority responsible. Using this benchmarking tool continuously, a reliable instrument for an ongoing animal welfare and health improvement instrument will come into being. Making use of this instrument intelligently will prevent animals in poor husbandry and poor animal care situations from being recognized early enough and will suffer much longer than is necessary (Dickhaus et al. 2009; Alt et al. 2010; Blaha and Richter 2011; Grandin 2017; Sandøe et al. 2020).

2.3 Animal Care Ratios for Larger Animal Herds and Flocks

The structural changes in agriculture towards sectoral concentration and larger farms and larger food animal herds and flocks (in the Eastern Block during the communistic period due to central planning, in the Western Free-Market Economy due to increasing competition forces) have led to the fact that the larger herds or flocks get, the more animals must be observed and cared for by one person. This has been regarded for several decades as a welcomed economic effect: apart from feed, labour causes the highest costs in livestock production. Thus, lowering labour costs per animal was seen as economically successful progress, since in the decades after World War II, making the then very small-scaled agriculture more efficient, lowering the prices of food, and making the physical work of the farmers and their employees less demanding was regarded in the 1950s and 1960s one of the long-awaited signals for "the slender after World War II diet is becoming better and affordable".

However, there was no stopping point for this growth process in terms of animal performance (producing more meat, milk, and eggs per animal). This means that the development went far beyond the threshold, where exploiting the animals and deteriorating the animals' quality of life began. The effect of this missing stopping point and the lack of consideration for the animals' needs resulted in an increase in farm efficiency and productivity, causing systemic animal welfare shortcomings of most food animals. Additionally, the availability of automation and computerization (feeding computers, computerized climatization of the barns and animal houses) had the consequence that the hours of physical presence of the farmers and caretakers in the animals' direct environment were reduced.

Now, with the changing understanding of man's responsibility for the sentient creatures in their custody, we need to counteract the obvious degradation of animal observation and the shortening of time to care for a single animal by the growth of herds and flocks. There is no reason, not to regulate by law a minimum "animal care ratio" for given herd and flock sizes. Appropriate research groups can be asked to determine for each animal species and each age group per species, how many animals per competent person can be sufficiently observed, twice a day, and appropriately be cared for if necessary.

Regulations on how many competent people must care for how many food animals in larger herds and flocks would also be a guide for auditing procedures in the framework of delegated self-controls and the frequency and intensity of the mandatory state controls of larger animal units by veterinary authorities.

2.4 The Use of Animal-Oriented Sensor Technologies and Automation

Automation, computer-based technical support, and artificial intelligence technologies are in general society perceived as being detrimental to the animals' well-being since they are regarded as "soulless". However, automation and sensor techniques can, of course, be of great value for the animals, if they are developed not only to save labour time but to support the animal caretakers in their responsibility to provide the animals with optimal feed and water and a healthy climate; and to help farmers to recognize early signs of disease (Benjamin and Yiks 2019; Kistanova et al. 2024). The sensor techniques allow for identifying very early stages of limping or coughing, increasing body temperature, and even ruminal disorders (Rutten et al. 2013; Ward et al. 2024). Additionally, they can record behavioural abnormalities such as aggressive ranking order fights in the absence of human observation.

It is an optimistic fact that in recent years more and more farmers have installed warning APPs on their smartphones to get warned if something in the barn is having adverse effects on the animals, which is considerably lowering the risk of ongoing stress and suffering of the animals.

With the next generation of farmers that are digital natives, the promising technological achievements including AI will definitively be used and, thus, make a considerable contribution to improving their care for the animals.

3 Conclusions

There is an urgent need to improve the well-being and health of animals in our current livestock production. The transition of livestock production into a direction where the animals' physical, mental, social, and emotional needs are met is a moral obligation of humankind (as we started to understand centuries ago about slavery). The improvement in the animals' health is, apart from our moral responsibility for the animals' well-being, a major contribution to optimizing the prevention of disease in humans, which we as an international community strive to achieve via the One Health High-Level Expert Panel (OHHLEP), which is supported by the UN organizations FAO, OIE (= WOAH), WHO, and UNEP.

Society must agree on small steps in the process of a consensual continuous improvement in the quality of life of our food animals rather than insisting on drastic ad-hoc systemic changes that result in immediate opposition. Over time, applying small improvement steps like implementing well-understood transition measures such as reconstructing animal housings according to the animals' needs, animal health databases to enable benchmarking systems that trigger continuous improvements, reasonable animal care ratios for large herds and flocks, as well as using sensor techniques and automation that make the animal observation and the timely care for animals in need better will lead to gradual systemic changes into the right direction.

References

AACTING-network (2018) Description of existing monitoring systems for collection, analysis, benchmarking and reporting of farm-level veterinary antimicrobial usage. Available from: https://aacting.org/monitoring-systems/. Accessed 2 Nov 2020

Aarestrup FM (2005) Veterinary drug usage and antimicrobial resistance in bacteria of animal origin. Pharmacol Toxicol 96(4):271–281

Alt M, Blaha T, Möbius G, Richter T, Schlenker G (2010) Animal-oriented animal welfare criteria in food animals. TVT-Nachrichten 1:41–44

Benjamin M, Yiks S (2019) Precision livestock farming in swine welfare: a review for swine practitioners. Animals 9(4):133

Blaha T (2005) The concept and epidemiological aspects of the risk-based meat inspection. In: 12th ISAH congress 2005. Warsaw, Proc

Blaha T, Koefer J (2009) The growing role of animal hygiene for sustainable husbandry systems. In: Aland A, Madec F (eds) (Hrsg.): Sustainable animal production: the challenges and potential developments for professional farming. Wageningen Acad. Publ, Wageningen, pp 23–32

Blaha T, Richter T (2011) Animal welfare in food animals—analysis of the status quo and where to go. Deutsch Tieraerztebl 8:1028–1038

Brambell R (1965) Report of the technical committee to enquire into the welfare of animals kept under intensive livestock husbandry systems, Cmd. (Great Britain. Parliament), H.M. Stationery Office, pp 1–84

Capper JL (2023) The impact of controlling diseases of significant global importance on greenhouse gas emissions from livestock production. One Health Outlook 5(17):1–18. https://doi.org/10.1186/s42522-023-00089-y

Cohen J (2024) Bird flu discovered in U.S. dairy cows is 'disturbing'. Science. https://doi.org/10.1126/science.z9sk33v

D'Eath RB, Niemi K, Vosough Ahmadi B, Rutherford KMD, Ison SH, Turner SP, Anker HT, Jensen T, Busch ME, Jensen KK, Lawrence AB, Sandøe P (2016) Why are most EU pig tails docked? Economic and ethical analysis of four pig housing and management scenarios in the light of EU legislation and animal welfare outcomes. Animal 10(4):687–699. https://doi.org/10.1017/S1751731115002098

De Briyne N, Berg C, Blaha T, Palzer A, Temple D (2018) (2018): phasing out pig tail docking in the EU – present state, challenges and possibilities. Porc Health Manag 4:27. https://doi.org/10.1186/s40813-018-0103-8

Dickhaus C-P, Meemken D, Blaha T (2009) Attempts to quantify the health status of pig herds: developing and validating a herd health score (HHS). In: Aland A, Madec F (eds) Sustainable animal production the challenges and potential developments for professional farming, pp 191–201

EFSA (2024) Avian influenza overview. December 2023–March 2024. https://doi.org/10.2903/j.efsa.2024.8754

Elbers ARW (1991) The use of slaughterhouse information in monitoring systems for herd health control in pigs. Doctoral thesis, Rijksuniv, Department of Agroscience, Utrecht

Fraser D (2008) Understanding animal welfare. Acta Vet Scand 50(Suppl 1)

Grandin T (2009) Improving animal welfare: a practical approach. Paperback: ISBN: 978-1-84593-541-2328 pages, ePDF: eISBN: 978-1-78064-062-4328 pages

Grandin T (2017) On-farm conditions that compromise animal welfare that can be monitored at the slaughter plant. Meat Sci 132:52–58

Grethe H (2007) High animal welfare standards in the EU and international trade—how to deal with potential 'low animal welfare havens'? Food Policy 32(3):315–333. https://doi.org/10.1016/j.foodpol.2006.06.001

Grethe H (2017) The economics of farm animal welfare. Ann Rev Resour Econ 9(1):75–94. https://doi.org/10.1146/annurev-resource-100516-053419

Huntington B, Bernardo TM, Bondad-Reantaso M, Bruce M, Devleesschauwer B, Gilbert W, Grace D, Havelaar A, Herrero M, Marsh TL et al (2021) Global burden of animal diseases: a novel approach to understanding and managing disease in livestock and aquaculture. Revue Scientifique et Technique de l'OIE 40(2):567–584

Jensen VF, Jacobsen E, Bager F (2004) Veterinary antimicrobial-usage statistics based on standardized measures of dosage. Prev Vet Med 64(2–4):201–215

Kistanova E, Yotov S, Zaimova D (2024) Intelligent animal husbandry: present and future. Animals 14(11). https://doi.org/10.3390/ani14111645

Kunzmann P (2013) Changing relationships with animals. TIERethik 5(6):55–77

McElwain TF, Thumbi SM (2017) Animal pathogens and their impact on animal health, the economy, food security, food safety and public health. Rev Sci Tech 36(2):423–433

Mellor DH (2017) Operational details of the five domains model and key applications to the assessment and Management of Animal Welfare. Animals 7. https://doi.org/10.3390/ani7080060

More SJ, Hanlon A, Marchewka J, Boyle L (2017) Private animal health and welfare standards in quality assurance programmes: a review and proposed framework for critical evaluation. Vet Rec 180(25):612

Navarre C, Daniels A, Johnston MO, Mathis C, Perrett T, Posey D (2020) Impact of recruitment and retention of food animal veterinarians on the U.S. In: Food Supply Council for Agricultural Science and Technology Issue Paper 67, p 16

Perry BD, Robinson TP, Grace DC (2018) Review: animal health and sustainable global livestock systems. Animal 12(8):1699–1708. https://doi.org/10.1017/S1751731118000630

Rittel HWJ, Webber MM (1973) Dilemmas in a general theory of planning. Pol Sci 4(2):155–169

Rutten CJ, Velthuis AGJ, Steeneveld W, Hogeveen H (2013) Sensors to support health management on dairy farms (invited review). J Dairy Sci 96:1928–1952

Sandøe P, Hansen HO, Halkjær Rhode HL, Houe H, Palmer C (2020) Benchmarking farm animal welfare—a novel tool for cross-country comparison applied to pig production and pork consumption. Animals 10(6):955. https://doi.org/10.3390/ani10060955

Sharan M, Vijay D, Prakash Yadav J, Singh Bedi J, Dhaka P (2023) Surveillance and response strategies for zoonotic diseases: a comprehensive review. Sience in One Health 2. https://doi.org/10.1016/j.soh.2023.100050

Soliman T, Barnes A, Slettemoen Helgesen I (2023) The hidden carbon impact of animal disease. PLoS One. https://doi.org/10.1371/journal.pone.0292659

Spiller A, Kühn S (2022) Pathways to finance a socially accepted animal husbandry. J Consum Prot Food Saf 17:109–111

Temple D, Dalmau A, de la Torre JLR, Manteca X, Velarde A (2011) Application of the welfare quality® protocol to assess growing pigs kept under intensive conditions in Spain. J Veterinary Behavior: Clin Appl Res 6(2):138–149

Van Boeckel TP, Glennon EE, Chen D, Gilbert M, Robinson TP, Grenfell BT, Levin SA, Bonhoeffer S, Laxminarayan R (2017) Reducing antimicrobial use in food animals. Consider user fees and regulatory caps on veterinary use. Science 357(6358):1350–1352. https://doi.org/10.1126/science.aao1495

Van Staaveren N, Doyle B, Manzanilla EG, Calderón Díaz JA, Hanlon A (2017) Validation of carcass lesions as indicators for on-farm health and welfare of pigs. J Anim Sci 95(4):1528–1536

Walsh TR, Gales AC, Laxminarayan R, Dodd PC (2023) Antimicrobial resistance: addressing a global threat to humanity. PLoS Med 20(7):e1004264. https://doi.org/10.1371/journal.pmed.1004264

Wang WH, Thitithanyanont A, Urbina AN, Wang SF (2021) Emerging and re-emerging diseases. Pathogens 10(7):827. https://doi.org/10.3390/pathogens10070827

Ward S, Pluske J, Plush K, Pluske J, Rikard-Bell C (2024) Assessing decision support tools for mitigating tail biting in pork production: current Progress and future directions. Animals 14:224

WHO (2024) Influenza at the human-animal interface summary and assessment, 3 May 202 Emergency Situational Updates

WOAH (World Organization of Animal Health) (2008) Terrestrial animal health code. http://www.oie.int/doc/ged/D5517.PDF

Zonderland JJ, van Riel JW, Bracke MBM, Kemp B, den Hartog LA, Spoolder HAM (2009) Tail posture predicts tail damage among weaned piglets. Appl Anim Behav Sci 121(3–4):165–170

Economic and Political Competitive Constraints to Improving Animal Welfare

9

Mariann Molnár

1 Introduction

Since the first criticisms of industrial animal production in the 1960s, the European Union (EU) has made efforts to ensure the welfare of farm animals. The established system of EU minimum standards has contributed to improving animal care practices and farming conditions; however, it has not been able to address the deeper factors that lead to the intensification of animal farming. Competitive pressures and imbalances in economic power have also induced a conflict of interest between key players of the animal industries (farmers, suppliers, and processors), reformers, and regulators, making it difficult to negotiate problems associated with animal farming and solutions to address these. While the priorities of the EU "European Green Deal", the "Farm to Fork Initiative", and the citizen-driven "End the Cage Age" initiatives propose a rapid phasing out of large-scale, intensive, cage-based farming systems, farmers need to sustain production and continue to sell their products on a highly competitive global market to stay alive. No wonder that 2024 has seen major farmers' protests in many parts of Europe, imposing what seems like an inescapable escalation of events.

Intensive farmers have been directly targeted by EU policies that aim at ensuring higher welfare standards in farming. Yet, these farmers are also tightly squeezed by the prices of production inputs and the market of their products. This eventually manifests in a situation where farmers are left with the only option to decrease input costs to make ends meet. The need to reduce input costs also means that farmers will be severely limited in their ability to put positive animal care values into practice. It is like a deadly circle: the more pressure we put on farmers, the less resilient they

M. Molnár (✉)
Animal Welfare Program, Faculty of Land and Food Systems, University of British Columbia, Vancouver, BC, Canada
e-mail: mariann.sz.molnar@ubc.ca

B. Grabkowsky, T. Blaha (eds.), *Intensive Livestock Production in Transition*,
https://doi.org/10.1007/978-3-031-97872-2_9

become, and the less likely they will be able to work on welfare. So, the question naturally emerges: How can we ensure that farmers make a decent living and—in turn—expect them to provide a good life for their animals?

This chapter seeks to find methods to ensure a truly effective transition in animal farming. It proposes that reform efforts need to go beyond a new set of revised regulations on farm animal welfare, one that addresses drivers that push production towards a level of confinement and cost-cutting that reduces animal welfare standards. It proposes that given the right socio-economic and political support, an increase in standards on intensive farms, along with a meaningful transition away from intensive farming methods, could be facilitated by mandatory certification of intensive production technologies along with incentives supporting farm diversification, on-farm processing, and alternative marketing strategies.

First, it reviews the evolution of the European pro-welfare reform up to the recent publication of the EU "Green Deal", "Farm to Fork Initiative", and the "End the Cage Age" citizen initiative. Second, it assesses the shortcomings of a principally legislative approach to welfare reform that the EU has adopted. It then turns to explore challenges and opportunities for transformative change in animal agriculture and ends by introducing real-life solutions that help farmers improve production standards.

2 From Silent Spring to EU Pro-Welfare Legislation and Back Again

Industrial methods of food production were first problematised by Carlson (2002) in the book *Silent Spring*, which critiqued the abundant use of DDT and other pesticides in intensive crop production in the United States. This book was soon followed by Ruth Harrison's book *Animal Machines* (1964) on the emerging large-scale, indoor animal agricultural practices in the UK and their effects on the health and welfare of farm animals. In a foreword that she wrote to Harrison's book, Rachel Carson advocated a Schweitzerian ethic based on reverence for life (Schweitzer 1971) and asked: "How can [farm] animals produce under such conditions safe or acceptable human food?"

Although ethical concern for animals in the United Kingdom dates at least from the eighteenth century (Harwood 1928), reforms in the twentieth century rested on emerging scientific evidence, the activities of professional non-governmental organizations, and a growing recognition of animal sentience. The UK's 1822 anti-cruelty law was supplemented in 1968 by legislation specifically targeting farm animal welfare in intensive systems. Provisions in the "Agriculture Miscellaneous Provisions Act" resulted from a complex set of ethical, economic, and public concerns for animals (Lawrence and Vigors 2020). In 1979 the Farm Animal Welfare Council (FAWC) publishing the well-known Five Freedoms (FAWC 1979), which has provided high-level guidance for reform.

Concern for farm animal welfare reached continental Europe in 1976 when the Council of Europe published the first Convention on "the protection of animals kept for farming purposes" (Eurogroup. 2010), opening a new phase of farm animal welfare reform efforts. Indeed, while the 1957 Treaty of Rome still identified animals as goods (Caporale et al. 2005), the 1979 Council of Europe Convention was the first major European legislation that aimed to protect the welfare of farm animals raised in intensive or confinement systems. Subsequently, the Treaty of Amsterdam (1997, later affirmed by the 2007 Treaty of Lisbon) recognized animals as sentient beings (Miele et al. 2010) and in 1998 the European Union adopted the 1976 Council of Europe Convention and passed Council Directive 98/58/EC concerning the protection of animals kept for farming purposes (Council Directive 1998), opening a new phase in the animal welfare movement.

These legal documents, along with the so-called Protocol on Animal Welfare under the Amsterdam Treaty, established the ruling principles "in key areas of European law and policymaking" (Camm and Bowles 2000 p. 197). In addition to the general requirements laid down in the Directive on issues such as appropriate staffing, inspection, and automatic and mechanical equipment used on farms, minimum standards were also prescribed that looked to address animal needs. These included standards on freedom of movement, the quality of buildings and accommodation, feed and water, mutilations, and breeding procedures (Council Directive 1998). Soon enough, specialized minimum standards were also introduced on slaughter and killing practices (1993), laying hens (1999), transport conditions (2005), chicken kept for meat production (2007), calves (2008), and pigs (2008). Minimum standards (Mellor and Webster 2014) were generally believed to ensure the application of "animal-friendly" production methods. These rules, however, principally target farmers and define basic standards of care and on-farm housing conditions at the expense of other, more complex needs of animals and other key players in the value chain that directly impact on-farm standards. Suppliers of intensive animal breeds, feed, and medication, and—most importantly—the industry of housing tools and technologies upstream from farmers in the value chain directly shape and define on-farm standards that the farmer must cope with. In addition, processors down the supply chain set the prices of the finished animal or other animal-derived products, greatly impacting profitability. This claim does not wish to underestimate the importance of the quality of care a farmer can provide, but it would like to emphasize that there is not a single piece of EU legislation that regulates any of these industries for welfare. In addition, although Member States are free to pass national rules that extend beyond the established EU legal framework, norms higher than common EU standards have rarely been adopted.

So, although legislative reforms have brought many benefits, such as obligatory pre-slaughter stunning, significant challenges remain. While the Council of Europe identifies animal welfare as an issue of common cultural heritage and acknowledges the duty to care (Caporale et al. 2005), the level of protection to be delivered to farm animals seems open to interpretation (Croney and Millman 2007; Miele et al. 2010), only visible when applying a critical approach to animal welfare reform. High on the EU legislative agenda are issues such as animal health and food safety (Blockhuis et al. 2010; McLeod-Kilmurray 2012), while low on the agenda are welfare problems associated with comfort standards or boredom. Other concerns that are not addressed include overproduction (Elinder 2005), food loss and waste (FAO 2019a), and the biodiversity impacts of intensive farming practices (WWF 2024). While many would argue that the intensification of agricultural production, the consolidation of animal farming and processing industries, and the widespread use of large-scale, indoor confinement farming methods are key areas where change is most needed (Fraser 2008) in practice, current EU policies do nothing to ensure a transition away from these methods. Standardized legislation also fails to account for the diversity of farm scales, technologies, and management practices that directly influence day-to-day animal welfare conditions (Fraser 2014; Molnár and Fraser 2021).

These challenges are now being acknowledged in the EU by the European Green Deal (European Commission 2019a, b) and the Farm to Fork Strategy (European Commission 2020), which present a much more complex understanding of the range of issues that need to be tackled. Within the context of sustainable development, the strategies aim to deliver significant improvements in farming conditions, including a substantial increase in the welfare of farmed animals and—for the first time—the assessment of farming technologies used. A momentum for change also seems to be present in the social arena, as 1.4 million EU citizens have signed the "End the Cage Age" initiative launched by Compassion in World Farming (European Commission 2021), which calls on the EU to phase out the use of confinement farming methods. Key goals of the EU and non-governmental initiatives are to review the EU's entire policy system for agriculture and reflect on environmental (sustainability, environmental health, biodiversity), animal health (food security), and animal welfare issues along the entire food chain, with a special emphasis on removing farm animals from confinement systems. This goal was proposed to "foresee better living and transport conditions, and enhanced protection of animals during slaughter" (European Commission n.d.), but after extensive consultations failed to meet the initial 2023 deadline (European Commission 2021). Since then, the Commission presented a Strategic Dialogue report on the future of agriculture, which proposes the full revision of animal welfare legislation by 2026 and the introduction of a comprehensive multi-tiered EU-wide animal welfare labelling scheme (European Commission 2024).

In summary, despite the substantial farm animal welfare reforms since Carlson and Harrison, intensive farming methods are again in the spotlight. Further transformative change in animal agriculture may be possible only if, in addition to legal protection, other important aspects such as the socio-economic context of animal farming and ethical principles guiding decision-making are also given sufficient attention (Anderson 2011).

3 Understanding the Need for Change

Effective EU policies on agriculture, food production, and farm animal welfare are thought to be challenged by a set of competing interests that often result in compromised arrangements (Rose 2010). Competing human and animal welfare interests and the contradictory animal welfare and economic mandates of the EU induce significant challenges to the animal welfare reform effort. Depending on the context of policy analysis, some would argue that the EU farm animal welfare reform efforts to date were able to act against industrialized forms of animal agriculture and achieve a "counter-commoditization strategy" (McLeod-Kilmurray 2012, p. 77; Vapnek and Chapman 2010). However, critics of the current legislative approach claim that pro-welfare legislation only focuses on "the irrational property owner" who inflicts harm to animals without reasonable human benefit (Ibrahim 2006, p. 187), and highlight the difficulty of an ever-increasing legislative burden on farmers (Miele et al. 2010). These views are supported by individual farmers or farmer associations, who express their concerns over the tension between EU regulations and production efficiency on a global market (Copa-Cogeca n.d.).

This tension between EU regulations and global trade in food has been understood as the outcome of World Trade Organization (WTO) rules of conduct (McLeod-Kilmurray 2012). WTO trade regulations necessitate the application of neo-liberal free trade principles (McLeod-Kilmurray 2012) offering limited chances for a consistent set of rules governing both trade interests and pro-welfare action. As Hobbs et al. (2002) note, under WTO regulations, the EU is obliged to facilitate the international trade of like products that cannot be distinguished from one another. Obligatory labelling of production methods is a clear farm animal welfare interest but cannot be carried out due to concerns over market distortion (Hobbs et al. 2002; Passatino et al. 2008; Swinbank 2006).

While free trade in its present form continues to be widely supported, a diverse set of data indicates that negative externalities caused by large, intensive industries cannot be sufficiently addressed on the market (Anderson 2011; Gruen 2011; Sandel 2013; FSEC 2024). A dominant assumption of free trade is that it is beneficial to consumers, who make choices consistently and according to their values (Cernat et al. 2018). This would mean that growing concern and awareness of farm animal welfare could—in theory—empower consumers to relieve animals from harmful production methods (Eurogroup. 2016). Most willingness-to-pay studies emphasize this premise (Lusk and Norwood 2011), yet other studies suggest that direct causality between public concern and consumer action cannot be found (European Commission 2016; Falk and Szech 2013; Harper and Henson 2001), as important self-protective mechanisms such as cognitive dissonance (Miele and Bock 2007; Wicks 2011), the complexity of the market (Falk and Szech 2013b), or price-sensitivity (Lusk and Norwood 2011) prevent consistent purchasing behaviours. Falk and Szech 2013 also find that moral decisions may be affected when participants feel that they have no direct influence on an immoral act and therefore have "a tendency to lower moral values, relative to individually stated preferences" (p. 710). Findings therefore suggest that moral concern alone is unable to counter-balance negative market externalities (Falk and Szech 2013), including any harm caused to wild or farm animals during food production.

Consistent animal advocacy is also needed. Since the 1960s, societal concern for the welfare of farm animals has evolved, and European animal welfare advocacy approaches have changed over time. Before the publication of "Animal Machines" (Harrison 1964), the United Kingdom was already debating the problem of farming becoming “more profitable, modern and efficient” (Woods 2012, p. 17), leading to concerns over farm animal stress, suffering, and cruelty (Sayer 2013; Woods 2012). After the publication of Harrison’s book, *Farm Animal Welfare*, a “fundamentally new language and concept” (Woods 2012, p. 14) emerged from an in-depth scientific, political, and public debate (Sayer 2013). Following recommendations by the Brambell Committee, animal welfare developed into an established scientific discipline (Fraser 2016; Webster 1994) and advocacy approach (Ibrahim 2006). Most pro-welfare policies were developed on animal welfare premises, acknowledging that the subjective state of welfare matters to animals, and that given the possibility they are ready to contribute to their state of well-being, including the achievement of pleasure and the avoidance of pain (Broom 1988a, b, 1991; Fraser 1993; Webster 1994, 2005). A welfarist perspective is therefore predominantly concerned with the quality of an animal’s life rather than its quantity (unless longevity is used as an indicator of welfare) and has evolved to strive for no less than to provide good lives for animals under human care irrespective of their use or value (Fraser 2012). An animal welfare approach therefore focuses on establishing ethical human-animal interactions via legislation, education, capacity-building, and good practice (Garner 2005).

However, other ethical frameworks have also been proposed that challenge the animal welfare approach. Most notable of these are the consequentialist, utilitarian animal liberationist ethic developed by Peter Singer (1995) and the categorical animal rights theory proposed by Tom Regan (1988), further developed into what is now called the abolitionist theory proposed by Gary L. Francione (1996, 2009). These theories rest on strong foundational principles and propose the application of “non-interference rights” (Fraser 2012) in the belief that human-animal interactions, such as farming animals and the consumption of food from animal origin, are inherently wrong (Singer 1995; Regan 1988; Francione 2009). Therefore, in its most extreme form, non-interference rights prescribe the total abolition of all human-animal interaction (Francione 1996, 2009; Regan 1988). Other approaches necessitate the abolishment of harmful interactions for all species of animals (Bekoff et al. 1992) or those that are comparable to humans in their capacity to suffer (Singer 1995). The animal welfare approach and its competing categorical, hegemonic discourses and ethical frameworks (Fraser 1999; Stibbe 2005) arguably lead to the fragmentation of scientific, social, and political reform efforts (Anderson 2011) solely based on moral disagreement. These ethical debates have not been sufficiently resolved since the 1960s (Woods 2011), and evidence suggests that an increasing number of animal advocacy approaches are based on non-interference rights in the United States as well as the EU (Braddock 2014; Morini 2018) making it increasingly difficult to reach a consensus on the future of farming, and work consistently to achieve an ethical outcome.

This is a very significant problem. A model on how powerless groups such as children or animals gain protection illustrates that meaningful change in agriculture will only be possible if there is a common ethical foundation and a clear understanding of the level of protection desired (Anderson 2011). Therefore, success of the ongoing EU reform may depend on the ability of reformers to establish a clear, united ethical imperative on what needs to be done for animals and why (Anderson 2011). To date, a comprehensive approach that could embrace actions beyond the review of existing policies has not been achieved. Some argue that "it will not be easy to reach a consensus on what animal welfare is and how it should be achieved/improved" (Miele et al. 2011, p. 116). It is likely that a significant broadening of reform efforts is needed, otherwise the present status quo of animal farming will prevail, ensuring only limited progress (Elzen et al. 2011) on those aspects of welfare that negatively impact the profitability of farming (Molnár and Fraser 2020; Molnár and Fraser 2021).

In the related field of environmental protection, alternative economic models seem to be developing. Circular economy, for example, presents a united moral and innovative economic framework that can function even within the established neoliberal free trade context (Murray et al. 2017). For farm animal welfare, a functional moral imperative (Fraser 2010) is still needed. Examples could include a Reverence for Life originally proposed by Albert Schweitzer (1971), Biosocial Communitarianism by J. Baird Callicott (1988, 1998), the One Health (Monath et al. 2010) and One Welfare (Colonius and Earley 2013) frameworks, and the Practical Ethic by David Fraser (2012). Anderson (2011) highlights that "once a new ethic is firmly established…it can be…as powerful as legal reform… Without this ethical shift mere legislative reform will probably be ineffective" (p. 62).

4 Challenges for Transformative Change in Animal Agriculture

While it may seem that farmers can freely decide how they keep their animals, in a case study of pig farming in Hungary, Molnár and Fraser (2020) found that producing for the ever-changing mainstream market severely constrains farmers' choices. With large-scale consolidation of food value chains, many farmers compete to sell almost identical (generic) products to a relatively small number of processors (Hendrickson and James 2005; Molnár and Fraser 2020). Larger, better-established companies have greater access to processors, leading to competitive advantages over smaller companies that sell fewer animals (Vernooij 2015). Competitive pressures on animal producers also seem to be induced by the inability of farmers to negotiate the price of their finished animals. As the price of pigs and other intensively reared animals usually go in cycles (European Commission 2019a, b), farmers either sell finished animals on a fixed price, which is defined in a long-term contract, or on the daily rate determined by the stock market (Molnár and Fraser 2020). In both cases, farmers will have to endure significant periods of losses (Fraser 2005, 2008) and will therefore need to economize to ensure continued production (Molnár and Fraser 2020). Free trade also provides processors with the opportunity to source finished animals internationally, resulting in further economic vulnerabilities for farmers. For example, if feed prices increase in one geographic region, processors will still be able to import animals from a distance and local farmers impacted by higher production costs will be unable to recover their investments (Molnár and Fraser 2020; Vernooij 2015).

With this imbalance in economic power between farmers and the processing industry, the rational adjustment of product supply and demand would be necessary. However, Molnár and Fraser (2020) discovered that, unlike other industries, animal farming cannot respond quickly enough to lower prices or demand, because production can only be adjusted with a significant time lag (in the case of pig farming at least six months or more) and in these periods the market might change significantly. This inability of the animal farming sector to increase or decrease production as the market dictates also induces pressure on farmers to either further intensify or abandon their farming operations, a feature that is already widely apparent in industrialized countries (Fraser 2005, 2008; Marquer et al. 2014). Consequently, those farmers who would like to remain competitive have to ensure a high level of production efficiency, which seems to be the most important reason why farmers who produce for the mainstream market build and then continue to use production methods that necessitate large-scale, automatized confinement systems (Molnár and Fraser 2020).

Most intensive confinement farming operations also specialize in producing a single commodity and invest in costly technologies that are suited for only one species of animal. These housing technologies are very difficult to adjust should regulations or consumer expectations change. Technological lock-in not only seems to prevent farmers from being able to enjoy the economic resilience that a mixed-farming enterprise would be able to offer (Molnár and Fraser 2020), but it also predisposes the continued use of these large, fixed investments (Foray 1997; Perkins 2003). The economic pressures that force farmers to decrease input costs also lead to decreased amenities for animals, including space, care staff, and labour costs, which further limit the possibility for farmers to put positive animal care values into practice (Hendrickson and James 2005; Fraser 2008; Molnár and Fraser 2020; Thompson 2001). While much research has emphasized that farmers are mostly concerned with animal health and productivity (Te Velde et al. 2002; Lund 2006; Vanhonacker et al. 2008; de Rooij et al. 2010; Spooner et al. 2014), this "entrepreneurial discourse" (Fraser 2008) was found to represent the beliefs of only the largest, most intensive farmers (≥1000 sow operations, Molnár and Fraser 2021). Farmers operating medium-scale enterprises (≈400–600 sows), who used the same methods, but at a lower scale, were more critical of intensive confinement systems, and in ideal circumstances would have provided their animals with increased welfare through higher standards of care and more natural living conditions (Molnár and Fraser 2021). Yet even these farmers struggled to transition their technologies and increase welfare standards (Molnár 2022). Therefore, the combined effects of intensive farming technologies and economic pressures significantly constrain the actions of farmers independent from their values (Molnár and Fraser 2021).

Challenges endured by farmers in EU Europe and beyond have significantly impacted the scale and intensity of farming operations (Fraser 2008), the health and well-being of farmers (Kouimintzis et al. 2007; Torske et al. 2016), the countryside and rural communities (Filson 2004), and overall farm animal welfare conditions (Eurogroup 2010). Yet, standard (top-down) problem-solving methods only address a limited subset of these issues. Inconsistent legislative enforcement presents a significant challenge (Caporale et al. 2005), along with the narrow focus of policies on the physical aspects of indoor environments (e.g. flooring, ventilation), without adequate reflection on other widely used practices that may lead to poor welfare (e.g. frequent group mixing, confinement, boredom), and unpredictable events that can also limit compliance (e.g. unusually high birth rates, Molnár and Fraser 2020). The economic status of farmers is another issue that can decrease the effectiveness of pro-welfare legislation, especially if farmers face prolonged periods of losses that eventually lead to inadequate standards of care (Molnár and Fraser 2020). The subsidy system also had its weaknesses. At times when the prices of finished animals were very low, subsidies compensated for major losses (Molnár and Fraser 2020), and during high-income periods, subsidies were not sufficient to finance major welfare-related investments (Molnár 2022). The payment system seemed to be designed for confinement farms only, indirectly incentivising a move to intensive systems (Molnár and Fraser 2020). Finally, a reliance on consumers purchasing high-welfare products seems also flawed, as the processing industry is mostly supplying retailers with uniform "commodity" products without clear and consistent labelling, which limits informed decision-making (Molnár and Fraser 2020).

Following global trends, the consolidation of animal agriculture is still ongoing within and beyond the EU (Molnár and Fraser 2020). This process seems to facilitate the continued expansion of already successful industries and farming operations, while others are forced to abandon production (Molnár and Fraser 2020, 2021). Challenges induced by an uneven distribution of economic pressures and power seem to affect those medium-scale farmers who, given the chance, may be willing to engage in farming practices that are more aligned with their values (Molnár and Fraser 2021) leading to a further increase in the scale and distribution of intensive practices (Fraser 2008; Molnár 2022; Perfecto et al. 2009; Rollins 2008).

5 Opportunities for Transformative Change in Animal Agriculture

Despite all challenges, the successful European child labour reform that liberated children from mines and factories indicates that major change is possible (Anderson 2011), and the momentum for change in agriculture seems ever-growing. Scholars are calling for a set of important objectives guiding reform efforts, including the transformation of the model of agriculture that leads to intensification (Yeates 2017), along with the responsible use of natural resources (Bansard and Schröder 2021), the conservation of biodiversity (WWF 2024), the reintegration of nature and natural processes into farming (Tree 2018; Tree and Burrell 2023; Salatin 2013), and the production of safe and wholesome food. In addition, production must ensure food security (FAO 2019b), and food distribution also needs to improve (FAO 2008). While this chapter primarily argues for the case of farm animal welfare, these goals are of equal importance and are inherently interconnected. Therefore, while the dominant form of agricultural production in the developed world is intensive confinement agriculture, alternative methods of production and retail present important models worth exploring.

Alternative agricultural production is a collective term representing a highly diverse set of animal farming methods (Verhoog et al. 2004), yet in comparison to confinement farming they are generally small-scale, extensive, or semi-intensive operations with a high level of personal involvement by the farmer. Animals are kept in loose group housing with access to indoor shelters and outdoor runs, and with longer (at least double) life-spans for both breeding and fattening animals (Molnár 2022). These farms are based on a low-capital model, as they do not invest in and rely on expensive automated technologies and are therefore free of most technological constraints (Molnár and Fraser 2020). Alternative farming also relies on diversification (mixed farming methods) to ensure resilience during periods of fluctuating prices and costs (Molnár and Fraser 2020). While in comparison to intensive farming production efficiency is lower and variable costs of production are higher on alternative farms, operations possess a greater level of stability as farmers rarely sell their finished animals to processors, but pursue niche markets, engaging directly in the processing and sale of their products (Molnár and Fraser 2020). As UNIDO (2020) also finds, a shorter food chain can secure a higher return for the farmer as well as other benefits to rural communities, biodiversity, and the natural environment (Tree 2018; Tree and Burrell 2023; Salatin 2013).

A transition away from intensive farming is assumed to compromise the food security of growing world populations due to a perceived inefficiency of alternative farming methods (Wax and Wanat 2021), their need for more land (Muller et al. 2017), and agricultural workers (Jansen 2000). Intensive confinement production is believed to provide a solution to these challenges. However, when examined in an interdisciplinary manner, accounting for all production externalities (Horrigan et al. 2002), the combined negative effects of overproduction (Elinder 2005), food loss and waste (Gustavvson et al. 2011), and the effect of automation on worker wages (Marinoudi et al. 2021), we are confronted with a completely different scenario. The ever-increasing rate of urbanization is taking its toll on rural communities (Beckersa et al. 2020), and several complex factors influence the willingness of people to engage in farming (Booth 2021). Again, contrary to mainstream thinking, the need for alternative farming methods—including agro-ecological (Tomich et al. 2011), regenerative (Rhodes 2017), organic farming (Vaarst 2010), permaculture (Ferguson and Lovell 2014), or silvopastoral systems (Cubbage et al. 2012)—are well understood by most farmers (Molnár 2022).

A transition away from using monoculture cropping is the first step to enable the production of a more diverse set of food on a single unit of land. To do this, farmers will once again need well-adapted plant varieties and heritage breeds of animals to be able to farm in a truly alternative way. Diverse farming systems, plants, and animals can also foster the recovery of soil health and even provide an important opportunity to restore biodiversity (Tscharntke et al. 2021; Tree 2018). It does not inevitably mean a complete transition away from the use of smart technologies; however, it does necessitate the preservation of rural communities and an influx of motivated, educated, and well-paid workforces. Some data already indicate a growing momentum for a new generation of farmers, who leave their urban, non-agrarian lives to engage in agricultural production (Milona and Ventura 2019). Other studies support the idea that consumers who recognise the problems of intensive confinement systems should reconnect with farmers to directly negotiate common concerns and mutually acceptable farming methods (Ventura et al. 2013, 2016). This is all possible if farmers engage in the development, processing, and trade of their own products on short value chains.

In our research, differences between quality-of-life considerations for animals in intensive versus alternative systems have been acknowledged by some small to medium-scale confinement farmers (Molnár and Fraser 2021), but not by those with very large operations. Hence, strategies to transition intensive production may require policies, incentives, and technological solutions that fit the priorities of producers working at different farm scales (Molnár and Fraser 2021). Large operations that are resistant to transformative change may require well-qualified staff and expertise to improve farm management (Molnár and Fraser 2021) and technological solutions (e.g. enabling the use of bedding, roughage, and environmental enrichment) that may improve animal welfare without compromising farm efficiency or competitiveness (Molnár 2022). On the other hand, small- and medium-scale confinement farmers seem most open and able to change, and given the right support, are most likely to transition away from indoor, intensive confinement systems, and incorporate more natural production methods (Molnár and Fraser 2021). In both cases, an increased level of transparency on farming methods will also be needed to ensure that all farm scales find societal approval and a stable market (Molnár 2022).

6 How Can Confinement Farming Be Transformed to Ensure a Meaningful Increase in Farm Animal Welfare Conditions?

The EU is highly committed to protecting the welfare of farm animals and is striving to further improve conditions, but it also appears to be in a difficult situation, because any transition of animal agriculture away from the use of intensive confinement methods must be balanced with other socio-economic interests. The next years will be crucial in determining the depth and breadth of EU involvement in pro-welfare action. The level of current commitments suggests that the EU will limit its reform efforts to a revision of animal welfare legislation and the introduction of a comprehensive animal welfare labelling scheme. Proposals on new regulations that could initiate the phasing out of cage-based systems are still to be seen. But all of these actions will likely continue to target farmers. To compensate farmers for changing production standards, the EU may provide legislative and financial incentives, for example, through targeted animal welfare payments. Yet, evidence suggests that this combined policy and payment approach may lead to the largest, most intensive confinement farmers or corporations (with the most lobby power) resisting a "forced" transition (Wax and Wanat 2021). Given that incentives are adequate in scale, and are widely available, higher EU standards may provide an opportunity for small- to medium-scale intensive confinement farmers, who already face severe economic pressures to change practices. If the EU pursues this route, European farmers will require a system of coherent protective measures to ensure that their investments will succeed, farmers will find a market, and that animal production will not simply be displaced to other jurisdictions (Grethe 2007) where lower standards prevail.

An alternative approach to the EU pro-welfare action is to go beyond existing commitments and extend the review to assess the influence of other direct and indirect factors—such as the impact of global free trade inducing continuous pressure to increase the scale and intensity of farms—and propose new regulations and incentives that address these other important drivers of confinement farming systems. Most importantly, this new policy system will have to acknowledge problems induced by (a) the unequal distributions of power and competitive pressures on the market, (b) the economic vulnerability of farmers, (c) the high level of specialization of large-scale intensive farms, (d) the problem of technological lock-in induced by expensive confinement technologies, and (e) the inability of the farming sector to quickly adjust production to demand. The current EU animal welfare reform effort may therefore need to do much more than prescribe the phasing-out of intensive confinement systems; it will have to challenge global agricultural production trends and provide a sound, workable alternative production and marketing strategy to farming communities. To ensure a transformative change in animal agriculture and a meaningful increase in farm animal welfare standards, farmers will also need to be involved in the process of reform (Molnár and Fraser 2020) and be empowered to put positive animal care values into practice (Molnár and Fraser 2021).

In addition to the legislative process, opportunities for transformative change in animal agriculture should also focus on preventing the further intensification of animal farming. This could be most easily carried out in those EU Member States, such as Hungary, that are in the process of intensifying their production methods or those that still depend on small- to medium-scale, semi-intensive, or extensive farming systems (those found in many EU accession states). While transitioning large-scale, intensive confinement farms is challenging, inspiration for change can be found in the example of alternative production and marketing systems. Well-targeted incentives to increase farm animal welfare conditions, the ability of producers to process their own products, and a workable solution on labelling (such as the EU egg labelling scheme) may improve market access and relieve farmers of those excessive economic pressures that predominantly drive intensification.

In the ideal scenario, the current EU reform effort should facilitate a sound transition away from intensive confinement farming methods, and enable the majority of animal production to be realised in small- to medium-scale, semi-intensive, and extensive, diverse, agro-ecological farming operations where animals are housed loosely, and where they can access both indoor and outdoor runs (Molnár 2022). A new era of farming may enable significantly improved farm animal welfare conditions as well as the increased resilience of farming communities. As ideals often provide momentum for reform efforts, it may be possible to assume that a meaningful transformation of intensive animal production is feasible (Anderson 2011; McCulloch 2013; Winter et al. 1998).

7 Conclusions

The greatest challenge of current EU farm animal welfare reform efforts—initiated by the EU Green Deal, the Farm to Fork Initiative, and the End the Cage Age citizen initiative—is the problem of how to realize transformative change in animal agriculture when many existing pro-welfare actions compete with conflicting mandates of the EU and contradict global free trade policies. Current commitments suggest that the EU will focus on a comprehensive review of the existing legal system to propose a new set of regulatory principles and the phasing out of caged-based farming methods. However, transformative change in animal farming appears severely constrained by the established system of intensive confinement production and trade practices.

This chapter suggests that a narrow, policy-based reform effort may increase the vulnerabilities of farmers and fail to address major forces that shape animal farming. To deliver meaningful change to farm animal welfare within and beyond the EU, reform efforts should extend the scope of legislative reform and engage in debating and devising an alternative system of animal production and trade within the current free-trade context. Given a carefully devised set of legislative, economic, and market-based incentives, farmers could break away from pressures that drive the intensification of animal agriculture and should transition from the use of specialised large-scale, intensive confinement farming methods to more environmentally and economically sustainable alternatives. The pro-welfare reform effort therefore requires a broad, holistic approach to ensure that actions offer a realistic path for transformative change in animal agriculture and a meaningful increase in farm animal welfare standards.

References

Anderson JL (2011) Protection for the powerless: political economy history lessons for the animal welfare movement. Stanford J Animal Law Policy 4:1–63. https://ssrn.com/abstract=1946337

Bansard J, Schröder M (2021) The sustainable use of natural resources: the governance challenge. In: IISD earth negotiations bulletin brief #16. International Institute for Sustainable Development: Winnipeg, Canada

Beckersa V, Poelmansc L, Van Rompaey A, Dendoncker N (2020) The impact of urbanization on agricultural dynamics: a case study in Belgium. J Land Use Sci 15:626–643. https://doi.org/10.1080/1747423X.2020.1769211

Bekoff M, Gruen L, Townsend SE, Rollin BE (1992) Animals in science: some areas revisited. Anim Behav 44:473–484. https://doi.org/10.1016/0003-3472(92)90057-G

Bentham J (1789) In: Payne T (ed) An introduction to the principles of morals and legislation. Son, London

Blockhuis HJ, Veissier I, Miele M, Jones B (2010) The welfare quality® project and beyond: safeguarding farm animal Well-being. Acta Agriculturae Scand Section A 60:129–140. https://doi.org/10.1080/09064702.2010.523480

Booth A (2021) Follow the food: the reason we're running out of farmers. BBC. https://www.bbc.com/future/bespoke/follow-the-food/the-reason-we-are-running-out-of-farmers/

Braddock K (2014) The utility of narratives for promoting radicalization: the case of the animal liberation front. Dynamics Asymmetric Conflict 1-22:38. https://doi.org/10.1080/17467586.2014.968794

Broom DM (1988a) Needs, freedoms and the assessment of welfare. Applied Animal Behavioural Sci 19:384–386. https://doi.org/10.1016/0168-1591(88)90023-8
Broom DM (1988b) The scientific assessment of animal welfare. Applied Animal Behavioural Science 20:5–19. https://doi.org/10.1016/0168-1591(88)90122-0
Broom DM (1991) Animal welfare: concepts and measurement. J Anim Sci 69:4167–4175. https://doi.org/10.2527/1991.69104167x
Callicott JB (1988) Animal liberation and environmental ethics: Back together again. Between the Species 4:163–169. https://doi.org/10.15368/bts.1988v4n3.1
Callicott JB (1998) 'Back together again' again. Environ Values 7:461–475. https://doi.org/10.3197/096327198129341672
Camm T, Bowles D (2000) Animal welfare and the treaty of Rome: a legal analysis of the protocol on animal welfare and welfare standards in the European Union. J Environ Law 12:197–205. https://doi.org/10.1093/jel/12.2.197
Caporale V, Alessandrini B, Dalla Villa P, Del Papa S (2005) Global perspectives on animal welfare: Europe. Sci Tech Rev Office International des Epizooties 24:567–577. ISSN: 0253-1933
Carson RL (2002) Silent Spring. Houghton Mifflin Harcourt Publishing Company, New York
Cernat L, Gerard D, Guinea O, Isella L (2018) Consumer benefits from EU trade liberalisation: how much did we save since the Uruguay round? Chief economist note. European Commission, Trade, Brussels
Colonius TJ, Earley RW (2013) One welfare: a call to develop a broader framework of thought and action. J Am Vet Med Assoc 242:309–310. https://doi.org/10.2460/javma.242.3.309
Council Directive (1998) Council directive 98/58/EC of 20 July 1998 concerning the protection of animals kept for farming purposes. Off J Eur Communities L221:0023–0027. https://eur-lex.europa.eu/legal-content/EN/ALL/?uri=CELEX:31998L0058
Copa-Cogeca (n.d.) Market and quality: what we call for. Copa-Cogeca, Brussels. https://www.copa-cogeca.eu/market
Croney CC, Millman ST (2007) Board-invited review: the ethical and behavioural bases for farm animal welfare legislation. J Anim Sci 85:556–565. https://doi.org/10.2527/jas.2006-422
Cubbage F, Balmelli G, Bussoni A, Noellemeyer E, Pachas AN, Fassola H, Colcombet L, Rossner B, Frey G, Dube F, de Silva ML, Stevenson H, Hamilton J, Hubbard W (2012) Comparing silvopastoral systems and prospects in eight regions of the world. Agrofor Syst 86:303–314. https://doi.org/10.1007/s10457-012-9482-z
De Rooij SJG, de Lauwere CC, van der Ploeg JD (2010) Entrapped in group solidarity? Animal welfare, the ethical positions of farmers and the difficult search for alternatives. J Environ Policy Plann 12:341–361. https://doi.org/10.1080/1523908X.2010.528882
Elinder LS (2005) Obesity, hunger and agriculture: the damaging role of subsidies. BMJ 331:1333–1336. https://doi.org/10.1136/bmj.331.7528.1333
Elzen B, Geels FW, Leeuwis C, van Mierlo B (2011) Normative contestation in transitions 'in the making': animal welfare concerns and system innovation in pig husbandry. Res Policy 40:263–275. https://doi.org/10.1016/j.respol.2010.09.018
Eurogroup (2010) Areas of concern: analysis of animal welfare issues in the European Union. Eurogroup for Animals, Brussels
Eurogroup (2016) Animal welfare top of mind: Eurobarometer 2016. Eurogroup for Animals, Brussels
European Commission (2016) Attitudes of Europeans towards animal welfare. Special Eurobarometer 442. European Commission, Brussels
European Commission (2019a) Communication from the commission to the European Parliament, the European Council, the Council, the European economic and social committee and the committee of the regions: the European Green Deal. COM 640 final. European Commission, Brussels, Belgium
European Commission (2019b) Annex to the communication from the commission to the European parliament, the European Council, the Council, the European economic and social committee and the committee of the regions: the European Green Deal. COM (2019), 640 final annex European Commission, Brussels, Belgium

European Commission (2020) A farm to fork strategy for a fair, health and environmentally friendly food system. European Commission, Brussels
European Commission (2021) Communication from the Commission on the European Citizens' Initiative (ECI) "End the Cage Age". C (2021) 4747 final. European Commission, Brussels
European Commission (n.d.) Video: animal welfare matters. European Commission, Brussels. https://ec.europa.eu/food/animals/animal-welfare_en#video-animal-welfare-matters
Falk A, Szech N (2013) Morals and markets. Science, 340:707–711. https://doi.org/10.1126/science.1231566
Falk A, Szech N (2013b) Organizations, diffused pivotality and immoral outcomes. In: DIW Berlin discussion paper no. 1305. https://doi.org/10.2139/ssrn.224393
FAO (2008) An introduction to the basic concepts of food security. In: Food security information for action practical guides. EC-FAO Food Security Programme, Rome
FAO (2019a) The state of food and agriculture 2019: moving forward on food loss and waste reduction. Food and Agriculture Organization of the United Nations, Rome
FAO (2019b) The future of food safety. In: First FAO/WHO/AU international food safety conference, Addis, 12–13 February 2019. Food and Agriculture Organization of the United Nations, Rome
FAWC (1979) Farm animal welfare council press statement: Surbiton, UK
Ferguson RS, Lovell ST (2014) Permaculture for agroecology: design, movement, practice, and worldview. A Rev Agronomy Sustain Develop 34:251–274. https://doi.org/10.1007/s13593-013-0181-6
Filson GC (2004) Social implications of intensive agriculture. In: Filson GC (ed) Intensive agriculture and sustainability: a farming system analysis. UBC Press, Vancouver
Foray D (1997) The dynamic implications of increasing returns: technological change and path dependent inefficiency. Int J Ind Organ 15:733–752. https://doi.org/10.1016/S0167-7187(97)00009-X
Fraser D (1993) Assessing animal Well-being: common sense, uncommon science. In: Baumgardt BR, Gray HG (eds) Food animal Well-being. Purdue University Press, West Lafayette
Fraser D (1999) Animal ethics and animal welfare science: bridging the two cultures. Applied Animal Behavioural Sci 65:171–189. https://doi.org/10.1016/S0168-1591(99)00090-8
Fraser D (2005) Animal welfare and the intensification of animal production: an alternative interpretation. Food and Agriculture Organization of the United Nations, Rome
Fraser D (2008) Animal welfare and the intensification of animal production. In: Thomson PB (ed) The ethics of intensification. Springer, Heidelberg, pp 167–189. https://doi.org/10.1007/978-1-4020-8722-6_12
Fraser D (2010) Towards a synthesis of conservation and animal welfare science. Anim Welf 19:121–124. https://www.wellbeingintlstudiesrepository.org/conbawel/1/
Fraser D (2012) A "practical" ethic for animals. J Agric Environ Ethics 25:721–746. https://doi.org/10.1007/s10806-011-9353-z
Fraser D (2014) Could animal production become a profession? Livest Sci 169:155–162. https://doi.org/10.1016/j.livsci.2014.09.017
Fraser D (2016) Animal welfare. In: Rowlinson MC, Ward C (eds) Routledge handbook of food ethics. Routledge, New York, pp 264–273
Fraser D, Weary DM, Pajor EA, Milligan BN (1997) A scientific conception of animal welfare that reflects ethical concerns. Anim Welf 6:187–205. https://www.wellbeingintlstudiesrepository.org/ethawel/1/
Fraser D, Mac Rae AM (2011) Four types of activities that affect animals: implications for animal welfare science and animal ethics philosophy. Anim Welf 20:581–590. https://www.wellbeingintlstudiesrepository.org/ethnawel/1/
Francione GL (1996) Animal rights and animal welfare. Rutgers Law Rev 48:397–469. https://heinonline.org/HOL/LandingPage?handle=hein.journals/rutlr48&div=16&id=&page=
Francione GL (2009) Animal welfare and the moral value of nonhuman animals. Law, Culture Humanities 6:1–13. https://doi.org/10.1177/1743872109348989

FSEC (2024) The economics of the food system transformation. In: Food Systems Economics Commission: Global Policy Report. https://foodsystemeconomics.org/wp-content/uploads/FSEC-Global_Policy_Report.pdf
Garner R (2005) Animal ethics. Polity, Cambridge
Grethe H (2007) High animal welfare standards in the EU and international trade—how to prevent potential 'low animal welfare havens'? Food Policy 32:315–333. https://doi.org/10.1016/j.foodpol.2006.06.001
Gruen L (2011) Ethics and animals: an introduction. Cambridge University Press, Cambridge
Gustavvson J, Cederberg C, Sonesson U, van Otterdijk R, Meybeck A (2011) Global food losses and food waste: extent, causes and prevention. Rome, Italy, Food and Agriculture Organization of the United Nations
Harrison R (1964) Animal machines: the new factory farming industry. Vincent Stuart Publishers Ltd., London
Harper G, Henson S (2001) Consumer concerns about animal welfare and the impact on food choice. EU FAIR CT98-3678, Final Report. Centre for Food Economics research (CeFER), Department of Agricultural and Food Economics. The University of Reading, Reading
Harwood D (1928, Republished 2002) Love for animals and how it developed in Great Britain. In Preece R, Fraser D (eds). Edwin Mellen Press, Lampeter
Hendrickson MK, James HS (2005) The ethics of constrained choice: how the industrialization of agriculture impacts farming and farmer behaviour. J Agric Environ Ethics 18:269–291. https://doi.org/10.1007/s10806-005-0631-5
Hume D (1987) In: Miller EF (ed) Essays: moral, political and literary. Liberty Fund, Indianapolis
Hobbs AL, Hobbs JE, Isaac GE, Kerr WA (2002) Ethics, domestic food policy and trade law: assessing the EU animal welfare proposal to the WTO. Food Policy 27:437–454. https://doi.org/10.1016/S0306-9192(02)00048-9
Horrigan L, Lawrence RS, Walker P (2002) How sustainable agriculture can address the environmental and human health harms of industrial agriculture. Research Review Environ Health Perspect 110:445–456. https://doi.org/10.1289/ehp.02110445
Ibrahim DM (2006) The anticruelty statute: a study in animal welfare. J Animal Law Ethics 1:175–202. https://scholarship.law.wm.edu/facpubs/1680
Jansen K (2000) Labour, livelihoods and the quality of life in organic agriculture in Europe. Biol Agric Horticul 17:247–278. https://doi.org/10.1080/01448765.2000.9754845
Lawrence AB, Vigors B (2020) Farm animal welfare: origins and interplay with economics and policy. In: Ahmadi B, Moran D, D'Eath R (eds) The economics of farm animal welfare. CABI International, Wallingford
Kouimintzis D, Chatzis C, Linos A (2007) Health effects of livestock farming in Europe. J Public Health 15:245–254. https://doi.org/10.1007/s10389-007-0130-4
Lund V (2006) Natural living – a precondition for animal welfare in organic farming. Livest Sci 100:71–83. https://doi.org/10.1016/j.livprodsci.2005.08.005
Lusk JL, Norwood FB (2011) Animal Welfare Economics. Appl Econ Perspect Policy 33:463–483. https://doi.org/10.1093/aepp/ppr036
Marinoudi V, Lampridi M, Kateris D, Pearson S, Sørensen CG, Bochtis D (2021) The future of agricultural jobs in view of robotization. Sustain For 13:1–15. https://doi.org/10.3390/su132112109
Marquer P, Rabade T, Forti R (2014) Pig farming in the European Union: considerable variations form one member state to another. Eurostat Statistics in Focus 15/2014, Agriculture Fisheries. https://ec.europa.eu/eurostat/web/products-statistics-in-focus/-/KS-SF-14-015
McCulloch SP (2013) A critique of FAWC'S five freedoms as a framework for the analysis of animal welfare. J Agric Environ Ethics 26:959–975. https://doi.org/10.1007/s10806-012-9434-7
McLeod-Kilmurray H (2012) Commoditizing nonhuman animals and their consumers: industrial livestock production, animal welfare, and ecological justice. Bull Sci Technol Soc 32:71–85. https://doi.org/10.1177/0270467612444585
Mellor DJ, Webster RJ (2014) Development of animal welfare understanding drives change in minimum welfare standards. Rev Sci Tech 33:121–130. https://doi.org/10.20506/rst.33.1.2258

Miele M, Bock BB (2007) Competing discourses of farm animal welfare and agri-food restructuring. International Journal of Sociology of Food and Agriculture 15:1–7. https://doi.org/10.48416/ijsaf.v15i3.275
Miele M, Murdoch J, Roe E (2010) Animals and ambivalence: governing farm animal welfare in the European food sector. In: Higgins V, Lawrence G (eds) Agricultural governance: globalization and the new politics of regulation. Routledge, Oxon
Miele M, Veissier I, Evans A, Botreau R (2011) Animal welfare: establishing a dialogue between science and society. Anim Welf 20:103–117. http://orca.cardiff.ac.uk/id/eprint/11640
Milona P, Ventura F (2019) New generation farmers: Rediscovering the peasantry. Journal of Rural Studies 65:43–52 https://doi.org/10.1016/j.jrurstud.2018.12.009
Molnár M (2018) What is the right thing to do? Perspectives on the "moral imperative" and the "ultimate goal" of farm animal welfare in Hungary. Doctoral dissertation. Department of Environmental Sciences and Policy, Central European University, Budapest
Molnár M (2022) How do farmers conceptualize the ideal and realistic future of animal farming? A critical analysis of trends in agriculture and their implications for farm animal welfare. Animals 12:2086. https://doi.org/10.3390/ani12162086
Molnár M, Fraser D (2020) Protecting farm animal welfare during intensification: Farmer perceptions of economic and regulatory pressures. Animal Welfare, 29(2):133-141. https://doi.org/10.7120/09627286.29.2.133
Molnár M, Fraser D (2021) Animal welfare during a period of intensification: the views of confinement and alternative pig producers. Anim Welf 30:121–129. https://doi.org/10.7120/09627286.30.2.121
Monath TP, Kahn LH, Kaplan B (2010) Introduction: one health perspective. ILAR J 51:193–198. https://doi.org/10.1093/ilar.51.3.193
Morini M (2018) Animals first!' The rise of animal advocacy parties in the EU: a new party family. Contemp Polit 24:418–435. https://doi.org/10.1080/13569775.2018.1434450
Muller A, Schader C, El-Hage Scialabba N, Brüggemann J, Isensee A, Erb KH, Smith P, Klocke P, Leiber F, Stolze M, Niggli U (2017) Strategies for feeding the world more sustainably with organic agriculture. Nat Commun 8:1290. https://doi.org/10.1038/s41467-017-01410-w
Murray A, Skene K, Haynes K (2017) The circular economy: an interdisciplinary exploration of the concept and application in a global context. J Bus Ethics 140:369–380. https://doi.org/10.1007/s10551-015-2693-2
Passatino A, Conte F, Russo M (2008) Animal welfare labelling and the approach of the European Union: an overview on the current situation. J Verbr Lebensm 3:396–399. https://doi.org/10.1007/s00003-009-0303-x
Perfecto I, Vandermeer J, Wright A (2009) Nature's matrix: linking agriculture, conservation and food sovereignty. Earthscan, London
Perkins R (2003) Technological 'lock-in'. In: Internet Encyclopaedia of ecological economics. International Society for Ecological Economics. http://isecoeco.org/pdf/techlkin.pdf
Regan T (1988) The case for animal rights. Routledge, London
Rhodes CJ (2017) The imperative for regenerative agriculture. Sci Prog 100:80–129. https://doi.org/10.3184/003685017X14876775256165
Rollins BE (2008) Animal production and the new social ethic for animals. Journal of Social Philosophy 25:71–83 https://doi.org/10.2527/2004.823955x
Rose M (2010) Challenges to the development and implementation of public policies to achieve animal welfare outcomes. Animals 1:69–82. https://doi.org/10.1111/ropr.12102
Sandel JM (2013) Market reasoning as moral reasoning: why economists should re-engage with political philosophy. J Econ Perspect 27:121–140. https://doi.org/10.1257/jep.27.4.121
Salatin J (2013) You can farm: the entrepreneur's guide to start and succeed in a farm enterprise. Chelsea Green Publishing, London
Sayer KA (2013) Animal machines: the public response to intensification in Great Britain, c.1960-c.1973. Agric Hist 87:473–501. https://doi.org/10.3098/ah.2013.87.4.473
Schweitzer A (1971) Reverence for life. Republished. In: Seymour P (ed) Albert Schweitzer: Reverence for life. Hallmark Editions, Kansas City

Singer P (1995) Animal liberation, 2nd edn. London, Pimlico
Spooner JM, Schuppli CA, Fraser D (2014) Attitudes of Canadian pig producers toward animal welfare. J Agric Environ Ethics 27:569–589. https://doi.org/10.1007/s10806-013-9477-4
Stibbe A (2005) Counter-discourses and the relationship between humans and other animals. Anthrozoös 18:3–17. https://doi.org/10.2752/089279305785594289
Swinbank A (2006) Like products, animal welfare and the World Trade Organization. J World Trade 40:687–711. https://heinonline.org/HOL/LandingPage?handle=hein.kluwer/jwt0040&div=44&id=&page=
Te Velde H, Aarts N, Van Woerkum C (2002) Dealing with ambivalence: farmers' and consumers' perceptions of animal welfare in livestock breeding. J Agric Environ Ethics 15:203–219. https://doi.org/10.1023/A:1015012403331
Thompson PB (2001) Animal welfare and livestock production in a post-industrial milieu. J Appl Anim Welf Sci 4:191–205. https://doi.org/10.1207/S15327604JAWS0403_03
Tomich TP, Brodt S, Ferris H, Galt R, Horwath WR, Kebreab E, Leveau JHJ, Liptzin D, Lubell M, Merel P, Michelmore R, Rosenstock T, Scow K, Six J, Williams N, Yang L (2011) Agroecology: a review from a global-change perspective. Annual Rev Environ Res 36:193–222. https://doi.org/10.1146/annurev-environ-012110-121302
Torske MO, Bjørngaard JH, Hilt B, Glasscock D, Krokstad S (2016) Farmers' mental health: a longitudinal sibling comparison—the HUNT study, Norway. Scandinavian J Work Environ Health 42:547–556. https://doi.org/10.5271/sjweh.3595
Tree I (2018) Wilding: the return of nature to a British farm. Bloomsbury Publishing, London
Tree I, Burrell C (2023) The book of wilding: a practical guide to rewilding big and small. Bloomsbury Publishing, London
Tscharntke T, Grass I, Wanger TC, Westphal C, Batáry P (2021) Beyond organic farming: harnessing biodiversity-friendly landscapes. Opinion Trends Ecol Evol 36:919–930. https://doi.org/10.1016/j.tree.2021.06.010
UNIDO (2020) Short food supply chains for promoting local food on local markets. United Nations Industrial Development Organization, Vienna
Vapnek J, Chapman M (2010) Legislative and regulatory options for animal welfare. In: Development law service, food and agriculture Organization of the United Nations, legal office, Rome
Vaarst M (2010) Organic farming as a development strategy: who are interested and who are not? J Sustain Develop 3:38–50. https://doi.org/10.5539/jsd.v3n1p38
Verhoog H, Lund V, Alrøe HF (2004) Animal welfare, ethics and organic farming. In: Vaarst M, Roderick S, Lund V, Lockeretz W (eds) Animal health and welfare in organic agriculture. CABI Publishing, Oxon, pp 73–94. https://doi.org/10.1079/9780851996684.0073
Vanhonacker F, Verbeke W, Van Poucke E, Tuyttens FAM (2008) Do citizens and farmers interpret the concept of farm animal welfare differently? Livest Sci 116:126–136. https://doi.org/10.1016/j.livsci.2007.09.017
Ventura BA, von Keyserlingk MAG, Schuppli CA, Weary DM (2013) Views on contentious practices in dairy farming: the case of early cow-calf separation. J Dairy Sci 96:6105–6116. https://doi.org/10.3168/jds.2012-6040
Ventura BA, Keyserlingk MAG, Wittman H, Weary DM (2016) What difference does a visit make? Changes in animal welfare perceptions after interested citizens tour a dairy farm. PLoS One 11:1–18. https://doi.org/10.1371/journal.pone.0154733
Vernooij A (2015) The EU pork industry: competitive power is key! Rabobank Food & Agribusiness Research and advisory. Rabobank Industry Note #509
Wax E, Wanat Z (2021) EU farm chief: badly implemented farm-to-fork plan could harm production. Strategy under fire over fears it could damage farmers' productivity, Politico. https://www.politico.eu/article/eu-farm-chief-badly-implemented-farm-to-fork-plan-could-harm-food-production/
Webster AJF (1994) Animal welfare: a cool eye towards Eden. A constructive approach to the problem of man's dominion over animals. Blackwell Science: Oxford-Cambridge, United Kingdom
Webster AJF (2005) The assessment and implementation of animal welfare: theory into practice. Sci Tech Rev Off Int des Epizooties 24:723–734. https://pubmed.ncbi.nlm.nih.gov/16358522/

Wicks D (2011) Silence and denial in everyday life: the case of animal suffering. Animals 1: 186–99. https://doi.org/10.3390/ani1010186

Winter M, Fry C, Carruthers SP (1998) European agricultural policy and farm animal welfare. Food Policy 23:305–323. https://doi.org/10.1016/S0306-9192(98)00036-0

Woods A (2011) From cruelty to welfare: The emergence of farm animal welfare in Britain, 1964–1971. Endeavour 2012:14–22 https://doi.org/10.1016/j.endeavour.2011.10.003

Woods A (2012) From cruelty to welfare: the emergence of farm animal welfare in Britain, 1964–71. Endeavour, Feature 36(1):14–22. https://doi.org/10.1016/j.endeavour.2011.10.003

WWF (2024) Living planet report 2024—a system in peril. WWF International, Gland. https://livingplanet.panda.org/en-GB/

Yeates JW (2017) How good? Ethical criteria for a 'good life' for farm animals. J Agric Environ Ethics 30:23–35. https://doi.org/10.1007/s10806-017-9650-2

Areas of Non-negotiable Achievements of Advanced Agri-food Systems: The Relevance of the Highest Possible Food Safety Standards in the Transformation Process

10

Diana Meemken and Thomas Blaha

As all chapters of this book make clear, governing a targeted transformation of established functional cycles requires mediating conflicts of interest and balancing conflicts of societal values. For both, a decision-making process is necessary, which considers the wants and needs of all affected stakeholder groups (also those that cannot express their interests themselves!). To find desirable decisions democratically, public discourse must comply with moral principles that guarantee outcomes that are just and fair to everybody and that are ethically justifiable (Habermas 1991). To achieve this, expert groups representing the major stakeholder groups should take over the moderation of the discourse so that all wants, needs, and interests are considered. In case of conflicts of societal values such as providing affordable food for all by efficient crop farming vs. maintaining biodiversity or keeping animals outdoors vs. the need to reduce emissions, etc., as a rule, decisions for balancing the contradicting areas need to be made. Just and fair trade-offs are the best way to solve these conflicts, i.e. none of the contradicting issues should be maximised. The best possible solutions are those where all interest groups that insist on maximising their demands are somewhat disappointed (Rittel and Webber 1973), which means that every interest group must agree on some trade-offs.

However, there are two major exceptions to this consensual approach to solving conflicts of societal values. The two areas in agri-food systems, where the

D. Meemken
Working Group Meat Hygiene, Institute of Food Safety and Food Hygiene, School of Veterinary Medicine, Freie Universität, Berlin, Germany
e-mail: diana.meemken@fu-berlin.de

T. Blaha (✉)
Former Head of the Field Station for Epidemiology (Bakum), University of Veterinary Medicine Hannover, Hannover, Germany
e-mail: thomas.blaha.ir@tiho-hannover.de

B. Grabkowsky, T. Blaha (eds.), *Intensive Livestock Production in Transition*,
https://doi.org/10.1007/978-3-031-97872-2_10

achievements of successful measures that protect the health of humans and animals are non-negotiable, are:

1. The achievements in providing safe food to all consumers, as every avoidable death due to contaminated food must be prevented.
2. The achievements in protecting animal herds and flocks from infections and preventing the spread of harmful pathogens from animals to humans, which potentially can lead to pandemics, as COVID-19 has demonstrated only recently.

1 A Short History of Food Safety

1.1 The Very Beginning

It is hard to debate that food availability has had the greatest impact on the existence and development of mankind (as for any other existing or emerging species on earth). Having access to food meant (and still means) simply the key to survival. Our early ancestors knew, of course, from their animal origin that not everything you can put in your mouth and swallow is eatable and wholesome. This means that the concept of "being aware of what you eat" has accompanied mankind ever since. In the beginning, the experience that some foods were not healthy (sometimes deadly) led to avoiding harmful foods. For unknown foods, we developed the concept of trial and error, which cost single lives, but taught us what to avoid and what to eat. Furthermore, the same principle was used to learn which guidelines for food production, preparation, and storage led to safe(r) resp. more digestible food. With developing more complex human societies, world religions like Hinduism, Buddhism, Judaism, Christianity, and Islam evolved and the associated religious scriptures that fundamentally defined coexistence in societies spread out. In addition, these scriptures include requirements for "clean" (kosher or halal) food like slaughtering without stunning to support sufficient bleeding of the animal to increase the shelf life of the produced meat or banning the consumption of pork. This was due to various reasons as pigs are considered unclean because they burrow or wallow in the ground, and they are food competitors. However, it can also be speculated that the rules were implemented to protect people from the health risks posed by consuming (undercooked) pig meat. The "religious" ban on pork may be based on observations of disease as a consequence of eating undercooked pig-derived food since the dominant health risks of meat at that time were possibly trichinella carried by pig meat.

That there was a continuous and increasing awareness of a potential threat due to toxic or otherwise unsafe food is proven by our knowledge about food tasters that had to eat or drink samples of what was served to, e.g., ancient kings or high priests (which also proves the fact that adulterated food was seen as a method of assassinations).

In the sixth to the fifteenth century, more than 90% of people in European medieval societies were farmers who slaughtered and processed the animals they kept on their own small farms and consumed the meat themselves.

In terms of animal welfare or animal hygiene, a very individualised and quite uninformed approach prevailed, with no further standards or legal requirements other than the religious guidelines described above.

With the formation of cities between the fifteenth and nineteenth centuries and the limited possibilities for preserving meat, urban abattoirs emerged in which butchers, organised in guilds, slaughtered animals and processed meat. According to their guild rules, they decided whether the meat is marketable at the normal prize or at a lower prize in cases of serious meat quality and possible safety deficits.

1.2 The Emergence of Science-Based Food Safety Measures

With the onset of the Industrial Revolution in the nineteenth century, food-associated health risks became targets of scientific developments. The connection between detectable pathological lesions in animal carcasses and the occurrence of disease in humans following the consumption of such meat has been recognised. Soon it was proven by scientists like Louis Pasteur, Andreas C. Gerlach, Rudolf Virchow, Robert Koch, and Robert von Ostertag that pathogens such as *Trichinella spiralis*, *Mycobacterium,* and *Bacillus anthracis* were the causative agents of the lesions in the carcasses and the consequent diseases in humans.

The veterinarian Robert von Ostertag was the lead author of the German Reichsfleischbeschaugesetz (German Meat Inspection Act), which came into force in 1900 and was the first of its kind globally. The core of it is the mandatory antemortem and postmortem inspection carried out by veterinarians paid by the state (to make them independent of the meat producers), including inspection, palpation, and incision, as well as the obligatory microscopical *Trichinella* examination for susceptible species such as pigs. Animals with apparent diseases were banned from being slaughtered, and meat unfit for human consumption due to pathological lesions was disposed of, significantly improving the safety of meat and meat products.

This "von Ostertag's" method, which was described in his handbook for meat hygiene and published in 1899, and soon became the blueprint for food safety regulations in other countries, was developed to reduce food-associated disease in humans, but it was also successful in the control of apparent animal diseases such as rinderpest or zoonoses like bovine tuberculosis or brucellosis that they were either eradicated or well controlled in the Global North until the end of the twentieth century (Edwards et al. 1997) (Fig. 10.1).

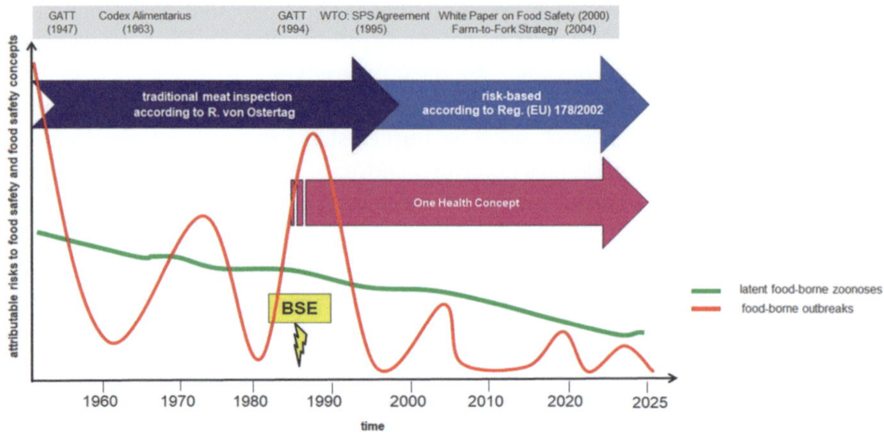

Fig. 10.1 Food-borne health risks and food safety concepts over time

2 International Trade and Food Safety

Trade and foreign policy have been intertwined throughout history, with foreign policy often tailored to promote trade interests. International trade is as old as humans were able to transport goods over long distances (the Silk Road travellers, and the trade ship armadas of Venice and Portugal as early protagonists of the international exchange of goods). The modern idea of international trade as a guarantee of peace and security was born in 1941 at a secret meeting of Churchill and Roosevelt on the Atlantic Ocean off the coast of Newfoundland, as they wanted to facilitate the global rules that underpin our multilateral economic system as a direct response to the Second World War and the desire to prevent its recurrence. The "Atlantic Charter" of 1941 became the basis of the "General Agreement on Tariffs and Trade" (GATT) and eventually the birth of the WTO. There were eight rounds of GATT negotiations, with the last GATT meeting in 1994 in Marrakesh being a cornerstone for the international trade of foodstuffs. Until then, food was not included in the free trade negotiations, since food safety concerns were seen as reason enough to exclude food from the free global trade. But in Marrakesh, the decision was made to make the free trade of foodstuffs also possible by setting special rules for international standards for harmonising the food safety regulations; this decision was adopted by the WTO (World Trade Organisation), which replaced 1995 the GATT negotiations. For this, the work of the Codex Alimentarius Commission, founded in 1963 by the UN Organisations FAO and WHO, was strengthened and declared as the basis for the safety of internationally traded food. Additionally, the WTO Agreement on the Application of Sanitary and Phytosanitary Measures (SPS Agreement) was issued (WTO 1995). This agreement is to both further the international trade of foodstuffs and to prevent the international spread of food-associated health risks by obliging all WTO member states to comply with the international food safety standards set by the Codex Alimentarius (ANON 1976) and to implement the HACCP Concept (Hazard Analysis Critical Control Point). The latter is a system that identifies, evaluates, and controls hazards along the production chain for foodstuffs, which aims at preventing food-borne physical, chemical, and microbiological risks from reaching the consumer.

3 The Development of the Modern Farm-to-Fork Strategy

As much as the Ostertag's meat inspection method was a real success story in terms of eradicating or at least controlling certain animal diseases and reducing food-associated risks due to the so-called classic meat-borne health risks to human health (Sexauer 1965; Edwards et al. 1997), it became clear in the 1990s that inapparent zoonoses such as salmonellosis, campylobacteriosis, hepatitis E, and toxoplasmosis could not be controlled using traditional meat inspection methods as there are no clinical signs and no macroscopically visible lesion.

The realisation that many, if not most, food-borne risks to human health originate in primary production led to the concept of "pre-harvest" food safety in the United States in the early 1990s. This concept was illustrated for farmers by coining the term "food animals are already food on hooves", which increased the awareness of farmers for their part of the responsibility for producing safe food.

In Europe, where the BSE crisis, originating at the end of the 1980s in Great Britain, and the *Salmonella* Enteritidis epidemic, as well as several dioxin scandals, severely shook the confidence in the reliability of the then-established state food safety system, the need to revise the existing food safety concept became clear. Therefore, the European Commission appointed an expert commission to come up with an analysis of the reasons for these kinds of food-associated health risks that originate in animal production. The result was the "White Paper on Food Safety" published in 2000 (Anon 2000). Important principles are the transfer of the responsibility for safe food to all stages of food production "from farm to fork" and to base all official controls on previously assessed risks (Berends and van Knapen 1999).

In just two years, the EU Commission founded the EFSA (European Food Safety Agency), which provides scientific advice on existing and emerging risks in the food sector. Furthermore, with the so-called basic regulation (Regulation (EC) No 178/2002), the EU Commission has initiated a genuine paradigm shift from mandatory end-product control to risk-based process optimisation. In addition, the regulation described an expansion of the objectives of the official tasks beyond the optimisation of food safety and the detection of notifiable animal diseases to the optimisation of animal health and animal welfare using feedback systems along the food chain. The core element of this feedback system is the food chain information, which provides information communicated by the farmers on potential risks that official veterinarians should use to determine the intensity of the meat inspection and on meat inspection results communicated by the official veterinarians that farmers should use to optimise herd health. By abstaining from the former mandatory lymph node incisions, cross-contamination of pathogens like *Salmonella* by the used knives is avoided, and communicated potential risks can be recognised through the targeted use of additional tests. In the almost two decades that followed, two generations of implementing regulations were created, each of which retained and refined the principles of the food chain and risk orientation. In practice, visual meat inspection has become very well established, while the other important part of the system, the targeted intensification of meat inspection using specific measures to control inapparent infectious agents, has remained below its potential and needs continuous improvement. However, all in all, the food safety standards achieved and maintained in the high-income countries have resulted in a food safety level that justifies the statement "food has never been as safe as today", which must not be compromised for whatever other benefit.

References

Anonymous (1976) Recommended international codes of hygienic practice for fresh meat for ante-mortem and post-mortem inspection animals and for processed meat products. Codex Alimentarius Commission, Rom, S. 13ff

Anonymous (2000) White paper on food safety of 12 January 2000 [COM/99/0719 final—Not published in the Official Journal

Berends BR, van Knapen F (1999) An outline of a risk assessment-based system of meat safety assurance and its future prospects. Vet Quarterly 21:128–134

Edwards DS, Johnson AM, Mead GC (1997) Meat inspection: an overview of present practices and future trends. Vet J 154:135–147

GATT (1947) General Agreement on Tariffs and Trade, Oct. 30, 1947, 61 stat. A-11, 55 U.N.T.S. 194

Habermas J (1991) Moral consciousness and communicative action. MIT Press, Cambridge

Rittel HWJ, Webber MM (1973) Dilemmas in a general theory of planning. Pol Sci 4(2):155–169

Sexauer RC (1965) Why state meat inspection? J Am Vet Med Assoc 147(S):1580–1583

WTO (1995) The WTO agreement on the application of sanitary and phytosanitary measures (SPS agreement), 10/11/20, http://www.wto.org/english/tratop_e/sps_e/spsagr_e.htm

Areas of Non-negotiable Achievements of Advanced Agri-food Systems: The Relevance of the Highest Possible Biosecurity Systems in the Transformation Process

11

Franz J. Conraths, Carolina Probst, Michael Bucherer, Jan Holzhausen, and Barbara Grabkowsky

1 Introduction

Biosecurity plays an important role in the transformation process that livestock farming is currently undergoing. It not only reduces the risk of introducing pathogens into animal holdings but can also help improve animal welfare by protecting animals from infectious diseases that can lead to pain, suffering, and death. Moreover, biosecurity is an essential component for sustainable and profitable livestock farming. It is important to note that it has recently been clarified, in the European Union, that the prime responsibility for biosecurity lies in the hands of the operator, i.e., the farmer (Regulation (EU) 2016/429).

Biosecurity measures are necessary to minimize avoidable risks and play a key role in the entire food production chain. They can be applied at different levels, including the choice of an appropriate location if a new farm is planned, constructional elements for farm buildings, and farm management practices. They address (i) the animal-human interface, e.g., protective clothing; (ii) the animal-animal

F. J. Conraths (✉) · C. Probst
Friedrich-Loeffler-Institut, Federal Research Institute for Animal Health, Greifswald-Insel Riems, Germany
e-mail: franz.conraths@fli.de; carolina.probst@bmz.bund.de

M. Bucherer · J. Holzhausen
Institute of Construction Design, Industrial and Health Care Building, Technical University of Braunschweig, Braunschweig, Germany
e-mail: m.bucherer@tu-braunschweig.de

B. Grabkowsky
Center for Sustainability Transformation Research in AgriFood Systems, Vechta Institute of Sustainability Transformation in Rural Areas, University of Vechta, Vechta, Germany
e-mail: barbara.grabkowsky@uni-vechta.de

B. Grabkowsky, T. Blaha (eds.), *Intensive Livestock Production in Transition*,
https://doi.org/10.1007/978-3-031-97872-2_11

interface, e.g., appropriate stocking density; and (iii) the animal-environmental interface, e.g., protection against contact with wild birds, rodents, or arthropods. Biosecurity planning involves different disciplines and specialties (human, animal/wildlife health; ecology; economics; behavioral sciences and communication; architecture, etc.) and points of view (traditional knowledge, environmental justice, financing, etc.). While biosecurity primarily aims at protecting livestock or poultry kept on a farm and the products derived from the animals, it may also help to avoid shedding pathogens and carriers of antimicrobial resistance into the environment. Biosecurity planners can therefore be important partners in implementing the One Health approach.

In this sense, the One Health approach should be an integral part of biosecurity planning. One Health, as defined by the One Health High-Level Expert Panel (OHHLEP), "is an integrated, unifying approach that aims to sustainably balance and optimize the health of people, animals and ecosystems" (Adisasmito et al. 2022). The One Health approach is particularly suitable to address zoonotic and vector-borne infections as well as challenges related to antimicrobial resistance (Anonymus 2022; Thal and Mettenleiter 2023). However, the One Health approach also implies that we should move away from an anthropocentric focus, also on animal husbandry, toward a more holistic view of health. To achieve the full integration of the environment—including biodiversity—with biosecurity concerns and measures, we need to accept the fact that humans are just one species within the large group of vertebrates and change our relationship with nature in general.

2 Biosecurity and Biosafety

The terms biosecurity and biosafety receive various definitions, depending on the discipline involved (veterinary science, food, environment, etc.) or the country where it is used (Burnette et al. 2013). As a result, both terms are frequently used without any agreed definition or scope. We use the term biosecurity as defined by the World Organisation for Animal Health (WOAH, former OIE) as a set of management and physical measures designed to reduce the risk of introduction, establishment, and spread of animal diseases, infections or infestations to, from and within an animal population (WOAH 2019). Regarding animal farms, biosecurity usually refers to *operational measures*, i.e., management practices designed to reduce the risk of the introduction, establishment, and spread of infections to, from, and within farms, including fisheries. Such measures may include personal hygiene, disease monitoring, medication regimes, applying the all-in all-out principle or a closed-herd policy, manure and carcass removal, controlling vermin and birds, cleaning, and disinfection (Dewulf and van Immerseel 2019; Grabkowsky et al. 2020). In addition, there is *constructional biosecurity*, which includes building measures, such as a quarantine stable or a washbasin (Bucherer et al. 2021). Both operational and constructional measures can be further divided into those directed at preventing pathogens from entering a farm (external biosecurity) and those preventing the spread within a farm (internal biosecurity) (Bucherer et al. 2021).

In the peer-reviewed literature, disease transmission routes and corresponding biosecurity practices from the individual farm to the international level are well documented but hardly led to recommendations on the construction of farms and stables. Most studies focus on specific diseases, e.g., avian influenza (Ssematimba et al. 2013; Conraths et al. 2016; Glass et al. 2019; Robertson 2020). Others concentrate on specific species, e.g., pigs (Filippitzi et al. 2018; Döring et al. 2019), poultry (van Steenwinkel et al. 2011; Carr and Howells 2018), or cattle (Brennan and Christley 2012; Renault et al. 2018a, b, 2020). Some publications focus more specifically on certain production types, e.g., dairy cattle (Sibley 2010), veal calves (Damiaans et al. 2019), fattening pigs (Boklund et al. 2004), foie-gras ducks (Delpont et al. 2018), or fish in aquaculture (Delabbio 2006).

According to the FAO, the three main elements of biosecurity are segregation, cleaning, and disinfection (FAO 2010). While cleaning and disinfection relate to management measures and therefore represent the "software" of biosecurity, segregation, i.e., the "creation and maintenance of barriers to limit the potential opportunities for infected animals and contaminated materials to enter an uninfected site" (FAO 2010), represents the "hardware" of biosecurity.

3 Who Is Responsible for Farm Biosecurity?

According to the Animal Health Act of the European Union, which is directly applicable in all EU member states, "operators" shall, as regards keeping animals and products under their responsibility, be responsible for:

(i) The health of kept animals.
(ii) Prudent and responsible use of veterinary medicines, without prejudice to the role and responsibility of veterinarians.
(iii) Minimising the risk of the spread of diseases.
(iv) Good animal husbandry (Regulation (EU) 2016/429, Article 10, Number 1, (a)).

Where appropriate, operators, including farmers, "take such biosecurity measures regarding kept animals, and products under their responsibility, as are appropriate for:

(i) the species and categories of kept animals and products,
(ii) the type of production; and,
(iii) the risks involved, taking into account: geographical location and climatic conditions; and—local circumstances and practices" ((Regulation (EU) 2016/429, article 10, paragraph 1, point (b)).

Where appropriate, operators, including farmers, must also "take biosecurity measures regarding wild animals" Regulation (EU) 2016/429, article 10, paragraph 1, point (c)). "Animal professionals," i.e., natural or legal persons having an occupational relationship with animals or products, other than operators or veterinarians, “shall take action to minimise the risk of the spread of diseases in the context of their occupational relationship with animals and products (Regulation (EU) 2016/429, article 10, paragraph 2).

The biosecurity measures referred to in point (b) of paragraph 1, article 10, Regulation (EU) 2016/429 “shall be implemented, as appropriate, through:

(a) Physical protection measures, which may include:
   (i) Enclosing, fencing, roofing, netting, as appropriate.
   (ii) Cleaning, disinfection and control of insects and rodents.
   (iii) In the case of aquatic animals, where appropriate: measures concerning the water supply and discharge;—natural or artificial barriers to surrounding water courses that prevent aquatic animals from entering or leaving the establishment concerned, including measures against flooding or infiltration of water from surrounding water courses.
(b) Management measures, which may include:
   (i) Procedures for entering and exiting establishments for animals, products, vehicles and persons.
   (ii) Procedures for using equipment.
   (iii) Conditions for movement based on the risks involved.
   (iv) Conditions for introducing animals or products into the establishment.
   (v) Quarantine, isolation or separation of newly introduced or sick animals.
   (vi) A system for safe disposal of dead animals and other animal by-products.”

While many farmers believed in the past that they fulfilled their duties, if they had followed the instructions of official veterinarians, they are now in charge of biosecurity on their premises. To make them capable of fulfilling their obligations regarding biosecurity, operators, a term that includes farmers and farm managers, and animal professionals must “have adequate knowledge of:

(a) Animal diseases, including those that are transmissible to humans.
(b) Biosecurity principles.
(c) The interaction between animal health, animal welfare and human health.
(d) Good practice of animal husbandry for the animal species under their care.
(e) Resistance to treatments, including antimicrobial resistance, and its implications” (Regulation (EU) 429/ 2016, article 11, paragraph 1).

Operators, animal professionals, and pet keepers are obliged to “cooperate with the competent authority and veterinarians in the application of the disease prevention and control measures” provided for in Regulation (EU) 429/ 2016.

In many countries, national legislation stipulates similar responsibilities or supplements supranational regulations at the national level. This means that farmers and farm managers are responsible for biosecurity on the premises where they keep livestock or poultry.

4 Farmers' Knowledge of Biosecurity

Farmers' knowledge and opinions about biosecurity have been assessed in various studies, many of which demonstrated room for improvement in sectors including pigs (Ribbens et al. 2008; Laanen et al. 2010), poultry (Gelaude et al. 2014), and cattle (Laanen et al. 2014). The following example may illustrate this: Although badgers are known to pose a risk of transmitting bovine tuberculosis (bTB) to cattle (Donnelly and Nouvellet 2013), commercial cattle farms that had experienced herd breakdowns of bTB had no biosecurity measures implemented several years after the outbreaks to prevent wild animals from entering the buildings and feed stores (Ward et al. 2008).

The motivation to implement biosecurity measures often seems to be associated with the type and size of the farm (Nöremark et al. 2010; Simon-Grifé et al. 2013; Sahlström et al. 2014), the current epidemiological situation, and the practicability of measures (Garforth et al. 2013; Shortall et al. 2017; Renault et al. 2018a). It has also been pointed out that on-farm conditions may represent obstacles to appropriate biosecurity measures (Nöremark and Sternberg-Lewerin 2014) or emphasize the need for training farmers in planning and implementing biosecurity measures (Kuster et al. 2015; Pritchard et al. 2015).

From a farmer's or veterinarian's perspective, informal recommendations on biosecurity have been provided by various institutions, associations, and stakeholders. Some apply to specific species, for example, the reports of the Food and Agriculture Organization (www.fao.org), the Terrestrial Animal Health Code of the World Organisation for Animal Health (WOAH 2019), national regulations such as the recommendations of the Federal Ministry of Food and Agriculture for keeping ruminants in Germany (BMEL 2014) or various guidelines of the German Association for Technology and Structures in Agriculture (https://www.ktbl.de/www.ktbl.de), and the Agricultural Construction Association (www.bfl-online.de). However, only a few of them specifically address constructional aspects of biosecurity. One of the few exceptions is the Guideline on biosecurity in cattle farms of the German Federal State of Lower Saxony (https://tierseucheninfo.niedersachsen.de/startseite/tierseuchenbekampfung/leitfaden_biosicherheit_in_rinderhaltungen/leitfaden-biosicherheit-in-rinderhaltungen-124077.html).

Due to the lack of coordinated guidance, an increasing number of checklists are available online for evaluating the biosecurity level of a farm. In Germany, the quality assurance system for food (QS-System) provides checklists for poultry, pig, and cattle (www.q-s.de), the Ministry for Agriculture and Environment of Sachsen-Anhalt issued checklists for different livestock, including fish and bees (mule.sachsen-anhalt.de), and the University of Vechta, Germany, has developed various online tools in German to assess and improve the biosecurity status of poultry, pig, and cattle farms together with the Friedrich-Loeffler-Institut, Federal Research Institute for Animal Health (https://risikoampel.uni-vechta.de), and large numbers of experts in the respective fields (Grabkowsky et al. 2020). Similar checklists are also available in English, for example, the risk-based "biocheck" scoring systems provided by Ghent University, Belgium (https://biocheck.ugent.be/en/worldwide).

5 Area- and Barrier-Management/ Constructional Biosecurity

Apart from several requirements, which also refer to animal health and animal welfare, the design of animal farms and the construction of farm buildings lie within the responsibility of the farmers, architects, or construction professionals. While plenty of regulations exist to warrant the constructional safety of farm buildings from statics to fire protection, the level of constructional biosecurity is ultimately left to the farmers' judgment.

Animal farms are complex and dynamic systems. Thus, several infrastructural features are relevant for biosecurity, from the geographic location to the properties of the constructional components used to build the farm. While the layout and construction of a farm can significantly reduce the risk of introducing pathogens into animal holdings, farmers and their advisors face the challenge of translating the individual protection requirements into the specifics of farm construction and building. This entails understanding constructional and functional options, assessing associated risks and benefits as well as balancing the implemented measures under consideration of influencing factors, such as economic impact, return on investment, and animal welfare. Building companies, engineers, and architects provide a large array of proposals for constructing animal farm buildings. While they usually comply with size specifications and other animal welfare standards, most of them fail to consider biosecurity issues. Definitions and nomenclature may vary between the different stakeholders involved in biosecurity planning. This may, in part, be the reason for the limited connections between them and the lack of interdisciplinary approaches to biosecurity planning in animal farms.

Within the Building Infection Protection Classification (BIPROC) project, architects and veterinary epidemiologists focused on the "hardware," i.e., the infrastructure of farms in the animal health context (Bucherer et al. 2021). This interdisciplinary approach to infection prevention aimed at analyzing livestock farms from both a construction-functional and a veterinary-epidemiological biosecurity perspective. In human medicine, a similar approach is successfully applied to the design of hospitals (Lateef 2009; Tabori and Dettenkofer 2017; Sunder et al. 2018). The project developed a theoretical framework that includes the key constructional elements in terms of biosecurity.

To gain mutual understanding between architects and veterinarians and to create a common basis of knowledge and terms, an interdisciplinary team visited conventional and organic farms, which kept cattle, poultry, and pigs and were in different parts of Germany. The following aspects were addressed: Physical farm design, workflows and routine management practices, numbers and types of biosecurity areas, their spatial layout, and physical separation. Based on these data, key constructional components concerning biosecurity were delineated and defined.

Based on the scientific literature and lessons learned from disease outbreak investigations on avian influenza (Conraths et al. 2016; Dietze et al. 2018; Globig et al. 2018), bovine viral diarrhoea (Gethmann et al. 2015), and infectious bovine rhinotracheitis, the available knowledge on pathogens, transmission routes, and potential pathways of entry into a farm was organized systematically. Potential vectors of pathogens (humans, animals) and fomites (e.g., working equipment) were characterized and classified according to their size, types of locomotion, and specific mobility abilities. Regarding their eligibility to pass through different biosecurity areas of a farm, they were differentiated into those that must be able to cross barriers in a controlled way (e.g., poultry, livestock, veterinarians, feed, equipment, etc.) and those that need to be kept out. Based on the findings, relevant constructional components were defined, and a common terminology was established. Constructional biosecurity aims to interrupt the pathways of pathogen transmission by controlling or stopping potential vectors (persons, animals) or fomites (materials) on their way into the farm. It consists of the quality of the layout, the construction quality, and the barrier quality and thus supports the area and barrier management. Constructional biosecurity is thus critical for achieving operational biosecurity.

Due to the large number of influencing factors, only a holistic approach can help establish an optimal farm-specific level of biosecurity. It requires farmers, architects, and veterinarians to work together as a transdisciplinary team from the planning stage onwards, although each of the disciplines has a different background, takes a different perspective and uses specific terminology, which may be a challenge for collaboration. They need to cover all three main elements of biosecurity, i.e., segregation, cleaning, and disinfection (FAO 2010), which requires knowledge of all spatial, functional, and operational interdependencies as well as the existence of locally active vectors and routes, by which pathogens can enter a farm. This means considering operational biosecurity and constructional biosecurity in conjunction.

Operational measures, such as the delivery of animal feed or cleaning the stable, need to be supported or made possible by constructional biosecurity measures, such as access control and provision of adequate space. Operational biosecurity measures are thus only made possible by the functional measures (farm layout) in conjunction with the space-forming and space-enclosing constructional measures (surfaces and farm building). Ideally, these measures are included in the site layout and the building design from the very beginning, leading to the farm being safe by design.

The following key components and basic design principles of physical infrastructure can be derived to support operational biosecurity. First, physical infrastructure components need to include biosecurity areas according to the protection requirements of the goods to be protected, i.e., animals, feed, or animal products. Along with the operational biosecurity management process, different functional areas can be set in a spatial relationship, and areas of the same biosecurity level can be grouped or combined to form a single biosecurity area (Fig. 11.1).

Then, a construction barrier must be planned for each biosecurity area. At the highest level of biosecurity, this barrier needs to be designed to perform uninterruptedly in all three dimensions (floor, walls, ceiling/roof) and is only allowed to be trespassed by locks. Locks allow the controlled passage of selected persons, animals, or objects through the barrier from one biosecurity area to the next. The transition between two biosecurity areas may be supported by a "conditioning process" (e.g., filtering the air, cleaning shoes) as an operational measure to achieve the required hygienic properties (Fig. 11.2).

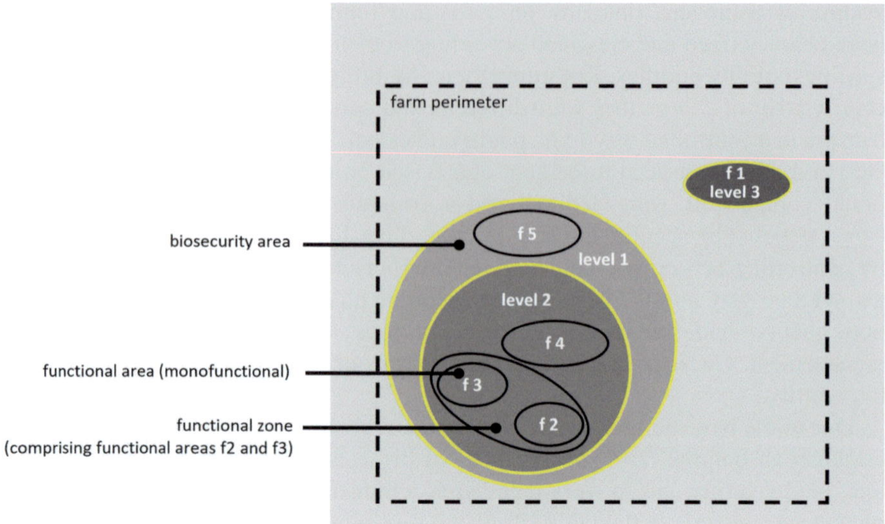

Fig. 11.1 Possible arrangements of areas with different biosecurity levels: independent, concentric

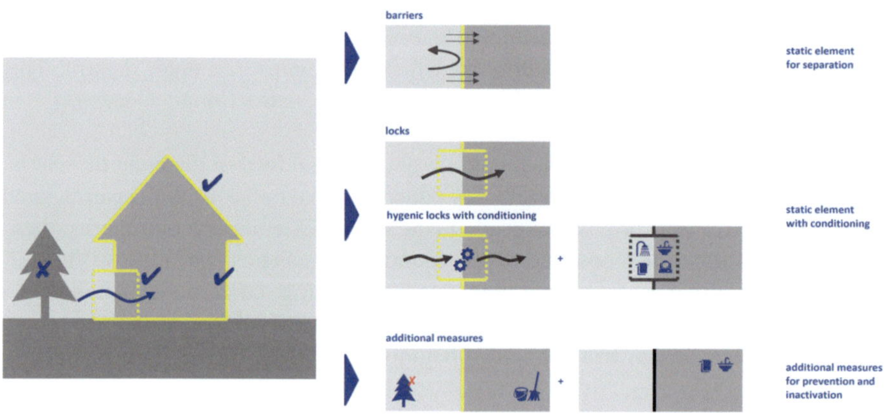

Fig. 11.2 Overview of levels of consideration

In addition, the constructional measures ("hardware") of the barriers are complemented by management practices ("software"). Zone demarcation and the one-way principle represent an intuitive example of a functional measure since it focuses on a linear process to avoid cross-contamination.

The design of the barrier itself is defined by the size and type of locomotion of the vector to be kept out. The outer shell of the biosafety area is therefore defined by its permeability and dimensionality and realized with a selected type of construction and materiality. Permeability determines the size of the smallest vector prevented from crossing a barrier, while allowing smaller vectors, that have been identified as innocuous, to pass freely. Dimensionality determines the type of locomotion prevented from crossing the barrier.

An important step in designing a biosecurity strategy is choosing a level between theoretical maximum protection ("zero risk"), possible implementation in practice with reasonable effort ("acceptable risk"), and keeping animals at an unacceptably high-risk level.

A risk assessment, i.e., the identification and quantification of risks associated with particular hazards (pathogens) and their entry pathways, including an estimate of the likelihood that a pathogen enters via the respective pathway and an assessment of the consequences of such an event's likelihood, may assist in determining an acceptable level of risk. Some risks can be acceptable or may not be fully known. Others may previously not have been identified in the region or suddenly occur, e.g., due to the spread of pathogens by human activity, wildlife or domestic animals, increasing population densities, destruction of natural wildlife habitats, or climate change. From a scientific point of view, consistent implementation is essential, while from a practical point of view, this consistent implementation may often appear impractical or even unfeasible. It is therefore necessary to decide on a case-by-case basis, which goals are to be achieved and where risks are consciously accepted by softening biosecurity criteria.

For the development of a holistic construction strategy for a farm, or for classifying the biosecurity status of an existing farm, the following three steps may be useful:

(i) Recording of the *PRESENT* status of a farm, in which the premise is divided into different biosecurity areas. This breakdown makes it possible to assign individual biosecurity levels and determine biosecurity areas.
(ii) Determining the *IDEAL* status to achieve maximum protection (zero risk) based on constructional infection prevention, and.
(iii) Choosing a *TARGET* status, which is based on an individual risk assessment. For this status, the farmer, supported by his advisors, decides which goals are to be achieved and in which areas risk levels deviating from "zero risk" are consciously accepted.

On the basis of these three scenarios, the existing level of protection can be compared with an ideal state. Vulnerabilities and thus potential for improvement can be identified, and a comprehensive implementation strategy can be developed.

Spatially, biosecurity starts at the perimeter fence of any premise where livestock or poultry is kept. It helps to ensure that only authorized personnel get access in a controlled way and may keep some unwanted animal species away. To control access of people and vehicles, gates must be closed and only opened as long as this is necessary for entering the premises. Disinfection points and devices at the entrances and gates may help to further reduce the risk of introducing pathogens, e.g., via wheels or shoes.

A plethora of materials, including metals (brick, stone, iron or steel, aluminum, timber, etc.), has traditionally been used for constructing stables and farm buildings. At least floors, walls and ceilings within the buildings must be easy to cleanse in disinfect. This holds also true for traffic areas in and outside buildings. Ventilation devices may represent a challenge for pathogens that are airborne or transmitted by wild birds. While wood or timber have been used as renewable construction material for farm buildings and stables in many countries for centuries, it has been argued that these materials may be more difficult to sanitize as compared to metal or concrete surfaces (Böhm et al. 2019). Yet, recent research suggests that wood and timber can be effectively disinfected (Oettler et al. 2024a). Moreover, at least some kinds of intact, finely sawn timber with a low depth of roughness had an inactivating effect on prototypes of non-enveloped (Enterovirus E) and enveloped viruses (Newcastle Disease Virus), while these pathogens remained infectious for a long period on moist steel (Oettler et al. 2024b). Although some standard concepts for stable construction have certainly proven their worth, it seems therefore necessary to take recent information regarding renewable building materials into account. Independent professional and science-based advice for stable construction can be obtained from the Kuratorium für Technik und Bauwesen in der Landwirtschaft e.V. (KTBL; https://www.ktbl.de/webanwendungen/baukost-investition-betriebsgebaeude) or the "Landgesellschaft" in German federal states, such as Lower Saxony or Mecklenburg-Western Pomerania (https://www.lgmv.de/bauen/landwirtschaftliches-bauen/).

6 Management and Biosecurity Plan

While farm and stable construction including the positioning of buildings in a way that avoids crossing routes and thus cross-contamination, each farm must adopt a defined set of rules to manage the activities of all persons that enter the farm premises. This comprehensive set of rules should be prepared in writing and revised at regular intervals to check each rule whether it is necessary, applicable, and regularly applied. All persons entering the farm premises must be informed at least about the rules that apply to them. This also includes veterinarians, professional farm visitors, guests, relatives, etc.

Operators, i.e., farmers or farm managers, must inform all farm workers about the general concept of the biosecurity plan, their role in implementing it, their responsibilities, and those of others working on the premises. All farm workers must be actively trained in biosecurity, in implementing the farm-specific biosecurity plan, and in the measures to be taken in case of any violation. They should be involved in the regular revisions of the biosecurity plan to ensure that living biosecurity becomes a matter of the heart for everybody employed on the farm.

7 Guidelines and Online Tools

Several guidelines and tools exist that may help to establish and improve farm-specific biosecurity measures, detect gaps, and prioritize measures to optimize farm-specific biosecurity plans. Ghent University operates the biocheck.ugent (https://biocheckgent.com/en), which is available in several languages including English, Chinese, French, Dutch and Spanish, for pigs, cattle, small ruminants, and poultry. The German "Risikoampel" (risk-traffic light) can be used for pigs (conventional stables and outdoor systems), also in an adaptation for Switzerland, for poultry (conventional holdings, a version for mobile chicken stables is under development), and soon also for cattle.

While existing tools are usually science-based and may also reflect legal requirements (e.g., the ASP-Risikoampel https://risikoampel.uni-vechta.de/plugins.php/aisurveyplugin/asp/survey/experts?disease_id=4), their application often lacks behind the expectations of the institutions that established them and of farmers' organizations that supported the development of the tools and recommend using them. The reasons for the limited acceptance of the tools and guidelines are manifold. Farmers argue that there are too many guidelines and tools and that applying them leads to additional bureaucracy. Online tools are usually questionnaire-based, and it is argued that they include too many questions, which may not all be equally relevant or applicable to a particular farm. Completing the questionnaire is regarded as time-consuming (usually 30–90 min). Revising and improving existing biosecurity concepts is seen as labor-intensive with little economic gain, at least at "peace times" regarding major animal diseases. Existing tools should therefore be comprehensive, but as short as possible, easy to understand, and ready to use. They should not only reveal existing gaps but also rank them from essential to less important and provide specific recommendations to overcome biosecurity shortages.

Low-threshold offers are needed for establishing, reviewing, and optimizing farm-specific biosecurity plans and procedures. Potential alternatives to guidelines, tools, etc. may thus be a confidential auditing by members of farmers' organizations trained in biosecurity, mutual visits and checks of farm managers with a focus on biosecurity issues, and discussion rounds among farmers involving biosecurity specialists and veterinarians.

8 Incentives for Improving Biosecurity at the Farm Level

To encourage farm managers to adopt a maximum of precautionary measures to warrant biosecurity, incentives may be useful. Animal insurance (including official compensation funds such as the "Tierseuchenkassen" of the German federal states) could, for example, reduce fees if the farm runs biosecurity measures at a level exceeding legal requirements.

On the other hand, insurance companies (including official schemes such as the "Tierseuchenkassen") may reduce compensations if a farm has violated legal biosecurity rules and if these violations caused or probably contributed to causing the disease outbreak that prompted the compensation claim. The Animal Disease Compensation Fund of the German Federal State of Lower Saxony (Niedersächsische Tierseuchenkasse) has successfully implemented such an approach in response to outbreaks of Highly Pathogenic Avian Influenza in recent years (U. Gerdes, personal communication).

In any case, improving farm biosecurity remains a constant challenge in agricultural transition processes and is embedded into a multifaceted setting that ranges from animal health over the protection of the environment to animal welfare, food safety, and protection from zoonotic infections and the spread of AMR.

9 Conclusions: Areas of Non-negotiable Achievements of Advanced Agri-food Systems

Thomas Blaha

thomas.blaha.ir@tiho-hannover.de

Former Head of the Field Station for Epidemiology (Bakum)
University of Veterinary Medicine
Hannover, Germany

This conclusion is applicable to both Chap. 10 and Chap. 11, which addresses areas of non-negotiable achievements in advanced agricultural and food systems, which emphasizes the importance of always implementing the highest possible food safety and biosecurity standards and must not be compromised in favour of e.g., animal welfare and other values in the transition process.

Of course, it is impossible to achieve totally risk-free food, but the stepwise implementation of the modern farm-to-fork food safety system has led to a food safety level that can be described as "food has never been as safe as it is today." Likewise, the current rules for biosecurity if complied with without avoidable breaches provide the best achievable minimization of the risks to animal health and the spread of zoonotic pathogens.

It should be taken into consideration that both are extremely important and valuable achievements of mankind, especially given the fact that nowadays food is usually produced from healthy animals in huge batches and distributed over long distances so that a single batch of contaminated or adulterated food could harm potentially Thousands and thousands of consumers in a very short time. This means that the maintenance of national and international high standards for the prevention of the spread of food-borne health risks to humans and the prevention of the spread of animal and zoonotic diseases must be protected against any change for the worse, and the standards and their implementation need continuous improvement and adaptation to emerging risks. Such adaptation is, for example, necessary due to ongoing climate change: the occurrence of mycotoxigenic fungi, so far only known in tropical environments, in Southern and Northern Europe has become a growing food safety concern. This change demands developing predictive models that account for climate change, incorporate agronomical and livestock variables, and enhance our monitoring and analytical capacities.

The efforts for the transition of the existing agri-food system are accompanied by slogans like "natural," "unprocessed," "no conservation stuff," "handcrafted," "homemade," "farm shops," and "organic," All of these characteristics suggest that food produced in this way is more sustainable and ethically better than industrially produced food. This may be true, but only as long as all food put on the market is as safe as it can be according to the state of the art of food hygiene and process optimization and as animal herds and flocks are optimally protected against the spread of disease.

No weakening of national and international food safety standards, which can lead to more illnesses and unnecessary deaths in humans as well as to considerable obstructions of international trade with agricultural primary products and food, can be ethically justified with whatever gain in sustainability and animal welfare.

Funding The Building Infection Protection Classification (BIPROC) project was funded as part of InfectControl 2020 by the Federal Ministry of Education and Research (BMBF) within the Zwanzig20—Partnership for Innovation Research Program (Grant 03ZZ0828A + B).

References

Adisasmito WB, Almuhairi S, Behravesh CB, Bilivogui P, Bukachi SA, Casas N et al (2022) One health: a new definition for a sustainable and healthy future. PLoS Pathog 18(6):e1010537. https://doi.org/10.1371/journal.ppat.1010537

Anonymus (2022) The need for a one health approach to zoonotic diseases and antimicrobial resistance: G7 science academies' statement 2022. https://doi.org/10.26164/leopoldina_04_00533

BMEL (2014) Bekanntmachung von Empfehlungen für hygienische Anforderungen an das Halten von Wiederkäuern, BAnz. AT 01.08.2014 B1, 1–17

Böhm R et al (2019) Reinigung und Desinfektion im Tierseuchenfall. Available online: https://umwelthygiene-tierhygiene.uni-hohenheim.de/fileadmin/einrichtungen/umwelthygiene-tierhygiene/April_2009/Nutztiersystmanagm_schwein/Gesetzestexte/Skriptum_Reinigung_und_Desinfektipon_im_Tierseuchenfall.pdf

Boklund A, Alban L, Mortensen S, Houe H (2004) Biosecurity in 116 Danish fattening swineherds: descriptive results and factor analysis. Prev Vet Med 66:49–62

Brennan ML, Christley RM (2012) Biosecurity on cattle farms: a study in north-West England. PLoS One 7:e28139

Bucherer M, Holzhausen J, Conraths F, Probst C (2021) Infrastructure of animal farms: key constructional elements in terms of biosecurity based on experience from Germany. Deutschland Berliner und Münchener Tierärztliche Wochenschrift 134:1–12. https://doi.org/10.2376/1439-0299-2020-37

Burnette RN, Hess JE, Kozlovac JP, Richmond JY (2013) Defining biosecurity and related concepts. In: Burnette R (ed) Biosecurity. Understanding, assessing, and preventing the threat. Wiley, Hoboken, pp 1–16

Carr J, Howells M (2018) Biosecurity on pig and poultry farms: principles for the veterinary profession. Practice 40:238–248

Conraths FJ, Sauter-Louis C, Globig A, Dietze K, Pannwitz G, Albrecht K, Höreth-Böntgen D, Beer M, Staubach C, Homeier-Bachmann T (2016) Highly pathogenic avian influenza H5N8 in Germany: outbreak investigations. Transbound Emerg Dis 63:10–13

Damiaans B, Renault V, Sarrazin S, Berge AC, Pardon B, Ribbens S, Saegerman C, Dewulf J (2019) Biosecurity practices in Belgian veal calf farming: level of implementation, attitudes, strengths, weaknesses and constraints. Prev Vet Med 172:104768

Delabbio J (2006) How farm workers learn to use and practice biosecurity. J Ext 44(6)

Delpont M, Blondel V, Robertet L, Duret H, Guerin J-L, Vaillancourt J-P, Paul MC (2018) Biosecurity practices on foie gras duck farms, Southwest France. Prev Vet Med 158:78–88

Dewulf J, van Immerseel F (2019) General principles of biosecurity in animal production and veterinary medicine. In: Dewulf J, van Immerseel F (eds) Biosecurity in animal production and veterinary medicine: from principles to practice. CABI, Wallingford, pp 63–76

Dietze K, Graaf A, Homeier-Bachmann T, Grund C, Forth L, Pohlmann A, Jeske C, Wintermann M, Beer M, Conraths FJ, Harder T (2018) From low to high pathogenicity-characterization of H7N7 avian influenza viruses in two epidemiologically linked outbreaks. Transbound Emerg Dis 65(6):1576–1587. https://doi.org/10.1111/tbed.12906

Donnelly C, Nouvellet P (2013) The contribution of badgers to confirmed tuberculosis in cattle in high-incidence areas in England. PLOS Curr 5

Döring S, Geisthardt N, Freitag H, Kobusch I, Boelhauve M, Mergenthaler M (2019) Animal hygiene indexes in relation to big-five personality traits of German pig farmers evaluated by self- and other-rating. Front Vet Sci 6:379

FAO (2010) Good practices for biosecurity in the pig sector. Issues and options in developing and transition countries 169. http://www.fao.org/3/i1435e/i1435e00.htm. Accessed 01 Apr 2020

Filippitzi ME, Brinch Kruse A, Postma M, Sarrazin S, Maes D, Alban L, Nielsen LR, Dewulf J (2018) Review of transmission routes of 24 infectious diseases preventable by biosecurity measures and comparison of the implementation of these measures in pig herds in six European countries. Transbound Emerg Dis 65:381–398

Garforth CJ, Bailey AP, Tranter RB (2013) Farmers' attitudes to disease risk management in England: a comparative analysis of sheep and pig farmers. Prev Vet Med 110:456–466

Gelaude P, Schlepers M, Verlinden M, Laanen M, Dewulf J (2014) Biocheck.UGent: a quantitative tool to measure biosecurity at broiler farms and the relationship with technical performances and antimicrobial use. Poult Sci 93:2740–2751

Gethmann J, Homeier T, Holsteg M, Schirrmeier H, Saßerath M, Hoffmann B, Beer M, Conraths FJ (2015) BVD-2 outbreak leads to high losses in cattle farms in Western Germany. Heliyon 1(1):e00019. https://doi.org/10.1016/j.heliyon.2015.e00019

Glass K, Barnes B, Scott A, Toribio J-A, Moloney B, Singh M, Hernandez-Jover M (2019) Modelling the impact of biosecurity practices on the risk of high pathogenic avian influenza outbreaks in Australian commercial chicken farms. Prev Vet Med 165:8–14

Globig A, Staubach C, Sauter-Louis C, Dietze K, Homeier-Bachmann T, Probst C, Gethmann J, Depner KR, Grund C, Harder TC, Starick E, Pohlmann A, Höper D, Beer M, Mettenleiter TC, Conraths FJ (2018) Highly pathogenic avian influenza H5N8 clade 2.3.4.4b in Germany in 2016/2017. Front Vet Sci 4:240. https://doi.org/10.3389/fvets.2017.00240

Grabkowsky B, Conraths FJ, Globig A, Wilke A, Denzin N (2020) A self-assessment tool to improve poultry farm biosecurity regarding avian influenza. A self-assessment tool to improve poultry farm biosecurity regarding avian influenza. Berl Münch Tierärztl Wochenschr, p 133

Kuster K, Cousin M-E, Jemmi T, Schüpbach-Regula G, Magouras I (2015) Expert opinion on the perceived effectiveness and importance of on-farm biosecurity measures for cattle and swine farms in Switzerland. PLoS One 10:e0144533

Laanen M, Beek J, Ribbens S, Vangroenweghe F, Maes D, Dewulf J (2010) Biosecurity on pig herds: development of an on-line scoring system and the results of the first 99 participating herds. Vlaams Diergeneeskd Tijdschr 79:302–306

Laanen M, Maes D, Hendriksen C, Gelaude P, de Vliegher S, Rosseel Y, Dewulf J (2014) Pig, cattle and poultry farmers with a known interest in research have comparable perspectives on disease prevention and on-farm biosecurity. Prev Vet Med 115:1–9

Lateef F (2009) Hospital design for better infection control. J Emerg Trauma Shock 2:175

Nöremark M, Sternberg-Lewerin S (2014) On-farm biosecurity as perceived by professionals visiting Swedish farms. Acta Vet Scand 56(1):28

Nöremark M, Frössling J, Lewerin SS (2010) Application of routines that contribute to on-farm biosecurity as reported by Swedish livestock farmers. Transbound Emerg Dis 57(4):225–236

Oettler MJ, Conraths FJ, Roesler U, Reiche S, Homeier-Bachmann T, Denzin N (2024a) Efficiency of Virucidal disinfectants on wood surfaces in animal husbandry. Microorganisms 12(5):1019. https://doi.org/10.3390/microorganisms12051019

Oettler MJ, Conraths FJ, Roesler U, Reiche S, Homeier-Bachmann T, Denzin N (2024b) Tenacity of animal disease viruses on wood surfaces relevant to animal husbandry. Viruses 16(5):789. https://doi.org/10.3390/v16050789

Pritchard K, Wapenaar W, Brennan ML (2015) Cattle veterinarians' awareness and understanding of biosecurity. Vet Rec 176:546

Regulation (EU) 2016/429 of the European Parliament and of the Concil of 9 March 2016 on transmissible animal diseases and amending and repealing certain acts in the area of animal health ('Animal Health Law'). Official Journal of the European Union, L 84/1

Renault V, Damiaans B, Sarrazin S, Humblet M-F, Dewulf J, Saegerman C (2018a) Biosecurity practices in Belgian cattle farming: level of implementation, constraints and weaknesses. Transbound Emerg Dis 65:1246–1261

Renault V, Damiaans B, Sarrazin S, Humblet M-F, Lomba M, Ribbens S, Riocreux F, Koenen F, Cassart D, Dewulf J, Saegerman C (2018b) Classification of adult cattle infectious diseases: a first step towards prioritization of biosecurity measures. Transbound Emerg Dis 65:1991–2005

Renault V, Lomba M, Delooz L, Ribbens S, Humblet M-F, Saegerman C (2020) Pilot study assessing the possible benefits of a higher level of implementation of biosecurity measures on farm productivity and health status in Belgian cattle farms. Transbound Emerg Dis 67:769–777

Ribbens S, Dewulf J, Koenen F, Mintiens K, de Sadeleer L, de Kruif A, Maes D (2008) A survey on biosecurity and management practices in Belgian pig herds. Prev Vet Med 83:228–241

Robertson ID (2020) Disease control, prevention and on-farm biosecurity: the role of veterinary epidemiology. Engineering 6(1):20–25

Sahlström L, Virtanen T, Kyyrö J, Lyytikäinen T (2014) Biosecurity on Finnish cattle, pig and sheep farms – results from a questionnaire. Prev Vet Med 117:59–67

Shortall O, Green M, Brennan M, Wapenaar W, Kaler J (2017) Exploring expert opinion on the practicality and effectiveness of biosecurity measures on dairy farms in the United Kingdom using choice modeling. J Dairy Sci 100:2225–2239

Sibley R (2010) Biosecurity in the dairy herd. In Pract 32:274–280

Simon-Grifé M, Martín-Valls GE, Vilar MJ, García-Bocanegra I, Martín M, Mateu E, Casal J (2013) Biosecurity practices in Spanish pig herds: perceptions of farmers and veterinarians of the most important biosecurity measures. Prev Vet Med 110:223–231

Ssematimba A, Hagenaars TJ, de Wit JJ, Ruiterkamp F, Fabri TH, Stegeman JA, de Jong MCM (2013) Avian influenza transmission risks: analysis of biosecurity measures and contact structure in Dutch poultry farming. Prev Vet Med 109:106–115

Sunder W, Holzhausen J, Gastmeier P, Haselbeck A, Dreßler I (2018) Bauliche Hygiene im Klinikbau. Planungsempfehlungen für die bauliche Infektionsprävention in den Bereichen der Operation, Notfall- und Intensivmedizin. Bundesinstitut für Bau-, Stadt- und Raumforschung im Bundesamt für Bauwesen und Raumordnung, Stand: Dezember 2017, Bonn

Tabori E, Dettenkofer M (2017) Baumaßnahmen und Krankenhaushygiene. Springer Medizin Verlag GmbH & Springer Verlag GmbH, Springer Nature

Thal DA, Mettenleiter TC (2023) One health – key to adequate intervention measures against zoonotic risks. Pathogens 12(3):415. https://doi.org/10.3390/pathogens12030415

van Steenwinkel S, Ribbens S, Ducheyne E, Goossens E, Dewulf J (2011) Assessing biosecurity practices, movements and densities of poultry sites across Belgium, resulting in different farm risk-groups for infectious disease introduction and spread. Prev Vet Med 98:259–270

Ward AI, Tolhurst BA, Walker NJ, Roper TJ, Delahay RJ (2008) Survey of badger access to farm buildings and facilities in relation to contact with cattle. Vet Rec 163:107–111

WOAH (2019) Chapter 6.5. Biosecurity procedures in poultry production. In: Terrestrial animal health code. World Organization for Animal Health Twenty-eighth edition, Paris

Transformative Change Towards Responsible Water Management and Sustainable Livestock Production

12

Claudia Pahl-Wostl and Johannes Halbe

1 Intensive Livestock Production and Water Management: Introduction

Livestock production has been an integral part of human food systems since the Neolithic Revolution about 11,000 years ago. Modern science fostered the systematic exploration of opportunities for increasing agricultural production, which paved the way for transformative change from diverse livestock integrated into mixed farming systems towards intensive livestock production starting with the Industrial Revolution in the eighteenth and nineteenth centuries (Hartung 2013). Since the 1950s, intensive livestock production has gained growing importance in industrial countries, which has been driven by populating growth, rising material wealth and dietary changes. Due to globalization, specialization as well as technological and organizational innovation, livestock production has become largely independent of local availability of fodder for animals since most fodder is imported. The density of animals in stables and thus the meat production per area continued to increase. This also led to a high volume of manure being dumped on the fields. Intensive agriculture has thus led to severe nutrient pollution of freshwater bodies with negative effects on ecosystems and drinking water supply. Eutrophication of lakes and drinking water pollution continue to be a major threat in many parts of the world (Sakadevan and Nguyen 2017).

Parts of Lower Saxony and North Rhine-Westphalia in Northwest Germany constitute a region that owes its prosperity to intensive livestock production. In the 1960s, laying hens and chickens as well as animal fattening achieved growing importance, resulting in a strong increase in animal stock. While farmers were able

C. Pahl-Wostl (✉) · J. Halbe
Institute of Geography and Institute of Environmental Systems Research, University of Osnabrück, Osnabrück, Germany
e-mail: cpahlwos@uni-osnabrueck.de; jhalbe@uos.de

B. Grabkowsky, T. Blaha (eds.), *Intensive Livestock Production in Transition*,
https://doi.org/10.1007/978-3-031-97872-2_12

to conduct specialization and growth in the field of pig fattening in the existing family farm structures, capital-intensive poultry farming resulted in new structures: vertically integrated corporations (Nischwitz 1996). The intensification process was further supported by the European Common Agricultural Policy that aimed at fostering agricultural production after the Second World War. In a self-sustaining circle, agro-food networks became ever more attractive for investors, but small-scale farms lost their competitiveness (Bäurle and Tamásy 2012). The consequences of these developments in the region resemble the general trends described in the previous paragraph. Intensive agriculture has led to severe nitrate pollution of freshwater bodies and has affected the drinking water supply. Most fodder is imported (mainly soy from South America), whereas maize produced in the region is predominantly used for electricity production. Policy incoherence induced a further worsening of the situation (Meergans and Lenschow 2018). A boom in the cultivation of energy crops, in particular maize, was created through subsidies that promoted electricity production from renewable energy sources. Climate change is expected to exacerbate the situation by increasing water scarcity and even more pressure to reduce groundwater pollution.

There is a clear need to address these complex challenges, now acknowledged in many regions of the world. The chosen kind of response is determined by how the challenge to be addressed is framed. Frames are understood as certain meanings attributed to a complex problem situation. Frames can depend upon different personal, cultural or disciplinary backgrounds (Isendahl et al. 2009). Regarding the impacts of intensive livestock production, we distinguish in this chapter between two framings that can be found in scientific, political and public discourses:

The problem to be addressed is nutrient pollution of freshwater bodies. Measures should prevent the disposal of excess nutrients/nitrate into the environment.

The problem is systemic and requires a systemic response. The production methods of intensive agriculture should be questioned. Agricultural production systems in particular and food systems in general must undergo a fundamental transformation.

These framings can be linked to different types of learning processes. The first framing corresponds to single-loop learning that is characterized by an incremental improvement of established practices without calling into question underlying paradigms and assumptions (Pahl-Wostl 2009). As phrased by Flood and Romm (1996): In single-loop learning, actors ask if they do things right without asking if they do the right things. We argue that the "nitrate problem" can only be solved by higher levels of learning, by calling into question dominating paradigms and business models. The challenge needs to be framed from a systemic perspective. This involves processes of double-loop learning (i.e. new solution strategies are considered and tested) and triple-loop learning (i.e. transformation of underlying paradigms, practices and governance structures).

The book chapter is structured as follows. First, we delineate the framing of the negative impacts of intensive livestock production on water resources as a nutrient problem. Second, we introduce a more systemic perspective on the problem and suggest transformative solution strategies. The chapter ends with a conclusion and an outlook on promising steps towards a systematic and transformative approach.

2 Framing the Challenge as a Nutrient Problem

The prevailing approach when dealing with environmental problems is to treat them in isolation. This also applies to nitrate pollution of water bodies from agriculture. Nitrate pollution is an externality for intensive livestock production. Its effects have no impact on and cause no direct costs for those who pollute. Hence, farmers have no economic incentive to address this problem. It must be dealt with by governmental intervention and/or direct negotiations between farmers and those negatively affected, such as water suppliers.

Regarding Germany, several attempts were made to regulate livestock densities and nutrient flows, such as setting upper limits of livestock per available land unit, regulating the nutrient balance of farms (fertilizer legislation) or introducing liquid manure markets. However, due to implementation deficits, these measures did not lead to satisfactory improvement regarding the nitrate pollution of groundwater (Meergans et al. 2020; Meergans and Lenschow 2018; Pahl-Wostl 2019). Regarding fertilizer legislation, effective monitoring and sanctioning mechanisms have been lacking for years. Water suppliers have been heavily affected by such developments. To rectify deficits in regulatory legislation, attempts have been made for years to achieve urgently needed progress in drinking water protection by using regional, voluntary instruments such as the Lower Saxony Cooperation Model for Drinking Water Protection (Aue 2017). Measures to remediate the worsening groundwater supply situation included voluntary cooperation with farmers. Voluntary agreements with farmers were financed with fees on drinking water to support financial compensation for farmers for restrictions on using their land in the vicinity of protection zones of wells. Furthermore, water suppliers voluntarily bought land for afforestation and offered training and education for professionals and the public at large. These programs showed some initial improvement and resulted in declining nitrate levels in groundwater in drinking water protection areas. However, in many wells, the nitrate level remained above the prescribed nitrate limit of 50 mg/l for drinking water. Since the EC Groundwater Directive, a Daughter Directive to the European Water Framework Directive, entered into force in 2006, this limit applies area-wide to all groundwater bodies. The threat of high fines due to violating these EU regulations has finally imposed severe pressure on agriculture in many regions of Europe characterized by intensive livestock production. However, attempts to finally enforce more stringent regulations have also led to violent protests by farmers in all parts of Europe.

In the Netherlands, the government concluded that the livestock needed to be reduced by one-third to meet nitrate regulations and to counteract the widespread environmental damage. To do so, the government intends to implement a buyout scheme and close down a significant number of livestock production sites. This led to violent protests by farmers and a polarized debate in society as a whole. What has become known as the "Dutch Nitrogen Crisis" attracted worldwide media attention (e.g. Tullis 2023). As in Germany, the need to address the surplus of nitrogen had been known for years, even decades. Due to years of inaction, the country finds itself now in an extremely difficult and conflict-ridden situation.

Attempts to address the nitrate pollution from intensive livestock production can be characterized as single-loop learning. Following this framing, the underlying structure of the current food system, including practices of intensive livestock production, economies of scale, an overall decline in the number of farms and globalized production networks, is assumed as given. Mainly end-of-pipe solutions are considered, such as manure markets or alternative usages of manure as substrate for biogas plants.

3 Framing the Challenge from a Systemic Perspective

Intensive agriculture is representative of the key characteristics of sustainability challenges, which have proven to be quite resistant to change: complex cross-scale dynamics, high uncertainties, combined with entrenched conflicts of interest, unequal power distributions and lock-in dynamics due to inflexible infrastructures and sunk costs. The various agricultural production networks contain numerous cases of systemic risks, some of which have already led to major disturbances of ecosystems or are likely to do so (e.g. Franz et al. 2018; de Raymond et al. 2021). This does not only concern biodiversity loss, soil degradation, water quality and water scarcity but also relates to climate change impact (e.g. Olesen and Bindi 2002), pest and disease outbreaks (e.g. Espinosa et al. 2020), supply chain vulnerabilities (e.g. Zhao et al. 2020) and their interactions. A systemic risk describes a risk to a component of a system with the potential of spreading throughout the system due to the connectedness between the system's constituent parts. Dealing with such risks requires a precautionary and systemic approach (Renn 2016). This also implies as well to the "nitrate" problem which cannot be solved by end-of-pipe solutions. It is inappropriate to reduce a complex sustainability challenge to a simple linear cause-effect relationship and one local impact.

From such a systemic perspective, intensive livestock production generates water problems not only at a local and regional scale but also in places far away due to the globalization of supply chains. The Brazilian Cerrado is a tropical savanna that covers more than 20% of the country's area. During the last decades, more than half of this vast landscape has been converted to agricultural land. Soy monocultures are particularly planted for export to China and Europe. Such intensive agricultural production has brought great profits to the agricultural industry, but at the expense of detrimental impacts on regional hydrology and climate and livelihoods (Raymond et al. 2023; Salmona et al. 2023; Gustafsson et al. in review). The increase in the frequency and severity of droughts poses a risk to the entire agribusiness model itself. The negative impacts of intensification are no longer an externality but weaken the production base. Transformative change seems to be the only way to ensure long-term economic, ecological and social sustainability (World Business Council for Sustainable Development 2023).

One driver for the expansion and intensification of soy production is the global increase in the demand for meat. On a global level, meat consumption amounts to about 320 million tons in 2018, which means a doubling in 20 years (Heinrich-Böll-Stiftung 2021). While China is the largest meat consumer globally, its per capita consumption in 2020 (62.2 kg) is still lower than in Germany (78.7 kg) (Destatis 2023). The annual meat consumption in Germany has risen from about 65 kg per capita in 1960 to about 105 kg per capita in 1988 (von Witzke et al. 2011). Since then, a reduction can be observed to around 86 kg per capita in 2005 (von Witzke et al. 2011) and about 70.6 kg in 2022 (Ahrens 2024). The indicators of Sustainable Development Goal (SDG) number 2 "Zero Hunger" account to the role of meat products in diets as well as related health impacts and nitrogen management. As a result, the status of SDG 2 for Germany indicates that "significant challenges remain" with a stagnating trend (Sachs et al. 2023, p. 250).

Currently, nearly 80% of the global soy production is used for animal feed (OWID 2021). However, numerous arguments have been put forward showing that an increase in meat consumption, particularly when produced from intensive livestock farming, cannot be sustained. The production of one kilo of beef uses about 15,000 litres of water, and that of pork is about 6000 litres. This refers to the water used in the whole production chain. Although the exact numbers depend on the details of the production process, water consumption in meat production is orders of magnitude higher than in grain production (Mekonnen and Hoekstra 2010). Food security for a much higher number of humans could be assured sustainably if soy were used primarily for human consumption rather than animal feed.

Klaura ct al. (2023) also point to the issue of meat wastage and loss along the supply chain, which is estimated to amount to 77.4 million tonnes globally in 2019. This means that nearly 18 billion animals were raised and killed without eventually being consumed by humans. However, the proportions of species differ considerably: chicken have the highest proportion of preventable animal death (93.6%) and cattle the lowest (0.4%). The origin of meat losses and wastage along the supply chain shows a diverse picture across world regions after Klaura et al. (2023). In Europe, most of the meat wastage appears in the consumption stage (~39.1%) followed by processing/packaging (~19.5%) and production (~19.4%). In Sub-Saharan Africa, losses mainly appear during the production stage (~65.8%), while the relevance of consumption losses is quite low (~4.1%).

In sum, environmental and ethical concerns of consumers are already observable in dietary choices that contain less meat consumption. Hence, one can expect major changes in the future in the modes of protein production and in patterns of consumption. Given these developments, it can be risky for livestock farming to stick with an outdated and unsustainable business model. Instead, it may be advisable to develop innovative and sustainable approaches that also increase resilience to climate change and uncertainties in global markets. Framing the challenge from such a systemic perspective implies higher levels of learning and transformative change. This means to ask: why are we not doing the right things and how can we do better? This involves a questioning of assumptions and current business models as well as a reframing that does not only focus on agricultural production but takes a whole system perspective, including input requirements and dietary choices on the demand side.

4 Transformative Change

Above we highlighted that sustainability challenges characterized by complex cross-scale dynamics, and high uncertainties, combined with entrenched conflicts of interest, unequal power distributions and inflexible, capital-intensive infrastructures have shown to be quite resistant to change. Indeed, supporting transformative change in such systems is by no means trivial (Pahl-Wostl 2015; Halbe and Pahl-Wostl 2019; Halbe et al. 2015). Pahl-Wostl (2009) introduced an evolutionary approach to transformative change based on the concept of triple-loop learning. Societal learning is assumed to be an exploratory process in which actors experiment with innovations until they encounter limitations and new frontiers. At the same time, the promotion and expansion of innovations also require political action to create supportive framework conditions. Transformative change thus embraces multi-level processes of purposeful action and of self-organization and emergence. Table 12.1 depicts processes at different levels that we consider essential for transformative change in livestock production in order to address water issues from a systemic perspective.

At the individual and household level, consumers have the ability to impact the food system through their dietary and consumption choices. As mentioned in the previous section, a current trend can be observed in Germany that involves less meat consumption (von Witzke et al. 2011) and higher market shares of vegan and vegetarian meat alternatives. For example, the market value of the total production of meat alternatives increased by 68% between 2019 and 2021; in the same timeframe, the market value of meat products decreased by 11.2% (Destatis, 2022). This trend is especially pronounced in the younger generation (15–29 years), which shows a twofold higher share of vegetarian and vegan diets (Heinrich-Böll-Stiftung 2021). This trend can be expected to continue or even intensify in the future, due to alarming environmental trends, including water quality or biodiversity issues that are likely to lead to higher pro-environmental awareness and behaviour. Today, 74% of consumers in Germany already take environmental and resource aspects of food production into account in their purchasing behaviour (forsa 2023).

At the farm level, more extensive, circular and diverse approaches to livestock production can help to sustainably include animals in farming systems and cultural landscapes. Several farming approaches already provide models to achieve this, such as ecological agriculture (Sanders and Heß 2019) or regenerative agriculture (von Meyer-Höfer et al., 2023). Besides food production, these farming approaches are connected to the protection and generation of further ecosystem services, such as water purification, soil quality and biodiversity. Nevertheless, this would imply higher prices and a changing role of meat from an everyday food item towards a food for special occasions. But one can argue that higher prices would better reflect the real value of meat. A more direct and regional marketing could be one approach to reduce margins from wholesale trade, connect consumers to regional farmers and increase consumers' appreciation of sustainably produced food. A diversification of farming businesses could also provide more resilience to changing markets and climate, and open up opportunities for new business models, for example, by linking

Table 12.1 Transformative change in livestock production: a multi-level process

Level	What	Why
Global	International regulations, trade agreements	Creating a level playing field for innovation at the international level
EU	EU agricultural policy CAP, food and environmental policy	Linking more financial payments with environmental goals Higher standards
National	Agricultural, food and environmental policies Professional standards	Implementing enabling regulations and financial programmes Assuring policy coherence Introducing labels
Landscape	Transformative landscape management Sustainable land use visions	Harness synergies between different land uses (water-energy-food nexus) Holistic perspective based upon ecohydrological flows and conditions
Regional (e.g. province) Local	Environmental policies Implementation strategies Networks for collective action	Creating enabling conditions for regional markets and innovation Sharing of experiences Supporting local food networks
Farm	Innovative, sustainable practices Diverse business models Regional food networks	Testing and further developing innovations Alternative (direct) marketing channels Provision of diverse ecosystem services
Individuals/ households	Dietary changes Consumption choices	Environmental consciousness Health concerns Ethical issues

agriculture to tourism or education (WBGU 2020). Community-supported agriculture is currently another trending model that actively involves consumers in farming activities and provides an economic model beyond markets (e.g. Boddenberg et al. 2017).

At the local and regional levels, networks between farmers and further food system actors (including consumers) support the development of regional cooperation, learning and economies. These networks can be initiated by farmers themselves or other actors, such as regional distribution networks (e.g. Partzsch 2019). Food councils are further bottom-up processes, often initiated by citizens, that bring diverse actors in the food system together, such as consumers, producers, commerce and regional policy (e.g. Sieveking 2019). Food councils can be an effective instrument to guide local food policy towards fostering sustainable consumption and farming practices, for example, by supporting canteens of public organizations to follow sustainability criteria and buy from regional producers. Such networks can support closer urban-rural exchanges, trust and the revitalization of rural areas by strengthening the embeddedness of agriculture in society and culture. Joint regulatory initiatives at the state level, such as the "Niedersächsische Weg" (Lower Saxony Approach), are also helpful approaches to develop momentum for sustainable development, based upon a participatory approach in policy-making (Box 12.1).

Box 12.1: Regenerative Agriculture

Regenerative agriculture is an alternative farming approach seeking the alleviation of climate change and the regeneration of soils, water bodies and biodiversity (e.g. Schreefel et al. 2020; Gordon et al. 2022). The term is not yet clearly defined, and regenerative farming projects show different thematic focuses. One of the focus areas is holistic livestock management, i.e. specific grazing systems aiming at carbon sequestration, soil health and drought resilience. Figure (12.1) shows grazing cows at a regenerative farm in the State of Brandenburg, whose founder was awarded the CERES "Farmer of the year" award in 2022. The cows are systematically moved between small plots that are confined by mobile fences.

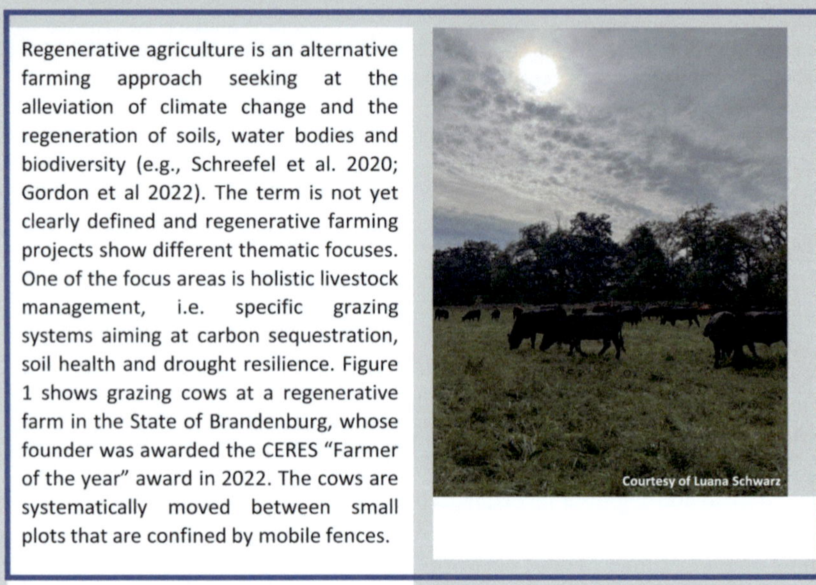

Fig. 12.1 Grazing cows at a regenerative farm

The spatial boundaries of a region can be defined based on administrative boundaries (e.g. districts or states), socio-economic criteria (e.g. historical/cultural factors or commuter flows) as well as physical (e.g. aquifers and river catchments) and ecological criteria (e.g. landscape types and natural regions). Intermediate scales are suitable spatial contexts to govern transformations towards more sustainable agriculture practices, as they allow the exploration of opportunities for upscaling of local innovation processes and their linkages to larger-scale processes. For example, natural and hydrogeological regions are suitable spatial units to explore the causes of environmental problems and envision sustainable land use futures. As another example, the choice of water bodies (catchments or aquifers) as spatial units recognizes the role of water as the connecting agent in a resource nexus. They provide thus an interesting scale for analysing and quantifying the effects of an upscaling of local innovations.

Transformative landscape management has been developed as a conceptual and methodological framework to transition processes at intermediate scales (Halbe and Ulbrich, 2024). Whatever specific intermediate spatial boundary is chosen (see previous paragraph), transformative landscape management has to deal with the socio-economic and ecohydrological complexities, that is, the interlinkages between different socio-economic and environmental levels and scales. It furthermore requires the screening of existing land use innovations and the analysis of their impacts as well as impeding and supporting factors. Such systems knowledge can be based upon literature reviews as well as case-specific empirical data from qualitative data from interviews or quantitative data from experiments and measurements. Target knowledge can be generated by envisioning the upscaling of these innovations towards sustainable land use futures. Such visioning processes require the close involvement of pioneers in order to generate transformative futures. However, the involvement of established food system actors is important to challenge transformative visions and foster their implementation. Transformation knowledge requires an analysis of suitable actions and multi-stakeholder platforms fostering the implementation of land use visions, including potential roles and responsibilities (Halbe and Pahl-Wostl 2019). Thus, some actions can be implemented by niche actors themselves, while others require the further development of networks and the involvement of regime actors.

At the national level, enabling regulations and financial programmes are required to support sustainable farming practices as well as regional food networks, markets and innovations (Feind et al. 2019). In particular, bureaucratic efforts need to be reduced for sustainability pioneers who prioritize the sustainability of farms and businesses over the generation of profit. Here, more openness towards innovation and experimentation is required from regulative authorities. Regulations and financial programmes are still tailored to an agricultural system that favours specialization, economies of scale and competitiveness in globalized markets. The recent efforts by the national government in Germany to adopt a real-world lab law illustrate a rising awareness of policymakers towards regulatory experimentation (cf. Parodi et al. 2023). Assuring policy coherence is another central task at the national level that requires additional efforts in the future (Meergans and Lenschow 2018; Platjouw et al. 2023). Agricultural, food and environmental policies need to be aligned in order to avoid detrimental side-effects of legislation (cf. detrimental effects of biogas plants on biodiversity and water quality). In particular, synergies between agricultural and environmental regulations should be identified and actively supported, for example, linked to biodiversity and water protection. Labelling initiatives at the national level are further impulses supporting consumers to make conscious decisions.

The EU level is certainly an important leverage point to introduce sustainable practice through the Common Agricultural Policy (CAP). Thereby, the potential detrimental effects of tighter regulatory standards at the national level on the competitiveness of farming businesses in the EU can be avoided. However, the past has shown that the reform of the CAP is a slow process, due to the complexity of the topic, the influence of lobbying groups and the diverging interests of EU member states (Feind et al. 2019; Pe'er et al. 2019). Despite claims that European agricultural policy implemented the principle of multi-functionality, in practice policies still give clear emphasis on agricultural export and market orientation (McMichael 2011).

At the global level, the exploitative approach to environmental resources is currently under pressure, which might intensify in the future for several reasons. First, soil degradation and water scarcity are increasing problems on a global scale, which weakens the stability of the global food system, including the high-input system of intensive livestock production. Second, accelerating climate change requires rapid and transformative action to reduce CO_2 emissions, but also to reduce land use change and foster regeneration—processes that are not compatible with intensive livestock production. Fourth, climate change and health concerns currently foster profound changes in the demand of consumers in the Global North, which also questions the current intensive livestock system. Finally, population growth and rising material wealth in the Global South increase the demand for meat and input factors for meat production. This will further increase the competition for inputs (e.g. fodder crops) in the global market. Finally, rising ethical concerns about livestock farming will put the current sector under pressure and foster alternative products. These trends can induce a rethinking and reconfiguration of global trade agreements as well as EU and national legislations.

5 Concluding Comments and Outlook

Systemic challenges require systemic solutions. Technical fixes will not solve problems related to intensive livestock production. However, they are still required to reduce the impact of the current system. We argue in this chapter that a systemic transformation is required in the long term. The question is not if transformative change will occur. The question is rather how and when it will occur. Will a smooth transition be possible driven by innovative action or will major crises and a partial or even complete breakdown of the current system be required to finally lead to radical change?

Intensive livestock businesses are faced with classical lock-in situations. Heavy investments in expensive and long-lasting infrastructure have led to a path dependency with little incentive for radical innovation. Specialization and a complex network of interdependencies make change even more difficult. Innovative action can be expected to originate in niches at the margin of the current food production system. Such pioneers need financial, logistic and organizational support to develop and explore the potential of innovations and to collaborate in larger networks. Networks can bring together producers, consumers, technology developers, environmentalists, investors and public administration representatives. Together, they can develop a critical momentum for transformative systemic change. They can develop a positive vision of a sustainable agriculture system and the role of livestock production.

Regions can be an appropriate intermediate scale to develop such a momentum, as they might comprise a critical mass of pioneers in different parts of the food system while still providing opportunities for bottom-up processes and concrete action based upon personal contacts and spatial proximity. These collaborative processes can be initiated by different actors, such as consumers, producers or municipalities. Transition governance analyses underline the multi-level complexity of the challenges but also diverse opportunities for action at individual, group, organizational and state levels (e.g. Halbe and Pahl-Wostl 2019). A key factor for transformative change is the development of an inspiring and coherent vision of a sustainable food system. A positive vision can bring together food system actors, animate others to join and provide a signpost for transformative challenges in the midst of the complexities of today.

Past transformation processes in the food sector, as well as other sectors and societal areas, can provide important lessons for today. An impressive example of the transformative power of a positive vision, leadership and collaboration in a region is given by the environmental and social restoration of the Emscher River in the North Rhine-Westphalia region (Tröltzsch et al. 2020). Due to mining and urbanization, the Emscher was used as an open sewer from the end of the nineteenth century onwards. It was considered "the cesspit of the Ruhr"—dead, smelly and people avoided it. The "Emscher Future Master Plan" became the driving force and a guiding narrative behind the transformation of an entire region during structural economic change. Today, large parts of the Emscher have been renatured, and sections of the river have been integrated into urban and regional planning. The region offers recreational and leisure areas, flood protection and biodiversity benefits from the renatured floodplains.

The current challenge of passing farms to the next generation (67% of individual farming businesses in Germany with a farm manager older than 55 did not yet have a successor in the year 2020, cf. DBV 2021) underlines the need to develop positive visions and development pathways for agriculture. A constructive dialogue between food system actors is needed to identify positive visions and gain momentum for collective action. It might be likely that such dialogues do not necessarily happen in the scope of conventional participatory formats (e.g. public roundtables) and media channels (e.g. social media or newspapers). Transformative learning and dialogues rather require trusting personal relationships and protected discursive spaces ranging from the private to the public sphere. Food councils are one example of such a setting that fosters constructive dialogues, trusting relationships and collaboration between diverse food system actors. Such food councils often arise in bottom-up processes. However, for long-term success, support from municipalities and local and regional authorities is required. Many initiatives develop innovative ideas with much creativity. To achieve their transformative potential, such processes at the local and regional levels need to be embedded in and better linked to dialogues and policy processes at the national and even international (EU) levels. Policies at these levels have so far tended to support the status quo and even presented obstacles to innovation.

References

Aue C (2017) Grenzen und Möglichkeiten des Grundwasserschutzes: Über steigende Nitratwerte und erfolgreiche Kooperationen zwischen Wasserwirtschaft und Landwirtschaft. Kritischer Agrarbericht, vol 25. München

Bäurle H, Tamásy C (2012) Regionale Konzentrationen der Nutztierhaltung in Deutschland. ISPA-Mitteilungen, vol Heft 79 ISPA, Vechta

Boddenberg M, Frauenlob MH, Gunkel L, Schmitz S, Vaessen F, Blättel-Mink B (2017) Solidarische Landwirtschaft als innovative Praxis – Potenziale für einen sozial-ökologischen Wandel. In: Jaeger-Erben M, Rückert-John J, Schäfer M (eds) Soziale Innovationen für nachhaltigen Konsum: Wissenschaftliche Perspektiven, Strategien der Förderung und gelebte Praxis. Springer Fachmedien Wiesbaden, Wiesbaden, pp 125–148. https://doi.org/10.1007/978-3-658-16545-1_6

de Raymond A, Bernard AA, Ben-Ari T, Daviron B, Nesme T, Tétart G (2021) Systemic risk and food security. Emerging trends and future avenues for research. Global. Food Secur 29:100547. https://doi.org/10.1016/j.gfs.2021.100547

Destatis (2022) Pressemitteilung Nr. N 025: Fleischersatz weiter im Trend. https://www.destatis.de/DE/Presse/Pressemitteilungen/2022/05/PD22_N025_42.html. Accessed 29 Apr 2024

Destatis (2023) Tierhaltung – Fleischkonsum. https://www.destatis.de/DE/Themen/Laender-Regionen/Internationales/Thema/landwirtschaft-fischerei/tierhaltung-fleischkonsum/_inhalt.html. Accessed 25 Apr 2024

Deutscher Bauernverband (DBV) (2021) Situationsbericht 2021/22: Trends und Fakten zur Landwirtschaft. https://www.bauernverband.de/fileadmin/berichte/2021/epaper/ausgabe.pdf. Accessed 22 Apr 2024

Distribution of soy production end uses worldwide in 2018 (2022) Stastista. https://www.statista.com/statistics/1254608/soy-production-end-uses-worldwide/. Accessed 26 Apr 2024

Doernberg A, Horn P, Zasada I, Piorr A (2019) Urban food policies in German city regions: an overview of key players and policy instruments. Food Policy 89:101782

Espinosa R, Tago D, Treich N (2020) Infectious diseases and meat production. Environ Resour Econ 76(4):1019–1044. https://doi.org/10.1007/s10640-020-00484-3

Feind P, Krämer C, Früh-Müller A, Heißenhuber A, Pahl-Wostl C, Purnhagen K, Thomas F, van Bers C, Wolters V (2019) Ein neuer Gesellschaftsvertrag für eine nachhaltige Landwirtschaft: Wege zu einer integrativen Politik für den Agrarsektor. Springer, Berlin, Heidelberg. https://doi.org/10.1007/978-3-662-58656-3

Fleischverbrauch in Deutschland pro Kopf in den Jahren 2010 bis 2023 (2024) Statista. https://de.statista.com/statistik/daten/studie/36573/umfrage/pro-kopf-verbrauch-von-fleisch-in-deutschland-seit-2000/

Flood R, Romm N (1996) Diversity management: triple loop learning, vol 25. Wiley, Chichester, p 154

forsa Gesellschaft für Sozialforschung und statistische Analysen mbH (2023) Ernährungsreport 2023: Ergebnisse einer repräsentativen Bevölkerungsbefragung. https://www.bmel.de/SharedDocs/Downloads/DE/_Ernaehrung/forsa-ernaehrungsreport-2023-tabellen.pdf?__blob=publicationFile&v=2. Accessed 25 Apr 2024

Franz M, Schlitz N, Schumacher KP (2018) Globalization and the water-energy-food nexus – using the global production networks approach to analyze society-environment relations. Environ Sci Pol 90:201–212. https://doi.org/10.1016/j.envsci.2017.12.004

Gordon E, Davila F, Riedy C (2022) Transforming landscapes and mindscapes through regenerative agriculture. Agric Hum Values 39(2):809–826

Gustafsson M, Schilling-Vacaflor A, Pahl-Wostl C (in review) Governing transnational water and climate risks in global supply chains. Earth System Governance

Halbe J, Pahl-Wostl C (2019) A methodological framework to initiate and design transition governance processes. Sustain For 11(3). https://doi.org/10.3390/su11030844

Halbe J, Ulbrich R (2024) Transformative landscape governance: upscaling of local innovations towards sustainable agrifood systems. International sustainability conference 2024 in Oslo, Norway

Halbe J, Pahl-Wostl C, Lange MA, Velonis C (2015) Governance of transitions towards sustainable development – the water–energy–food nexus in Cyprus. Water Int 40:1–18. https://doi.org/10.1080/02508060.2015.1070328

Hartung J (2013) A short history of livestock production. In: Livestock housing: modern management to ensure optimal health and welfare of farm animals, pp 21–34. https://doi.org/10.3920/978-90-8686-771-4_01

Heinrich-Böll-Stiftung (ed) (2021) Der Fleischatlas 2021. Berlin

Isendahl N, Dewulf A, Brugnach M, François G, Möllenkamp S, Pahl-Wostl C (2009) Assessing framing of uncertainties in water management practice. Water Resour Manag 23(15):3191–3205. https://doi.org/10.1007/s11269-009-9429-y

Klaura J, Breeman G, Scherer L (2023) Animal lives embodied in food loss and waste. Sustain Prod Consump 43:308–318. https://doi.org/10.1016/j.spc.2023.11.004

McMichael P (2011) Food system sustainability: questions of environmental governance in the new world (dis)order. Glob Environ Chang 21(3):804–812. https://doi.org/10.1016/j.gloenvcha.2011.03.016

Meergans F, Lenschow A (2018) Die Nitratbelastung in der Region Weser- Ems: Inkohärenzen in Wasser-, Energie- und Landwirtschaftspolitik. Neues Archiv Für Niedersachsen 2:105–117

Meergans F, Aue C, Knieper C, Kochendörfer S, Lenschow A, Pahl-Wostl C (2020) Overcoming coordination gaps between water, energy and agriculture: future paths to water protection in Weser-Ems. Briefing Paper. German Development Institute Bonn, Germany. https://doi.org/10.23661/bp25.2020

Mekonnen MM, Hoekstra AY (2010) The green, blue and grey water footprint of farm animals and animal products. Value of Water Research Report Series, vol 48. UNESCO-IHE, Delft, the Netherlands

Nischwitz G (1996) Veredlungswirtschaft in Südoldenburg unter dem Einfluss sich wandelnder sozioökonomischer und politischer Rahmenbedingungen. Vechtaer Studien zur Angewandten Geographie und Regionalwissenschaft vol 17. Vechta, Vechta

Olesen JE, Bindi M (2002) Consequences of climate change for European agricultural productivity, land use and policy. Eur J Agron 16(4):239–262. https://doi.org/10.1016/S1161-0301(02)00004-7

OWID (2021) Distribution of soy production end uses worldwide in 2018. Statista, https://www.statista.com/statistics/1254608/soyproduction-end-uses-worldwide Accessed: June 26, 2025.

Pahl-Wostl C (2009) A conceptual framework for analysing adaptive capacity and multi-level learning processes in resource governance regimes. Glob Environ Chang 19:354–365

Pahl-Wostl C (2015) WAter governance in the face of global change – from understanding to transformation. Water governance: concepts, methods and practice, vol 1. Springer International Publishing, Cham. doi:https://doi.org/10.1007/978-3-319-21855-7

Pahl-Wostl C (2019) Governance of the water-energy-food security nexus: a multi-level coordination challenge. Environ Sci Pol 92:356–367. https://doi.org/10.1016/j.envsci.2017.07.017

Parodi O, Schwichtenberg R, Stelzer F, Rhodius R, Schreider C, von Wirth T, Lang D, Marg O, Wagner F, Egermann M, Bauknecht D, Wanner M (2023) Stellungnahme des Netzwerks Reallabore der Nachhaltigkeit zur Reallabore-Gesetz-Initiative. GAIA-Ecol Perspect Sci Soc 32(4):399–401

Partzsch L (2019) Food localization and agency: the cases of Regionalwert AG and Luzernenhof in Freiburg, Germany. In: Kalfagianni A, Skordili S (eds) Localizing global food. Routledge, London, pp 55–70. https://doi.org/10.4324/9780429449284-5

Pe'er G, Zinngrebe Y, Moreira F, Sirami C, Schindler S, Müller R, Bontzorlos V, Clough D, Bezák P, Bonn A, Hansjürgens B, Lomba A, Möckel S, Passoni G, Schleyer C, Schmidt J, Lakner S (2019) A greener path for the EU common agricultural policy. Science 365(6452):449–451. https://doi.org/10.1126/science.aax3146

Platjouw FM, Nesheim I, Enge C (2023) Policy coherence for the protection of water resources against agricultural pollution in the EU and Norway. Rev Eur Comp Int Environ Law 32(3):485–500. https://doi.org/10.1111/reel.12509

Raymond CM, Anderson CB, Athayde S, Vatn A, Amin AM, Arias-Arévalo P, Christie M, Cantú-Fernández M, Gould RK, Himes A, Kenter JO, Lenzi D, Muraca B, Murali R, O'Connor S, Pascual U, Sachdeva S, Samakov A, Zent E (2023) An inclusive typology of values for navigating transformations towards a just and sustainable future. Curr Opin Environ Sustain 64:101301. https://doi.org/10.1016/j.cosust.2023.101301

Renn O (2016) Systemic risks: the new kid on the block. Environ Sci Policy Sustain Dev 58(2):26–36. https://doi.org/10.1080/00139157.2016.1134019

Sachs JD, Lafortune G, Fuller G, Drumm E (2023) Implementing the SDG stimulus. Sustainable Development Report 2023. Dublin University Press, Dublin

Sakadevan K, Nguyen ML (2017) Chapter four – livestock production and its impact on nutrient pollution and greenhouse gas emissions. In: Sparks DL (ed) Advances in agronomy, vol 141. Academic Press, pp 147–184. https://doi.org/10.1016/bs.agron.2016.10.002

Salmona YB, Matricardi EAT, Skole DL, Silva JFA, Coelho Filho OA, Pedlowski MA, Sampaio JMC et al (2023) A worrying future for river flows in the Brazilian Cerrado provoked by land use and climate changes. Sustain For 15(5):4251. https://doi.org/10.3390/su15054251

Sanders J, Heß J (2019) Leistungen des ökologischen Landbaus für Umwelt und Gesellschaft. 2. überarbeitete und ergänzte Auflage. Thünen Reports, vol 65. Johann Heinrich von Thünen-Institut, Braunschweig. https://doi.org/10.3220/REP1576488624000

Schreefel L, Schulte RP, De Boer IJM, Schrijver AP, Van Zanten HHE (2020) Regenerative agriculture–the soil is the base. Glob Food Sec 26:100404

Sieveking A (2019) Food policy councils as loci for practising food democracy? Insights from the case of Oldenburg, Germany. Politics Gov 7(4):48–58

Tröltzsch J, Gerner N, Meergans F, Stein U, Sutcliffe R (2020) Coordination and Cooperation of Water Management, NatureConservation and Open Space Development in the Emscher Restoration. Briefing Paper. German Development Institute Bonn, Germany. doi:https://doi.org/10.23661/bp22.202

Tullis P (2023) Nitrogen wars: the Dutch farmers' revolt that turned a nation upside-down. The Guardian, 16 Nov 2023

von Meyer-Höfer M, Gerlach R, Jumshudzade Z, Paulsen H M (2023) Klimaschonende Landwirtschaft angehen–Dialog zwischen Verbrauchern und Landwirten fördern. https://www.thuenen.de/media/institute/ol/Aktuelles_Veranstaltungen/Dokumente/Klimaschonende_Landwirtschaft_angehen_TI_OLMA_280421.pdf. Accessed 22 Apr 2024

Von Witzke H, Noleppa S, Zhirkova I (2011) Fleisch frisst Land: Ernährung, Fleischkonsum, Flächenverbrauch. WWF Deutschland

World Business Council for Sustainable Development (2023) Deep dive: Soy production in the Cerrado, Brazil. World Business Council for Sustainable Development, Geneva

WBGU (2020) Landwende im Anthropozän: Von der Konkurrenz zur Integration. Berlin

Zhao G, Liu S, Lopez C, Chen H, Lu H, Mangla SK, Elgueta S (2020) Risk analysis of the agri-food supply chain: a multi-method approach. Int J Prod Res 58(16):4851–4876. https://doi.org/10.1080/00207543.2020.1725684

Potential of Functional Plant Diversity for Climate Resilience of Grasslands

13

Manfred Kayser, Martin Komainda,
and Johannes Isselstein

1 Introduction

Worldwide, grassland covers more than 70% of the agriculturally utilized land (Soussana and Lüscher 2007); in Europe, this proportion is about 30%, varying from 10% to 90% among the different countries. Grasslands can roughly be categorized into permanent and temporary grassland—temporary grasslands are sown mixtures used in rotation with other crops on arable land (Allen et al. 2011). In northwestern parts of Europe, forage production is often intensively managed (high nutrient input, frequent defoliation) and coupled with dairy farming (Reheul et al. 2017). In southern parts of Europe, apart from dairy, a range of extensive grassland systems, providing more biodiversity, have been preserved as an adaptation strategy to an often mountainous or drier environment (Zamora et al. 2007). In Germany, the grassland area is approximately 4.7 million ha (28% of the agriculturally utilized

M. Kayser (✉)
Division of Grassland Science, Department of Crop Sciences, Georg-August-University Göttingen, Göttingen, Germany

GeoLab, University of Vechta, Vechta, Germany
e-mail: manfred.kayser@agr.uni-goettingen.de

M. Komainda
Division of Grassland Science, Department of Crop Sciences, Georg-August-University Göttingen, Göttingen, Germany
e-mail: martin.komainda@uni-goettingen.de

J. Isselstein
Division of Grassland Science, Department of Crop Sciences, Georg-August-University Göttingen, Göttingen, Germany

Campus Center of Biodiversity and Sustainable Land Use, University of Göttingen, Göttingen, Germany
e-mail: jissels@gwdg.de

B. Grabkowsky, T. Blaha (eds.), *Intensive Livestock Production in Transition*,
https://doi.org/10.1007/978-3-031-97872-2_13

land), with an additional 0.66 million ha used as sown temporary grassland (DESTATIS 2022).

Maintaining agricultural grassland is based on certain conditions, among which are regular defoliation (mowing or grazing), water supply, and nutrient input. On soils with a heavy texture, with a high water table, or on shallow soils with high stone content, grassland will often be the only feasible form of agricultural land use. The role of grasslands in the landscape context is complex, as grasslands are found and managed along a strong gradient from very intensive agricultural use to sites of extensive use and low input.

Grasslands have different functions in agricultural production and within the landscape and provide a range of important ecosystem services, among which are agricultural production, biodiversity, water retention, erosion control, groundwater protection, soil conservation, carbon sequestration, aesthetics and leisure activities (Schils et al. 2022; Isselstein and Kayser 2014).

In agricultural production, grassland provides a basis for keeping livestock such as ruminants and equines or provides biomass for other purposes such as energy generation. The intensity of grassland management is mainly determined by the intended livestock husbandry and the demands for forage quality, most importantly, energy and protein. Ruminants are able to utilize the cell wall–rich biomass of grassland herbage, which implies that a mainly grass-based milk and meat production is, in this regard, not a direct competitor for human-edible plant products (Wilkinson and Lee 2018; Peyraud and Peeters 2016).

Grasslands play a major role in the biodiversity of agricultural landscapes in Europe. About one-third of the domestic vascular plants are found in grassland, and in Central Europe, grassland vegetation consists of four times more plant species than arable land (Leuschner and Ellenberg 2017; Dengler et al. 2014). This is partly due to the existence of a range of traditionally more extensive management systems, along with intensive dairy farming, that use and preserve a variety of soil and site conditions (Isselstein et al. 2005; Gilhaus et al. 2017).

There are potential trade-offs between agricultural production and biodiversity. In recent decades, a more species-rich grassland vegetation has been replaced by a few highly productive species, mainly grasses, with a high nutritive value (Isselstein et al. 2005; Krause et al. 2014; Leuschner et al. 2014). In the northern parts of Central Europe, the abundance of traditionally managed species-rich grasslands and the number of plant species have drastically declined compared to the situation in the middle of the twentieth century (Krause et al. 2011; Wesche et al. 2012). There is increasing pressure from society to at least halt these losses and take action to maintain and restore biodiversity.

The remaining highly diverse grasslands are scarce and can only be utilized by extensive livestock systems (sheep, grass-based beef production), and their maintenance requires financial support by agri-environmental schemes (Batáry et al. 2015; Hammes et al. 2016). Currently, introducing or maintaining biodiversity is often perceived as a constraint to productive grassland farming (Isselstein et al. 2005).

Recently, such a perception has elicited scrutiny and led to further discussion about the role and importance of biodiversity in grassland farming (Soussana and

Lüscher 2007; Gauly et al. 2013; Ergon 2023). Climate change, socio-economic pressures, the need to reduce protein import from overseas, the need for higher efficiency of nutrient use, especially N, and the food-feed discussion (beef and dairy production are expected to rely less on human-edible feed components) are the main drivers behind this change of perception.

The aim of this chapter is to explore the potential for maintaining and increasing biodiversity in a transformation towards achieving a more sustainable and resilient agricultural grassland management under conditions of climate change and increasing constraints on resources. This involves the integration and use of biodiversity to support a range of production systems, from low-input systems to livestock systems that rely on sufficient amounts of high-quality forage.

The review adopts two basic approaches to the potential significance of biodiversity for agriculturally utilized grassland under climate change in Central Europe (temperate humid climate): (i) the field level: how can the production level and the resilience of production of intensively managed grassland as the basis of productive livestock systems (such as dairy) be supported by more diverse grass swards, i.e., a higher forage plant diversity, and (ii) the landscape level: exploring the role of an increased habitat heterogeneity within the grassland dominated landscape for herbage production, i.e., enriched field margins and silvo-pastoral systems. Questions reamin as to the extent with which these measures can contribute to production and to the reduction of external inputs; and how they could support yield stability and facilitate a better forage and nutrient use.

In Chap. 2, we describe the challenges of and mitigation options for climate change for grassland production. In Chap. 3, we consider possible definitions and concepts for using biodiversity in agricultural production. Chapters 4 and 5 explain and give examples of the integration of biodiversity at the field and landscape level. Finally, Chap. 6 concludes with a summarizing outlook.

2 Challenges

Climate change causes a rise in temperatures and an increased variability of weather events (Trnka et al. 2011; Ergon et al. 2018). Compared to 1881, in 2021, the German average annual air temperature had increased statistically by 1.6 °C (DWD 2023a). In fact, the five warmest years have occurred after the year 2000. The number of hot days per year with an average temperature of at least 30 °C increased from three in 1950 to currently 9 days per year. While the precipitation sum has remained stable during summer, the average precipitation sum in winter has increased. In summer, however, the duration of precipitation-free days has increased, leading to more frequent drought periods (DWD 2023b). Campbell et al. (1997) report that within different grassland types, the amount and timing of precipitation are the major determinants of plant community structure and function and that, typically, 90% of the variance in primary production of grassland systems can be accounted for by annual precipitation.

Some implications of climate change for dairy production have been reviewed by Gauly et al. (2013). The main influences on plant growth and thus forage quantity in relation to climate change are a prolonged growing season, increased temperatures, altered precipitation patterns leading to extreme rainfall events and droughts and effects of increased CO_2 concentrations (Myneni et al. 1997; Ciais et al. 2005; Taube and Herrmann 2009). A higher level of CO_2 concentration in the atmosphere amplifies the nitrogen (N) limitation for grass growth and gives the N-fixing plants an advantage over grasses (Picon-Cochard et al. 2004; Lüscher et al. 2005). The effects of rainfall deficiency will vary greatly and are dependent on the botanical composition, soil conditions, season, growth stage and fertilization (Thumm and Tonn 2010; Carlsson et al. 2017).

Periods of drought are expected to be more severe and occur more often in future. For instance, the summer of 2018 saw some of the most severe droughts ever recorded in Central and Northern Europe (Buras et al. 2019). A study by Hofer et al. (2017) found that water shortage reduced white clover (*Trifolium repens* L.) productivity by up to 88% compared to a rain-fed control. The most widely sown companion-grass species in productive grass-clover swards is perennial ryegrass (*Lolium perenne* L.), which suffered significantly in productivity by −65% compared to the rain-fed control (Hofer et al. 2017).

In a scenario of 3 °C global warming in 2100, yield losses could be five times higher compared to today (Camalleri et al. 2020). This makes adaptation to drought conditions a priority for future grassland farming.

2.1 General Mitigation Options

Adaptation of grass and forage management to climate change should consider differences among agro-climatic zones, in seasonality and in regional and local conditions (Lüscher et al. 2022; Ergon 2023). Some factors like soil and site conditions and the main climatic factors, namely precipitation and temperature, can clearly not be controlled by farmers apart from the possibilities of land melioration and irrigation. Irrigation of grazed fields can yield high biomass and improved forage quality, but whether it is economically feasible is debatable (Hack-Ten Broeke 2001; Smith et al. 2010).

Possible adaptation strategies could include: Firstly, the management and structure of the whole farm could change and adapt, for example, by moving from intensive dairy production to beef production, or by adopting arable-based ley farming and maize production instead of permanent grassland. Secondly, at the field level, the use of adapted varieties and new species can help to synchronize with changed and extended growth periods, with milder winters and with drought periods in spring and summer. Thirdly, breeding new varieties and improving the agricultural use of less common species provides the basis and material for these mitigation and adaptation options.

3 Conceptual Background

The interactions between grassland farming and utilization, biodiversity and climate change are complex. Ecological processes and political and management decisions often do not occur at the same scale. Farming decisions need to be made in the context of the farm scale, which is, in turn, a part of the landscape scale and subject to regional, national and EU politics. The farm scale usually includes a number of fields, which are often managed according to the main livestock production of the farm. Still, within-field variability of soil and water conditions offers opportunities to increase plant diversity. A concept that can be used in this context is that of functional traits, which form the basis for the development of adaptation strategies which make use of diverse swards and multi-species mixtures. This also requires adaption to new patterns of climatic seasonality. It is thus relevant to understand the interactions between phenology, seasonal acclimation processes, stress resistance and recovery after stress (Ergon 2023).

3.1 The Concept of Functional Traits

The functional traits of plants enable them to occupy certain niches and survive challenging growing situations. Schellberg and da Pontes (2012) have reviewed the concept of plant functional traits and nutrient gradients in grassland. Violle et al. (2007) define functional traits as morpho-physio-phenological properties that affect plant fitness indirectly via their impact on growth, reproduction and survival, the three components of individual performance. The combination of functional traits of individual plants is related to their role within the community and the reaction to biotic and abiotic stress factors (Schellberg and da Pontes 2012). Understanding the role of functional traits and related species interaction, species compatibility, phenology and seasonal acclimation processes will help to plan and manage plant mixtures and plant community compositions for purposes like stress prevalence, resistance and recovery (Ergon, 2023; Lüscher et al. 2022). Ecological theory has claimed that species-rich mixtures with their variety of potential functional traits maximize diversity effects, such as the production of herbage biomass during drought (Craven et al. 2016).

3.2 Reactions of Plants to Abiotic Stress

Tolerance and resistance are different expressions of responses to abiotic stress resulting from, e.g. drought, flooding, temperature and heavy metals (Taiz and Zeiger 2010). Drought is the meteorological term for a period of insufficient precipitation that results in plant water deficit. Drought tolerance is the ability by which a plant or plant community maintains its biomass production during arid or drought conditions. Drought resilience is the capacity of a system (ecosystem, plant and plant community) to respond to a period of water deficit by resisting damage and subsequently recovering. More resilient grasslands have, therefore, a higher stability of production or forage quality over time (seasons, years).

3.3 Plant Functional Groups

Grassland swards and sown grassland mixtures consist of an array of plant species of distinct functional groups. Functional groups are defined as species with similar general morpho-physiological characteristics (Hunt et al. 2004). Each functional group may possess certain functional traits, although plant species from different functional groups may express similar functional traits (e.g. deep rooting in certain grasses and in herbs). An overview of some typical grassland forage plants belonging to different functional groups and their respective functional traits is given in Table 13.1.

Usually one (e.g. pure grass mixtures), sometimes two (grasses + legumes) and in rare cases three (grasses + legumes + herbs) functional groups are sown in grassland mixtures. Each functional group may be represented by several species, and a

Table 13.1 Some plant species with their respective functional groups and functional traits

	Functional groups			Functional traits					
				Deep	Tillering[a]			N_2	
Species	Grass	Legume	Herb	rooting	Below	Above	Tussock	fixation	Persistence
Dactylis glomerata	x						x		x
Lolium perenne	x					(x)			x
Festuca arundinacea	x			x			x		x
Plantago lanceolata			x	x			x		(x)
Taraxacum officinale			x	x					x
Cichorium intybus			x	x			x		(x)
Achillea millefolium			x	x	x				x
Trifolium repens		x				x		x	x
Trifolium pratense		x		x			x	x	
Medicago sativa		x		x	(x)		x	x	
Lotus corniculatus		x			(x)			x	x

Adapted from Dierschke and Briemle (2008), Senghas and Seybold (2003) and Hunt et al. (2004)
[a]Tillering refers to the ability for lateral growth of tillers above- (stolons) and below-ground (rhizomes), while other species form tussocks

higher species number will increase species richness and inter-specific variability (Prieto et al. 2015). Species-rich swards could use more niches and utilize water and nutrient resources more efficiently (Scherer-Lorenzen et al. 2003; Niklaus et al. 2006). A number of investigations have confirmed that species-rich systems have an increased yield stability and are better adapted to varying environmental conditions and stress factors (Knops et al. 1999; Tilman et al. 2006; Nölke et al. 2021). Plant community composition can influence stress resistance or the recovery of individual plants in the sward through, firstly, an effect on stress levels and, secondly, through an effect on resistance levels or the recovery ability of the individual plant (Ergon 2023). In a review of the importance of multi-species grassland leys to enhance ecosystem services in crop rotations, Malisch et al. (2023) state that (a) the identity of the plant species is crucial, and that (b) more balanced mixtures tend to maximize the benefits of synergistic interspecific interactions and deliver the strongest yield responses (Lüscher et al. 2019). A mixture may consist of a blend of varieties within a single species, multiple species or a combination of multiple species and multiple varieties within species (Malisch et al. 2023). Considering the effects of plant diversity in permanent grassland on resource utilization, adaptation to seasonality and pedo-climatic conditions, climate change and, generally, abiotic stress, the concept of functional traits can be extended to utilize planned mixtures in ley cropping for multiple functions. These include, among others, yield stability, pollination and weed suppression (e.g. Thielecke et al. 2024).

4 Grassland Biodiversity at the Field Level

At the farm or regional level, Lüscher et al. (2022) have identified two ways of increasing the diversity of genotypes and/or species: either 'within fields' by growing the same complex plant community on all sites, or 'among fields' by growing different (simpler) plant communities on different sites. Both these strategies are feasible in sown grassland leys through the targeted composition of the seed mixture. The selection of a mix of forage species and varieties with specific properties (or traits) enables the designing of mixtures with various functions and types of utilization and for certain pedo-climatic conditions (Lüscher et al. 2019; Malisch et al. 2023).

4.1 Species Diversity and Plant Production

Ecological research over the last two decades has established that there is a positive correlation between productivity and biodiversity for grasslands that are sown and extensively managed (e.g. Hector et al. 1999; Tilman et al. 2001; Fig. 13.1a). This claim is at least partly based on results generated from manipulated environments focusing on ecological functions with limited applicability to agricultural production conditions where intensive management and frequent defoliation are typical (Wrage et al. 2011). Under such more intensive grassland management, a positive

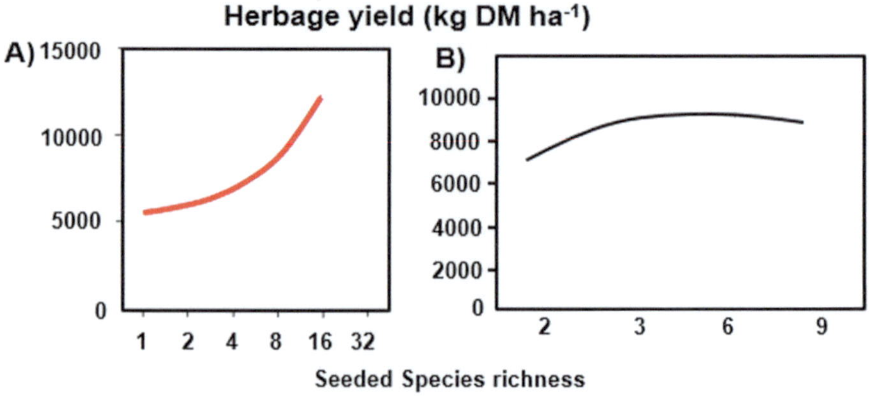

Fig. 13.1 Species richness-productivity relationship in (**a**) low-input grassland. (Adapted from Hector et al. 1999) and (**b**) under productive intensively managed grazing conditions in the USA. (Adapted from Sanderson et al. (2005))

response in biomass yield to species richness can generally be confirmed, but this response function already reaches a plateau with only a few sown species (Fig. 13.1b). Herbage biomass yields from different management intensities can be at a similar level (Fig. 13.1). Under extensive management, as shown in Fig. 13.1a, and at peak biomass yield, the yield of digestible organic matter is low due to a late harvest date that is far beyond the date of maximum digestibility. At an increased management intensity, as shown in Fig. 13.1b, the digestible organic matter yield will be higher, which is an important measure for livestock production.

The forage quality of grasslands, characterized by energy content, protein content and digestibility, needs to fit the requirements of the associated form and intensity of livestock production. The relationship between species richness and forage quality is even less clear than with biomass yield and is often negative. Usually, a high forage quality is achieved by frequent and regular defoliation, often in combination with higher inputs of nitrogen and a high proportion of grasses in the sward. Cell contents such as water-soluble carbohydrates or crude protein (CP) increase the digestibility and, hence, the metabolizable energy (ME) of forage, whereas cell wall constituents such as the acid detergent fibre (ADF), which consists of cellulose and lignin, reduce the ME concentrations. Under conditions of a lower management intensity with less frequent defoliations, swards tend to have higher ADF, reduced digestibility and lower ME. However, species identity is important to understand the influence of species richness on forage quality (Fig. 13.2). Generally, legumes lead to higher CP concentrations and reduced ADF, whereas dicotyledonous herbs often have higher ADF concentrations and thus lower ME concentrations.

Experimental data from Jerrentrup et al. (2020; Fig. 13.3) show a significantly positive relationship between species richness and CP, a slightly negative relationship for species richness and ADF, but no relationship with ME. Although dicotyledonous species mostly have lower cell wall proportions than grasses, the in vivo digestibility of herbs is lower than that of perennial ryegrass (Derrick et al. 1993).

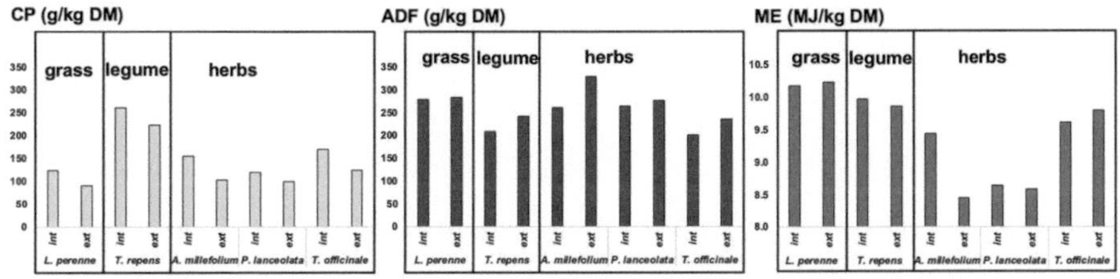

Fig. 13.2 Effects of species identity on forage quality under a 3-cut (ext) and a 6-cut (int) defoliation system. (Adapted from Isselstein (1994)) as averages of two full production years; concentrations of crude protein (CP), acid-detergent fibre in the organic matter (ADF) and the metabolizable energy (ME)

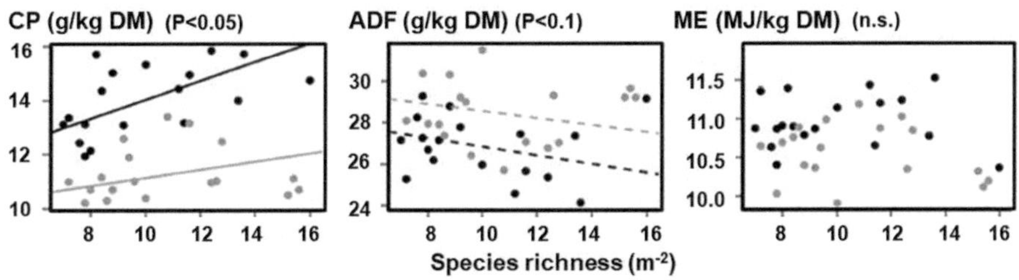

Fig. 13.3 Relationship between species richness and forage quality (adapted from Jerrentrup et al. (2020)); concentrations of crude protein (CP), acid-detergent fibre in the organic matter (ADF) and the metabolizable energy (ME); values for 2 years (grey and black); P-values indicate significance of the relationship

This is related to the higher lignification of the cell wall (Hummel et al. 2006). The resulting lower digestibility of the cell wall is the reason why lower ADF concentrations in species-rich swards do not lead to higher ME concentrations (Fig. 13.3).

The consequences of the relationship between species richness and forage quality for the practice can be explained by the pictures in Fig. 13.4. In dicotyledonous species, flower stems develop fast and at a high rate during the summer period, while most common forage grasses produce flower stems only in the primary growth in spring. For herbs, the emergence of flower stems with higher fibre content is related to reduced digestibility and, consequently, lower ME concentrations. A higher defoliation frequency would reduce the proportion of flower stems in the sward and lead to higher ME and, generally, better forage quality (Figs. 13.2 and 13.3).

4.2 Species Richness and Drought

Water has direct and indirect effects on the occurrence of certain plant communities and the sward composition, and on the processes of dominance and competition due to adaptation to water deficit or conditions of water logging or flooding (Opitz von

Fig. 13.4 (**a**) Tall fescue (*Festuca arundinacea*) in the vegetative growth stage in summer with low ADF and high ME concentration, (**b**) Timothy (*Phleum pratense*) in the generative stage, (**c**) Ribwort plantain (*Plantago lanceolata*) and (**d**) Caraway (*Carum carvi*) during the reproductive stages with flower stems and low ME concentrations during summer

Boberfeld 1994). During wet periods, differences in water use among species are smaller compared to periods when water deficit occurs (Nippert and Knapp 2007). More drought-resistant plants in temperate climates often exhibit a greater plasticity in rooting depth, wider root–shoot ratio and finer root architecture (Evans 1977; Herndl et al. 2010).

Ecological studies have found that the richness of functional groups within a mixture significantly influences productivity (Hector et al. 1999). According to data from a mesocosm study by Lanta and Lepš (2006), the relationship between biomass production and functional group richness per sward is much stronger than the effect of species richness (p-value and slope; Fig. 13.5). It follows that for an adapted composition of grassland mixtures, the diversity in functional groups should be maximized to benefit from the diversity of functional traits and the inclusion of forage herbs is suggested (e.g. *C. intybus*, *P. lanceolata*, Table 13.1). However, adding herbs to mixtures of grasses and legumes could reduce the proportional yield contribution of grasses while the share of legume is not affected.

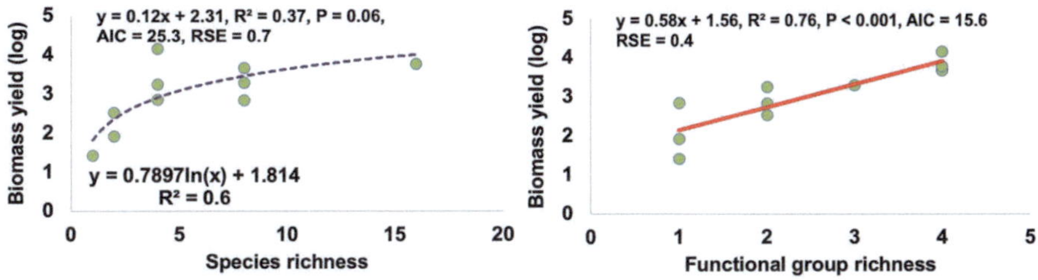

Fig. 13.5 Relationship between sown species richness (left) and functional group richness (right) and the log-transformed biomass production based on outputs of linear models. The logarithmic function produced a higher R^2 in the left figure, and we, therefore, included the prediction line. (Adapted from Lanta and Lepš 2006. *RSE* residual standard error, *AIC* Akaike information criterion)

The effects of species diversity in grassland on resistance to drought stress and subsequent recovery have been reviewed by Lüscher et al. (2022). They conclude that plant species diversity can stabilize forage production in two ways. Firstly, growth reduction under stress is significantly smaller for diverse compared to simple plant communities as diverse swards offer the opportunity to include drought-resistant (or drought-surviving) species (Finn et al. 2018). Secondly, positive interactions among species in more diverse plant communities maintain the ecosystem functioning under moderate drought, allowing these communities to compensate for drought-induced yield reductions (Lüscher et al. 2022).

In a study by Komainda et al. (2020) on the effect of species diversity in an open-sided greenhouse, water supply was varied (control vs. induced temporary drought periods) for one-, three- and five-species mixtures containing different combinations of functional groups and functional traits. The study showed that it is not the most species-rich mixture that has the highest drought tolerance. Rather, it is primarily the mixtures where the functional traits of several species are optimized (Mahaut et al. 2019) and where the traits do not overlap (Fig. 13.6) that cope best with drought. For instance, a mixture of five species with the highest potential trait diversity, but also high trait overlap, was less drought tolerant than a mixture with three species where functional traits were already maximized but where they did not overlap (Fig. 13.6) with consequently lower inter-specific niche competition. Under drought, mixtures which were optimized according to traits (without potential overlap) produced 3% more herbage biomass than the five-species mixture with potential trait overlap. This yield increase seems small but actually is in the range of perennial ryegrass breeding, where a yield gain of 4.3% was achieved per decade (Reheul et al. 2017). It can therefore be concluded that (low-input) grassland mixtures need to be optimized with respect to functional trait diversity. However, knowledge of the functional traits that certain plant species provide and where these are needed is required for improved grassland mixture design under future climate conditions.

The challenge under future climate conditions is to provide stable herbage biomass and forage quality over several years in order to avoid costly grassland renewal.

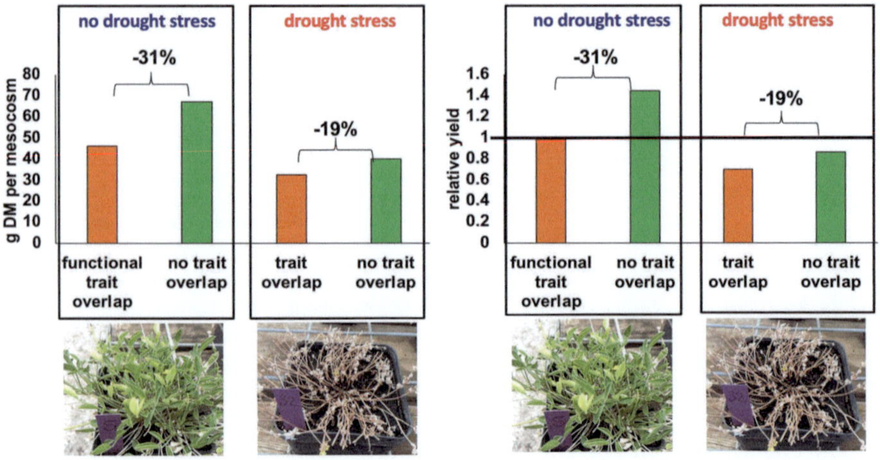

Fig. 13.6 Effects of functional diversity and trait redundancy in three-species swards on total and relative biomass yield (relative to the average biomass yield) under optimal water supply and drought conditions. Swards consisted of three-species combinations of *Trifolium repens*, *Dactylis glomerata*, *Lolium perenne*, *Plantago lanceolata* and *Taraxacum officinale*; averaged across three drought stress periods. (Adapted from Komainda et al. 2020)

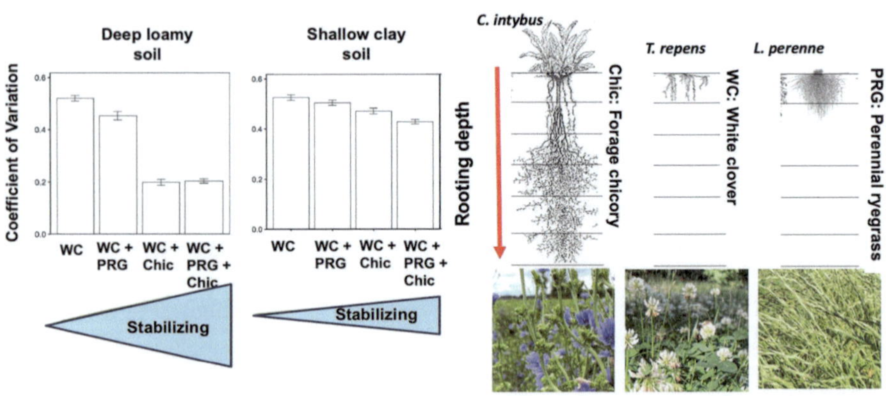

Fig. 13.7 Coefficient of variation (CV) of annual yield = yield stability of different species-rich stands at two locations, over a 5-year period, near Göttingen, Germany. *WC* white clover, *PRG* perennial ryegrass, *Chic* chicory. The wedges (blue) show the degree of stabilization of the more species-rich stands. (Adapted from Nölke et al. 2021. Rooting patterns are taken from Kutschera 1960)

Resilient grassland swards are often regarded as stable systems that react less to stress situations and show smaller fluctuations in yield over time. Temporal stability is a measure of climate resilience as expressed in terms of the intra- or inter-annual coefficients of variation (CV). The smaller the CV, the smaller the fluctuations and, therefore, the more stable or resilient the stands and swards. The following graph (Fig. 13.7) shows how the CV developed in a field experiment over 5 years depending on the crop combinations (white clover, white clover + ryegrass, white clover + forage chicory and white clover + ryegrass + forage chicory). Thus, one to three species from three functional groups were combined (Nölke et al. 2021). The experimental design included the following traits: deep-rooting of the forage herb chicory (*Cichorium intybus* L.) and biological N fixation of the legume white clover (*Trifolium repens* L.); the traits, however, were not combined. The effective inclusion of deep-rooting herbs in forage production may, therefore, depend on the N-fixing partner species. From a trait perspective, this partner would need to have a shallow root in order to avoid niche competition. Chicory was, thus, combined with white clover as it is the most common grassland legume. The experiment was conducted from 2015 to 2020 and included at least 2 years (2018 and 2019) with notable drought periods. The results indicated that supplementing white clover and ryegrass with the deep-rooted forage herb chicory leads to a significant stabilization effect. However, this effect is dependent on external factors. Only where the location allows deep root formation does the inclusion of a deep-root component lead to an increase in resilience. This example illustrates how climate resilience in grassland production can be achieved by sowing a herb-rich mixture.

5 Grassland Biodiversity at the Landscape Level

A higher structural heterogeneity of an agricultural landscape will support and increase the level of biodiversity. Using a concept of the effect of landscape structure on biodiversity, Fahrig et al. (2011) distinguish between compositional heterogeneity, which is the number of cover types in a given landscape, and configurational heterogeneity, which is the size and spatial arrangement of cover types in that landscape. Putting this in context with ecosystem services and biodiversity would lead to a description of ecological functioning of a landscape and a concept of functional landscape heterogeneity. During the prolonged dry periods in recent years, it has been observed that the growth and biomass production of agricultural crops and grasslands were less affected in structured landscapes with trees and hedges than in largely open landscapes.

Meteorologists are interested in how the structure of the earth's surface interacts with weather and climate. In a review paper, Pielke and Avissar (1990) report that even small changes in the surface condition can have significant effects on the local and regional climate. It is, therefore, useful to examine how the spatial arrangement of (different) plant types affects the microclimate and growth conditions within this arrangement. Agroforestry systems are examples of spatial arrangements that have been investigated well; these combine planted trees with either arable crops (silvoarable) or with grasslands (silvopastoral). In a recent compilation of experimental data from various sites in temperate climates (Central Europe), Veldkamp et al. (2023) outlined the multiple benefits of silvoarable and silvopastoral systems for a range of ecosystem services. They found that, among others, soil biological activity, structure of soil biota and soil organic carbon stocks increased compared to conventional agricultural production. Other measures to increase the heterogeneity of the land surface, like establishing hedgerows, coppices, late growing grassland margins, and generally, a greater variety of land use types and of agricultural crops, can have similar effects.

In the context of demands for a transformation in agricultural production, it is of interest to investigate whether climate change adaption measures at the landscape level can be combined with promoting biodiversity. In order to answer this question, the results of two examples are considered: (i) silvopastoral systems and (ii) enriched grass margins.

5.1 Silvopastoral Systems

Tree lines in agroforestry systems increase the surface roughness and alter the microclimatic conditions in the alleys between the tree lines which creates a range of different growth conditions depending on distance to the tree line and orientation in relation to the cardinal point: wind speed is reduced, extreme temperatures are buffered, and erosion is reduced. There is higher relative humidity and lower temperatures close to the trees, potential hydraulic redistribution of water, and possible lower water vapour pressure—meaning that water losses via evaporation and transpiration are reduced (Quinkenstein et al. 2009; Castillo et al. 2020; Gerjets et al. 2021; Jacobs et al. 2022). Consequently, agroforestry systems have the potential to mitigate expected adverse climate change effects, such as heat waves and drought periods. The mitigating effects of agroforestry systems occur more frequently under climatic conditions with higher temperatures and less water availability (Torralba et al. 2016). For temperate silvopastoral systems, Sutterlütti et al. (2023a, b) found that under dry conditions, tree lines improved the water status of grassland plants close to the tree lines and that light availability and temperatures were modified by the trees.

An analysis of global data on the effect of silvopastoral systems on biodiversity (Perez-Alvarez et al. 2023) revealed mainly beneficial effects across taxa as well as across climatic regions, with no evidence of a negative effect (Table 13.2). There appear to be interactions between climate region and taxa; birds and mammals do not seem to be markedly affected, while dung beetles show positive effects. For agroforestry systems in temperate climates, the effects are less clear (Table 13.2).

To evaluate the effect of silvopastoral systems on the biodiversity of higher plants, findings from research in old wooded and orchard grasslands (*Streuobstwiesen*) under extensive agricultural management can be used. These habitats are characterized by a high level of plant species richness with a high nature conservation value (Garbarino and Bergmeier 2014). Detailed vegetation patterns of orchard grasslands were studied by Paesel et al. (2019) in Lower Saxony, Germany. Their hypothesis was that trees would increase spatial heterogeneity and that the number of ecological niches would also increase, leading to a higher number of plant species. In the survey, 32 orchard grasslands were analysed, and three different small-scale habitats were identified within every orchard, representing a gradient of competition between the trees and the grass sward (Fig. 13.8). The results showed that the

Table 13.2 Results of a meta-analysis of globally available data on the effect of silvopastoral systems on the diversity of different taxa groups; biodiversity response either positive (green arrow) or neutral (yellow arrow); (---) no sufficient data available

Climates/regions	Tropics	Sub-tropics	Mediterranean	Temperate	Across climates
Birds	→	→	---	→	→
Dung beetles	↗	↗	---	---	↗
Insects	→	---	→	→	↗
Mammals	→	---	---	---	→
Higher Plants	→	---	---	---	↗
Across taxa	↗	↗	→	→	↗

From: Perez-Alvarez et al. (2023)

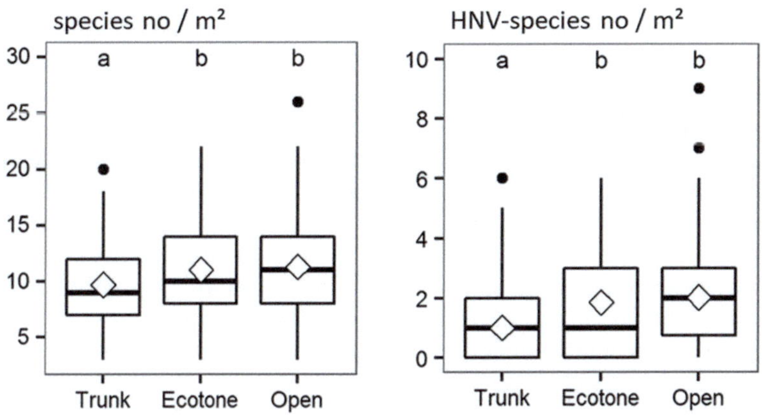

Fig. 13.8 Diversity of higher plants of different habitats within orchard grasslands ($n = 32$ sites in Lower Saxony, Germany) in relation to the distance to trees: Trunk, beneath the tree canopy; Ecotone, transition zone between canopy and open grassland; Open, open grassland barely affected by trees. Treatments with the same letter are not significantly different at $p < 0.05$ (Tukey test). (From Paesel et al. 2019)

species richness increased from the area directly affected by the tree canopy ('trunk') to the open grassland area ('open'), which was only marginally affected by the trees. The more important finding, however, was that the number of species occurring in all three areas was only 40%, meaning that the total species number of the orchard grassland (including high nature value species) was not only an effect of the extensive management but also of the spatial heterogeneity.

While the effects of trees as structural elements on species richness seem to be positive, their effects on biomass production and potential for agricultural use are more complex. Schmiedgen et al. (2021) investigated the herbage production potential of silvopastoral systems in Germany and found that herbage accumulation close to tree lines was reduced by 41% compared to the control, a part of the field with no trees. The herbage utilization efficiency, that is, the proportion of gross live-green herbage production which is utilized before entering senescence, was not affected by the distance to the tree line. This implies that a more intensive grassland management with frequent defoliation or fertilization close to the tree lines will not improve production and that this area could be better used for improving plant diversity by extensive management (Klimek et al. 2007). Schmiedgen et al. (2021) have shown that there is considerable potential to follow that finding: by employing remote sensing technology, they found that 4.4% of the German grassland area is directly affected by woody vegetation and could be included in management schemes to increase biodiversity.

5.2 Enriched Grass Margins

Extensive grassland systems with no history of agricultural intensification show a considerably high heterogeneity of small-scale environmental conditions accompanied by high biodiversity. In contrast, intensively managed grasslands for dairy production are often quite species-poor (Isselstein et al. 2005). However, even intensively managed permanent grasslands have some potential for higher plant diversity. Lowland grasslands often exhibit considerable microtopographic heterogeneity with a mosaic of depressions and elevated patches on peat bog grassland or of drainage ditches alternating with slightly elevated areas on heavy mineral marsh soil. Mroncz et al. (2024) showed that this kind of within-field variability affects soil water availability, forage production and forage quality. They concluded that a higher variability in soil and site conditions may contribute to a more resilient herbage growth under conditions of climate change with more frequent drought or heavy rain events. It can also be assumed that a higher gradient of soil water conditions and soil nutrient status within a field leads to more diverse vegetation. In a botanical survey of 110 intensively managed grassland fields on marsh and peat soils in North-West Germany, Leuschner et al. (2024) recorded details of the botanical composition of the field centre, the field margin (2–8 m from edge) and the field edge, usually bordering a drainage ditch. The total number of species and the number of high-nature-value species did not differ between the field centre and margin, but sward composition was significantly different between the centre and edge and between margin and edge (Fig. 13.9, Leuschner et al. 2024). The edges contained species that were adapted to a higher water table and lower nutrient status. This finding was further confirmed by investigating the soil seed bank. The seed bank at the edge of the field contained a higher number of plant species (particularly wetland species) but was species-poor in the centre and the margin. Despite these findings, the biodiversity status and nature conversation potential of intensively managed permanent grassland sites in North-West Germany are comparatively low. Establishing more species-rich vegetation would require the introduction of suitable plant

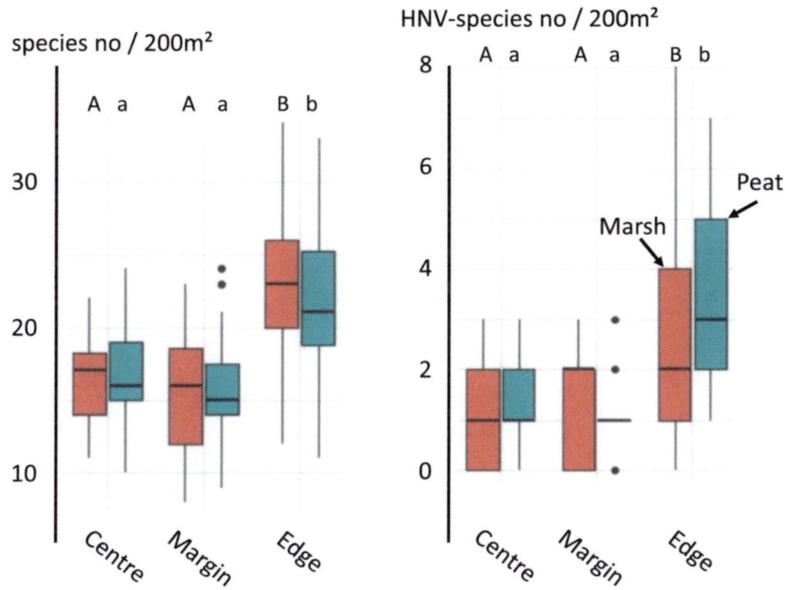

Fig. 13.9 Box plots of the plant species richness (total species number and high-nature value (HNV) species number) of intensively managed marsh and peat grasslands ($n = 55$ per soil type) in North-West Germany in relation to the position within the field: Centre, interior of a site; Margin, within 2–8 m distance from the field edge; Edge, drainage ditch (grassland edge including the ditch slope); means with the same letter (a, b for Peat and A, B for Marsh) within a soil type are not significantly different (Tukey test, $p < 0.05$). (After Leuschner et al. 2024)

species by enriching field margins, which would then require that these edges and margins are extensively managed to provide niches for longer-lasting effects. In a field experiment on a range of grassland sites, Schmitz et al. (2022) tested to what extent grassland margins can be diversified by introducing new species either by sowing seed mixtures (oversowing after shallow harrowing) or by transferring hay from species-rich grasslands. This attempt was compared to a control area which was not enriched. Management of the margins was changed to leave out any N fertilization and from frequent cutting for silage to a less frequent two-cut defoliation system. This approach proved to be successful: the species number of the aboveground vegetation in areas with oversowing seed mixtures and in areas with transfer of hay from species-rich grassland treatment was up to five times higher than in the control, and the effect was still observable in the second year (Fig. 13.10; Schmitz et al. 2022).

These results show that even in landscapes with intensive ruminant husbandry accompanied by high intensity of grassland management, there is potential to increase biodiversity at the landscape scale without compromising the production of herbage of a high quality at the farm level.

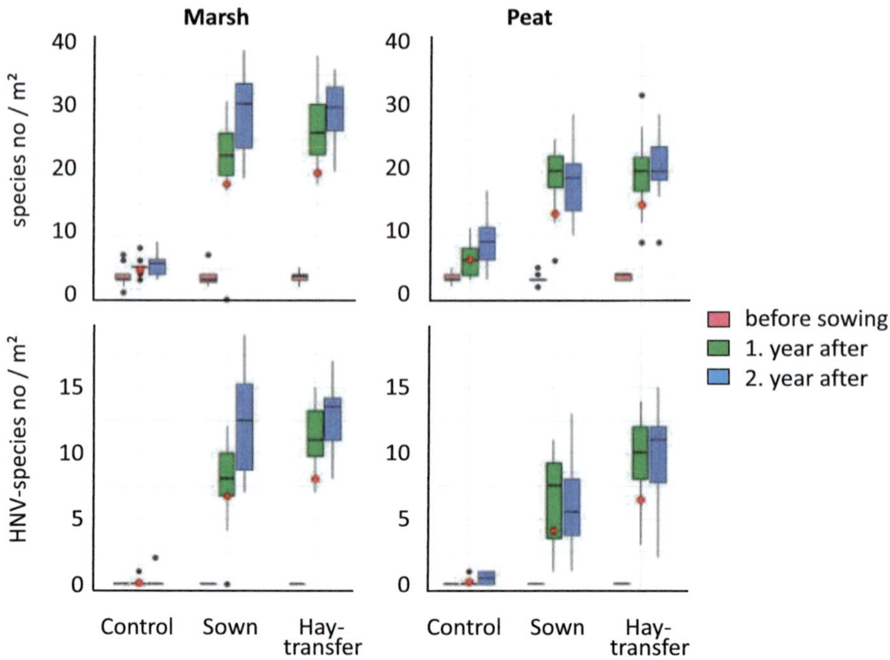

Fig. 13.10 The effect of different ways to improve field margins of marsh and peat grassland sites in North-West Germany on the plant species, and the high nature value (HNV-) plant species number. Treatments consisted of sowing a species-rich seed mixture and of a species-rich hay transfer compared to an untreated control; in both intervention measures, the soil was prepared by harrowing. (After Schmitz et al. 2022)

6 Conclusions

In a transformation towards more climate-resilient grasslands, plant diversity can play a crucial role. In this review, we have shown that there is a range of management options to increase biodiversity, but that knowledge of underlying processes at different scales and of concepts such as functional traits and ecosystem services is needed.

The expected effects of climate change on grassland production are manifold and interrelated. The interactions create a very complex picture where, in some cases, the effects of climate change are enhanced, while in others, they are compensated.

The question of the potential role of biodiversity in mitigating adverse climate change effects on grassland production cannot, as yet, be fully answered. This review has illustrated that there is evidence for potential at both the plot and field and the landscape scale. Still, further knowledge is required to cope with the challenges of climate change in grassland farming practice. New plant material from breeding efforts must be productive and has to have the potential to adapt to abiotic stress with changing seasonality and the occurrence of stress periods. Furthermore, there is a need to search for novel plant species with special traits which can be included in future grass swards and sowing mixtures. As a start, pre-breeding programmes should be set up for novel species to investigate their functional traits and how these traits can be included into existing concepts for breeding and for mixtures. This requires the consideration of the interactions among different species in a grass sward as well as their interaction with environmental conditions. Common forage species will still be the backbone of agricultural grassland production. However, an improved resilience to adverse climate change effects needs to be included in breeding activities.

At the landscape scale, there is even less scientific knowledge on the relationship between biodiversity, climate change and grass production. This is surprising as in biodiversity research, spatial scales have for some time been considered to be of utmost importance. Future grassland research should, therefore, extend the focus from the plot and field to the landscape scale. The view of a farmer is always to maintain the functioning of the production system, and management decisions at the field level are governed by the overriding requirements of that system. There is a need for a transdisciplinary analysis of current grassland use systems, including nature conservation. This ultimately means integrating agricultural management into biodiversity research.

References

Allen VG, Batello C, Berretta EJ, Hodgson J, Kothmann M, Li X, McIvor J, Milne J, Morris C, Peeters A, Sanderson M (2011) An international terminology for grazing lands and grazing animals. Grass Forage Sci 66:2–28. https://doi.org/10.1111/j.1365-2494.2010.00780.x

Batáry P, Dicks LV, Kleijn D, Sutherland WJ (2015) The role of Agri-environment schemes in conservation and environmental management. Conserv Biol 29:1006–1016. https://doi.org/10.1111/cobi.12536

Buras A, Rammig A, Zang CS (2019) Quantifying impacts of the drought 2018 on European ecosystems in comparison to 2003. Biogeosciences 17:1655–1672

Camalleri C, Naumann G, Mentaschi L, Formetta G, Forzieri G, Gosling G, Bisselink B, De Roo A, Feyen L (2020) Global warming and droght impacts in the EU. JRC PESETA IV project – Task 7. https://joint-research-centre.ec.europa.eu/system/files/2020-05/pesetaiv_task_7_drought_final_report.pdf

Campbell BD, Statford Smith DM, McKeon GM (1997) Elevated CO2 and water supply interactions in grasslands: a pastures and rangelands management perspective. Glob Chang Biol 3:177–187

Carlsson M, Merten M, Kayser M, Isselstein J, Wrage-Mönnig N (2017) Drought stress resistance and resilience of permanent grasslands are shaped by functional group composition and N fertilization. Agric Ecosyst Environ 236:52–60

Castillo MS, Tiezzi F, Franzluebbers AJ (2020) Tree species effects on understory forage productivity and microclimate in a silvopasture of the southeastern USA. Agric Ecosyst Environ 295:106917. https://doi.org/10.1016/j.agee.2020.106917

Ciais P, Reichstein M, Viovy N, Granier A, Ogee J, Allard V, Aubinet M, Buchmann N, Bernhofer C, Carrara A, Chevallier F, De Noblet N, Friend AD, Friedlingstein P, Grunwald T, Heinesch B, Keronen P, Knohl A, Krinner G, Loustau D, Manca G, Matteucci G, Miglietta F, Ourcival JM, Papale D, Pilegaard K, Rambal S, Seufert G, Soussana JF, Sanz MJ, Schulze ED, Vesala T, Valentini R (2005) Europe-wide reduction in primary productivity caused by the heat and drought in 2003. Nature 437:529–533

Craven D, Isbell F, Manning P, Connolly J, Gruelheide H, Ebeling A, Roscher C, van Ruijven J, Weigelt A, Wilsey B, Beierkuhnlein C, de Luca E, Griffin JN, Hautier Y, Hector A, Jentsch A, Kreyling J, Lanta V, Loreau M, Meyer ST, Mori AS, Naeem S, Palmborg C, Polley HW, Reich PB, Schmid B, Siebenkäs A, Seabloom E, Thakur MP, Tilman D, Vogel A, Eisenhauer N (2016) Plant diversity effects on grassland productivity are robust to both nutrient enrichment and drought. Philos Trans R Soc B 371:20150277. https://doi.org/10.1098/rstb.2015.0277

Dengler J, Janišová M, Török P, Wellstein C (2014) Biodiversity of Palaearctic grasslands: a synthesis. Agric Ecosyst Environ 182(2014):1–14. https://doi.org/10.1016/j.agee.2013.12.015

Derrick RW, Moseley G, Wilman D (1993) Intake, by sheep, and digestibility of chickweed, dandelion, dock, ribwort and spurrey, compared with perennial ryegrass. J Agric Sci 120:51–61

DESTATIS (2022) Statistisches Bundesamt. https://www.destatis.de/DE/Home/_inhalt.html

Dierschke H, Briemle G (2008) Kulturgrasland: Wiesen, Weiden und verwandte Staudenfluren, 2nd edn. Eugen Ulmwer KG, Stuttgart

DWD (2023a) German Weather Service: Temporal development in temperature and precipitation in Germany. https://www.dwd.de/DE/klimaumwelt/klimawandel/ueberblick/ueberblick_node.html. Accessed 18 Feb 2024

DWD (2023b). German Weather Service: Temporal variability in precipitation. https://www.dwd.de/DE/leistungen/zeitreihen/zeitreihen.html;jsessionid=4A7BEF53B37C0A13AB4D392ED2B1399B.live11052?nn=344934. Accessed 18 Feb 2024

Ergon A (2023) Combined adaptation to climate and mixed cropping in perennial forages. Grassland Sci Eur 28:185–191

Ergon A, Seddaiu G, Korhonen P, Virkajärvi P, Bellocchi G, Jørgensen M, Østrem L, Reheul D, Volaire F (2018) How can forage production in Nordic and Mediterranean Europe adapt to the challenges and opportunities arising from climate change? Eur J Agron 92:97–106. https://doi.org/10.1016/j.eja.2017.09.016

Evans PS (1977) Comparative root morphology of some pasture grasses and clovers. N Z J Agric Res 20:331–335

Fahrig L, Baudry J, Brotons L, Burel FG, Crist TO, Fuller RJ et al (2011) Functional landscape heterogeneity and animal biodiversity in agricultural landscapes. Ecol Lett 14:101–112

Finn JA, Suter M, Haughey E, Hofer D, Lüscher A (2018) Great gains in annual yields from increased plant diversity than losses from experimental drought in two temperate grasslands. Agric Ecosyst Environ 258:149–153. https://doi.org/10.1016/j.agee.2018.02.014

Garbarino M, Bergmeier E (2014) Plant and vegetation diversity in European wood-pastures. In: Hartel T, Plieninger T (eds) European wood-pastures in transition. A socio-ecological approach. Routledge, Oxon, pp 113–125

Gauly M, Bollwein H, Breves G, Brugemann K, Danicke S, Das G, Demeler J, Hansen H, Isselstein J, Konig S, Loholter M, Martinsohn M, Meyer U, Potthoff M, Sanker C, Schroder B, Wrage N, Meibaum B, von Samson-Himmelstjerna G, Stinshoff H, Wrenzycki C (2013) Future consequences and challenges for dairy cow production systems arising from climate change in Central Europe – a review. Animal 7:843–859

Gerjets R, Richter F, Jansen M, Carminati A (2021) Hydraulic redistribution by hybrid poplars (Populus nigra x Populus maximowiczii) in a greenhouse soil column experiment. Plant Soil 463(1–2):145–154. https://doi.org/10.1007/s11104-021-04894-0

Gilhaus K, Boch S, Fischer M, Hölzel N, Kleinebecker T, Prati D, Rupprecht D, Schmitt B, Klaus VH (2017) Grassland management in Germany: effects on plant diversity and vegetation composition. Tuxenia 37(1):379–397. https://doi.org/10.14471/2017.37.010

Hack-ten Broeke MJD (2001) Irrigation management for optimizing crop production and nitrate leaching in grassland. Agric Water Manag 49:97–114

Hammes V, Eggers M, Isselstein J, Kayser M (2016) The attitude of grassland farmers towards nature conservation and agri-environment measures – a survey-based analysis. Land Use Policy 59:528–535

Hector A, Schmid B, Beierkuhnlein C, Caldeira MC et al (1999) Plant diversity and productivity experiments in European Grasslands. Science 286:1123–1127

Herndl M, Bohner A, Krautzer B, Graiss W, Kandolf M (2010) Arten- und Sortenscreening von Gräsern, Kräutern und Leguminosen als Grundlage für Grünlandbewirtschaftung auf trockenen Standorten. Species and variety screening of grass, herbs and legumes as a basis for grassland management at dry sites. Project Report IFZ Raumberg-Gumpenstein http://www.raumberg-gumpenstein.at/c/index.php?option=com_fodok&Itemid=100033&task=detail&filter_publnr=8157. Accessed 2 Mar 2011

Hofer D, Suter M, Buchmann N, Lüscher A (2017) Nitrogen status of functionally different forage species explains resistance to severe drought and post-drought overcompensation. Agric Ecosys Environ 236:312–322. https://doi.org/10.1016/j.agee.2016.11.022

Hummel J, Südekum K-H, Streich WJ, Clauss M (2006) Forage fermentation patterns and their implications for herbivore ingesta retention times. Funct Ecol 20(6):989–1002. https://doi.org/10.1111/j.1365-2435.2006.01206.x

Hunt R, Hodgson JG, Thompson K, Bungener P, Dunnett NP, Askew AP (2004) A new practical tool for deriving a functional signature for herbaceous vegetation. Appl Veg Sci 7:163–170

Isselstein J (1994) Zum futterbaulichen Wert verbreiteter Grünlandkräuter. Habilitationsschrift, Gießen

Isselstein J, Kayser M (2014) Functions of grassland and their potential in delivering ecosystem services. Grassland Sci Eur 19:199–214

Isselstein J, Jeangros B, Pavlu V (2005) Agronomic aspects of biodiversity targeted management of temperate grasslands in Europe–a review. Agron Res 3:139–151

Jacobs SR, Webber H, Niether W, Grahmann K, Lüttschwager D, Schwartz C et al (2022) Modification of the microclimate and water balance through the integration of trees into temperate cropping systems. Agric For Meteorol 323:109065

Jerrentrup JS, Komainda M, Seither M, Cuchillo-Hilario M, Wrage-Mönnig N, Isselstein J (2020) Diverse swards and mixed-grazing of cattle and sheep for improved productivity. Front Sustain Food Syst 3:125. https://doi.org/10.3389/fsufs.2019.00125

Klimek S, Kemmermann AR, Hofmann M, Isselstein J (2007) Plant species richness and composition in managed grasslands: the relative importance of field management and environmental factors. Biol Conserv 134:559–570. https://doi.org/10.1016/j.biocon.2006.09.007

Knops JMH, Tilman D, Haddad NM, Naeem S, Mitchell CE, Haarstad J, Ritchie ME, Howe KM, Reich PB, Siemann E, Groth J (1999) Effects of plant species richness on invasion dynamics, disease outbreaks, insect abundances and diversity. Ecol Lett 2:28–293

Komainda M, Küchenmeister F, Küchenmeister K, Kayser M, Wrage-Mönnig N, Isselstein J (2020) Drought tolerance is determined by species identity and functional group diversity rather than by species diversity within multi-species swards. Eur J Agron 119:1–10. https://doi.org/10.1016/j.eja.2020.126116

Krause B, Culmsee H, Wesche K, Bergmeier E, Leuschner C (2011) Habitat loss of floodplain meadows in North Germany since the 1950s. Biodivers Conserv 20:2347–2364

Krause B, Wesche K, Culmsee H, Leuschner C (2014) Diversitätsverluste und floristischer Wandel im Grünland seit 1950. Natur und Landschaft 89(H 9+10):399–404

Kutschera L (1960) Wurzelatlas mitteleuropäischer Ackerunkäuter und Kulturpflanzen. Frankfurt/M., Germany

Lanta V, Lepš J (2006) Effect of functional group richness and species richness in manipulated productivity–diversity studies: a glasshouse pot experiment. Acta Oecol 29:85–96

Leuschner C, Ellenberg H (2017) Ecology of Central European non-forest vegetation: coastal to alpine, natural to man-made habitats: Vegetation Ecology of Central Europe, Volume II (Vol 2). Springer

Leuschner C, Krause B, Meyer S, Bartels M (2014) Strukturwandel im Acker- und Grünland Niedersachsens und Schleswig-Holsteins. Natur und Landschaft 89(H 9+10):386–391

Leuschner C, Lott S, Schmitz A, Wellinghoff J, Komainda M, Isselstein J (2024) Plant species richness in intensively managed temperate wet grasslands: current state and the importance of edge habitats. Biodivers Conserv 33:3275–3294. https://doi.org/10.1007/s10531-024-02912-y

Lüscher A, Fuhrer J, Newton PCD (2005) Global atmospheric change and its effect on managed grassland systems. In: McGilloway DA (ed) Grassland: a global resource. Academic Publishers, Wageningen, pp 251–264

Lüscher A, Grieder C, Huguenin-Elie O, Klaus V, Reidy B, Schneider MK, Schubiger F, Suter D, Suter M, Kölliker R (2019) Grassland systems in Switzerland with a main focus on sown grasslands. Grassland Sci Eur 24:3–16

Lüscher A, Barkaoui K, Finn JA, Suter D, Suter M, Volaire F (2022) Using plant diversity to reduce vulnerability and increase drought resilience of permanent and sown productive grasslands. Grass Forage Sci 77:235–246. https://doi.org/10.1111/gfs.12578

Mahaut L, Fort F, Violle C, Freschet G (2019) Multiple facets of diversity effects on plant productivity: species richness, functional diversity, species identity and intraspecific competition. Funct Ecol 34:287–298. https://doi.org/10.1111/1365-2435.13473

Malisch CS, Eriksen J, Loges R, Finn JA, Brophy C, Huguenin-Elie O (2023) The importance of multi-species grassland leys to enhance ecosystem services in crop rotations. Grassland Sci Eur 28:192–206

Mroncz A, Isselstein J, Komainda M, Leuschner C (2024) Microtopography causes small-scale variation in harvest and forage quality of high yielding silage grassland in northern Germany. Grassland Res 3:33–42. https://doi.org/10.1002/glr2.12072

Myneni RB, Keeling CD, Tucker CJ, Asrar G, Nemani RR (1997) Increased plant growth in the northern high latitudes from 1981 to 1991. Nature 386:698–702

Niklaus PA, Wardle DA, Tate KR (2006) Effects of plant species diversity and composition on nitrogen cycling and the trace gas balance of soils. Plant Soil 282:8–98

Nippert JB, Knapp AK (2007) Linking water uptake with rooting patterns in grassland species. Oecologia 153:261–271

Nölke I, Komainda M, Tonn B, Feuerstein U, Isselstein J (2021) Including chicory and selecting white clover varieties as strategies to improve temporal stability of forage yield and quality in white-clover-based temporary grassland. Eur J Agron 130:126362. https://doi.org/10.1016/j.eja.2021.126362

Opitz von Boberfeld W (1994) Grünlandlehre. Ulmer, Stuttgart, Germany

Paesel HK, Schmitz A, Isselstein J (2019) Heterogeneity and diversity of orchard grassland vegetation in Central Germany: role of tree stock, soil parameters and site management. Agrofor Syst 93:825–836

Perez-Alvarez R, Chara J, Snyder LD, Bonatti M, Sieber S, Martin EA (2023) Global meta-analysis reveals overall benefits of silvopastoral systems for biodiversity. bioRxiv, 2023–07. https://doi.org/10.1101/2023.07.30.551160

Peyraud J-L, Peeters A (2016) The role of grassland based production system in the protein security. Grassland Sci Eur 21:29–43

Picon-Cochard C, Teyssonneyre F, Besle JM, Soussana JF (2004) Effects of elevated CO2 and cutting frequency on the productivity and herbage quality of a semi-natural grassland. Eur J Agron 20:363–377

Pielke RA, Avissar R (1990) Influence of landscape structure on local and regional climate. Landsc Ecol 4:133–155

Prieto I, Violle C, Barre P, Durand J-L, Ghesquire M, Litrico I (2015) Complementary effects of species and genetic diversity on productivity and stability of sown grasslands. NaturePlants 1:15033. https://doi.org/10.1038/nplants.2015.33

Quinkenstein A, Wöllecke J, Böhm C, Grünewald H, Freese D, Schneider BU, Hüttl RF (2009) Ecological benefits of the alley cropping agroforestry system in sensitive regions of Europe. Environ Sci Policy 12(8):1112–1121. https://doi.org/10.1016/j.envsci.2009.08.008

Reheul D, Cougnon M, Kayser M, Pannecouque J, Swanckaert J, De Cauwer B, van den Pol-van Dasselaar A, De Vliegher A (2017) Sustainable intensification in the production of grass and forage crops in the Low Countries of north-west Europe. Grass Forage Sci 72:369–381

Sanderson MA, Soder KJ, Muller LD, Klement KD, Skinner RH, Goslee SC (2005) Forage mixture productivity and botanical composition in pastures grazed by dairy cattle. Agron J 97:1465–1471. https://doi.org/10.2134/agronj2005.0032

Schellberg J, da Pontes L (2012) Plant functional traits and nutrient gradients on grassland. Grass Forage Sci 67:305–319. https://doi.org/10.1111/j.1365-2494.2012.00867.x
Scherer-Lorenzen M, Palmborg C, Prinz A, Schulze ED (2003) The role of plant diversity and composition for nitrate leaching in grasslands. Ecology 86:1539–1552
Schils RL, Bufe C, Rhymer CM, Francksen RM, Klaus VH, Abdalla M, Milazzo F, Lellei-Kovács E et al (2022) Permanent grasslands in Europe: land use change and intensification decrease their multifunctionality. Agric Ecosyst Environ 330:107891. https://doi.org/10.1016/j.agee.2022.107891
Schmiedgen A, Komainda M, Kowalski K, Hostert P, Tonn B, Kayser M, Isselstein J (2021) Impacts of cutting frequency and position to tree line on herbage accumulation in silvopastoral grassland reveal potential for grassland conservation based on land use and cover information. Ann Appl Biol 179:75–84
Schmitz A, Lott S, Leuschner C, Isselstein J (2022) Mehr Farbe ins Grün: Maßnahmen zur Förderung des Artenreichtums in intensiv bewirtschaftetem Grünland. Mitteilungen der Arbeitsgemeinschaft Grünland und Futterbau 21:30–33
Senghas K, Seybold S (2003) Flora von Deutschland und angrenzender Länder, 92 edition. Quelle Meyer, Wiebelsheim, Germany
Smith PC, Calanca P, Fuhrer J (2010) Modelling irrigation demand from two grasslands in Switzerland under contrasting climatic conditions and soil properties. In: Swayne DA, Yang W, Voinov AA, Rizzoli A, Filatova T (eds) Modelling for environment's sake international environmental
Soussana JF, Lüscher A (2007) Temperate grasslands and global atmospheric change: a review. Grass Forage Sci 62:127–134
Sutterlütti R, Komainda M, Kayser M, Isselstein J (2023a) Tree lines do not reduce grassland productivity and herbage quality in alley cropping under drought. J Agron Crop Sci 209:761–771
Sutterlütti R, Kayser M, Komainda M, Isselstein J (2023b) Alley cropping tree lines alter temperature and light quantity and influence daily growth rates of grassland. Grass Forage Sci 78(402):415
Taiz L, Zeiger E (2010) Plant physiology, 5th edn. Sunderland
Taube F, Herrmann A (2009) Relative benefit of maize and grass under conditions of climate change. Landbauforschung 331:115–125
Thielecke D, Isselstein J, Komainda M (2024) Multi-species for multifunctional grassland: evaluation of minor and novel forage species for temperate humid grasslands. Ann Appl Biol 185:91–107. https://doi.org/10.1111/aab.12920
Thumm U, Tonn B (2010) Effect of precipitation on dry matter production of a meadow with varied cutting frequency. Grassland Sci Eur 15:90–92
Tilman D, Reich PB, Knops J, Wedin D, Mielke T, Lehman C (2001) Diversity and productivity in a long-term grassland experiment. Science 294:843–845
Tilman D, Reich PB, Knops JMH (2006) Biodiversity and ecosystem stability in a decade-long grassland experiment. Nature 441:629–632
Torralba M, Fagerholm N, Burgess PJ, Moreno G, Plieninger T (2016) Do European agroforestry systems enhance biodiversity and ecosystem services? A meta-analysis. Agric Ecosyst Environ 230:150–161. https://doi.org/10.1016/j.agee.2016.06.002
Trnka M, Olesen JE, Kersebaum KC, Skjelvag AO, Eitzinger J, Seguin B, Peltonen-Sainio P, Rötter R, Iglesias A, Orlandini S, Dubrovsky M, Hlavinka P, Balek J, Eckersten H, Cloppet E, Calanca P, Gobin A, Vucetic V, Nejedlik P, Kumar S, Lalic B, Mestre A, Rossi F, Kozyra J, Alexandrov V, Semeradova D, Zalud Z (2011) Agroclimatic conditions in Europe under climate change. Glob Chang Biol 17:2298–2318. https://doi.org/10.1111/j.1365-2486.2011.02396.x
Veldkamp E, Schmidt M, Markwitz C et al (2023) Multifunctionality of temperate alley-cropping agroforestry outperforms open cropland and grassland. Commun Earth Environ 4:20. https://doi.org/10.1038/s43247-023-00680-1
Violle C et al (2007) Let the concept of trait be functional. Oikos 116:882–892. https://doi.org/10.1111/j.0030-1299.2007.15559.x

Wesche K, Krause B, Culmsee H, Leuschner C (2012) Fifty years of change in central European grassland vegetation: large losses in species richness and animal-pollinated plants. Biol Conserv 150:76–85

Wilkinson JM, Lee MRF (2018) Use of human-edible feeds by livestock. Animal 12:1735–1743. https://doi.org/10.1017/S175173111700218X

Wrage N, Strodthoff J, Cuchillo MH, Isselstein J, Kayser M (2011) Phytodiversity of temperate permanent grasslands: ecosystem services for agriculture and livestock management for diversity conservation. Biodivers Conserv 20:3317–3339. https://doi.org/10.1007/s10531-011-0145-6

Zamora J, Verdú JR, Galante E (2007) Species richness in Mediterranean agroecosystems: spatial and temporal analysis for biodiversity conservation. Biol Conserv 134:113–121. https://doi.org/10.1016/j.biocon.2006.08.011

14 Upcycling Food Loss and Waste to Animal Feed Is Essential for a Circular and Sustainable Food System

Gerald Shurson

1 Introduction

Despite numerous technological advances that have enabled the production of more food with fewer resources, 33–40% of all food that is produced globally is lost or wasted at all stages of food supply chains without being consumed (Wang et al. 2023). There is no single factor in our global food system that has a greater detrimental effect on the social, economic, and environmental sustainability than food loss and waste (FLW). This abysmal inefficiency of our global food system is a major contributing factor to our inability to provide food security for nearly one-third (29.6%) of the hungry and malnourished people in the world because of poor access, inadequate distribution, and poor food preservation capabilities (UNEP 2024). Unless this current amount of FLW is drastically reduced, our future ability to provide adequate amounts of nutritious high-quality food to an increasing number of people in the world will not be possible because of the declining amounts of finite water, land, and ecosystem resources that exist on planet Earth. Not only are we failing to adequately nourish our current global population, but this enormous amount of FLW represents more than 750 billion USD in economic losses on an annual basis based on producer prices (FAO 2013). The annual global amount of 1.6–2.5 billion tonnes of FLW has also created a tremendous environmental burden which wastes 25% of freshwater resources (FAO 2023), 30% of agricultural land area, and 38% of total energy consumption, while contributing about 8% of total greenhouse gas (GHG) emissions (FAO 2019). In fact, about half of the total GHG emissions from our global agri-food system are attributed to FLW, with approximately 30% of these emissions originating from end-of-life food waste disposal

G. Shurson (✉)
Department of Animal Science, College of Food, Agriculture, and Natural Resource Sciences, University of Minnesota, St. Paul, MN, USA
e-mail: shurs001@umn.edu

B. Grabkowsky, T. Blaha (eds.), *Intensive Livestock Production in Transition*,
https://doi.org/10.1007/978-3-031-97872-2_14

(Zhu et al. 2023). As a result, the goal of halving global FLW by 2030 has been a key focus of climate and food security agendas (Crippa et al. 2021; Zurek et al. 2022), but it is highly unlikely that it will be achieved.

To do this, government policies need to be enacted that encourage the implementation of effective practices that can dramatically reduce FLW and improve access, distribution, and preservation of food to meet the nutritional needs of all societies. Improving our ability to feed hungry people continues to be the most important sustainability goal that must be achieved throughout all food supply chains from production to consumption. The next most important sustainability priority is to divert non-consumed FLW away from disposal in landfills, open dumps, and incineration, which are environmentally detrimental disposal options, and toward a more resource-efficient circular food system that includes using appropriate sources of FLW in animal feed (Papargyropoulou et al. 2014). By recycling energy and nutrients from various FLW sources into animal feed, significant improvements in environmental sustainability of food animal production can be achieved including reduced land and water use (Govoni et al. 2021; Govoni et al. 2022), along with less nitrogen (Uwizeye et al. 2019) and phosphorus (Mokjatturas et al. 2024) fertilizer use, while also reducing the overall environmental footprint of food production (Pinotti et al. 2021). The remaining next best circularity options for using FLW not suitable for use in animal feed are to recycle these nutrients into biogas using anaerobic digestion, and into nutrient-rich soil amendments using aerobic composting, which are far superior to the detrimental environmental consequences of disposal in landfills (Dou et al. 2024). Collective evidence from field studies representing a wide range of conditions around the world have shown that processing and recycling FLW into animal feed would reduce GHG emissions by 1,059 kg of CO_2 equivalent per tonne of food waste that is diverted from landfills, which is comparable to the reduction of 939 kg and 1,157 kg of CO_2 equivalent per tonne of food waste recycled using aerobic composting and anaerobic digestion, respectively (Wang et al. 2025). However, recycling FLW into animal feed provides additional environmental benefits resulting from sparing enormous amounts of land, water, agricultural fuel and fertilizer use compared with recycling through composting and biogas production (Wang et al 2025). Therefore, upcycling FLW into animal feed represents one of the most significant ways of achieving multiple economic, social, and environmental sustainability development goals of our global food system, but it will require transforming our current linear "take, make, consume, dispose, pollute" agri-food system to a circular food system.

2 What Is a Circular Agri-Food System?

There are different definitions of circularity and its application to the economy, agriculture, and food systems. In general, circularity means creating a "closed loop" of production and consumption that focuses on manufacturing more durable goods, minimizing waste, and recycling materials that have reached their "end of life" into the next generation of products so that further extraction of scarce natural resources are minimized. There are seven pillars commonly used to describe the ideal features of a circular economy, which prioritize recycling, renewable energy use, and improving the efficiency of using natural resources to minimize detrimental environmental impacts. These circularity pillars also include improving the human experience by ensuring health and well-being for all. Unfortunately, the human element is often omitted in discussions for creating circular economies, food, and agricultural systems, but it is an essential component that must be included. Therefore, the core components of a circular economy should include:

1. *Materials* are cycled at a continuous high value.
2. All *energy* is based on *renewable* sources.
3. *Water (stewardship)* is extracted at a sustainable rate and resource recovery is maximized.
4. *Biodiversity* is structurally supported and enhanced through all human activities.
5. Human *society and culture* are preserved through social governance.
6. *The health and well-being* of humans and other species are structurally supported.
7. *Value* is generated by human activities beyond financial and includes aesthetic, emotional, and ecological value.

Creating a circular agri-food system must include recycling energy and nutrients from FLW into animal feed in order to transform relationships between ecological and economic systems and improve human society and well-being. Accomplishing this will enable reducing our reliance on consumption of fossil fuels, use of water, land, nitrogen and phosphorus fertilizer, and reduce many of the adverse effects of climate change caused by food animal production.

Unfortunately, there is a lack of government policies and regulations in most countries that provide significant incentives to recycle FLW into high-value purposes, such as use in animal feed, to reduce environmental burdens and maintain economic viability (Velasco-Muñoz et al. 2021; Marinova and Bogueva 2022; Nath et al. 2023). Immediate attention needs to be devoted to reviewing and revising outdated institutional and legal frameworks (Shurson et al. 2023) that will enable designing a circular food system that includes recycling FLW into animal feed as one of the core components (Shurson and Urriola 2022).

3 Terminology, Definitions, and Classification of Food Loss and Waste Sources

Historically, many terms have been used to describe various sources of food loss and waste fed to livestock and poultry. Although terms such as "swill" and "garbage" feeding, which were used many decades ago, may have been appropriate to describe the practice of feeding household food waste to pigs on a small-scale, backyard farms at that time, they no longer accurately represent today's state-of-the-art processing technology, high feed safety requirements, and economically and environmentally valuable feeding applications. More appropriate terms for FLW should be "residual foods," "food discards," "food scraps," "food surplus," and "former foods."

Many definitions have also been used to describe FLW, but they generally refer to edible foods intended to be consumed by humans (Parfitt et al. 2010; FAO 2013; HLPE 2014; Buzby et al. 2014). In contrast, although many types of agri-industrial by-products such as oilseed meals and bioethanol co-products have been widely recycled and utilized in animal feeds for many years (Coffey et al. 2016), these types of materials are not included in FLW definitions because they are secondary by-products of agri-industrial processes that are unsuitable or not originally intended for human consumption. "Food loss" generally refers to a reduction in the quantity or quality of food in the early stages of food supply chains (i.e., production, harvest) that causes a decrease in the amount of food available or suitable for human consumption (Pinotti et al. 2021). "Food waste" is generally used to describe human edible food materials produced or remaining after processing, manufacturing, preparing, or after consumption which occurs at the mid-to-late stages of food supply chains (Pinotti et al. 2021). An alternative definition of FLW has also been proposed which distinguishes among FLW and agri-industrial by-products based on low opportunity cost (produced from inputs that do not compete with human nutrition) versus high opportunity cost (could have been consumed by humans or were grown on arable land; Cottee et al. 2022). However, in general, the main differences among FLW definitions are the: (1) stage of the food supply chain where it occurs, (2) inclusion or exclusion of human inedible food parts, and (3) end-of-life or discard options (Spang et al. 2019).

Although understanding terms and definitions of FLW is important, the criteria used to classify FLW sources that meet feed safety and regulatory requirements and are nutritionally suitable for use in animal feeds are most important. Nath et al. (2023) proposed an FLW classification system based on (1) the natural resource from which it was derived (i.e., plant vs. animal), (2) the extent of cooking (i.e., cooked, semi-cooked, uncooked), (3) the stage of the food supply chain where it was produced, and (4) physical state (i.e., solid, semi-solid, liquid). Shurson et al. (2023) proposed a more extensive approach for classifying FLW for potential use in animal feeds based on feed safety and nutritional value characteristics which include country of origin, presence of animal tissue, species of animal tissue, nutritional composition and density, complexity, physical form, preservation and quality, consistent supply of significant quantities, presence of impurities, and extent of thermal processing. This classification system can be used as a general guide for assessing the suitability of various FLW sources for use in animal feeds.

4 Historical Use and Regulations of Food Loss and Waste in Animal Feed

For many centuries in most countries around the world, post-consumer food waste from households was primarily fed to pigs in a semi-liquid form known as "swill," or in a semi-solid form known as "garbage feeding" on small-scale farms. This widespread practice was an early recognition of the importance of creating a circular food system that recycles and repurposes nutrients from waste streams for high-value purposes. Although the practice of swill feeding of food waste to pigs still occurs on some small-scale pig farms in Asia, Eastern Europe, and parts of Central and South America, it is not practiced in large-scale, intensive swine production systems in Western Europe, North America, and other major pork producing countries. This divergence in the evolution of feeding post-consumer food waste to pigs in various countries has had several pivotal implications on the current regulations and extent of using all types of FLW in animal feeds.

First, the historical terms "swill" and "garbage" feeding may have been appropriate decades ago, but they have created negative perceptions and images among consumers, livestock producers, veterinarians, and practitioners in the feed industry which has been a barrier for greater use of FLW in animal feed. For some consumers, the concept of re-feeding animal-derived foods (i.e., meat) back to animals for subsequent food production is viewed negatively (Amato et al. 2023).

Second, feeding uncooked or inadequately heat-processed animal-derived food waste to animals has been a historical human food safety concern (e.g., *Trichinella spiralis*, *Toxoplasma gondii*, *Salmonella*, *Escherichia coli*), has also been attributed to devastating national animal disease outbreaks (e.g., foot and mouth disease), and can be a potential route for spreading Bovine Spongiform Encephalitis (BSE) in ruminants if they are fed prion contaminated ruminant-derived by-products (Shurson et al. 2022a, 2023). These historical health concerns occurred as a result of feeding uncooked or inadequately heat-processed FLW to animals, and led to a wide variation in the current feed safety regulations among various countries for using processed (rendered) animal proteins in animal feeds (Shurson et al. 2023). Some countries (i.e., Japan, Taiwan, and S. Korea) have extensive feed safety, quality control programs, processing facilities, and infrastructure in place to promote and successfully utilize more than 65% of food waste in animal feeds without animal disease outbreaks and public food safety concerns (Shurson et al. 2023). Upcycling FW to animal feed is more restrictive in the United States, and because of a lack of infrastructure and economic incentives, only about 10% of surplus food is processed into animal feed (Shurson et al. 2023). The use of processed animal proteins in animal feeds is the most restrictive and heavily regulated or prohibited in the European Union (Shurson et al. 2023). The main driver of the ban on using rendered animal by-products in feed for all animal species in the United Kingdom in 1989, and subsequently in the European Union in 2001, was the BSE epidemic which resulted in a major transformation in the rendering industry to improve feed safety (Woodgate 2023). Although food and feed safety are legitimate considerations for upcycling animal-derived FLW sources into animal feeds, all endoparasites, bacteria, and viruses, except for African swine fever virus, are effectively inactivated in animal tissues when heated up to up to 80 ° C for 30 min (Shurson et al. 2022a, 2023). Unfortunately, prions associated with BSE require greater time, temperature, and pressure for inactivation than that required for inactivating endoparasites, bacteria, and viruses. Taylor (2000) reported that heating ruminant carcass tissue to 132 °C for 4.5 h at 3 bar pressure (45 psi) is completely effective for inactivating prions, whereas Auvermann et al. (2004) reported that heating at 136–138 °C for 18 min at 2 bar pressure (29.4 psi) was also effective. If these conditions can be applied to animal carcass rendering facilities, feeding ruminant-derived by-products back to ruminants could be safely achieved once this safety concern is validated through appropriate research. Until then, unless the processing of animal tissues achieves these processing conditions, existing laws prohibiting the re-feeding of ruminant-based processed animal protein to ruminants should be maintained. However, feeding these materials to pigs and poultry is safe and of no concern because prions are not transmitted to these species (Jahns et al. 2006; Wells et al. 2003). Furthermore, several technologies are available to accurately identify species-specific DNA and prion proteins in raw materials, which can be used for monitoring prion contamination if they are implemented in rendering and animal-derived food waste processing facilities (Shurson et al. 2022a).

Third, the ongoing widespread African swine fever (ASF) epidemic in all continents except North and South America and Australia has been a barrier to using FLW in swine feed in these countries. The main concern is the unlikely possibility of feeding uncooked or improperly heated pig meat imported from countries with ASF-infected pigs to pigs in countries that have no ASF-infected pigs (Gogin et al. 2013). One of the main drivers of this concern is the recent reports inferring that feeding uncooked household food waste (i.e., swill) to pigs raised in small-scale, backyard pig farms contributed to the spread of ASFV in various countries where this devastating disease is present (Costard et al. 2009; Kagira et al. 2010; Phengsavanh et al. 2010; Heilman et al. 2020; Bellini et al. 2016). However, it is important to recognize that these events occurred in countries where ASF is endemic and where no heat treatment of household food waste is required or enforced before feeding household food waste to pigs. The other main concern is that the ASF virus, unlike other animal viruses, is extremely resilient to high temperatures and survives for many months in fresh, processed, and frozen pork products and tissues (Mazur-Panasiuk et al. 2019). In fact, recent results from Balestreri et al. (2023) indicate that viable ASF virus particles are still present after heating at 100 °C for 20 min, which suggests that food products containing pig meat and produced in countries with endemic ASF should not be fed to pigs but can be fed to other food-producing animals.

5 Why Is Food Loss and Waste Such a Problem?

If FLW were a country, it would rank third after the United States and China for contributions to global carbon emissions (FAO 2013). Food loss and waste occurs in all countries and at all stages of the food supply chain from agricultural production, post-harvest handling and storage, processing, distribution, and consumption for all types of food. However, the amount of FLW across the entire food chain varies by geographic region and economic status, with Europe and North America wasting about 280–300 kg/capita annually compared with sub-Saharan Africa and South and Southeast Asia wasting about half of this amount (120–170 kg/capita/year; FAO 2011). The proportion of FLW that occurs at various stages of food supply chains also varies based on the economic status of countries, with about 40% of food losses occurring at post-harvest and processing levels in low-income countries, while more than 40% of food waste occurs at the retail and consumer levels of high-income countries (FAO 2011). Even more remarkable, is that the amount of food waste at the consumer level (222 million tonnes) in high-income countries is nearly the same as the total annual net food production (230 million tonnes) in sub-Saharan Africa (FAO 2011). The per capita food waste by consumers in Europe and North America (95–115 kg/year) is more than 10 times greater than per capita food waste in sub-Saharan Africa and South and Southeast Asia (6–11 kg/year; FAO 2011). Most of the FLW that occurs in low-income countries is due to economic, managerial, and technical limitations associated with harvesting technologies, inadequate storage and cooling facilities, adverse climatic conditions, and lack of adequate infrastructure, packaging, and marketing systems (Table 14.1). Overcoming these challenges by encouraging small farmers in these regions to organize into cooperatives, diversify production, and adopt current technologies, along with increasing public and private investment in infrastructure, transportation, processing, and packaging, could alleviate much of the hunger and malnutrition experienced by

Table 14.1 Various causes of food loss and waste between low- and high-income countries (FAO 2011)

Low-income countries	High-income countries
Premature harvesting	Excess production exceeds demand
Poor storage facilities cause spoilage	Non-conforming to "appearance quality standards"
Lack of adequate transportation and infrastructure	Unsafe food and failure to comply with food safety standards
Unsafe food and failure to comply with food safety standards	Disposing is less expensive than using or re-using
Lack of adequate food processing facilities	Large quantities, types, and brands of foods on display in retail stores
Inadequate market systems with too few wholesale and retail markets that are often small, unsanitary, and lack cooling equipment	Premature removal from retail shelves due to exceeding freshness or "sell-by" dates
Poverty and limited household income make it unacceptable to waste food	Abundance and price incentives for volume purchases encourage waste

people living in these parts of the world (FAO 2011). In contrast, most of the FLW that occurs in high-income countries is a result of careless and affluent consumer behaviors and a lack of coordination among different segments of food supply chains (FAO 2011). Overproduction, discarding fresh and safe fruits and vegetables because of blemishes and failure to meet appearance standards, inadequate planning of food purchases, and disposing of retail food products that exceed "best-before dates" are major contributors to excessive food losses (Table 14.1).

Global environmental hotspots caused by FLW also vary among geographic regions and have a substantial detrimental effect of natural and societal resources (FAO 2013). Globally, cereals represent the greatest type of FLW based on calories (energy), while fruits and vegetables represent the greatest loss by weight. Therefore, cereal wastage in Asia is a major environmental problem due largely to the high carbon intensity (methane emissions) from rice production and rice wastage, along with significant burdens on arable land and surface and groundwater water use. Although the amount of meat wastage is relatively low in all geographic regions compared with cereals, roots and tubers, and fruits and vegetables, it has a substantial impact on carbon footprint and land use, especially in high-income and Latin American countries that waste high amounts of meat produced. High amounts of fruit wastage in Asia, Latin America, and Europe cause excessive burdens on ground and surface water, while high amounts of vegetable waste in industrialized Asia, Europe and South and Southeast Asia contribute to a high carbon footprint. These regional hotspots should be the focus and deemed high priorities for diverting these specific types of FLW to animal feed in order to achieve an immediate environmental benefit.

Unfortunately, the extent of FLW recycling in various countries is often, but not exclusively associated with economic status, with low-income countries generally having minimal government policies and regulations to prevent FLW and promote recycling (Shekdar 2009). For example, Nigeria and Ghana have limited policies for food waste management and recycling which results in most FLW being disposed of in landfills (De Clercq et al. 2017). India produces enormous amounts of FLW but almost all of it is disposed of in landfills due to a lack of recycling policies and infrastructure (Thi et al. 2015). However, even more affluent countries like Brazil and Mexico have inadequate legislation and government policies, low private sector participation, and limited funding and financial incentives which have led to minimal food waste recycling (Hoornweg and Gianelli 2007). The United States and European Union allow recycling FLW into animal feed under specific conditions, but have strict feed safety laws and regulations associated with preventing BSE and pathogen transmission that have severely restricted use of animal-derived foods in certain animal feeds. Shekdar (2009) suggested using an integrated sustainable solid waste management approach and action plan based on Asian country-specific scenarios that includes public participation, policy and legal framework, institutional arrangement, operations management, use of appropriate technology, and financial management involving the entire process from collection, transportation, processing, and disposal. Likewise, Pasarin and Viinikainen (2022) outlined seven key national regulatory measures for creating an effective framework for FLW prevention and reduction at all stages of the food supply chain which include (1) establishing an adequate institutional framework, (2) adopting a food material hierarchy, (3) setting binding legal targets to reduce FLW, (4) establishing a food donation system, (5) ensuring food safety and quality, (6) defining clear date labelling requirements, and (7) adopting appropriate fiscal measures. However, despite these many useful recommendations, very few meaningful actions have been implemented in most countries around the world.

6 Why Has Not More Food Loss and Waste Been Diverted to Animal Feed?

There are several reasons why the diversion of FLW into animal feed has been minimal relative to the amount produced in most countries:

1. General lack of societal awareness of the magnitude of the environmental, economic, and social significance of the FLW problem.
2. Differences in economic wealth, attitudes, and cultural beliefs among societies in various countries toward the importance of minimizing waste and associated sustainability benefits that could be realized.
3. Negative perceptions and images of feeding "garbage" and "swill" to pigs and subsequent consumption of food products from animals fed food waste.
4. Abundant, low-cost feed crops, oilseed meals, and agri-industrial by-products in high-income countries with large-scale intensive livestock production systems.
5. Often challenging transportation and handling logistics to preserve the nutritional value of high moisture foods and avoid spoilage before processing.
6. Variability in supply and nutritional composition of most sources to optimize their use in precision livestock feeding programs.
7. Government laws and regulations that restrict or prohibit the feeding of certain types of food waste to specific animal species due to food safety and animal disease transmission concerns based on devastating economic consequences from historical animal disease epidemics.
8. Current concerns in major pig production countries that do not have African swine fever (ASF), about the possibility of transmission of ASF from feeding food waste containing imported pig meat from countries with ongoing African swine fever in their domestic pig population.
9. Lack of economic and government policy incentives to encourage entrepreneurship and ease of business entry to collect, process, and distribute FLW to commercial feed manufacturers and farms.
10. Lack of infrastructure to collect, process, store, distribute, and utilize different types and forms of FLW in animal feeds.

Of all of these factors, the greatest barrier involves government laws and regulations prohibiting or significantly restricting the use of various types of FLW, especially FLW containing animal tissues, in animal feeds. Most of the laws and regulations involving feeding FLW to animals were developed decades ago to prevent animal disease transmission from uncooked or ruminant animal-derived FLW sources, but that was before effective heat processing technologies were developed and implemented in the rendering and animal plasma industries to inactivate endoparasites, bacteria, and viruses of concern (Shurson 2020; Shurson et al. 2022a). However, there is a continued need to regulate FLW containing ruminant tissues re-fed to ruminant animals because of the risk of transmission of BSE, which is a fatal degenerative disease in ruminants caused by consuming prions (abnormal proteins associated with nervous tissue) from ruminant-based tissues and foods derived from infected animals (Prince et al. 2003). However, governments in Japan, South Korea, and Taiwan have been pioneers in establishing laws, regulations, economic incentives and subsidies, and infrastructure to mandate the collection and recycling of all sources of FLW which have led to utilizing more than 65% of total FLW generated into safe animal feed (Shurson et al. 2023). In Japan, national laws were enacted that require increasing self-sufficiency in animal feed production by reducing reliance on imported feed ingredients to support domestic animal production (Sugiura et al. 2009). The Japanese system has been very effective in preventing the spread of animal diseases, including BSE, because it involves a certification system for collecting, transporting, storing, and processing specific types of food waste into "Ecofeed" (Sugiura et al. 2009), and diverting food waste containing animal-derived foods toward other recycling options such as composting. In contrast, because of current feed safety laws in the United States and European Union, and lack of a comprehensive feed safety certification system for using FLW in animal feeds, only about 5–10% of the total FLW produced in these regions is used in animal feeds.

7 Can the Use of FLW in Animal Feeds Support Optimal Animal Performance?

All feed ingredients, including FLW sources, are "packages" of energy and digestible nutrients in various proportions and amounts that can partially contribute to meeting the nutrient requirements of animals when combined with specific amounts of other feed ingredients to provide a nutritionally balanced diet. Results from some studies have shown that several sources of thermally processed pre-consumer food waste have nutritional value for pigs that is comparable, and even exceeds that provided by conventional feed ingredients such as corn and soybean meal (Fung et al. 2019a, b). The key to avoiding underfeeding and overfeeding of energy and nutrients to animals is to have accurate nutrient composition and digestibility data for all ingredients that are included in an animal's diet during the feed formulation process.

This is the greatest challenge when using new or atypical feed ingredients that have not been previously evaluated through animal feeding trials. As a result, animal nutritionists are often reluctant to use various types of poorly characterized FLW sources in animal diets because of the desire to have consistent and predictable nutritional composition of the ingredients being fed to avoid overfeeding and underfeeding nutrients to animals.

Therefore, a legitimate question to consider is whether the use of various types of FLW in animal feeds can maintain animal productivity and feed efficiency. To answer this question, Wang et al. (2024) conducted a critical review and analysis of global primary data from previous studies to collectively determine the effects of feeding FLW to food-producing animals on animal productivity. Their systematic literature review included 102 published studies from 31 countries involving all major food-producing animal species that were fed various types of FLW, and common measures of animal productivity were determined. Results showed that animal productivity was maintained or improved in 75–77% of the studies when livestock and poultry were fed diets containing various amounts and types of FLW, compared with productivity and efficiency responses from animals fed conventional control diets. In the 23–25% of studies that showed decreased animal productivity from feeding FLW, these reductions could have been avoided by correcting dietary nutrient deficiencies through making diet formulation adjustments and using appropriate nutritional supplements. These findings provide compelling evidence that animal productivity (measured by growth rate, feed intake, and gain efficiency) can be maintained at optimal levels when feeding all types of FLW to food-producing animals if accurate information on the energy and nutrient composition of the FLW source is available, and if appropriate supplements and formulation adjustments are made to current any nutrient deficiencies or imbalances. However, not all sources of FLW are suitable for direct feeding to certain food animal species based on high moisture, dietary fiber and lignin content, imbalances in amino acids, presence of antinutritional factors (e.g., mycotoxins, phytate), or potential contamination of specific disease-causing agents (i.e., prions, African swine fever virus) in animal-derived sources. However, the energy and nutrients in these types of FLW can still be upcycled to animal feed using several emerging technologies.

8 Emerging Technologies to Upcycle Nutrients in FLW in Animal Feed and Other Valuable Uses

8.1 Co-ensiling and Fermentation

Many types of FLW possess suboptimal nutritional characteristics limiting their diet inclusion rates and feeding applications for certain animal species. Bioprocessing various types of FLW using feed-grade fungi, yeast, bacteria, and enzymes are low-cost, long-term preservation technologies that can enhance their nutritional value and facilitate their use in animal feeds in a circular agri-food system. Several studies have shown that bioprocessing agro-industrial by-products and FLW can reduce the carbon, nitrogen, and phosphorus footprints of food animal production by improving protein quality, degrading lignin, improving dietary fiber utilization, increasing phosphorus digestibility, and reducing anti-nutritional factors such as mycotoxins (Sun et al. 2023). Fruit and vegetables comprise about 50% of the total annual global FLW from harvest through retail sales (Wang et al. 2023). Because of the high moisture and relatively high dietary fiber content of fruit and vegetable (FV) waste, it is best suited for use in cattle feed, but very little is fed directly to ruminants because of the relatively short shelf-life and handling challenges. However, FV waste can be directly used as part of a total mixed ration with other high moisture and fiber forages or ensiled with other forages for long-term storage and preservation for later use (Baker et al. 2024). Results from a recent study showed that co-fermenting crop residue (i.e., wheat straw) with fresh FV discards using two food-grade fungi strains was effective in degrading lignin and improving the energy and protein value as a ruminant feed ingredient (Sun et al. 2024). Results from a related study showed that feeding discarded fresh or ensiled citrus fruit to lactating dairy cows supported optimal milk production and quality, reduced feed cost, and provided significant reductions in several environmental burdens (Baker et al. 2024). Although these bioprocessing technologies are promising solutions for upcycling the nutritional value of low value plant-based biomass and foods, further studies are needed to determine optimal fermentation conditions, assess their relative effectiveness when used to bioprocess various types of FLW sources, develop scalable technologies for farm-level applications, conduct techno-economic evaluations, and evaluate the environmental benefits of using these technologies (Sun et al. 2024).

8.2 Substrate for Insect Production

There is considerable interest in producing and using insect meals as an environmentally sustainable energy and protein ingredient in animal feeds. Insect meals are derived from five major insect species and are marketed as a low environmental footprint replacement for soybean meal and fish meal (Makkar et al. 2014). In addition to energy and protein, insects also contain bioactive compounds such as lauric acid, antimicrobial peptides, and chitin which may enhance immune responses when fed to food-producing animals (Sogari et al. 2019). Most insect production systems use low-value animal manure or food waste to upcycle and concentrate protein and lipids into highly palatable feed ingredients for use in ruminant, swine, poultry, and aquaculture diets (Makkar et al. 2014). Black soldier fly larvae (BSFL) are the most popular type of insect meal being commercially produced using FLW as a substrate, mainly because it requires minimal space and energy to efficiently produce a nutrient-dense insect meal (i.e., high in protein, fat, and minerals) with pharmacologically beneficial compounds (Mlambo et al. 2023). Although insect meal has been promoted as an environmentally sustainable feed ingredient for use in food animal production, very few studies have focused on determining LCA environmental impacts when feeding it to livestock and poultry (DeFoliart and Paoletti

2005; Ooninex and de Boer 2012; van Zanten et al. 2015; Salomone et al. 2017). In addition, there are several challenges that must be overcome for widespread implementation of this strategy for upcycling nutrients in FLW to animal feed, including (1) assessing the nutritional suitability of using various FLW sources for BSFL production, (2) determining optimal diet inclusion rates of insect meal in various types of animal diets, (3) determining the causes and overcoming the inconsistent effects on yield and quality of food products produced from feeding diets containing insect meal to poultry, (4) consumer acceptance of food products produced from feeding BSFL diets, and (5) regulatory and policy frameworks in various countries (Mlambo et al. 2023).

8.3 Microalgae Production

Although agri-food wastewater is not considered part of FLW, perhaps it should be because many sources contain relatively high concentrations of valuable nitrogen, phosphorus, carbon, and other nutrients that should be captured and upcycled into more useful products in a truly circular agri-food system. Upcycling nutrients in wastewater to microalgae can potentially become an economical source of protein and oil in animal feeds, human diets, and as a feedstock for biofuels production. Microalgae strains vary in their capability to grow and utilize various nutrients in diverse wastewater streams, but because of their high productivity, microalgae production can potentially be part of a sustainable food production system if large-scale commercial cultivation systems can be successfully developed. Wastewater streams from agriculture and food processing industries are often rich sources of carbon and nitrogen which are primary nutrients required for the growth of microalgae. A few studies have been conducted to show that selected microalgae strains can efficiently and economically use nutrients in dairy wastewater (Lu et al. 2016) and meat processing (Lu et al. 2015) streams to remove nitrogen and increase biomass yield, while concentrating nitrogen and oil in the microalgae biomass. More research is needed to explore producing microalgae from nutrients in agri-food wastewater streams and subsequent use in animal feeds and other high-value applications.

8.4 Anaerobic Digestion and Aerobic Composting

Although not all FLW sources are suitable for use in animal feed, disposal of FLW in landfills must be avoided because it violates the principles of a circular and sustainable food system by wasting valuable nutrients that could be re-purposed for productive uses, and because of the significant amount of GHG emissions that occur from disposing of FLW in landfills. Therefore, FLW sources that are not suitable for use in animal feed should be recycled to biogas as a renewable energy source through anaerobic digestion, or recycled using aerobic composting to produce nutrient-rich soil amendments to support a circular agri-food system. Dou et al. (2024) provided recent estimates indicating that FLW disposal in landfills is the dominant end-of-life method used in the United States (44%), China (60%), and the European Union (33%). The lesser amount of FLW disposed in landfills in the European Union is because of using incineration, which is also not environmentally sustainable. To put this in perspective, if equal one-third proportions of the total FLW produced in the United States was diverted to animal feed, anaerobic digestion, and aerobic composting so that none was disposed of in landfills, the reduction in GHG emissions from landfill disposal would be equivalent to the amount of enteric methane produced by about 9 million dairy cows (Dou et al. 2024). This magnitude of reduction in GHG emissions is truly remarkable and is even greater when accounting for emissions from China, the European Union, and the rest of the world. Although recycling FLW through animal feed, anaerobic digestion, and aerobic composting provided comparable marked reductions in GHG emissions compared with landfill disposal, recycling FLW into animal feed can provide additional environmental benefits by sparing enormous amounts of land, water, fossil fuels, fertilizer, herbicides, and pesticides currently used to produce feed crops such as corn and soybeans (Dou et al. 2024).

Similar to the wide variation in regulations and use of FLW in animal feeds in various countries (Shurson et al. 2023), there are unique strengths and weaknesses in governmental policies among countries regarding policies to promote the conversion of FLW to biogas (De Clerq et al. 2017). Specifically, China and South Korea have highly centralized policies incentivizing the rapid development and use of food waste in the biogas industry; France and the United Kingdom have developed and implemented policies that provide incentives to produce multiple outputs from FLW treatment facilities; South Korea has implemented waste management systems to reduce the volume of food waste produced at the source; and Germany, France, and South Korea have established several co-digestion projects to process food waste together with other waste sources (De Clerq et al. 2017). Although aerobic composting is the lowest value FLW recycling option, it provides a valuable function in a circular food system by recycling FLW sources that are not suitable for use in animal feeds or biogas from anaerobic digestion into nutrient rich soil amendments which requires the least amount of capital investment.

9 Feeding FLW to Livestock and Poultry Dramatically Reduces the Environmental Footprint of Food Animal Production

Several studies have shown that diverting FLW toward animal feed not only enhances the recovery of nitrogen and phosphorus (Kerr et al. 2017, 2019) and reduces the environmental footprint of swine diets (Shurson et al. 2022b), but also provides a more environmentally sustainable option than diverting it toward biogas production, composting, and landfills (Padeyanda et al. 2016; zu Ermgassen et al. 2016; Salemdeeb et al. 2017; Shurson 2020; Dou et al. 2021). Uwizeye et al. (2019) evaluated the nitrogen flows, losses, and use in global pork supply chains and the environmental benefits associated with feeding food waste to pigs produced in intensive production systems. Their results showed nitrogen use efficiency was improved, and nitrogen losses were reduced at the feed production stage when feeding food waste to pigs. As a result, the amount of soybeans used in swine diets could be reduced by 31 million tonnes, the amount of feed grains could be reduced by 20 million tonnes, and 16 million hectares of land could be spared and used for other purposes. Shurson et al. (2022b) showed that feeding diets containing only 8.6% thermally processed supermarket food waste to growing-finishing pigs had the least impact on GHG emissions, land use, and water use in all major US pork production regions compared with conventional feeding programs consisting of corn, soybean meal, corn distillers dried grains with solubles, and phytase, and even when using a low environmental footprint feeding program comprised of low protein, synthetic amino acid supplemented corn-soybean meal diets. Similarly, a recent study conducted by Baker et al. (2024) showed that feeding fresh citrus waste or co-ensiled citrus and hay to lactating dairy cows not only supported optimal milk production and quality, but it also reduces feed cost, spares land, water, fertilizer, and herbicide use associated with feed production, and decreases GHG emissions by avoiding disposal in landfills. Therefore, greater use of FLW in animal feeds is needed to overcome the ecosystem costs associated with intensive livestock and poultry production.

10 Conclusions

Food systems are responsible for one-third of global GHG emissions and require the use of the majority of available land and freshwater resources on Earth. Reducing food loss and waste at all stages of food supply chains, improving access and distribution of surplus food for feeding hungry people, and recycling the remaining food loss and waste into animal feeds, production of biogas, and compost in a circular agri-food system would dramatically reduce the social, economic, and environmental burdens of our global food system. Urgent action is needed to build circular agri-food systems that include recycling a much greater proportion of food loss and waste into animal feeds than is currently practiced in most countries. If this can be accomplished, the environmental burdens of intensive livestock production can also be dramatically reduced.

References

Amoto M, Demartini E, Gaviglio A, Marescotti ME, Verneau F (2023) Consumer's preferences for chicken fed different processed animal proteins; a best-worst analysis in Italy. Nutrients 15:1800. https://doi.org/10.3390/nu15071800

Auvermann B, Kalbasi A, Ahmed A (2004) Rendering, Chapter 4, in: Carcass disposal: a comprehensive review. National Agricultural Biosecurity Center Consortium, USDA APHIS Cooperative Agreement Project, Carcass Disposal Working Group, Kansas State University, Manhattan

Baker L, Bender J, Ferguson J, Rassler S, Pitta D, Chann S, Dou., Z. (2024) Leveraging dairy cattle to upcycle culled citrus fruit for emission mitigation and resource co-benefits: a case study. Resour Conserv Recycl 203:107452. https://doi.org/10.1016/j.resconrec.2024.107452

Balestreri C, Schroeder DC, Sampedro F, Marqués G, Palowski A, Urriola PE, van de Ligt JLG, Yancy HF, Shurson GC (2023) Unexpected thermal stability of two enveloped megaviruses, Emiliania huxleyi virus and African swine fever virus particles as measured by viability PCR. Virol J 21:1. https://doi.org/10.1186/s12985-023-02272-z

Bellini S, Rutili D, Guberti V (2016) Preventive measures aimed at minimizing the risk of African swine fever virus spread in pig farming systems. Acta Vet Scand 58:82. https://doi.org/10.1186/s13028-016-0264-x

Buzby JC, Wells HF, Hyman J (2014) The estimated amount, value, and calories of postharvest food losses at the retail and consumer levels in the United States. Economic Research Service, USDA, EIB-121, Washington, DC

Coffey D, Dawson K, Ferket P, Connolly A (2016) Review of the feed industry from a historical perspective and implications for its future. J Appl Anim Nutr 4:e3. https://doi.org/10.1017/jan.2015.11

Costard S, Porphyre V, Messad S, Rakotondrahanta S, Vidon H, Roger F, Pfeiffer D (2009) Multivariate analysis of management and biosecurity practices in smallholder pig farms in Madagascar. Prev Vet Med 92:199–209. https://doi.org/10.1016/j.prevetmed.2009.08.010

Cottee J, McCormack C, Hearne E, Sheane R (2022) The future of feed: how low opportunity cost livestock feed could support a more regenerative UK Food system. World Wildlife Fund, U.K. 39 pp. https://www.wwf.org.uk/sites/default/files/2022-06/future_of_feed_full_report.pdf

Crippa M, Solazzo E, Guizzardi D, Montforti-Ferrario F, Tubiello FN, Leip A (2021) Food systems are responsible for a third of global anthropogenic GHG emissions. Nat Food 2:198–209. https://doi.org/10.1038/s43016-021-00225-9

De Clercq D, Wen Z, Gottfried O, Schmidt F, Fei F (2017) A review of global strategies promoting the conversion of food waste to bioenergy via anaerobic digestion. Renew Sust Energ Rev 79:204–221. https://doi.org/10.1016/j.rser.2017.05.047

DeFoliart GR, Paoletti MG (2005) Overview of role of edible insects in preserving biodiversity. In: Ecological implications of Minilivestock: potential of insects, rodents, frogs, and snails, pp 123–140

Dou Z, Galligan D, Shurson G (2021) Food supply chain and waste in climate mitigation. In: The Role of Agricultural Science and Technology in Climate 21 Project Implementation. Council for Agricultural Science and Technology, Ames, IA, QTA2021-1. pp 14–17. https://www.cast-science.org/publication/the-role-of-agricultural-science-and-technology-in-climate-21-project-implementation/

Dou Z, Wang Y, Ying H, Stefanovski D, Shurson GC, Chen T, Yin Y, Zheng H, Nakaishi T, Cui Z (2024) Food waste treated as resources can achieve meaningful reductions in the climate and resource burdens of agrifood systems. Nature Food

FAO (2011) Global food losses and food waste – extent, causes and prevention. Rome

FAO (2013) Food Wastage footprint: impacts on natural resources. Summary report. Food and Agriculture Organization of the United Nations, Rome. https://www.fao.org/3/i3347e/i3347e.pdf

FAO (2019) The State of Food and Agriculture 2019. Moving forward on food loss and waste reduction. Food and Agriculture Organization of the United Nations, Rome. https://www.fao.org/3/ca6030en/ca6030en.pdf

FAO (2023) The State of Food and Agriculture 2023 – Revealing the true cost of food to transform agrifood systems. Food and Agriculture Organization of the United Nations, Rome.https://doi.org/10.4060/cc7724en

Fung L, Urriola PE, Shurson GC (2019a) Energy, amino acid, and phosphorus digestibility of thermally-treated food waste sources for swine. Transl Anim Sci 3:676–691. https://doi.org/10.1093/tas/txz028

Fung L, Urriola PE, Baker L, Shurson GC (2019b) Estimated energy and nutrient composition of different food waste sources and their potential for use in sustainable swine feeding programs. Transl Anim Sci 3:143–152. https://doi.org/10.1093/tas/txy099

Gogin A, Gerasimov V, Malogolovkin A, Kolbasov D (2013) African swine fever in the North Caucasus region and The Russian Federation in years 2007-2012. Virus Res 173:198–203. https://doi.org/10.1016/j.virusres.2012.12.007

Govoni C, Chiarelli DD, Luciano A, Ottoboni M, Perpelek SN, Pinotti L, Rulli MC (2021) Global assessment of natural resources for chicken production. Adv Water Resour 154:103987. https://doi.org/10.1016/j.advwatres.2021.103987

Govoni C, Chiarelli DD, Luciano A, Pinotti L, Rulli MC (2022) Global assessment of land and water resource demand for pork supply. Environ Res Lett 17:074003. https://doi.org/10.1088/1748-9326/ac74d7

Heilmann M, Lkhagvasuren A, Khishgee T, Bold B, Ankhanbaatar B, Fusheng G, Raizman E, Dietze K (2020) African swine fever in Mongolia: course of the epidemic and applied control measures. Vet Sci 7:24. https://doi.org/10.3390/vetsci7010024

High Level Panel of Experts (HLPE) (2014) Food losses and waste in the context of sustainable food systems. Report HLPE Food Security Nutrition Committee World Food Security, Rome. https://openknowledge.fao.org/server/api/core/bitstreams/b1949fae-23d4-473c-8b87-8c4359b74d6c/content

Hoornweg D, Gianelli N (2007) Managing municipal solid waste in Latin America and the Caribbean. Gridlines Note No. 28. https://openknowledge.worldbank.org/server/api/core/bitstreams/a391c044-ffb9-5888-a189-34f56523c274/content

Jahns H, Callanan JJ, Sammin DJ, McElroy MC, Bassett HF (2006) Survey for transmissible spongiform encephalopathies in Irish pigs fed meat and bone meal. VetRecord 159:137–142. https://doi.org/10.1136/vr.159.5.137

Kagira JM, Kanyari PW, Maingi N, Githigia SM et al (2010) Characteristics of the smallholder free-range pig production system in western Kenya. Trop Anim Health Prod 42:865–873. https://doi.org/10.1007/s11250-009-9500-y

Kerr BJ, Jha R, Urriola PE, Shurson GC (2017) Effects of animal protein meal composition on digestible and metabolizable energy value and prediction in growing pigs. J Anim Sci 95:2614–2626. https://doi.org/10.2527/jas.2016.1165

Kerr BJ, Urriola PE, Jha R, Thomson J, Curry SM, Shurson GC (2019) Amino acid composition and digestible amino acid content in animal protein by-product meals fed to growing pigs. J Anim Sci 97:4540–4547. https://doi.org/10.1093/jas/skz294

Lu Q, Zhou W, Min M, Ma X, Chandra C, Doan YT, Ma Y, Zheng H, Cheng S, Griffith R, Chen P, Chen C, Urriola PE, Shurson GC, Gislerød HR, Ruan R (2015) Growing *chlorella* sp. on meat processing wastewater for nutrient removal and biomass production. Bioresour Technol 198:189–197. https://doi.org/10.1016/j.biortech.2015.08.133

Lu Q, Zhou W, Min M, Ma Y, Chen P, Zheng H, Doan YT, Liu H, Chen C, Urriola PE, Shurson GC, Ruan R (2016) Mitigating ammonia nitrogen deficiency in dairy wastewaters for algae cultivation. Bioresour Technol 201:33–40. https://doi.org/10.1016/j.biortech.2015.11.029

Makkar HPS, Tran G, Heuzé V, Ankers P (2014) State-of-the-art on use of insects as animal feed. Anim Feed Sci Technol 197:1–33. https://doi.org/10.1016/j.anifeedsci.2014.07.008

Marinova D, Bogueva D (2022) Circular agriculture. In: Food in a planetary emergency. Springer, Singapore. https://doi.org/10.1007/978-981-16-7707-6_5

Mazur-Panasiuk N, Żmudzki J, Woźniakowski G (2019) Africa swine fever virus – persistence in different environmental conditions and the possibility of its indirect transmission. J Vet Res 63:303–310. https://doi.org/10.2478/jvetres-2019-0058

Mlambo V, Dibakoane SR, Mashiloane T, Mukwevho L, Wokadala OC, Mnisi CM (2023) Rethinking food waste: exploring a black soldier fly larvae-based upcycling strategy for sustainable poultry production. Resour Conserv Recycl 199:107284. https://doi.org/10.1016/j.resconrec.2023.107284

Mokjatturas S, Chinwetkitvanich S, Patthanaissaranukool W, Polprasert C, Polprasert S (2024) Phosphorus mass flows and economic benefits of food waste management: the case study of

selected retail and wholesale fresh markets in Thailand. Clean Techn Environ Policy 27:219. https://doi.org/10.1007/s10098-024-02847-6
Nath PC, Ojha A, Debnath S, Sharma M, Nayak PK, Sridhar K, Inbaraj BS (2023) Valorization of food waste as animal feed: a step towards sustainable food waste management and circular bioeconomy. Animals 13:1366. https://doi.org/10.3390/ani13081366
Oonincx DGAB, de Boer IJM (2012) Environmental impact of the production of mealworms as a protein source for humans – a life cycle assessment. PLoS One 7:e51145. https://doi.org/10.1371/journal.pone.0051145
Padeyanda Y, Jang Y-C, Ko Y, Yi S (2016) Evaluation of environmental impacts of food waste management by material flow analysis (MFA) and life cycle assessment. J Mater Cycles Waste Manag 18:493–508. https://doi.org/10.1007/s10163-016-0510-3
Papargyropoulou E, Lozano R, Steinberger JK, Wright J, Ujang N, Bin Z (2014) The food waste hierarchy as a framework for the management of food surplus and food waste. J Clean Prod 76:106–115. https://doi.org/10.1016/j.jclepro.2014.04.020
Parfitt J, Barthel M, MacNaughton S (2010) Food waste within food supply chains: quantification and potential for change to 2050. Philos Trans R Soc B 365:3065–3081. https://doi.org/10.1098/rstb.2010.0126
Pasarin, V., & Viinikainen, T. (2022): Enabling a legal environment for the prevention and reduction of food loss and waste. Legal Brief, No. 9, Rome, FAO. https://doi.org/10.4060/cc2278en
Phengsavanh P, Ogle B, Stür W, Frankow-Lindberg BE, Lindberg JE (2010) Feeding and performance of pigs in smallholder production systems in northern Lao PDR. Trop Anim Health Prod 42:1627–1633. https://doi.org/10.1007/s11250-010-9612-4
Pinotti K, Luciano A, Ottoboni M, Manoni M, Ferrari L, Marchis D, Tretola M (2021) Recycling food leftovers in feed as opportunity to increase the sustainability of livestock production. J Clean Prod 294:126290. https://doi.org/10.1016/j.jclepro.2021.126290
Prince MJ, Bailey JA, Barrowman PR, Bishop KJ, Campbell GR, Wood JM (2003) Bovine spongiform encephalopathy. Rev Sci Technol 22:37–60. https://doi.org/10.20506/rst.22.1.1389
Salemdeeb R, Vivanco DF, Al-Tabbaa A, zu Ermgassen EKHJ (2017) A holistic approach to the environmental evaluation of food waste prevention. Waste Manag 59:442–450. https://doi.org/10.1016/j.wasman.2016.09.042
Salomone R, Saija G, Mondello G, Giannetto A, Fasulo S, Savastano D (2017) Environmental impact of food waste bioconversion by insects: application of life cycle assessment to process using *Hermetia illucens*. J Clean Prod 140:890–905. https://doi.org/10.1016/j.jclepro.2016.06.154
Shekdar AV (2009) Sustainable solid waste management: an integrated approach for Asian countries. Waste Manag 29:1438–1448. https://doi.org/10.1016/j.wasman.2008.08.025
Shurson GC (2020) "What a waste" – can we improve sustainability of food animal production systems by recycling food waste streams into animal feed in an era of health, climate, and economic crises? Sustain For 12:7071. https://doi.org/10.3390/su12177071
Shurson GC, Urriola PE (2022) Invited review: sustainable swine feeding programs require the convergence of circular agriculture and food systems with one health. Anim Front 12:30–40. https://doi.org/10.1093/af/vfac077
Shurson GC, Urriola PE, van de Ligt JLG (2022a) Can we effectively manage parasites, prions, and pathogens in the global feed industry to achieve one health? Transbound Emerg Dis 69:4–30. https://doi.org/10.1111/tbed.14205
Shurson GC, Pelton REO, Yang Z, Urriola PE, Schmitt J (2022b) Environmental impacts of eco-nutrition swine feeding programs in spatially explicit geographic regions of the United States. J Anim Sci 100:1–17. https://doi.org/10.1093/jas/skac356
Shurson GC, Dierenfeld ES, Dou Z (2023) Rules are meant to be broken – rethinking the regulations on the use of food waste as animal feed. Resour Conserv Recycl 199:107273. https://doi.org/10.1016/j.resconrec.2023.107273
Sogari G, Amato M, Biasato I, Chiesa S, Gasco L (2019) The potential role of insects as feed: a multi-perspective review. Animals 9:119. https://doi.org/10.3390/ani9040119

Spang ES, Moreno LC, Pace SA, Achmon Y, Donis-Gonzalez I, Gosliner WA, Jablonski-Sheffield MP, Momin MA, Quested TE, Winans KS, Tomich TP (2019) Food loss and waste: measurement, drivers, and solutions. Annu Rev Environ Resour 44:117–156. https://doi.org/10.1146/annurev-environ-101718-033228

Sugiura K, Yamatani S, Watahara M, Onodera T (2009) Ecofeed, animal feed produced from recycled food waste. Vet Ital 45:397–404

Sun X, Dou Z, Shurson GC, Hu B (2023) Fungal bioprocessing of wheat straw with fruit and vegetable discards to produce cattle feeds for enhanced sustainability. Resour Conserv Recycl 199:107251. https://doi.org/10.1016/j.resconrec.2023.107251

Sun X, Dou Z, Shurson GC, Hu B (2024) Bioprocessing to upcycle agro-industrial and food wastes into high-nutritional value animal feed for sustainable food and agriculture systems. Resour Conserv Recycl 201:107325. https://doi.org/10.1016/j.resconrec.2023.107325

Taylor DM (2000) Inactivation of transmissible degenerative encephalopathy agents: a review. Vet J 159:10–17. https://doi.org/10.1053/tvjl.1999.0406

Thi NBD, Kumar G, Lin C-Y (2015) An overview of food waste management in developing countries: current status and future perspective. J Environ Manag 157:220–229. https://doi.org/10.1016/j.jenvman.2015.04.022

UNEP (2024). Food Waste Index Report 2024: https://wedocs.unep.org/20.500.11822/45230

Uwizeye A, Gerber PJ, Opio CI, Tempio G, Mottet A, Makkar HPS, Falcucci A, Steinfeld H, de Boer IJM (2019) Nitrogen flows in global pork supply chains and potential improvement from feeding swill to pigs. Resour Conserv Recycl 146:168–179. https://doi.org/10.1016/j.resconrec.2019.03.032

Van Zanten HHE, Mollenhorst H, Oonincx DGAB, Bikker P, Meerburg BG, de Boer IJM (2015) From environmental nuisance to environmental opportunity: housefly larvae convert waste to livestock feed. J Clean Prod 102:362–369. https://doi.org/10.1016/j.jclepro.2015.04.106

Velasco-Muñoz JF, Mendoza JMF, Aznar-Sánchez JA, Gallego-Schmid A (2021) Circular economy implementation in the agricultural sector: definition, strategies and indicators. Resour Conserv Recycl 170:105618. https://doi.org/10.1016/j.resconrec.2021.105618

Wang X, Dou Z, Feng S, Zhang Y, Ma L, Zou C, Bai Z, Lakshmanan P, Shi X, Liu D, Zhang W, Deng Y, Zhang W, Chen X, Zhang F, Chen X (2023) Global food nutrients analysis reveals alarming gaps and daunting challenges. Nature Food 4:1007–1017. https://doi.org/10.1038/s43016-023-00851-5

Wang Y, Rassler S, Stefanovski D, Bender J, Deutsch J, Chen T, Cui Z, Dou Z (2024) Evidence of animal productivity outcomes when fed diets including food waste: a systematic review of global primary data. Resour Conserv Recycl 203:107411. https://doi.org/10.1016/j.resconrec.2024.107411

Wang Y, H Ying D Stefanovski, G C Shurson, T Chen, Z Wang, Y Yin, H Zheng, T Nakaishi, J Li, and Z Cui, Z Dou (2025) Food waste treated as resources can achieve meaningful reductions in the climate and resource burdens of agrifood systems. Nature Food 6, 478-490. https://doi.org/10.1038/s43016-025-01140-z

Wells GAH, Hawkins SAC, Austin AR, Ryder SJ, Done SH, Green RB, Dexter I, Dawson M, Kimberlin RH (2003) Studies of the transmissibility of the agent of bovine spongiform encephalopathy to pigs. J Gen Virol 84:1021–1031. https://doi.org/10.1099/vir.0.18788-0

Woodgate SL (2023) Meat industry by-products: a bio-refinery approach to the production of safe, value added products for sustainable agriculture applications. Front Anim Sci 4:1259200. https://doi.org/10.3389/fanim.2023.1259200

Zhu J, Luo Z, Sun T, Li W, Zhou W, Wang X, Fei X, Tong H, Yin K (2023) Cradle-to-grave emissions from food loss and waste represent half of total greenhouse gas emissions from food systems. Nat Food 4:247–256. https://doi.org/10.1038/s43016-023-00710-3

zu Ermgassen EKHJ, Balmford A, Salemdeeb R (2016) Reduce, relegalize, and recycle food waste. Science 352:1526. https://doi.org/10.1126/science.aaf9630

Zurek M, Hebinck A, Selomane O (2022) Climate change and the urgency to transform food systems. Science 376:1416–1421. https://doi.org/10.1126/science.abo2364

By-products from Agriculture and the Agri-Food Industry for Livestock Nutrition

15

Christian Visscher and Julia Gickel

1 Introduction

From a global perspective, the growing world population seems to require a further increase in the production of food of animal origin (Jia et al. 2023; Rauw 2022; van Zanten et al. 2016). Scientists propose an increase in demand by 14% per person and 38% in total between the years 2020 and the year 2050 (Komarck et al. 2021) before declining by 2100 (Jia et al. 2023). However, livestock husbandry is one of the most disputed production chains of human food due to the large ecological footprint (Van Raamsdonk et al. 2023).

Animal feeds are responsible for the use of up to 40% of all arable land and more than 30% of cereal crop production (Govoni et al. 2023; Sandström et al. 2022). In total, more than 2.5 billion ha are used for feed production all over the world (Mottet et al. 2017). Additionally, approximately 23% of all captured fish mainly become part of fish and livestock feeds (Govoni et al. 2023; Sandström et al. 2022).

In the context of a continually growing world population, agriculture is challenged worldwide by two major incentives: minimizing negative environmental impacts (reducing greenhouse gas emissions, cropland use, freshwater use, N and P fertilizer application) and maximizing food security (Sun et al. 2024; Van Raamsdonk et al. 2023). This development also increases the need to align existing processes

C. Visscher (✉)
Institute for Animal Nutrition, University of Veterinary Medicine Hannover, Foundation, Hannover, Germany
e-mail: christian.visscher@tiho-hannover.de

J. Gickel
WING (Science for Innovative and Sustainable Poultry Farming), University of Veterinary Medicine Hannover, Foundation, Hannover, Germany
e-mail: julia.gickel@tiho-hannover.de

B. Grabkowsky, T. Blaha (eds.), *Intensive Livestock Production in Transition*,
https://doi.org/10.1007/978-3-031-97872-2_15

and growth with efficiency standards. In the European context, it is the desire for greater sustainability that is driving this further development.

2 Historical Role and Development of Livestock Nutrition

Historically, the development of a stable food supply from animals allowed the world's population to grow, residential centers to develop, and cities to emerge (Coffey et al. 2016). Keeping the animals indoors also led to the professionalization of breeding and feeding because the cause (genetics and feed) and effect (performance) could be better linked. Providing livestock with a "nutritionally balanced" diet became necessary (Coffey et al. 2016), and nutrition became a scientific discipline primarily during the last 200 years. Albrecht Daniel Thaer, a German scientist, developed the first feed standards by comparing potential feedstuffs to meadow hay and assigning a "hay value" as a comparative measure in 1810 (Coffey et al. 2016; Johnson 2007). Approximately 50 years later, the "Weende method," developed at an Experiment Station in Germany, became apparent as the first comprehensive system for the analysis of animal feed (Coffey et al. 2016; Severe 2022; Van Soest and McQueen 1973). Based on this mathematical system to assess the nutritional value, the first by-products were also used in animal diets in the second half of the nineteenth century (Coffey et al. 2016).

In the USA, leftovers from grinding grain for human consumption were thought to have potential feeding value for livestock (Coffey et al. 2016). The first corn gluten, a by-product from the corn starch industry, was manufactured in 1882 (Jiao et al. 2022). In 1890, one of the first animal-based by-products (meat scraps) was recognized for its superior nutritional value and adopted by the emerging commercial feed industry, as brewing by-products were incorporated into animal feed (Coffey et al. 2016). In Europe, during that time, the value of nutrition began to be recognized in increasing animal performance and the profitability therein (Coffey et al. 2016). Oilseed crushing and, therefore, oil cake became a large constituent of animal feeds early on, and mass quantities of other by-products found their way to the feed industry (Coffey et al. 2016).

Nowadays, livestock nutrition is based on a wide range of expertise, including animal physiology, feed technology, agricultural science, biochemistry, and animal health. The aim is to keep animals healthy and maximize their performance while minimizing environmental impact (Perry et al. 2018). As companies continue to be more consumer-oriented, the focus on sustainability and their environmental footprint will become increasingly important. Greenhouse gas emissions and environmental impacts of animal production will be considered, and there is a need to push feed mill and farm efficiency forward (Coffey et al. 2016; Van der Poel et al. 2020). What is essential, however, is to reduce the proportion of crops used directly as animal feed. In 2022, 67% of cereals, 58% of pulses, and 93% of soybeans in the EU-28 were used for animal feed (Detzel et al. 2022). Nevertheless, fodder crops are needed to feed the number of animals required to meet current demand; this means there is insufficient non-food-competing feedstuff available (van Zanten et al. 2023).

Food production has to be optimized in terms of environmental impacts and food security for the world's growing population. Therefore, changes in food production inevitably affect the feed production chain. The ecological footprint of food differs depending on the type, process, and geographic region of the production (Van Raamsdonk et al. 2023). In this context, livestock production is one of the most controversial human food production chains due to its large environmental impact.

3 Feedstuffs for Livestock and the Role of By-products

In the past, there was a stricter, economically driven focus on only a few crops for animal feed production. To mark its fiftieth anniversary, the European Feed Manufacturers' Association (FEFAC) compared how the composition of compound feed changed between 1970 and 2008 (FEFAC 2009). The association concluded that there had been no dramatic changes in the proportion of components. The proportion of grain was the largest overall and increased from 41.0% to 46.9%. However, according to the authors of that report, a detailed analysis would reveal that changes have occurred within the categories (e.g., partial replacement of soya extraction meal by rapeseed extraction meal). Certain categories of feed materials have almost disappeared (e.g., tapioca or meat and bone meal), and some others such as pulses have gained some importance (FEFAC 2009). The wide range of residual by-products from food production products should increasingly be incorporated into rations.

By-products may be classified into different categories according to their origin. One category comprises crop residues, i.e., the plant material remaining after harvesting, including parts such as straw, leaves, stalks, roots, and stover (Sandström et al. 2022; Shinde et al. 2022). In 2020, more than 5000 Mt. of crop residues were produced worldwide, with an upward trend since 1960 (Shinde et al. 2022). Another category includes all by-products that come from different processing systems. This includes by-products from crop processing like cereal bran and distiller's grains (including distiller's grains from biofuel and brewer's grains from barley beer industries) as well as by-products from sugar production (including sugarcane and sugar beet molasses and sugar beet pulp), oilseed meals (including rapeseed, soybean, sunflower seed, palm kernel, sesame seed, cottonseed, groundnut, and other oilseed meals), and citrus pulp (Sandström et al. 2022). Many of these products in this highly variable category are well suited for the nutrition of animals and are therefore widely used in their feed (Pan et al. 2019; Sandström et al. 2022). The category of livestock by-products from non-ruminant origins includes processed animal protein from pig and poultry production like blood meal, hydrolyzed feather meal, meat and bone meal, poultry by-product meal, and poultry oil (Sandström et al. 2022). In the commercial slaughtering process, a substantial part of the whole animal is not used as meat and may therefore be classified as a by-product (Irshad and Sharma 2015). Many of them have been used for animal nutrition ever since

(Leoci 2014). In addition to their use in feed production, animal by-products are widely used in the pharmaceutical, wool, and leather industries (Leoci 2014). Another category comprises by-products from fisheries and aquaculture processing as direct fish by-products and those processed into fishmeal and fish oil (Sandström et al. 2022).

Direct feed-food competition occurs when food-grade or human-edible ingredients are used in feed (Chary et al. 2024). About 15% (940 million tons in dry matter) of the total feedstuffs (6100 million tons) used in livestock and aquaculture production consisted of food-competing feedstuff that could be directly used as human food (Sandström et al. 2022). Globally, up to 49% of feed use in aquaculture (total feed use, 67 million tons), 68% in poultry (total 421 million tons), and 38% in pork meat production (total of 1200 million tons) consisted of food-competing feedstuff. For cattle meat (total of 5200 million tons) and dairy (total of 1920 million tons), the share (in quantities of feed in dry matter) was only 3–4% (Sandström et al. 2022).

4 Measures to Improve Circularity in Animal Nutrition

The global agricultural, food, and animal production systems should be re-designed to be more circular and to function within “planetary boundaries” (Springmann et al. 2018). This requires practices in feed production and livestock husbandry that reduce greenhouse gas emissions, cropland use, freshwater use, and N and P fertilizer application (Sun et al. 2024). Upcycling non-conventional feeding materials from agricultural by-products and food waste into animal feed for the sustainable production of animal-derived food products is one of the most promising options (Dou et al. 2024; Sun et al. 2024).

4.1 Global Perspective

Meat and dairy products from livestock husbandry contribute to 16–18% of the global food supply and 25–33% of the global protein consumption (Govoni et al. 2023; Mottet et al. 2017). Animals therefore consume about 980 Mt. of energy-rich crops as feed per year (Govoni et al. 2023). The production of these feed crops requires the use of valuable natural resources such as freshwater (81 km^3) and fertile land (185 Mha) suitable for human food production (Govoni et al. 2023). The virtual share of arable land used for feed production was estimated to be 37–42% of total arable land depending on the methodology (Manceron et al. 2014). In fact, substituting human edible crops with available by-products could potentially make 17.6–25.4 Mha of fertile land available and provide about 8.5–13.6 km^3 of freshwater (Govoni et al. 2023). An additional volume of 87–124 km^3 of green water would be available for the growth of other food crops (Govoni et al. 2023). Therefore, livestock species, including aquaculture species, should not consume human edible biomass but instead convert by-products from crops, livestock, and fisheries that are inedible for humans into edible biomass. Additionally, animals in circular food systems can also convert plant-based food waste and grass resources into food (van Riel et al. 2023).

Ideally, up to 11–12% of the food-competing feedstuff could be globally replaced with definite by-products without negatively affecting animal productivity (Sandström et al. 2022). The share of replacement increases to 25–29%, allowing for some decrease in animal productivity to tolerate the replacement of cereals with crop residues. For high-value products, the share of potential replacement is the highest for fishmeal and fish oil made from whole fish (Sandström et al. 2022). Nearly all whole fish used to produce fishmeal and fish oil could be replaced with by-products (Sandström et al. 2022). For this maximum replacement potential of food-competing feedstuff with by-products, there is a potential for making available a total of 430–654 × 10^{12} kcal and 17–26 million tons of protein for humans, corresponding to 5–7% and 7–11% of the current global food supply, respectively (based on data from 2016 to 2018). Almost 10% of cereal is used for feed and up to 3.8–6.0 million tons of vegetable oils from oilseeds (~31–42% of their feed use), 8–19 million tons of pulses (~50–88% of their feed use), and 2.9–3.9 million tons of fishmeal made from whole fish, corresponding to more than 17 million tons of whole fish (~11% of the current seafood supply) could be directed to human food use (Sandström et al. 2022). The replacement would free up 31–54 million ha of cropland when applying average global yields for the replaced feed crop groups. If also considering crop residues as replacement materials, the replacement would free up more than 100 million ha of cropland, corresponding to 7% of the world's total arable land use in 2018 (Sandström et al. 2022).

4.2 European Perspective

Land requirements associated with the food consumption of the European Union (EU) and United Kingdom (UK) equated to 115 Mha of cropland and 74 Mha of pasture, with around 60% (68.7 Mha) of the cropland being used in animal agriculture (Sun et al. 2022). The share of cropland feeding livestock is larger than the global average (~40%), since the EU and the UK together consume more animal products per capita than the global average and have a strong export market in high-value animal products (Sun et al. 2022). Implementing circularity principles in a self-sufficient European food system could theoretically reduce agricultural land use by up to 71% and per capita agricultural greenhouse gas emissions by 29% while producing enough macro- and micronutrients for a healthy diet on the continent. The freed-up land could also be used for producing food for export and supplying food to a further 767 million people outside Europe at reduced per capita greenhouse gas emissions (Sandstrom and Kummu 2023). Future feed consists of food losses and waste, grass, by-products, and fodder crops. The share of food losses and waste in animal feeds largely differs across animal types from zero in ruminant feed to large shares in fish, chicken, and pig feeds. Although feed-food competition decreases, animals are still largely fed with fodder crops in all circular scenarios. Without fodder crops, there is insufficient non-food-competing feedstuff available to feed the number of animals needed to meet the current supply of animal protein or to meet all nutritional requirements (for example, vitamin B12) (van Zanten et al. 2023).

4.3 Specific Perspective: Food Waste

In circular food systems, waste, for example, food waste and overconsumption of nutrients, is minimized and, if unavoidable, utilized (recycled) in the most sustainable way (Dou et al. 2024; van Zanten et al. 2023). If by-products are used as feed for farm animals, inedible biomass for humans is transformed into valuable food (van Zanten et al. 2023). With the publication of "Guidelines on the use of foodstuffs no longer intended for human consumption as animal feed (2018/C 133/02)," the EU intends to give much higher priority to the conservation of resources (European Commission 2018).

Food waste is diverse and varies in type (plant- vs animal-based), form (solid, semi-solid, liquid), source (stage of the supply chain), and nutritional attributes (chemical composition, digestibility, bioavailability), with fruit and vegetables having the highest wastage among all food groups (Dou et al. 2024). Rejected or discarded residues occur at harvesting, handling, packaging, and processing (Dou et al. 2024).

While feeding municipal food waste to livestock is currently illegal in the EU, it is common practice in many parts of the world, and there is growing interest in its possible use as a substitute for high-quality, low-cost conventional pig feed (Salemdeeb et al. 2017). Initial findings, including EU studies on pigs, indicate that the conversion of this product range into pig feed would have a lower environmental and health impact than, for example, the processing of waste by composting or anaerobic digestion (Salemdeeb et al. 2017). Without wanting to repeat former mistakes, the use of unproblematic materials offers considerable potential. Currently, only around 3 million tonnes of manufacturing by-products are used for animal feed in the EU, compared to an estimated 102 million tonnes of food waste (Dou et al. 2018).

Former food products (FFPs) have lost their suitability for human nutrition for different reasons but are still valuable as animal feed. When used in complete feed to replace traditional feed ingredients, FFPs do not decrease animal growth performance (Luciano et al. 2020; Pinotti et al. 2023). Commonly used are leftovers from the confectionery and bakery industry, showing a high energy content due to the richness of sugars, oils, and starch (Luciano et al. 2020; Pinotti et al. 2023). Pigs, as omnivorous animals, are particularly suitable for upgrading these ingredients to a higher quality of animal protein (Giromini et al. 2017; Pinotti et al. 2023). To ensure human food safety, all involved partners (farmers, nutritionists, industries, and governments) need to give serious attention to avoid an unsuitable load of pathogenic organisms (Pinotti et al. 2023). Reducing food waste, increasing recycling, or enhancing ex-food management will potentially reduce the environmental impact of feed and livestock production and develop them to a higher circularity (Pinotti et al. 2023; Srikanthithasan et al. 2024).

5 Veterinary Aspects of By-product Utilization

Many studies have documented animal productivity parameters (growth, yield, feed use efficiency) when fed food waste, but their findings remain fragmented. A systematic literature review identified 102 studies, which originated from 31 countries and encompassed all major food-producing animal species fed food waste of various types and rates (Wang et al. 2024). Animals in 75–77% of all cases performed as optimally or increased productivity compared to those fed a control diet; the remaining 23–25% showed decreased performance. Meta-analysis of a data subset indicated that pig growth did not differ between control and treatment diets of varying food waste rates (Wang et al. 2024). Ultimately, balancing major nutrients from all feed ingredients is the key to maintaining optimal productivity while pursuing sustainability (Wang et al. 2024). However, contamination of by-products with mycotoxins presents a drawback for their reliable valorization for food and feed, hampering their application range, especially as food ingredients (Lopes et al. 2023). A total of six by-products were assessed for mycotoxin content: these included distiller-dried grain with solubles, brewer's spent grain, brewer's spent yeast, cocoa shell, grape pomace, and sugar beet pulp. Common mycotoxins in these by-products are aflatoxin B1, ochratoxin A, fumonisins, deoxynivalenol, and zearalenone. The high prevalence of contaminated samples, which surpasses the limits established for human consumption, thus limits their valorization as ingredients in the food industry (Lopes et al. 2023).

Some fruits and oil extraction by-products (e.g., prickly pears, olive, and hemp cake) could improve the fatty acid profile of milk and/or meat (Vastolo et al. 2022). In addition, the gut microbiota and animal health status antioxidant may be positively affected by these by-products (Vastolo et al. 2022). The substitution of traditional feedstuffs with by-products could be an interesting prospect for animal nutrition. As all fruit and vegetable residues vary in their chemical composition, it is important to evaluate the right dosage of supplementation for the specific by-product in the animal diet (Campos et al. 2020; Vastolo et al. 2022).

6 Conclusion

Nutrients from by-products and other sources like food waste can contribute to circular and sustainable food production systems when the food waste materials are properly handled and converted to safe and nutritious feeds for animals. Established scientific understanding firmly supports this notion, as does field evidence from numerous feeding studies that have documented the productive and beneficial use of feeds derived from a variety of food waste materials. Continued success in Japan in upcycling more than one-third of its food waste to animal feed attests to the feasibility of this approach while providing an aspiring working example.

References

Campos DA, Gómez-García R, Vilas-Boas AA, Madureira AR, Pintado MM (2020) Management of fruit industrial by-products - a case study on circular economy approach. Molecules 25(2):320. https://doi.org/10.3390/molecules25020320

Chary K, van Riel A-J, Muscat A, Wilfart A, Harchaoui S, Verdegem M et al (2024) Transforming sustainable aquaculture by applying circularity principles. Rev Aquac 16(2):656–673. https://doi.org/10.1111/raq.12860

Coffey D, Dawson K, Ferket P, Connolly A (2016) Review of the feed industry from a historical perspective and implications for its future. J Appl Anim Nutr 4:e3. https://doi.org/10.1017/jan.2015.11

Detzel A, Krüger M, Busch M, Blanco-Gutiérrez I, Varela C, Manners R et al (2022) Life cycle assessment of animal-based foods and plant-based protein-rich alternatives: an environmental perspective. J Sci Food Agric 102(12):5098–5110. https://doi.org/10.1002/jsfa.11417

Dou Z, Toth JD, Westendorf ML (2018) Food waste for livestock feeding: feasibility, safety, and sustainability implications. Glob Food Sec 17:154–161. https://doi.org/10.1016/j.gfs.2017.12.003

Dou Z, Dierenfeld ES, Wang X, Chen X, Shurson GC (2024) A critical analysis of challenges and opportunities for upcycling food waste to animal feed to reduce climate and resource burdens. Resour Conserv Recycl 203:107418. https://doi.org/10.1016/j.resconrec.2024.107418

European Commission (2018) Commission Notice – guidelines for the feed use of food no longer intended for human consumption Brüssel, 16.04.2018: European Commission

FEFAC (2009) FEFAC 50th Anniversary 1959–2009. Retrieved from Bruxelles

Giromini C, Ottoboni M, Tretola M, Marchis D, Gottardo D, Caprarulo V et al (2017) Nutritional evaluation of former food products (ex-food) intended for pig nutrition. Food Addit Contam Part A 34(8):1436–1445. https://doi.org/10.1080/19440049.2017.1306884

Govoni C, D'Odorico P, Pinotti L, Rulli MC (2023) Preserving global land and water resources through the replacement of livestock feed crops with agricultural by-products. Nature Food 4(12):1047–1057. https://doi.org/10.1038/s43016-023-00884-w

Irshad A, Sharma B (2015) Slaughter house by-product utilization for sustainable meat industry: A review. J Anim Prod Adv 5:681. http://dx.doi.org/10.5455/japa.20150626043918

Jia JW, Dawson TP, Wu F, Han QM, Cui XF (2023) Global meat demand projection: quo Vadimus? J Clean Prod 429:139460. https://doi.org/10.1016/j.jclepro.2023.139460

Jiao Y, Chen HD, Han H, Chang Y (2022) Development and utilization of corn processing by-products: a review. Food Secur 11(22). https://doi.org/10.3390/foods11223709

Johnson DE (2007) Contributions of animal nutrition research to nutritional principles: energetics. J Nutr 137(3):698–701. https://doi.org/10.1093/jn/137.3.698

Komarek AM, Dunston S, Enahoro D, Godfray HCJ, Herrero M, Mason-D'Croz D et al (2021) Income, consumer preferences, and the future of livestock-derived food demand. Global Environ Change 70:102343. https://doi.org/10.1016/j.gloenvcha.2021.102343

Leoci R (2014) Animal by-products (ABPs): origins, uses, and European regulations. Universitas Studiorum

Lopes P, Sobral MMC, Lopes GR, Martins ZE, Passos CP, Petronilho S, Ferreira IMPLVO (2023) Mycotoxins' prevalence in food industry by-products: a systematic review. Toxins 15(4). https://doi.org/10.3390/toxins15040249

Luciano A, Tretola M, Ottoboni M, Baldi A, Cattaneo D, Pinotti L (2020) Potentials and challenges of former food products (food leftover) as alternative feed ingredients. Animals 10(1):125. https://doi.org/10.3390/ani10010125

Manceron S, Ben-Ari T, Dumas P (2014) Feeding proteins to livestock: global land use and food vs. feed competition. OCL Oilseeds Fats Crops Lipids 21(4):10. http://dx.doi.org/10.1051/ocl/2014020

Mottet A, de Haan C, Falcucci A, Tempio G, Opio C, Gerber P (2017) Livestock: on our plates or eating at our table? A new analysis of the feed/food debate. Glob Food Sec 14:1–8. https://doi.org/10.1016/j.gfs.2017.01.001

Pan Z, Zhang R, Zicari S (2019) Integrated processing technologies for food and agricultural by-products. Academic Press

Perry BD, Robinson TP, Grace D (2018) Animal health and sustainable global livestock systems. Animal 12(8):1699–1708. https://doi.org/10.1017/s1751731118000630

Pinotti L, Ferrari L, Fumagalli F, Luciano A, Manoni M, Mazzoleni S et al (2023) Review: pig-based bioconversion: the use of former food products to keep nutrients in the food chain. Animal 17(Suppl 2):100918. https://doi.org/10.1016/j.animal.2023.100918

Rauw WM (2022) 188 prospects of swine production in the context of climate change. J Anim Sci 100(Supplement_2):90–91. https://doi.org/10.1093/jas/skac064.148

Salemdeeb R et al (2017) Environmental and health impacts of using food waste as animal feed: a comparative analysis of food waste management options. J Clean Prod 140:871–880. https://doi.org/10.1016/j.jclepro.2016.05.049

Sandstrom V, Kummu M (2023) Towards circular food systems in Europe. Nat Food 4(4):279. https://doi.org/10.1038/s43016-023-00732-x

Sandström V, Chrysafi A, Lamminen M, Troell M, Jalava M, Piipponen J et al (2022) Food system by-products upcycled in livestock and aquaculture feeds can increase global food supply. Nat Food 3(9):729. https://doi.org/10.1038/s43016-022-00589-6

Severe JH (2022) A brief and comprehensive history of the development and use of feed analysis: a review. Open Agric 7(1):837–846. http://dx.doi.org/10.1515/opag-2022-0141

Shinde R, Shahi DK, Mahapatra P, Singh CS, Naik SK, Thombare N, Singh AK (2022) Management of crop residues with special reference to the on-farm utilization methods: a review. Ind Crop Prod 181:114772. https://doi.org/10.1016/j.indcrop.2022.114772

Springmann M, Clark M, Mason-D'Croz D, Wiebe K, Bodirsky BL, Lassaletta L et al (2018) Options for keeping the food system within environmental limits. Nature 562(7728):519–525. https://doi.org/10.1038/s41586-018-0594-0

Srikanthithasan K, Giorgino A, Fiorilla E, Ozella L, Gariglio M, Schiavone A et al (2024) Former foodstuffs in feed: a minireview of recent findings. Environ Sci Pollut Res 31:1–12. http://dx.doi.org/10.1007/s11356-024-32695-2

Sun Z, Scherer L, Zhang Q, Behrens P (2022) Adoption of plant-based diets across Europe can improve food resilience against the Russia–Ukraine conflict. Nature Food 3(11):905–910. https://doi.org/10.1038/s43016-022-00634-4

Sun X, Dou ZX, Shurson GC, Hu B (2024) Bioprocessing to upcycle agro-industrial and food wastes into high-nutritional value animal feed for sustainable food and agriculture systems. Resour Conserv Recycl 201:107325. https://doi.org/10.1016/j.resconrec.2023.107325

Van der Poel A, Abdollahi M, Cheng H, Colovic R, Den Hartog L, Miladinovic D et al (2020) Future directions of animal feed technology research to meet the challenges of a changing world. Anim Feed Sci Technol 270:114692. https://doi.org/10.1016/j.anifeedsci.2020.114692

Van Raamsdonk L, Meijer N, Gerrits E, Appel M (2023) New approaches for safe use of food by-products and biowaste in the feed production chain. J Clean Prod 135954:135954. https://doi.org/10.1016/j.jclepro.2023.135954

van Riel AJ, Nederlof MAJ, Chary K, Wiegertjes GF, de Boer IJM (2023) Feed-food competition in global aquaculture: current trends and prospects. Rev Aquac 15(3):1142–1158. https://doi.org/10.1111/raq.12804

Van Soest P, McQueen R (1973) The chemistry and estimation of fibre. Proc Nutr Soc 32(3):123–130. https://doi.org/10.1079/pns19730029

van Zanten HH, Mollenhorst H, Klootwijk CW, van Middelaar CE, de Boer IJ (2016) Global food supply: land use efficiency of livestock systems. Int J Life Cycle Assess 21(5):747–758. https://doi.org/10.1007/s11367-015-0944-1

van Zanten HHE, Simon W, van Selm B, Wacker J, Maindl TI, Frehner A et al (2023) Circularity in Europe strengthens the sustainability of the global food system. Nat Food 4(4):320–330. https://doi.org/10.1038/s43016-023-00734-9

Vastolo A, Calabrò S, Cutrignelli MI (2022) A review on the use of agro-industrial CO-products in animals' diets. Ital J Anim Sci 21(1):577–594. https://doi.org/10.1080/1828051x.2022.2039562

Wang YC, Rassler S, Stefanovski D, Bender J, Deutsch J, Chen T et al (2024) Evidence of animal productivity outcomes when fed diets including food waste: a systematic review of global primary data. Resour Conserv Recycl 203:107411. https://doi.org/10.1016/j.resconrec.2024.107411

Future Livestock Systems Based on Circularity

16

Wilhelm Windisch

1 Introduction

Livestock production affects the environment and climate. According to FAO, the agrifood system causes around 30% of total anthropogenic greenhouse gas (GHG) emissions, with livestock being a major source (12–16% of total anthropogenic GHG) (FAO 2023). Within livestock, enteric methane (CH_4) from ruminants and, to a lesser extent, CH_4 from manure storage significantly contribute to GHG emissions when expressed as CO_2 equivalents. Simultaneously, global demand for food protein is expected to rise by at least 20% in the upcoming decades. Since plant-based alternatives, cell-cultured proteins, etc. do not seem capable of filling the gaps in demand, future livestock production will be driven to deliver even more meat, milk, and eggs (FAO 2023). This pushes livestock production into the conflict of targets to significantly reduce emissions at higher yields of products.

Further pressure on livestock arises from the severely shrinking availability of agricultural areas due to the continued growth of the global human population and losses of valuable lands, along with urbanization, erosion, and desertification. Besides the climate crisis, this is considered to be one of the greatest challenges facing humanity (Smith 2018), particularly as climate change accelerates the loss of agricultural areas. On the other hand, around 40% of global cropland is currently used to produce livestock feed (Mottet et al. 2017). However, it has to be kept in mind that this practice started in the middle of the last century, along with enormous progress in plant production due to fertilizers, phytosanitary tools, plant breeding, mechanization, etc. Edible products from agriculture became available to livestock feeding at virtually unlimited quantities, and livestock production systems were optimized to upgrade these materials into economically superior goods such as

W. Windisch (✉)
Technical University Munich (TUM), TUM School of Life Sciences, Munich, Germany
e-mail: wilhelm.windisch@tum.de

B. Grabkowsky, T. Blaha (eds.), *Intensive Livestock Production in Transition*,
https://doi.org/10.1007/978-3-031-97872-2_16

meat, milk, eggs, etc. This linear system of livestock economy has been driven for decades by cereals, maize, and soybean as globally available commodities. The production systems evolved toward maximum efficiency in upgrading these commodities. Along with this development, poultry meat production has been most successful, followed by pig production. But nowadays, these systems face increasing limitations from several directions: the rising global population has been progressively claiming edible feed commodities for own consumption, uncertainty about future progress in global plant productivity, the agricultural area becoming scarce, and the need for livestock production to reduce emissions. Altogether, limitations on resources to produce enough food for humans will force current livestock production to move from linear concepts toward circularity. The following contribution will illustrate major concepts and consequences for future livestock systems.

2 Plant-Based Biomass Is the Limiting Resource of Food Production

Planetary biomass originates almost entirely from plants conducting photosynthesis. This biomass nurtures all heterotrophic organisms in the trophic levels above, including humans. Indeed, building blocks of nutritious biomass (e.g., glucose, amino acids) could also be synthesized chemically, and in some cases, it is already current practice (e.g., methionine, a limiting essential amino acid fed to livestock). However, chemical synthesis consumes large amounts of energy. Therefore, the provision of biomass through chemical synthesis will probably remain a niche production for highly specialized purposes, while the vast majority of biomass will continue to originate from plants mainly grown in agriculture.

The dependence on plants as donors of biomass has implications for its use. Our primary interest in biomass focuses on its material functionality (e.g., as food). But in any case, we first harvest the plant in its native form. At this stage, it does not fully match our primary interest, and therefore, we must continue extracting the valuable subfractions while generating residues (see the next chapter). This principle applies to all kinds of production of plant-based food for humans. The large quantities of biomass unsuitable for the primary purpose may be further utilized by either generation of energy (combustion, biogas, etc.), or by channeling it into a secondary system of conversion into a usable form, e.g., feeding it to livestock and harvesting animal-based food. Indeed, such biological conversion systems are incomplete and entail further losses (e.g., livestock dung). However, they significantly expand the yield of human food from the same plant-based biomass that was originally produced on a given spot of agricultural area.

Using biomass as an energy source, however, irreversibly destroys its material functionality and reverses the initial process of collecting energy from sunlight via photosynthesis. It must be kept in mind that growth and maintenance of plant tissues consume almost all the collected sunlight energy. In the end, the combustion energy of totally harvested plant biomass accounts for only around 1–3% of sunlight energy that was shed to the respective agricultural area along the complete cycle of vegetation. On the other hand, any use of biomass for food, feed, or other functional purposes leaves considerable amounts of residues which cannot be utilized anymore for functional purposes (e.g., excrements from livestock). These residues still contain combustion energy, and it is reasonable to exploit them as the terminating step of utilizing biomass.

In total, the shortage of global agricultural area limits the capacity to produce plant-based biomass and hence raises its value as a direct as well as an indirect resource of food. Consequently, future use of biomass must be arranged along a cascade that maximizes the overall yield of high-quality materials. This cascade may be summarized as follows:

fork > *forage* > *fuel*

The first target of agriculture and downstream processing of harvested plant materials is the direct generation of plant-based food for humans (fork). Non-edible biomass inevitably occurring along this process should be fed to livestock (forage) to convert it into additional food. Energy exploitation (fuel) should terminate the utilization of biomass and should be limited to materials that resist feasible conversion into human food.

3 Most of Plant-Based Biomass Is Non-edible

Agriculture does not produce human food directly; it rather accesses entire plant materials that further need to be processed. This generates non-edible biomass at the site of plant cultivation (coproducts, e.g., straw), as well as along downstream processing in the food industry (by-products, e.g., bran from milling). The left side of Table 16.1 presents the distribution of total agricultural biomass accessible with harvesting machines into main products and coproducts for major plant cultivation systems. Accordingly, most plant production systems on cropland generate at least the same amounts of non-edible coproducts compared to main products.

Downstream processing of harvested goods from agriculture in the food industry further releases significant amounts of biomass as by-products (see Table 16.1, right side). As these materials are not consumed by humans, they have retrospectively been considered as non-edible biomass, even though they still may contain some edible subfractions. Most of these by-products have high feeding values. This explains why it is common practice to recirculate by-products of food processing into livestock feeding.

Summing up the entire production line from agriculture until final processing, cultivation of food-delivering plants on cropland entails around double as much non-edible as edible biomass. In addition, maintenance of soil health and fertility forces plant production to avoid repetitive cultivation of crops on the same spot and to follow a multi-annual system of rotation of different plants (crop rotation). This includes plants that accumulate nitrogen and humus without delivering food to humans (e.g., clover, alfalfa). Particularly in organic farming, these "green

Table 16.1 Appearance of coproducts along agricultural plant production and by-products along downstream processing of harvested biomass in the food industry (on the basis of dry matter)

Agricultural plant production			Non-edible residues along downstream processing of harvested biomass in the food industry		
Plant culture	Coproduct	% of total harvest	Technology	By-product	% of input
Cereals	Straw	45	Cereal milling	Bran	20–25
Maize	Straw	50	Starch processing	Pulp	25–30
Rapeseed	Straw	65	Sugar extraction, beets	Sugar beet pulp	40–45
Soybean	Straw	50	Oil mills, rapeseed	Cake, expeller, extracts	55–60
Sugar beets	Leaves	40	Oil mills, soybean	Cake, expeller, extracts	65–80
Potato	Straw	20	Brewery, bioethanol	Draff, DDGS[a]	25–30

[a]*DDGS* dried distillers' grain with solubles

fertilizers" may occupy up to one-third of total cropland, thus increasing the generation of non-edible biomass at the expense of plant-based food for humans.

Another significant source of non-edible biomass is permanent grassland. On a global scale, one-third of ice-free land is permanently used for agricultural purposes, with a share of 32% as croplands, while the majority (68%) comprises meadows and pastures (FAO 2021). Most of these permanent grasslands cannot be converted into cropland mainly for topographical and climatic reasons (> 80%, Mottet et al. 2017). In Alpine areas, such as Austria and Switzerland, the respective share ranges at 60%, and even regions with intensive crop production comprise significant areas of grasslands (e.g., 30% in Germany).

Grasslands are often considered as competitors to forests and an artificial product of agriculture. Indeed, forests would succeed grasslands in many areas once agricultural utilization would be terminated. But this is also because humans extinguished the large herds of wild herbivores (the so-called megafauna) in almost all areas where permanent agriculture has been practised. Under fully natural conditions, this megafauna would create a mosaic of open grasslands and spots of forest (Thompson et al. 2023). Therefore, well-managed agricultural grasslands might be considered as an option to maintain a semi-natural habitat rich in biodiversity.

Summing up biomass from croplands and grasslands, total agriculture generates far more non-edible biomass than edible biomass. As an example of intensive agricultural regions of Central Europe, Figure 16.1 depicts the current situation in Germany (Vorndran et al. 2024). Most of the total annual biomass production of around 120 Mio MT of dry matter is non-edible and comprises mainly biomass from grassland (20%), maize silage (16%), and coproducts (29%). The harvested main products (cereals, other seeds, sugar beets, potatoes, etc.) represent 32% of total biomass dry matter. Assuming complete use of these materials for the downstream generation of human food, this section would further divide into final food

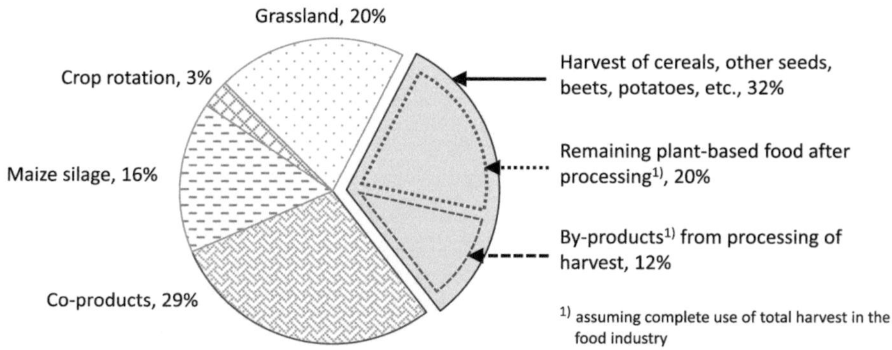

Fig. 16.1 Distribution of biomass from German agriculture among sectors of plant production and downstream processing of harvests (based on dry matter) (Data from Vorndran et al. (2024))

(20%) and by-products (12%). Accordingly, each kg of final food would account for at least 4 kg of non-edible biomass. Indeed, this ratio may considerably vary on a local scale depending on the distribution of grassland vs. cropland and the productive traits of the respective areas. Furthermore, maize silage could theoretically be harvested as maize kernels plus maize straw, hence raising the yield of edible biomass to some extent. However, this would not change the overall message that most agricultural biomass is non-edible but suitable for livestock feeding. Also on a global scale, around 85% of total biomass consumed by livestock was estimated to be non-edible (Mottet 2017).

The most challenging aspect of estimating the provision of non-edible biomass from agriculture is the definition of edibility itself. To assess the gray zone between edible vs non-edible biomass, the concept of the *human edible fraction* (*hef*) has been proposed (Ertl et al. 2015) (Table 16.2). It classifies the edibility of plant-based biomass based on nutrient analyses as well as process technologies to separate nutritious from inferior materials. *hef*-values below around 20% denote materials that are categorially non-edible, such as highly fibrous biomass like straw. Intermediate *hef*-values characterize materials that humans usually do not consume as such. Typically, they undergo technical processing to separate nutritious components from inferior subfractions, as in barley (pot barley vs bran) or in rapeseed (oil vs extracts).

Top *hef*-values above around 70% denote materials that are categorically edible, such as wheat, maize (kernels), and soybean. Therefore, feeding such materials to livestock would introduce food competition and hence contradict the principle of using biomass along the cascade *fork* > *forage* > *fuel*. However, this concept is valid only for a priori considerations, such as when deciding the type of plant to be cultivated on a given spot of cropland. Final edibility, however, results a posteriori from economic, technical, and sanitary quality traits that decide whether the harvested material finally ends up in human consumption. For example, unfavorable weather conditions during harvest may turn edible biomass into livestock feed. Another grey area is the cultivation of barley, triticale, etc. on low-quality croplands for the

Table 16.2 Human edible fraction (hef) of materials used as livestock feed (% based on dry matter) (adapted from Windisch and Flachowsky (2022))

Feedstuff	*hef* (%)	Minimum-maximum
Wheat	80	60–100
Maize (kernels)	80	70–90
Barley	65	40–80
Soybean	92	50–93
Rapeseed	59	30–87
Wheat bran	10	0–20
Maize silage	29	19–45
Grass and products derived thereof, straw, pulp, DDGS[a]), …	0	0

[a]*DDGS* dried distillers' grain with solubles

purpose of livestock feeding. Indeed, these materials are edible in theory but are hardly consumed under current practical conditions. Also, from the viewpoint of indirect food competition, producing feed-grade cereals on low-quality cropland does not necessarily displace food-producing crops, as those require high-quality soils. Consequently, there will always exist a certain proportion of cereals that finally turn out to be non-edible.

The process technology of harvesting and downstream processing is decisive to the edibility of biomass by humans. This even applies to materials which are usually considered to be categorially non-edible, such as grass. It contains valuable protein inside the plant cells, but it is hardly accessible for digestion by monogastric livestock (e.g., pigs, poultry) as well as by humans due to the caging effect of the indigestible cell wall. However, pressing out and precipitating the cell juice could provide valuable protein from grassland at amounts that are comparable to the protein yield of soybeans grown on cropland (Brugger et al. 2016). Simultaneously, the large quantities of fibrous residues still remain valuable feed materials to ruminants. This example demonstrates that novel process technology may upgrade the future use of biomass and may significantly change the differentiation between edible vs non-edible subfractions.

In total, the production of plant-based food along with agriculture and downstream processing of harvested materials always produces much more non-edible than edible biomass. This biomass arises without any food competition to humans and hence may be considered as a valuable and sustainable feed resource for livestock for conversion into human food.

4 Feeding Non-edible Biomass to Livestock Creates a Win-Win Situation

Non-edible biomass contains large quantities of plant nutrients (nitrogen, phosphorus, etc.) that must be returned to the soils to maintain fertility. In principle, three strategies of recirculation are available: rotting in the field, fermentation in biogas plants, using the residues as fertilizer, or feeding the non-edible biomass to livestock and returning the dung to the soils. It should be noted that livestock dung originating solely from recirculation of non-edible biomass does not produce excessive emissions of nitrogen, phosphorus, etc., since the latter occurs only in the case of extra import of feedstuffs into the system. In any case, the non-edible biomass will almost completely disaggregate, release the same plant nutrients, and end up in virtually the same humus particles in the soils irrespective of the pathway of circulation.

To prevent leaching into the environment, it is important to synchronize the recirculation of plant nutrients from non-edible biomass with the needs of the plants in the next cycle of vegetation as closely as possible. While rotting releases plant nutrients unsynchronized and hence causes comparably high emissions, both the biogas residues and livestock dung may be stored and applied in a targeted way. This significantly reduces leaching and improves fertilizing efficiency. The most important effect, however, is an increased production of biomass by around a factor of two along the entire crop rotation period as compared to rotting, and hence a doubling of human food production from the same agricultural spot (Bryzinski 2020). From the viewpoint of agronomy, the efficiency of generating plant-based food is comparable to the two routes of utilizing non-edible biomass, namely biogas production or livestock feeding. However, considering the emerging scarcity of agricultural area and biomass-derived thereof, the generation of food through livestock will get priority over energy according to the cascade *fork* > *forage* > *fuel*.

Livestock may produce significant amounts of extra food from non-edible biomass. Assuming 4 kg of non-edible biomass per kg of plant-based food, the quantities of extra animal-based protein may be estimated to range at the same magnitude as the basal production of plant-based protein. In terms of kilocalories, livestock may add around half of the basal level. In other words, the inclusion of livestock into the cascade of using plant-based biomass increases the number of humans that may be fed from the same spot of agricultural area by at least one-half compared to an agricultural system without livestock. Since this effect occurs without food competition, no inherent antagonism exists between plant-based vs animal-based food. On the contrary, both production lines are coupled synergistically, provided the utilization of plant-based biomass follows the cascade *fork* > *forage* > *fuel*. Inclusion of livestock into agriculture not only fosters plant production by delivering an efficient fertilizer but also supports food security through the provision of significant quantities of high-quality food on top of the basal plant production (see also Windisch and Flachowsky 2022; Thompson et al. 2023).

5 "Climate-Killer-Cow" May Be a Misleading Narrative

Ruminants are the most efficient converters of non-edible biomass into human food. However, the carbon footprints of milk and ruminant meat, as calculated by standard metrics (GWP100), are among the highest among animal-derived products. The main reason is CH_4, which is emitted mainly during microbial fermentation in the forestomachs. Indeed, CH_4 is a strong greenhouse gas, but it degrades quickly (atmospheric half-life time around 12 years). In contrast, CO_2 is a long-lived greenhouse gas that accumulates in the atmosphere once released, e.g., through the combustion of fossil energy sources or the degradation of soil humus after land use change. Although CO_2 is a weaker greenhouse gas compared to CH_4, any emission linearly heats up the atmosphere, while the climate effect of the short-lived CH_4 depends on the change in emission rate rather than on the emission itself (Allen et al. 2018). This leads to different climate effects of ruminants depending on the local dynamics of population and production intensity (Manzano et al. 2023).

At constant emission rates (e.g., constant ruminant head counts), the atmospheric concentration of CH_4 reaches an equilibrium between newly formed CH_4 and the degradation of formerly emitted CH_4. In this situation, the current CH_4 emissions do not further heat up the atmosphere (Neu 2022). Consequently, the emission of CH_4 from ruminants kept at such conditions is virtually climate neutral. The rather climate effect is of an indirect nature and is characterized by the theoretical cooling effect, which could be introduced by stopping CH_4 emissions, e.g., through the elimination of ruminating livestock. Rising headcounts and/or rising production intensities, however, would introduce fresh CH_4 into the atmosphere, which could initially express its full warming potential because the equilibrium with degradation is not yet reached. Consequently, the short-term climate effects of increasing ruminant production are much more severe than reductions would bring climate benefits. Currently used metrics of CO_2 equivalents (GWP100) do not sufficiently consider the different behavior of CH_4 compared to CO_2, as they linearly assess the absolute emissions of CH_4 (factor 28 compared to CO_2) rather than the emission rates. This may lead to significant overestimations or underestimations of the climate impact on local ruminant production systems (Manzano et al. 2023). For the situation in Austria, for example, it has been estimated that current metrics (GWP100) overestimate the CO_2 equivalents of milk and beef meat by around factor 2 (Hörtenhuber et al. 2022).

Global ruminant production and, consequently, CH_4 emission from livestock are particularly concentrated in Latin America and South Asia, where both head counts and production intensities are rising. In contrast, figures from Europe are comparatively small. For example, in Germany, the head counts of ruminants expressed as cattle units and the corresponding emissions of CH_4 have been massively decreasing during the past decades, now ranging below the pre-industrial level (Schulze 2014; Kuhla and Viereck 2022). Figure 16.2 schematically summarizes the cumulative climate impact (radiative forcing) of total national emissions of CO_2, CH_4, and N_2O from Austria, considering the different degradation behaviors of these long-, short-, and medium-lived greenhouse gases (Guggenberger et al. 2022). While continuous accumulation of the long-lived CO_2 caused by far the most heating effects over time, the short-lived CH_4 has been producing a comparably small band rather than an accumulation. Considering that only around half of the CH_4 band is attributable to agriculture, even the complete elimination of ruminant production would entail only minor cooling effects on the atmosphere.

Of course, CH_4 emissions from ruminant livestock must be carefully watched and minimized through measures such as feed additives and feeding strategies. Limiting livestock feeding to inevitably occurring, non-edible biomass would

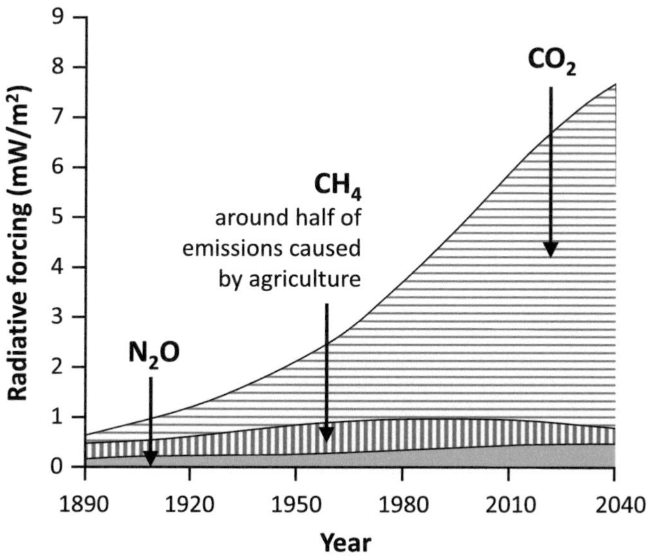

Fig. 16.2 Cumulative climate impact (radiative forcing, mW/m^2) of national emissions of carbon dioxide (CO_2), methane (CH_4), and nitrous oxide (N_2O) from Austria (adapted from Guggenberger et al. (2022))

further constrain ruminant production and keep the equilibrium between CH_4 emission and atmospheric degradation at a lower level than now. Under such conditions, CH_4 emissions from ruminants may be considered fairly climate neutral as this CH_4 would not contribute further to atmospheric heating, and the cooling effect in case of completely eliminating ruminants from livestock production would be relatively small.

6 Too Much as Well as Too Little Livestock Harm the Environment and Climate

The circularity of non-edible biomass releases nitrogen, phosphorous, and other plant nutrients bound therein, virtually independent of the recycling pathway (rotting, biogas, livestock feeding). Hence, feeding non-edible biomass to livestock does not pose a principal risk of environmental pollution, as the release of plant nutrients via dung remains within the boundaries of natural agricultural matter circulation. Furthermore, all emissions caused by both edible and non-edible biomass are inevitably coupled within the common framework of plant cultivation, harvest, and downstream processing. If the non-edible biomass would not be used for livestock feed or energy purposes, the respective share of emissions would not disappear. Therefore, it would be necessary to add it on top of the share of edible biomass, hence significantly raising the overall footprints of plant-based food. These emissions must be offset against those emissions which are specifically introduced by livestock, e.g., for housing,

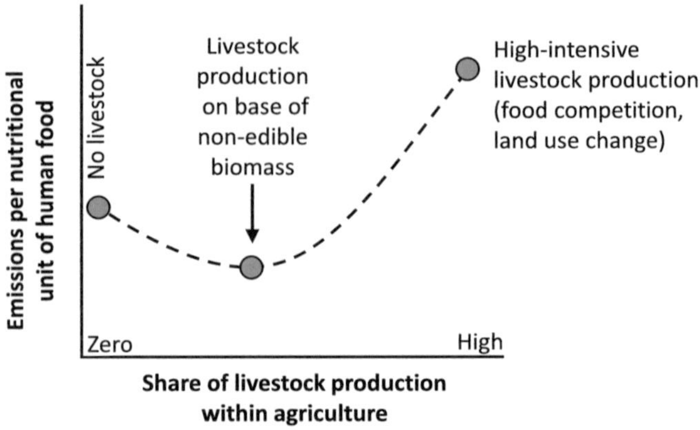

Fig. 16.3 Schematic response of emissions per nutritional unit of human food (kilocalories, protein, etc.) to rising share of livestock production within agriculture

technical equipment, grassland management, etc. Considering that enteric CH4 from ruminants is fairly climate neutral under the condition of circularity (see above), the footprints of animal-based food become significantly smaller once the use of total biomass follows the cascade *fork > forage > fuel*.

Abstinence from livestock would eliminate the extra meat, milk, etc. produced with non-edible biomass and would thus reduce total food production from a given spot of agricultural area. Therefore, agricultural systems without livestock would be forced to compensate for these food losses either by significantly intensifying plant production and/or by expanding the use of cropland (Van Zanten et al. 2018). Furthermore, such an agricultural practice would inevitably raise the generation of non-edible biomass. As depicted schematically in Fig. 16.3, this would increase the emissions per nutritional unit of human food (kilocalories, protein, etc.). In other words, completely discharging livestock from agriculture would raise the pressure on the environment and climate to nourish a given number of humans in comparison to a mixed agricultural system, where livestock is balanced with plant production according to the cascade fork > forage > fuel. But emissions would also rise on the other side of the imbalance once livestock production is intensified above the threshold given by the circularity of non-edible biomass. In this situation, extra feed must be produced at the expense of human food, and the extra animal-based products generated thereof must take full responsibility for additional emissions arising along the entire pathway of production, particularly if associated with land use change. Consequently, the minimum impact of food production on the environment and climate may be achieved by the smart inclusion of livestock into the entire agricultural system, based on the circularity of inevitably occurring, non-edible biomass. The future challenge will be to identify this minimum according to local conditions of agriculture and downstream food processing.

Figure 16.4 exemplifies the effect of rising animal performance on absolute emissions ((a), left side) and corresponding footprints of animal-based products,

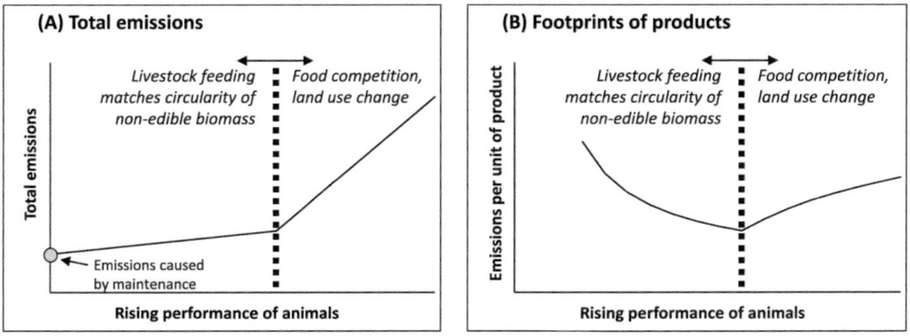

Fig. 16.4 Schematic effect of rising animal performance on (**a**) total emissions and (**b**) footprints of animal-based products (emissions per unit)

calculated by total emissions per unit of product ((b), right side). The emissions respond to rising performance according to a broken line function: starting from the maintenance level, rising performance results in a comparably small increase in emissions. The corresponding footprints show an asymptotic decrease due to the progressive dilution of maintenance emission over a rising gain of products. However, once animal performance exceeds the productive potential of the non-edible biomass, food competition must be introduced to cover the extra feed requirements, possibly entailing also land use change. Consequently, the absolute emissions experience a much steeper increase, and the corresponding footprints may start to rise depending on the extent of change in the steepness of absolute emissions. In total, the minimum footprints are achieved when animal performance matches the maximum feeding potential of the non-edible biomass.

Indeed, current intensive livestock production is characterized by high footprints due to the presence of food competition and, partially, land use change. These high footprints are often considered to be a constant value, seen as an inherent burden of livestock on the environment and climate. Accordingly, recommendations for more sustainable nutrition systems usually assume linear reductions in emissions with restricted consumption of animal-based food. Indeed, such strategies may, to a certain extent, reduce environmental and climate impacts compared to the current situation. However, they overlook the synergistic potential to further minimize the footprints of animal-based food by specifically avoiding food competition. In addition, they ignore the risk that overall footprints per unit of human food start to rise again if livestock production becomes too extensive or weak to make full use of converting inevitably occurring, non-edible biomass into extra human food. Therefore, the future goal is to minimize the overall footprint for the entire food production system, including both plant-based and animal-based food, by balancing the proportion of livestock within agriculture (Fig. 16.3) and optimizing the performance levels within livestock (Fig. 16.4, right side). Unfortunately, food production scenarios considering the dynamic, non-linear nature of emissions and footprints are still missing (Manzano et al. 2023).

7 Feed Efficiency Solves the Trade-Off Between Footprints and Productivity

Feeding livestock solely with inevitably occurring, non-edible biomass will limit livestock production. This limitation particularly affects poultry and, to a lesser extent, pigs, as these monogastric livestock species require feed qualities equal to or close to human food. Ruminants, however, could largely continue with already existing non-edible biomass. Corresponding model scenarios for Switzerland suggest significant reductions in poultry products (meat, eggs) and pork meat by more than 90% and 50%, respectively, while milk and beef meat production could maintain around two-thirds of current levels (Baur and Flückiger 2018). For German conditions, recent model scenarios indicate that around half of the current milk and beef meat production could be maintained solely based on national grassland, and almost two-thirds when non-edible by-products from food processing are further included (De Luca and Müller 2024). Therefore, ruminants will become the most important livestock species in production systems using plant-based biomass according to the cascade *fork > forage > fuel*.

In systems with limited availability of feed biomass, both in terms of quantity and nutritional quality, the efficiency of converting non-edible biomass into animal-based food becomes pivotal (e.g., Windisch and Flachowsky 2022). A first step to improve overall feed efficiency is avoiding losses of biomass during harvest and storage, and in the case of grassland also proper pasture management. Furthermore, access to non-edible biomass from cropland as well as from downstream processing must be established as far as possible. The next step is raising the quality of non-edible biomass for feed purposes. Techniques have already been established in the past to improve the feed value of highly fibrous materials such as straw, e.g., through degradation of lignocellulose. But the most promising route to fundamentally improve feed quality should be plant breeding. Indeed, this is a novel task since until now plant breeding has been focusing mainly on food-related traits rather than on the use of the entire plant biomass along the cascade *fork > forage > fuel*. Finally, livestock feeding systems should avoid unproductive feed consumption as much as possible. Precision feeding in combination with the use of feed additives such as essential amino acids and feed enzymes (e.g., phytases) is a well-established technique to improve feed efficiency. Rising performance of individual animals may further improve feed efficiency due to dilution of maintenance costs over a larger quantity of products (see above, Fig. 16.4). However, this approach is limited by the feed value of the non-edible biomass that is currently available. Vice versa, improvements in feed quality extend the limits to raising the performance of individual animals. Finally, amending livestock management practices to promote robust offspring development, undisturbed production cycles, and better animal health and welfare can significantly improve feeding efficiency. All these measures also reduce the CH_4 burden per unit of ruminant products (meat, milk).

8 "Alternatives" to Animal-Based Food Should Also Follow the Cascade *Fork > Forage > Fuel*

Vegan products, despite being advertised as sustainable and climate-friendly alternatives to animal-based food, inevitably generate large amounts of non-edible biomass from agricultural cultivation of plants until processing into the final products. For example, oat drinks utilize only one-third of biomass introduced into the production process via oat kernels. Following the cascade *fork > forage > fuel*, these by-products should be fed to livestock, especially considering that oat pulp is a valuable feedstuff. Metaphorically speaking, any glass of oat drink entails another glass of cow milk, demonstrating the symbolic relationship between vegan products and animal-based food. Therefore, vegan products should be addressed as synergistic partners to animal-based food rather than alternatives.

Cellular meat from muscle cells is advertised to provide sustainable and climate-friendly proteins without posing animal welfare problems. Furthermore, the harvested material may be completely used for human consumption, while slaughtering livestock animals always generates non-edible residues. Indeed, there are still various (bio)technological and biological constraints inhibiting upscaling to industrial scales of production (e.g., Wood et al. 2023). But the most striking aspect is the fact that the muscle cells must be fed with a culture media containing solely pure nutrient molecules (glucose, amino acids, fatty acids, etc.). Such cultural media represent the highest possible degree of nutritional quality that human food ever could achieve. Furthermore, it is produced out of edible biomass by elaborate and energy-consuming technical procedures mimicking the process of digestion that would occur by nature in livestock animal. Exploitation of nutrients provided by the culture media is limited due to the lack of the biological functionality of the liver and kidney in such technical systems. Therefore, unused nutrients must be disposed of when changing the culture media to get rid of metabolic residues from the cell culture. In addition, muscle cells consume considerable amounts of nutrients along with unproductive maintenance metabolism and tissue turnover. Overall, cellular meat poses a strong food competition to humans with limited capacity to convert input nutrients into harvested food.

Precision fermentation with, e.g., genetically modified microorganisms offers an alternative method of protein production compared to livestock animals. Indeed, the proposed microorganisms are autotrophic in view of nitrogen and may generate highly specified proteins out of soluble nitrogen. Therefore, precision fermentation seems to share similarities with ruminants, which derive their demand for protein mainly from the digestion of *N*-autotrophic microbes growing in the forestomachs. Furthermore, the protein expressed by precision fermentation is equivalent to the native protein produced by livestock (e.g., microbes genetically modified with genes for milk casein can produce pure casein). But there are crucial differences: Most precision fermentation microorganisms require dietary energy sources that are in principle edible (glucose, sucrose). An exception is the methylotrophic yeasts like *Pichia pastoris*, which grow on methyl alcohol. Furthermore, all the yeast systems deliver only protein out of the biomass fed to the fermenter, while the input of dietary energy is largely lost via CO_2 emissions along the microbial metabolism. Ruminants, on the other hand, feed the microbes in their forestomachs primarily with non-edible materials. These microbes can utilize only around one-quarter of fermented energy due to the anaerobic environment in the forestomach. The end-products of fermentation (mainly acetic, propionic, and butyric acid) are absorbed by the host animal, thus providing the majority of their dietary energy requirements. While precision fermentation systems can produce edible protein only, the exploitation of both the anaerobic microbial protein and energy metabolism enables ruminants to produce a multitude of matter out of the same quantity of biomass. For example, whole milk contains four times the edible dry matter and dietary energy compared to the sole production of milk protein in microbial systems. This is why precision fermentation will probably not be able to quantitatively compete with ruminants for bulk production of dietary protein for humans. Nevertheless, the high specificity and purity of the proteins derived from precision fermentation might attract future interest for special purposes in the food industry. Furthermore, methylotrophic yeast may expand the overall spectrum of food production as they can utilize extremely degraded biomass that is even unsuitable to ruminants.

Fungi present an interesting principle to convert strictly non-edible material into human food. These organisms not only synthesize protein from soluble nitrogen but also thrive on materials very rich in lignocellulose. However, fungi are strictly aerobic organisms and therefore primarily degrade processed biomass into CO_2, with only small amounts of harvestable material remaining. On the other hand, materials rich in lignocellulose can hardly be degraded in the forestomachs of ruminants. While fungi may excel in the niche of processing highly ligneous materials, they cannot rival ruminants in efficiently converting the majority of non-edible biomass into human food.

9 Conclusion and Outlook

Livestock plays a crucial role in utilizing non-arable areas, maintaining non-edible biomass in circulation, providing high-quality organic fertilizers, and generating additional high-quality food without food competition. These benefits are maximized when the intensity of livestock production is balanced within the circularity of non-edible biomass and follows the cascade *fork > forage > fuel*.

In essence, the necessary adaptations in future livestock production systems are comparable to the ongoing transition from fossil energy to renewable energy sources. Similarly, the non-edible biomass can be seen as a renewable feed resource that can be utilized without automatically harming environment and climate. However, gaining and storing these resources can be challenging and labor-intensive, and limitations in quantity and feed quality restrict the production volume of animal-based food and raise production costs. In such a system, non-edible biomass is a valuable resource that must be utilized with maximum efficiency.

Non-edible biomass composition varies widely, both temporally and spatially, and encompasses a wide range from highly fibrous materials like straw to top-quality feedstuffs such as feed-grade cereals. Optimally utilizing these heterogeneous feed materials requires a proper mix of different livestock species. For example, feed-grade cereals and high-quality by-products should be fed to poultry and pigs, while more fibrous materials, coproducts from cropland, and biomass from grassland are suitable for ruminants. Livestock breeds of the future need to be flexible and resilient against temporal fluctuations in quantity and quality of feed biomass, while achieving the highest possible zootechnical performance with the available non-edible biomass.

The *fork > forage > fuel* approach will drive advancements in crop rotation systems, plant breeding, and technologies for harvest and downstream processing, leading to higher yields of plant-based food from the total plant-based biomass. This will result in further reductions in the quantity of non-edible biomass and an increase in its fiber content. Consequently, the available feed resources for livestock production will decrease, particularly for poultry and to a lesser extent for pigs, while the importance of ruminants in effectively utilizing non-edible biomass will increase. In the end, sustainable food production systems will still incorporate livestock, but to a significantly lesser extent compared to the current system. This shift reflects the need to optimize resource utilization and minimize environmental and climate impacts while still meeting the demand for high-quality food for humans.

References

Allen MR, Shine KP, Fuglestvedt JS, Millar RJ, Cain M, Frame DJ, Macey AH (2018) A solution to the misrepresentations of CO_2-equivalent emissions of short-lived climate pollutants under ambitious mitigation. Npj Clim Atmos Sci 1:16. https://doi.org/10.1038/s41612-018-0026-8

Baur P, Flückiger S (2018) Nahrungsmittel aus ökologischer und tiergerechter Produktion. In: Eine Studie im Auftrag von Greenpeace Schweiz Wädenswil: ZHAW Institut für Umwelt und natürliche Ressourcen. https://doi.org/10.21256/zhaw-1411

Bryzinski T (2020) Erträge, Energieeffizienz und Treibhausgasemissionen ökologischer und konventioneller Pflanzenbausysteme: Methodische Einflüsse und feldexperimentelle Ergebnisse

Brugger D, Nadler C, Windisch WM, Bolduan C (2016) Feed protein value of acidic precipitates obtained from press juices of three types of green forage leaves. Anim Feed Sci Tech 222:236–241. https://doi.org/10.1016/j.anifeedsci.2016.10.017

Dissertation an der Technischen Universität München, Germany. ISBN: 979–8574395912/

De Luca K, Müller A (2024) Das Potenzial einer Grünlandbasierten Milchproduktion in Deutschland. Forschungsinstitut für biologischen Landbau FiBL und Greenpeace Deutschland e.V

Ertl P, Zebeli Q, Zollitsch W, Knaus W (2015) Feeding of by-products completely replaced cereals and pulses in dairy cow and enhanced edible feed conversion ratio. J Dairy Sci 98(2):1225–1233. https://doi.org/10.3168/jds.2014-8810

FAO (2021) Land use statistics and indicators statistics. Global, regional and country trends 1990–2019. Vol. FAOSTAT Analytical Brief Series No 28. Rome

FAO (2023) Pathways towards lower emissions—a global assessment of the greenhouse gas emissions and mitigation options from livestock agrifood systems, Rome. https://doi.org/10.4060/cc9029en

Guggenberger T, Terler G, Herndl M, Fritz C, Grassauer F (2022) Langzeitbewertung von Treibhausgasemissionen in Österreich. Forschungsbericht der HBLFA Raumberg-Gumpenstein, Irdning-Donnersbachtal

Hörtenhuber SJ, Seiringer M, Theurl MC, Größbacher V, Piringer G, Kral I, Zollitsch WJ (2022) Implementing an appropriate metric for the assessment of greenhouse gas emissions from livestock production: a national case study. Animal 16(10):100638. https://doi.org/10.1016/j.animal.2022.100638

Kuhla B, Viereck G (2022) Enteric methane emission factors, total emissions and intensities from Germany's livestock in the late 19th century: a comparison with the today's emission rates and intensities. Sci Total Environ 848:157754. https://doi.org/10.1016/j.scitotenv.2022.157754

Manzano P, Rowntree J, Thompson L, del Prado A, Ederer P, Windisch W, Lee MRF (2023) Challenges for the balanced attribution of livestock's environmental impacts: the art of conveying simple messages around complex realities. Anim Front 13(2):35–44. https://doi.org/10.1093/af/vfac096

Mottet A, de Haan C, Falcucci A, Tempio G, Opio C, Gerber P (2017) Livestock: on our plates or eating at our table? A new analysis of the feed/food debate. Glob Food Sec 14:1–8. https://doi.org/10.1016/j.gfs.2017.01.001

Neu U (2022) Climate effect and CO_2 equivalent emissions of short-lived substances. Swiss academies. Communications 17(5). https://doi.org/10.5281/zenodo.6343974

Schulze E (2014) Deutsche Agrargeschichte; 7500 Jahre Landwirtschaft in Deutschland; ein kurzer Abriss. Shaker Verlag. Aachen. ISBN 978-3-8440-2636-8

Smith P (2018) Managing the global land resource. Proc R Soc B 285:20172798. https://doi.org/10.1098/rspb.2017.2798

Thompson L, Rowntree J, Windisch W, Waters SM, Shalloo L, Manzano P (2023) Ecosystem management using livestock: embracing diversity and respecting ecological principles. Anim Front 13(2):28–34. https://doi.org/10.1093/af/vfac094

Vorndran AM, Steinhoff-Wagner J, Windisch WM (2024) Ermittlung des Aufkommens an nicht-essbarer Biomasse aus dem Agrar- und Verarbeitungssektor Deutschlands. Züchtungskunde 69:56–76

Windisch W, Flachowsky G (2022) Livestock-based bioeconomy. In: Thrän D, Moesenfechtel U (eds) The bioeconomy system. Springer, Berlin, Heidelberg, pp 68–83. https://doi.org/10.1007/978-3-662-64415-7

Wood P, Thorrez L, Hocquette J-F, Troy D, Gagaoua M (2023) "Cellular agriculture": current gaps between facts and claims regarding "cell-based meat". Anim Front 13(2):68–74. https://doi.org/10.1093/af/vfac092

Specificities, Changes and Challenges of Pig Production in France

17

Christine Roguet and Elisa Husson

1 Introduction

With 22 million pigs produced in 2023, France is the third largest producer behind Spain and Germany. The pig sector accounts for 5% of the French final agricultural production in 2023, versus 29% in Denmark, 18% in Spain and 12% in Germany. The very dynamic growth of its pig production in the 1980s and 1990s enabled France to fill its domestic deficit in 1994. Production has become concentrated, both geographically (56% of pigs are in Brittany in 2023) and structurally, taking advantage of economies of scale and agglomeration (Gaigné and Ben Arfa 2011; Roguet et al. 2015). However, this growth and concentration ended up impacting the environment (degradation of water quality, green algae, odours, ammoniac, etc.) and led to growing societal opposition and very restrictive environmental regulations in the most intensive areas (areas with structural nitrate surpluses). Growth was stopped in the early 2000s, and pig production has since struggled to maintain its level. With a self-sufficiency rate close to 100% (compared to 410% in Denmark, 240% in the Netherlands, 195% in Spain and 130% in Germany in 2022), the French pig sector fears a return to a supply deficit and imports. As in other countries, pig production in France faces a wide range of challenges: food sovereignty, competitiveness, farmers' income, renewal of the workforce, transmission of farms and agroecological transition.

After a brief overview of French pig production and its evolution in the first part, the second part describes the specific characteristics of French pig farms. The final part of this chapter offers some insight into the main societal issues in France and illustrates the solutions implemented by industry players to address them.

C. Roguet (✉) · E. Husson
IFIP-institut du porc, Pacé, France
e-mail: christine.roguet@ifip.asso.fr; elisa.husson@ifip.asso.fr

B. Grabkowsky, T. Blaha (eds.), *Intensive Livestock Production in Transition*,
https://doi.org/10.1007/978-3-031-97872-2_17

2 Current Situation and Trends of the French Pig Production

2.1 Supply and Demand: A 100% Self-Sufficiency Rate

Pork production and consumption in France have evolved interdependently over time (Fig. 17.1). Between 1960 and 1985, the boom in consumption widened France's pork deficit, with 15% of French demand for pork covered by imports, mainly from Belgium and the Netherlands. This shortfall in animal production prompted the government to enact the 1966 Livestock Law to encourage the development of technical support, with public aid "designed to encourage the creation and modernization of the buildings to rear cattle, pigs, sheep and goats" and then to implement a "Pig Production Rationalization Plan" (1970–1981). As a result, pork production increased by 45% between 1985 and 2000, and France became self-sufficient in 1994. This very dynamic growth provided France with modern and technically among the most efficient livestock farms in the world. However, when implementing the Nitrates Directive in 1991, "the French government rather focused its control measures on the creation and expansion of farms rather than on the balance of fertilisation" (Le Goffe 2013). Any expansion or creation of livestock farms has been banned in Brittany, leading over the years to a lack of investment (Roguet et al. 2007; Bertin and Ramonet 2016), a lack of consistency and ultimately a deterioration of performances compared to competing countries (Duflot 2015). Pig production started to decline, and the number of pigs slaughtered in France decreased from 27 to 22 million heads between 1999 and 2023. Slaughtering in tons remained relatively stable until 2010 thanks to the significant increases in sow productivity and carcass weights. But, since 2010, these productivity and carcass weight gains

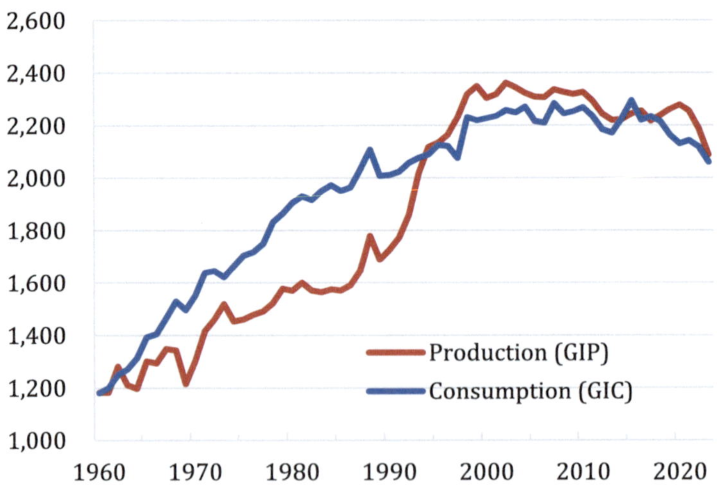

Fig. 17.1 French supply balance sheet since 1960 (1000 tons cwe). (Source: IFIP from Agreste, Eurostat)

have not been sufficient to offset the decline of herds: between 2010 and 2023, slaughtering decreased by 13.5% in heads and by 8.8% in tons.

At the same time, carcass grading scales have encouraged farmers to produce increasingly lean pigs to meet nutritional recommendations and consumer expectations (less visible fat). Progress has been rapid, leading to a standardization of the pigs produced. Simultaneously, hygiene practices, controls and traceability have been strengthened, significantly reducing health risks and reassuring consumers. Thanks to significant productivity gains at all levels of the industry, pork has also become more affordable for consumers.

In France, a quarter of pork is consumed as fresh meat and three-quarters as processed products, a distribution that changes little. More than 70% of pork dedicated to home consumption is sold in hyper and supermarkets, 13% in hard discounters and the rest in butchery and cured-meat shops, on markets, through direct sales or online.

2.2 A Stable but Fragile Balance: France Imports and Exports a Quarter of Its Production

Throughout its development, France became a major player in world markets. The country exports 27% of its production and imports about a quarter of its consumption. However, the balance between supply and demand is highly complex. Ham demand from French consumers is significant, and the national production only covers part of it (72%), compared to 127% for shoulders, 120% for loins and 128% for belly (Diot 2021). Imports of pork cuts help to fulfil the supply deficit, such as hams, and to meet the needs of the processing sector. Exports allow slaughterhouses to sell certain products for which there is a market surplus, such as loins and offal.

French pork exports are mainly destined for the European market (64% in 2023). In the intra-EU market, Italy has always been France's main customer, but the value of its products is lower there. Over the years, France has also tried to diversify its markets. An increasing share of its exports has gone to third countries, benefiting from the strong growth in demand from China. In 2023, China is still France's biggest customer.

For imports, Spain and Germany are the main suppliers to the French market. They provide mainly deboned cuts and processed meat (delicatessen, sausages and other preparations) (Fig. 17.2). French consumer expectations, business relationships along the supply chain and labour shortage in the slaughterhouses and processing plants encourage imports of value-added pork products from more competitive markets (Husson and Van Ferneij 2022). This difficult supply-demand balance spearheads a trade deficit in value (−272 million euros in 2022). In volume, the French trade balance remains in surplus (+170,000 tonnes in 2022), but it has deteriorated significantly over the last 2 years (Fig. 17.3) because of weaker demand from China.

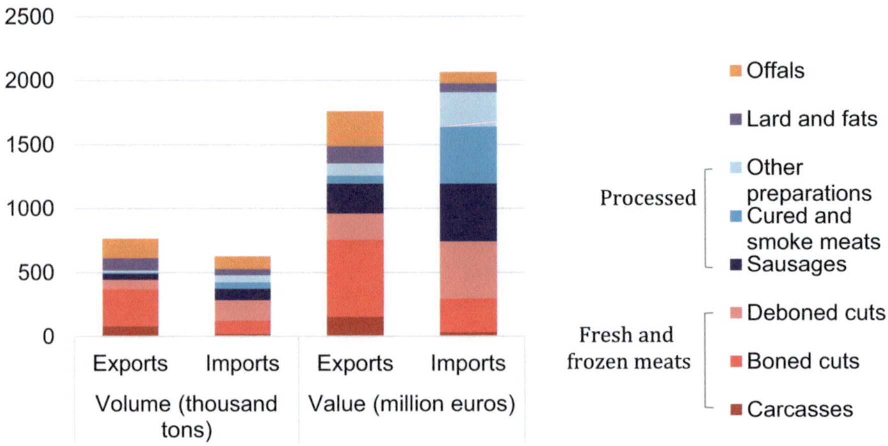

Fig. 17.2 Typology of products imported into and exported from France, by volume and by value. (Source: IFIP from customs (Eurostat))

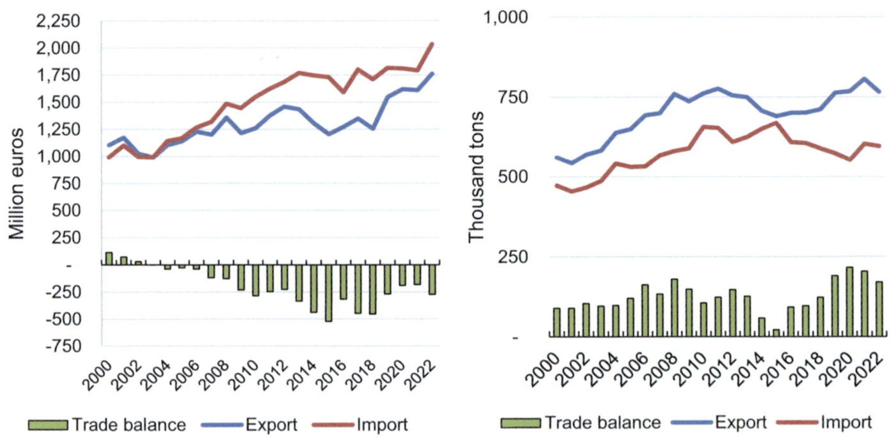

Fig. 17.3 French pork trade balance in volume (1000 tons) and in value (million euros) (excluding live animals trade). (Source: IFIP from customs (Eurostat))

2.3 Spatial Concentration: 56% of Pigs in Brittany on 6% of UAA

Brittany is France's leading livestock region, with 56% of pigs (Fig. 17.4), 33% of poultry and 20% of dairy cows in mainland France concentrated on only 6% of the country's utilized agricultural area (UAA). This colocation results from a trade-off between the positive effects of economies of scale and agglomeration (sharing marketing infrastructures, upstream and downstream industrial equipment, information and innovation) and the negative effects of environmental regulations and competition between farms in different sectors, particularly for land and labour (Roguet

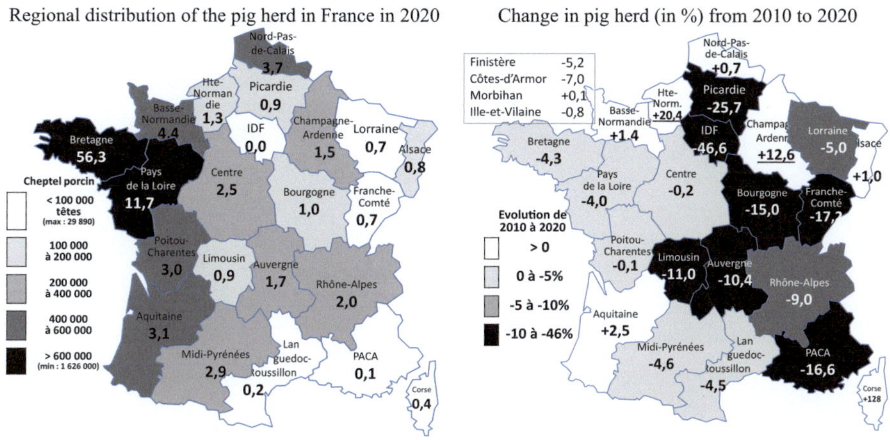

Fig. 17.4 Regional distribution of the pig herd in France in 2020 and change (in %) from 2010 to 2020. (Source: IFIP from Agreste-SSP, Agricultural census of 2020 and 2010)

et al. 2015). The strength of these economies is such that environmental regulations, however restrictive they may be, have not been able to counteract this spatial agglomeration (Gaigné et al. 2011).

3 Special Features of French Pig Farms

The number of farms with at least one pig in France fell from 597,000 in 1970 to 13,048 in 2020. This decline is mainly due to the disappearance of a very large number of small pig farms, often without sows (Fig. 17.5). In 2020, almost the entire French pig herd (13.4 million head) were raised in around 8400 farms with more than 100 pigs or 20 sows, compared with twice as many in 2000, of which 4400 held almost all the sows. Pig herd remained quite stable over the decades 1970 and 1980 (+7.1%), increased steadily in the 1990 (+26.3%) and decreased by 22.2% over 2000–2023 (Fig. 17.6). At the same time, sow population decreased by 15.2% between 1970 and 1990, increased by 17.4% between 1990 and 2000 and then dropped by 38.1% from 2000 to 2023. French pig farms are very different from those in other countries. This is a legacy of the 1970 production rationalization plan.

3.1 A Farrow-to-Finish Farm Type, on a Family Scale, Tied to the Land

France has made the choice of a farrow-to-finish production model because of its technical and sanitary skills and its economic and financial solidity, both at the farm and national level (France imports and exports very few piglets, 199,900 and 4600, respectively, out of 23 million pigs produced in 2022). More than 80% of sows and

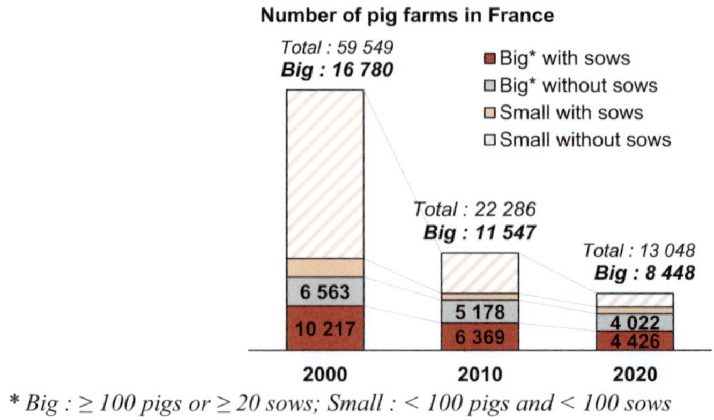

Fig. 17.5 Evolution of the number of pig farms according to their size over the last three agricultural censuses. *Big: ≥100 pigs or ≥20 sows; Small: <100 pigs and <100 sows. (Source: IFIP from Agreste, SSP, Agricultural censuses of 2020, 2010 and 2000)

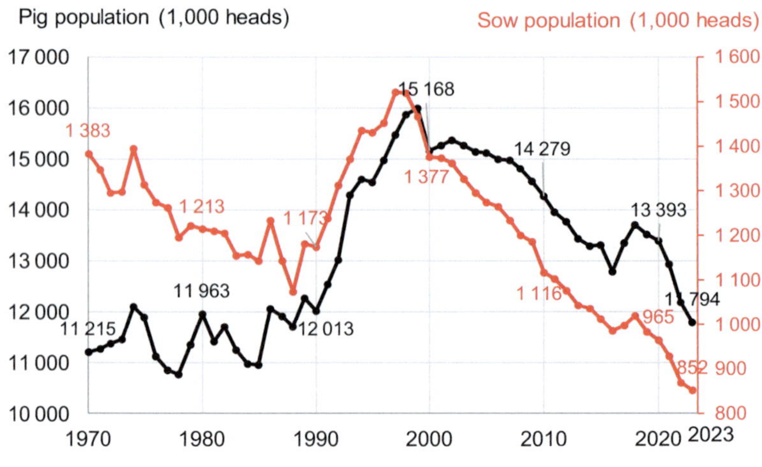

Fig. 17.6 Evolution of pig and sow population in France from 1970 to 2023. (Source: IFIP from Agreste-SSP, pig census of December)

60% of fattening places are in farrow-to-finish farms. However, this model could be weakened in the future by the choice of farmers to stop farrowing in order to reduce workloads, particularly at the end of their careers, or due to lack of labour or performance (Roguet et al. 2022). Faced with the risk of a shortage of piglets, some cooperatives are investing together with farmers in collective farrowing units (Roguet et al. op. cit.). A collective farrowing unit is a company that produces piglets (usually until weaning) on behalf of its members, who fatten them on their own farms.

Between 2000 and 2020, the average size of farms with more than 20 sows (99% of the herd) increased from around 140 to 230 sows. In 20 years, the number of farms with less than 100 sows will have decreased by a factor of 3.2 (Fig. 17.7). But

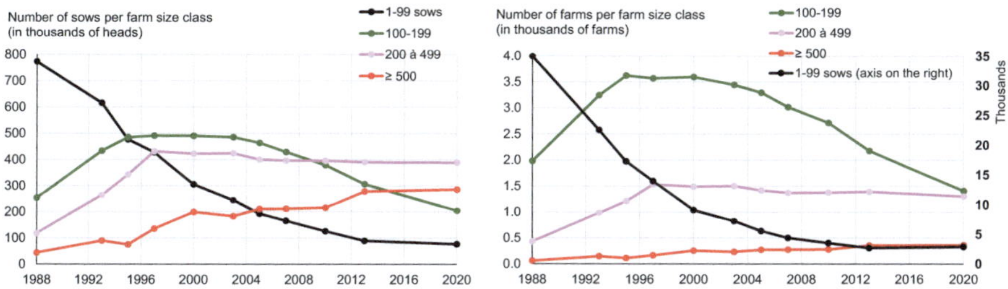

Fig. 17.7 Evolution of the distribution in number of sows and farms by size class. (Source: IFIP from Agreste-SSP, Agricultural censuses and farm structure surveys)

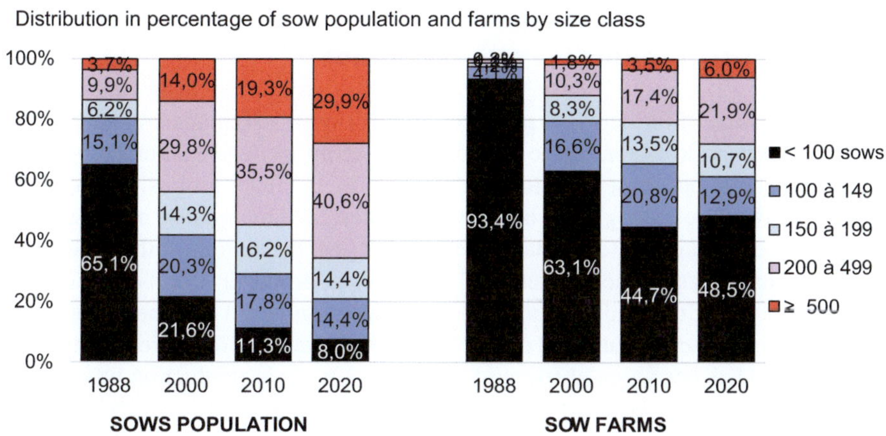

Fig. 17.8 Evolution of the distribution of sows and farms by size class. (Source: IFIP from Agreste-SSP, Agricultural censuses)

they still accounted for almost half of all farrowing farms and 8% of the sow herd in 2020. By contrast, 30% of sows were in farms with more than 500 sows in 2020, compared with 14% in 2000 (Fig. 17.8).

The typical French pig farm is specialised (with no other animal production), farrowing and fattening, with 258 sows on average, 80 ha of UAA and 3.07 full-time equivalent (FTE) of which 55% is hired in 2020 (Roguet 2023). Farms combining pigs and herbivores are declining but still accounting for 29% of pig farms and 21% of pig herds in 2020, with a smaller pig activity. Finally, mixed crop-livestock farms, where a pig unit and a crop unit make a balanced contribution to gross output, account for 20% of pig farms and 11% of pigs in 2020.

In France, structural changes on pig farms are influenced by farmers' quest for autonomy in fattening and land, for efficiency and better working conditions and for additional income (Roguet 2017; Roguet et al. 2022). The quest for autonomy leads farmers to stop selling piglets or fattening them under contract, even if this means reducing their number of sows, increasing their land area, investing in renewable energy, on-farm processing and/or in direct selling. The average UAA of French pig farms has increased from 66 ha in 2000 to 102 ha in 2020.

Ultimately, despite the trend towards specialisation and expansion, pig farms in France remain very diverse in terms of activity and size of pig units, combinations of productions, workforce and farmer profiles, according to the latest Agricultural Census of 2020. This diversity may act as an attractive lever. According to farmers, there is no ideal model of pig farming, as long as there is consistency in farrowing and fattening, enough land for feed production and environmental security and enough workforce (Roguet et al. 2022). Finally, regarding the perception by society, the French pig farming model can rely on its specific features: its family nature in terms of size, labour and capital, its autonomy in farrowing and fattening, which limits the transport of piglets, and its link to the land which makes it possible to use the cereals produced in the feed of pigs and the slurry in the fertilization of crops.

3.2 An Organization in Producer Groups

The organization of farmers into producer groups is also a significant characteristic of the French pork sector. Some groups were set up by feed manufacturers and slaughterhouses, but most were formed by farmers. Producer groups were defined in a 1962 law: "Producer groups lay down rules designed to organise and discipline production or marketing, to regulate prices and to gear the actions of their members to the requirements of the market. […] Recognised producer groups may benefit from special priorities and advantages in the allocation of aid that the government may grant for the organisation of production." The organisation of farmers into producer groups has been a great success: in 2022, 89% of pig production was carried out by farmers who were members (6870 in total) of one of the country's 32 producer groups. The initial role of producer groups is twofold: (1) to disseminate technical and genetic progress in order to improve the competitiveness of farms and better meet the needs of the downstream market, and (2) to market pigs on behalf of their members, thereby strengthening their bargaining power by pooling supply (Duflot et al. 2021).

In addition, the groups have extended their services: animal health and financing (guarantees, loans). They have also extended their vertical range of action with strategic or profitable participation, downstream (slaughtering and cutting to guarantee outlets, processing), and/or upstream (feed, genetics, livestock equipment and facilities, manure management, etc.). In addition, pig producer groups play a role in maintaining production, supporting farmers in their projects and stimulating local dynamism. In 2021, the pig AOP (Association des Organisations de Producteurs) du Grand Ouest was created and officially recognized. Ten pig producer groups from the Great West, representing 75% of production in the Great West (nearly half of national production), have decided to join forces to strengthen the market power of their member farmers, respond to market needs and the expectations of consumers and citizens, and obtain operational programmes enabling them to co-finance projects, in particular for the modernization of pig farms.

4 Societal Issues and Their Impact on the National Pig Production

Beyond the key objectives of any economic activity, i.e. to meet market demands and remain competitive, French pig production faces multiple societal issues. The environmental and health impacts of spatial and structural concentration, together with the population's growing sensitivity to how animals are treated, have led to increasingly radical criticism of certain farming methods and to strong local opposition (Roguet et al. 2020).

4.1 Four Main Themes of Debate on Livestock Farming in France

From 2014 to 2018, the research project ACCEPT ("Acceptability of livestock farming by the society in France: the cartography of the controversies, collective mobilisations and prospective"), led by Ifip, analysed the controversy on livestock farming in France, its actors, its audience and its sociological mechanisms, with the aim of helping the animal sectors to renew the dialogue and reconcile with society. The controversy on livestock farming reflects the lack of common understanding among society on the role of livestock farming and its treatment of animals (Van Tilbeurgh 2017). More specifically, the debates on animal husbandry were grouped into four themes after an analysis of the discourses collected in over 40 interviews with stakeholders in France. These four registers reflect the disagreements within society about the impact of human activities on the environment and health, the way animals should be treated, and development models (Fig. 17.9). Each register of debates covers a wide range of topics, as "micro-controversies". A controversy is a particular type of conflict which involves three categories of stakeholders: the two opposing parties (the farming community and NGOs, in this case to simplify) and a

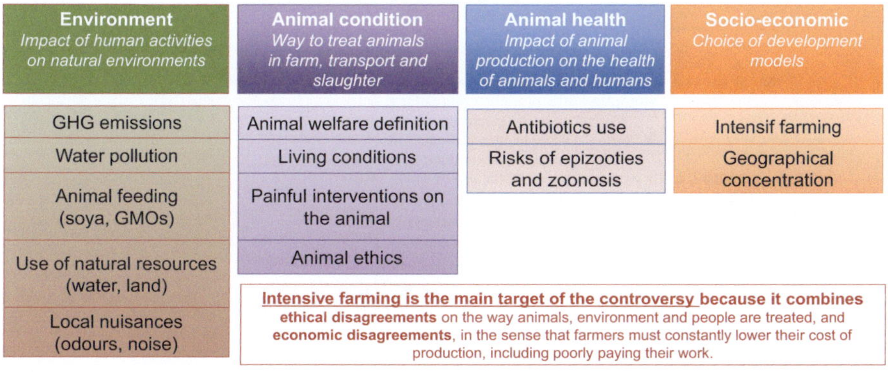

Fig. 17.9 Four main themes of debates around livestock farming in French society today. (Source: Delanoue and Roguet 2015)

public that will decide between them. The public is made up of people who have the capacity for action and sufficient knowledge to assert a point of view and enforce it (Van Tilbeurgh 2017). The challenge for each opponent is to rally the public to his cause, in particular through arguments justifying their points of view and positions. Regarding animal husbandry, the debate also takes place in front of "onlookers", an audience whose reaction will impact that of the public or insiders. The convergence of biocentric, anti-speciesism and fair-trade ethics empowers today the controversy on animal husbandry and challenges intensive livestock farming. These historic ethical guidelines used to be distinct in the past. Intensive livestock farming, however, is particularly targeted due to new ethical concerns regarding the treatment of animals, the environment and humans, as well as economic considerations. However, due to market forces, farmers are under pressure to reduce production costs, which can result in low wages for workers, in order to offer affordable meat and animal products to consumers (Van Tilbeurgh 2017).

The opinion poll conducted in 2016 among 2000 citizens representative of the population quantified the different views of French society on livestock farming (Delanoue et al. 2017, 2018). Among a list of items (Fig. 17.10), ensuring access to the outdoors for all animals and strengthening animal welfare regulations were identified by French people as the top two actions to be carried out as a priority. This statement confirms the high level of concern for animal welfare, commonly associated with free-range farming. Among the purchasing criteria proposed, the French origin of the products was most often cited first, but overall, outdoors farming came up most frequently. Moreover, 97% of persons surveyed expressed support for the labelling of "free-range" products. These statements need to be put into perspective. In another survey among 1003 people representative of the French population carried out in 2018 by Ifop on behalf of LIT Ouesterel, 84% of French people affirmed they were in favour of introducing an animal welfare label. However, only 64% of respondents were ready to pay more. When purchasing ham, they prioritized price, origin and use-by date, with the farming method ranking tenth out of 13 criteria, cited by only 1 in 8 people.

Fig. 17.10 Actions to be carried out as a priority for livestock farming according to citizens. (Source: opinion pool in 2016, 2000 people representative of the French population, Delanoue et al. (2017))

The 2016 opinion poll identified five types of French citizens. “Competitors” (10% of the French population) are individuals who are satisfied with current farming practices and the living conditions of animals on French farms. They support the intensive model, including its configuration and practices, because it allows to produce food at affordable prices, to be competitive and to export. “Progressives” (50% of the population) are individuals who want improvements in farming conditions, particularly for environmental reasons or because they are sensitive to animal welfare, without advocating for radical changes to French farming systems or their consumption practices. This population is the key audience in the controversy, as their opinions are less radical and more susceptible to being influenced. “Alternatives”, which constitute 25% of the population, advocate for the replacement of intensive livestock farming with alternative, extensive or official quality and origin systems (organic farming, Label Rouge). They do not oppose the rearing of animals for human consumption but rather advocate for a reduction in the consumption of animal products. They themselves consume relatively little. “Abolitionists” (less than 2% of the population), 80% of whom are women, consume little or very little animal products, appear very shocked by the condition of animals on farms and have a very negative opinion of all aspects of livestock farming. Finally, 3% of the population has no opinion or expectations regarding livestock farming.

These challenges must be dealt with because the debates about livestock farming have become very virulent, with intrusions into the farms and illegally produced videos of farms or slaughterhouses. Anger rose in the farming community. Consumption of meat and cold cuts is decreasing in France. The agricultural profession, and in particular the profession of pig farmer, lacks attractiveness. The generations of farmers are changing, and spouses often work outside; the agricultural world is not outside society. The challenges therefore lie in reconciling livestock farming with society, and meat with consumers, in order to halt the decline in consumption and to attract new livestock farmers, as well as in meeting the demand for animal products across diverse markets (domestic, export to Europe and third countries) and consumer profiles.

4.2 Market Segmentation in Response to Animal Health and Welfare Expectations

Since the early twentieth century, France has differentiated animal products based on their production system and quality. The Label Rouge, typically French, was born in 1960 from the desire of poultry farmers to offer a different breeding method addressing both tradition, yield and superior quality products. However, despite the guarantees offered and the efforts to raise “consumers” awareness, pig production under the official quality labels has remained underdeveloped. Less than 1% of pigs are produced under organic farming labels and only 4% under Label Rouge in 2023. As a result, numerous segmentation approaches have been developed between conventional production and official quality labels. In the ACCEPT project, around 40 of these approaches were analysed and classified into three main categories (Roguet

	Reassurance	Improvement	Breakthrough
Objectives	**Reassure** the other segment of the chain (B to B) or the consumer (B to C) on compliance with regulations, on origin and quality Promotion of **existing good practices**	**To make all farmers progress** to an "improved standard" or only some of them to segment the market. Multidimensional quality: health, animal welfare, environment... **Small steps on large volumes**	**Propose a real alternative** to the conventional livestock farming with specifications setting requirements much higher than the regulations Clear break (organic, label, straw, etc.) or gradual (multi-level labels) **Big steps on small volumes**
Collectives and Interprofessional initiatives	Origine France **Reassurance**	Guides to good farming practice	Quality and origin certification (SIQO such as Organic or Farmhouse Label Rouge)
Individual and Private initiatives		Specifications carried out upstream (cooperatives) or downstream (processors, distributors) **Différenciation and segmentation**	Specifications carried out by farmers, NGOs, distributors

Fig. 17.11 Three main types of segmentation approaches in France to meet societal demands. (Source: Roguet et al. 2018)

et al. 2018; Fig. 17.11): reassurance (no change in production methods compared to the regulatory standard), progress (towards an improved standard of production) and breakthrough (alternative modes of production) approaches. Reassurance approaches, such as the "Viandes de France" (French meats) label, aim to communicate with other actors in the industry (B-to-B) or with consumers (B-to-C) to reassure them about the regulatory standards, the origin of the product, its taste quality, etc. Progress approaches aim to encourage farmers to adopt practices that go beyond the regulatory standards. In France, the specifications proposed by cooperatives or retailers are generally multidimensional: animal health and welfare, environmental impact and producer remuneration. The most common approaches—reassurance or progress—propose changes to farming conditions, taking small steps but involving as many animals and farms as possible, while the few breakthrough approaches aim to offer a real alternative to production standards (big steps with small volumes).

The controversy over the use of antibiotics has been calmed by the fact that there has been a sharp reduction in their use in France, driven by the national EcoAntibio plan and the creation of an "antibiotic-free" supply chain. The first EcoAntibio plan (2012–2016) was a success, with antibiotic use in animals falling by 37% against an initial target of −25% (Anses-ANMV 2020). In the second EcoAntibio plan (2017–2021), the specific target of a 50% reduction in colistin exposure in the cattle, pig and poultry sectors over 5 years has also been exceeded (−64%). In the pig sector, changes in the regulation of critical antibiotics were anticipated by a joint initiative of farmers and veterinarians in 2010 calling for a "moratorium on the use of C3G and C4G" (third and fourth generation cephalosporins), a decisive action to change behaviour (Paul et al. 2023). At the same time, in the early 2010s, economic actors in the pork sector began to develop private "antibiotic-free" (AF) standards in response to demand from certain distributors and to communicate the efforts made by farmers (Roguet and Hémonic 2022). In the absence of a collective response and a legal definition of the AF claim, unlike "GMO-free", specifications have proliferated (Fig. 17.12) as a way for companies to stand out from their competitors and to improve their image. The specifications are evolving as part of a progress-based

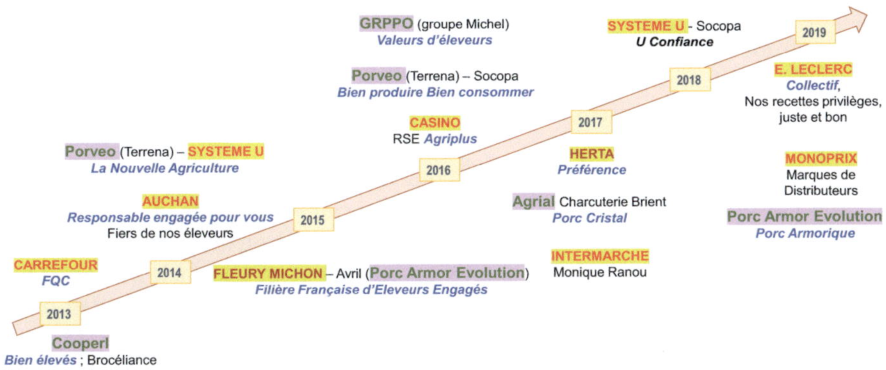

Fig. 17.12 Multiple "antibiotic-free" pig supply schemes in France. (Source: Roguet and Hémonic 2022)

approach: the AF claim "from 42 days of life" is gradually replaced or supplemented by "since birth". This criterion is always included in a more global communication approach (nutrition, environment, animal welfare, etc.). In 2020, around 15% of pigs in France were produced in the AF chain, satisfying the demand which mainly concerns cooked ham. The largest AF chains in terms of volume are "Bien élevé" of Cooperl cooperative (three million AF-produced pigs, including one million since birth, out of a total of six million), "La Nouvelle Agriculture" produced by the Terrena cooperative (200,000 pigs out of a total of 825,000) and "Engagé dans l'élevage" of Avril and the Fleury-Michon (120,000 pigs out of 1.9 million). The example of antibiotic-free production illustrates the effective and rapid impact that the structuring of the French pig industry (see Sect. 2.2) can have on an issue through incentives or actions undertaken at the level of producer groups.

4.3 Developing Animal Welfare Innovative Farming Models

In response to societal demand for a less intensive farming model, more in line with the natural behaviour of pigs, a number of research projects have been carried out in France in recent years. Let us take three examples, supported by different players (regional interprofessional organisations, academic research, applied research and cooperatives).

From 2017 to 2021, the GOPEI Occitanie project, funded by Europe and led by MIDIPORC, the pork interprofession of Occitanie region, with the support of IFIP, co-developed a pig farming model for the Occitanie region that jointly meets the expectations of farmers, industry, consumers and society. A consultation involving around a dozen farmers, supply chain stakeholders, agricultural chamber advisors, and consumers first established compromises to manage conflicting objectives — for example, improving animal welfare through bedding or open buildings versus the resulting increase in uncontrolled gas emissions. An example of antagonism was improving animal welfare by adding litter or opening buildings and increasing uncontrollable gas emissions. Then, the technical specifications for the farm were drawn up jointly, proposing larger surface areas per animal at all physiological stages (especially during fattening), differentiated spaces for various activities, limited-duration restraint (insemination, farrowing), provision of straw and, for finishing pigs, outdoor access.

The LIT OUESTEREL project (Laboratoire d'Innovation Territoriale Ouest Terre d'Elevage), led by INRAE and winner of a call for projects in 2019 (France 2030 Territoires d'Innovation), works over 8 years to improve animal health and welfare and the income, working and living conditions of livestock farmers (One Welfare). Created in 2020, the LIT OUESTEREL association brings together more than 60 partners from research (INRAE), technical institutes (IFIP, IDELE, ITAVI), producer organisations, distributors, slaughterhouses, veterinarians, animal welfare associations and innovation specialists in a process of co-construction and open innovation. A participatory design process has been carried out to develop prototypes of innovative pig breeding buildings in terms of animal health and welfare, involving breeding professionals and the general public at all physiological stages (farrowing, gestation, lactation, post-weaning and fattening) (Leroux et al. 2024). Members of the public (via a competitive game) and professionals (ideation workshop) were asked to come up with solutions to meet the expectations of the French population. According to a representative survey in 2018, these expectations regarding pig housing were, in descending order, access to the outdoors for the animals (very important for 52% of respondents), an end to individual confinement, an increase in space per animal and an increase in natural light and floor litter. Due to their ignorance of pig farming, citizens offered very few ideas (15). There were considerably more ideas from professionals (185), but they were also more conventional. Putting these ideas into practice resulted in four innovative approaches to pig housing. They all include more space per pig, more natural light, different types of floors, an enriched environment and an end to cages and mutilations (castration, teeth grinding, tail docking). The main difference is the choice of whether or not to provide access to the outdoors (yard or run). Thus, two concepts correspond to an "improved standard" (closed buildings, 1 m^2/pig for fattening, additional enrichment materials, etc.) for a mass market, while the other two represent breakthrough buildings for a niche market (runs, 1.5 m^2/pig, large groups, differentiated living areas or, in the extreme case, a single open building supplemented by mobile huts for pregnant sows). The prototypes were sketched out and then rendered in 3D for immersive virtual tours. As in the GOPEI Occitanie study (Espagnol et al. 2021), the expert multi-criteria assessment highlighted antagonisms between the improvement of animal welfare, animal health, environmental impacts and zootechnical performance.

In response to growing social concerns about animal welfare, Le Gouessant cooperative (4500 members and 800 employees in Brittany) launched Physior in 2019, an alternative rearing method based on specifications between conventional and organic. All animals have access to open-air runs, including the maternity section, and have access to straw, more floor space (1.20 m^2 per fattening pig), three living areas (bedding on solid floor, feeding, outdoor dunging) and straw as a manipulating material. Sales are guaranteed by a contract between the farmer and his group (Syproporcs) and Le Gouessant cooperative (a tripartite contract over 12 years guaranteeing the margin after feed costs with a pig price indexed to the feed price). Two fattening farms with 1200 and 672 places were built in 2020, followed by a farrow-to-finish farm with 270 sows in 2021. This farrow-to-finish farm also serves as a support for the "Elevage Porc Pilote" (EPP) research project, labeled by two competitiveness clusters (Valorial and Images et Réseaux) and funded by the Brittany Region. The pens are equipped with numerous sensors, probes and cameras to monitor technical and economic performance in real time. By 2025, Le Gouessant aims to have between 15 and 20 farms committed to the Physior approach, producing around 50,000 pigs per year, or 8% of pigs sold by Syproporcs (Chambre régionale d'agriculture de Bretagne 2021).

4.4 Meeting the Environmental Challenge Through the Circular Economy: Example of Cooperl

The transition to a circular economy is a national objective and one of the pillars of sustainable development. The French law relating to the energy transition for green growth of 18 August 2015 gave the following definition: "The transition to a circular economy aims to go beyond the linear economic model consisting of extracting, manufacturing, consuming and throwing away by calling for a sober and responsible consumption of natural resources and primary raw materials as well as, in order of priority, the prevention of waste production, in particular through the reuse of products, and, following the hierarchy of waste treatment methods, the reuse, the recycling or, failing that, the waste recovery." The Cooperl producer group has initiated this transition since the beginning of the 1990s.

With 5.4 million pigs marketed in 2022, a quarter of French production, Cooperl is the leading organisation of pork producers in the country. Born in 1966 from the association of 24 pig farmers from the Lamballe region in Brittany, the cooperative has 2950 member farmers and 7700 employees (including 4900 in Brittany). From a simple group of pig farmers at the beginning, its development followed a sector logic to cover animal nutrition, breeding materials and equipment, technical-economic support for breeders, slaughtering-cutting-processing, local distribution and the environment.

Cooperl's circular economy and industrial and territorial ecology approach consists of transforming waste from the pig industry (effluent surplus, stercorary materials, fats, meat flour, treatment station sludge) into value-added co-products while saving resources (water, energy). Its strategy has evolved from controlling waste management costs to seeking autonomy for the cooperative and reducing the carbon impact, with an increasing use of digital technology to manage the growing complexification of the interdependence of flows. Cooperl's ecological approach, embodied by Cooperl Environnement (160 employees in 2020), consists of developing cutting-edge technologies to turn the sector's environmental issues into "solutions that create value for all" (Cooperl 2019). It includes water recycling, valuation

1993	1994	1998	2003	2016	2019	2021	forthcoming
Fertival Valorization of hygienized and dried solid organic matter from slurry into natural fertilizers	**Dénitral** Technical monitoring of pig manure treatment stations. Collection of solid materials	Plant for the valorization of slaughter by-products (blood, bones, bristles, casings) into high value-added proteins and fats	**Wastewater treatment plant** Recycling of 60% of wastewater By-products recovered into biofuels and fertilizers	**Production of hot water** from the waste energy of production tools Heat supply to installations and administrative buildings	**Methanization station** Biomethane injection into the Lamballe domestic gas network Separation of digestate phases: pure recycled water, liquid fertilizer, fertilizers and amendments	**Production of biofuel** from wastewater greasy waste	**Production of microalgae** recovering waste energy and BioCO_2 from methanization Sale of pigments, lipids, polymers and finished products (food supplements, cosmetics, animal feed, plant biostimulants)

Fig. 17.13 History of Cooperl Environnement innovations

of by-products, production of renewable energy and reduction of energy consumption. Cooperl Environnement has four sub-divisions: water (clean water), ingredients (proteins, fats, microalgae), bioenergy (hot water, renewable energy, biogas, biofuel) and natural fertilizers (organic breeding materials, fertilizers and amendments) (Cooperl 2021). The main achievements of Cooperl Environnement over time are presented in Fig. 17.13.

In 1994, to limit competition for land and its increase in price, a consequence of regulatory constraints on nitrogen in Brittany, Cooperl created DENITRAL in Lamballe (Côtes d'Armor), a company specialised in the installation and maintenance of livestock effluent treatment stations (67 stations in 2023). DENITRAL collects the solid fraction produced by on-farm slurry treatment stations. It is then dried at the recovery centre CEDEV. Thermal drying uses steam energy produced by a biomass (including animal meal) cogenerator, reducing its cost. The hygienized and dried organic matter constitutes the basic ingredient used by FERTIVAL, a subsidiary of Cooperl since 1997. FERTIVAL's activity consists of formulating and packaging natural fertilizers for viticulture, arboriculture, field crops and market gardening. 50,000 tonnes are sold each year, a third of which is exported.

The steam requirement of the industrial site, 145,000 tonnes per year, is covered by 100,000 tonnes by the biomass cogenerator and by 45,000 tonnes by the COMBIOVAL tool created in 2005. Since 2008, COMBIOVAL has been recovering the flotation grease from the water in the treatment plant (greasy waste from the site's factories or partner companies) to obtain, after treatment, "COMBIOVAL grease", concentrated at 95%. This is then valued in the form of biofuel, which supplies the site's steam production tools or is used to supply the CARBIOVAL biofuel production line.

In 1998, Cooperl created a factory to transform meat and products not intended for consumption (bones, blood, intestines, bristles), collected in the cooperative's Breton slaughterhouses and cutting tools, into proteins (for pet food, aquaculture, etc.) and fats (for animal feed, oleochemistry, biofuels, etc.). The objective is to recover all of the Cooperl's non-food meats and fats towards value-added outlets, on the one hand, and to use the renewable energy produced during processing, on the other hand.

In 2003, Cooperl created on the Lamballe site a wastewater treatment plan to treat water from the site's various activities (slaughterhouse, washing station, freezing, etc.) and reduce drinking water consumption. Of the 850,000 m^3 of industrial effluent produced on the Lamballe site, 100,000 m^3 are treated by the public treatment plan. Of the 750,000 m^3 treated on site, 350,000 m^3 can be reused; the rest is released into the neighbouring river because French legislation only authorises the use of drinking water from a public station for uses in contact with food. The biological treatment process, ultrafiltration and then reverse osmosis, can recycle up to 60% of wastewater. In practice, for regulatory reasons, only 30% of treated wastewater is reused in the site's tools. The by-products are recovered and valued as fertilizer and biofuels to power the steam production tools.

Since 2016, the unavoidable waste energy from Cooperl's production tools has been recovered by a network of heat exchangers and used to heat a 4 km network of pipes covering all sites in the Lamballe area with water at 85 °C. It supplies heat to production processes and administrative buildings. This heat is used in particular to maintain the methanization unit described below at a mesophilic temperature (38°).

In 2019, DENITRAL started a methanization plant "Emeraude bioenergy" (15,000 m^3 including two digesters and a post-digester of 5000 m^3 each) to increase the valuation of organic matter from member farms and the slaughterhouse and to further reduce the need for land to spread slurry. With 156,000 tonnes per year of digestate produced and treated, it is the largest French methanization unit without a land-spreading plan (Levasseur et al. 2023). It operates with two inputs: pig faeces (38,000 t/year) from around one hundred farms equipped with TRAC technology, which separates urine (spread directly on the land) from solid matter, and swill from the slaughterhouse (70,000 t/year). These inputs are diluted by adding water from on-site recycling (evapoconcentrator). The biogas produced is purified of its carbon dioxide to become biomethane (79 million kWh per year), which is injected into the domestic gas network of the town of Lamballe (i.e. 3,100 housings). The 5,000 tons of $BioCO_2$ produced cover 75% of Cooperl's needs for anesthetizing pigs, producing cold and conservation of trays used for food packaging. The digestate is not spread but transported to FERTIVAL and separated by decantation-centrifugation into liquid and solid phases. The liquid phase is treated by evaporation to produce pure water, which is reused at the head of the methanization unit, and a liquid fertilizer (ammonium sulphate) after passing through a sulfuric acid washing tower. The solid phase is dried to produce fertilizers and amendments. Thus, 100% of the NPK from the digestate is preserved to supply the organic fertilizer production sector.

Since 2021, fatty waste has been transformed into biofuel to power the cooperative's trucks. The process consists of extracting free fatty acids then transforming them by esterification into fatty acid methyl esters, a component of biodiesel. All co-products of the process (odours, wastewater and residues) are treated on-site using existing tools. This pilot project is expected to reach industrial deployment in 2025.

Finally, an industrial production of micro-algae (e.g. spirulina) is planned, making it possible to recover waste energy and $BioCO_2$ generated from methanization, while also enabling the commercialisation of molecules with high added value (pigments, lipids, polymers) and finished products from it on the food supplements, cosmetics, animal feed and plant biostimulant markets.

5 Conclusion

Through its organisation and innovation, the French pork industry has successfully met the challenges it has faced over the years: meeting market demand in terms of both quantity and quality, reducing environmental impact, dropping the use of antibiotics, improving animal welfare, etc. Changes in practices have been driven by regulatory standards (mostly European), private standards and voluntary commitments. Cooperatives such as Cooperl, presented in this chapter, have invested in advanced circular economy approaches and industrial ecology initiatives.

Regionally, production growth in Brittany has been halted by the implementation of the Nitrates Directive, with no compensation from other regions. As a result, pig production in France has been eroding since the early 2000s, and in recent years, concerns about France's ability to maintain its pork sovereignty have been growing. Indeed, production is struggling to attract new farmers, while in 2020, one third of pigs were held by farmers aged over 55. Since 2020, the succession of health (Covid, ASF), geopolitical (war in Ukraine) and economic (inflation) crises has increased the pressure on prices and their volatility, at a time when major investments would be needed to modernise or upgrade livestock buildings. In this context, the French pork industry should be able to count on its farm model (farrow-to-finish, land-based, family-run), which is sound in technical, health, economic and social terms, its unique organisation in producer groups and its limited dependence on external markets and their uncertainties. Many challenges remain for the future, such as the demography of farmers, climate change (adaptation and mitigation), health risks, improvement of animal welfare, value sharing and decline in consumption and production.

References

Anses-ANMV (2020) Suivi des ventes de médicaments vétérinaires contenant des antibiotiques en France en 2019, rapport annuel, 97p

Bertin C, Ramonet Y (2016) Etat des lieux des bâtiments d'élevage de porcs en Bretagne chez les naisseurs-engraisseurs en 2015. Journées Rech Porcine 48:1–7

Chambre régionale d'agriculture de Bretagne (2021) Physior, un mode d'élevage entre le conventionnel et le bio. Segmentation dans la filière porc, décembre 2021. doc_num.php (chambres--agriculture.fr), 2p

Cooperl (2019) Cooperl environnement—Hautes technologies environnementales. Plaquette de présentation, 19p

Cooperl (2021) Bilan d'activité – exercice 2020, 67p

Delanoue E, Roguet C (2015) Acceptabilité sociale de l'élevage en France: recensement et analyse des principales controverses à partir des regards croisés de différents acteurs. INRAE Prod Anim 28(1):39–50

Delanoue E, Dockès AC, Chouteau A, Philibert A, Magdelaine P, Roguet C (2017) Points de vue et attentes des consommateurs et citoyens vis-à-vis de l'élevage. Une étude quantitative auprès de 2 000 personnes en France. Journées Rech Porc 49:295–300

Delanoue E, Dockès AC, Chouteau A, Roguet C, Philibert A (2018) Regards croisés entre éleveurs et citoyens français: vision des citoyens sur l'élevage et point de vue des éleveurs sur leur perception par la société. INRA Prod Anim 31:51–68

Diot V (2021) Cartographie de la consommation et de la distribution des produits du porc en France en 2018. doc_num.php (FranceAgriMer.fr), 14p

Duflot B (2015) Coûts de revient internationaux, les écarts entre bassins de production se réduisent. Tech PORC, mars-avril 2015 22, pp. 4–6

Duflot B, Oudin B, You G (2021) Place des outils d'organisation économique dans les filières et impacts sur les revenus des agriculteurs. Centre d'études et de prospective, n°168-Août 2021, 4p

Espagnol S, Rousselière Y, Guingand N, Courboulay V, Alibert L, Levasseur P, Roguet C, Durand C (2021) Quels bâtiments d'élevage porcin pour demain ? Enseignements du projet GOPEI Occitanie. « Elevages porcins: des solutions pour demain ». Editions Ifip-Institut du porc, 42p

Gaigné C, Ben Arfa N (2011) Environnement et concentration géographique des productions animales: Quels effets sur la compétitivité de l'Ouest de la France ? Projet PSDR CLAP, Grand Ouest, Série Les Focus PSDR3, Focus CLAPn°2, 4p

Gaigné C, Le Gallo J, Larue S, Schmitt B (2011) Does manure management regulation work against agglomeration economies? Evidence from French hog production. Am J Agric Econ 94:116–132

Husson E, Van Ferneij JP (2022) Le commerce intra-européen des viandes et coproduits du porc: tendances et évolutions récentes des principaux flux et identification des marchés d'intérêt. Journées Rech Porcine 54:55–30

Le Goffe P (2013) The nitrates directive, incompatible with livestock farming? The case of France and northern European countries. Notre Europe, J. Delors Institute, Policy paper, 28p

Leroux M, Kergoulay F, Rousselière Y, Girma O (2024) De la réflexion de groupe à l'immersion dans des bâtiments porcins co-construits innovants en santé et bien-être animal. Journées Rech. Porcine 56:229–234

Levasseur P, Blazy V, Langlois AS, Bourrin S, Lenouvel J, Jarrige de la Sizeranne L, Espagnol S (2023) Filières collectives de gestion des déjections animales et des digestats. Brochure réalisée dans le cadre du projet GESTE et mise en forme par L'IFIP, 30p

Paul M, Leblanc-Maridor M, Rousset N, Hémonic A, Marguerie J, le Coz P, le Normand B, Hercule J, Roguet C, Chauvin C, Belloc C, Ducrot C (2023) Réduction de l'usage des antibiotiques en filières monogastriques: état d'avancement et perspectives. INRAE Prod Anim 35(4):293–306

Roguet C (2017) Evolution des modèles d'exploitations porcines en France sous l'effet des contraintes économiques et réglementaires depuis 2008. Journées Rech Porcine 49:277–282

Roguet C (2023) Evolution, diversité et typologie des exploitations porcines en France: enseignements du recensement agricole de 2020, comparaison aux recensements de 2010 et 2000. Journées Rech Porcine 55:1–6

Roguet C, Hémonic A (2022) Les filières « porcs élevés sans antibiotiques » en France: caractéristiques, atouts, limites et perspectives. Projet européen ROADMAP. Journées Rech Porcine 54:321–326

Roguet C, Massabie P, Gourmelen C, Douguet G (2007) Les bâtiments d'élevage de porcs en France: quel besoin d'investissement au regard de leur état ? TechniPorc 30(6):1–7

Roguet C, Gaigné C, Chatellier V, Cariou S, Carlier M, Chenut R, Daniel K, Perrot C (2015) Spécialisation territoriale et concentration des productions animales européennes: état des lieux et facteurs explicatifs. INRA Prod Anim 28(1):5–22

Roguet C, Marion C, Magdelaine P, Dockes AC (2018) Les démarches mises en œuvre par les filières animales en France en réponse aux attentes sociétales en termes de bien-être animal: typologie et perspectives NESE n° 44, Décembre 2018, pp. 7–35

Roguet C, Delanoue E, Dockès AC, Magdelaine P, van Tilbeurgh V, Grannec ML (2020) ACCEPT – Acceptabilité des élevages par la société en France: cartographie des controverses, mobilisations collectives et prospective. Innovations Agronomiques 79:315–329

Roguet C, Lecuyer B, Le Clerc L (2022) Trajectoires individuelles d'évolution des élevages de porcs en France: mécanismes, déterminants et perspectives de la concentration structurelle. Journées Rech Porcine 54:43–48

Van Tilbeurgh V (2017). Enjeux sociétaux et adaptation des filières aux controverses sur l'élevage. Douzièmes Journées de la Recherche Avicole et Palmipèdes à Foie Gras, Tours, 5 et 6 avril 2017. Hal-01580518

A Dichotomous Approach to Extensive and Intensive Animal Farming 18

Kai Frölich

The present agriculture strategy ensures the food supply for the constantly growing global human population and increases biodiversity in agricultural landscape. This strategy is currently facing one of its greatest challenges: On the one hand, ever higher productivity must be achieved in order to feed an increasing number of people; on the other hand, the agricultural landscape has a special responsibility for the preservation of biodiversity. In Germany, the arable land represents around 50% of the total land area. Furthermore, in developed countries like Germany, production costs are rising to ensure modern, animal-friendly husbandry, which is demanded by many consumers. Almost everyone agrees that a transformation of current production methods is necessary. But how this should be done and at what speed it must be accomplished is the subject of controversial debates.

If we consider the planetary boundaries defined by Rockström et al. (2009), it is apparent that the extinction of species, in developed countries with highly industrialised agricultural production, especially over-fertilization (i.e., the excess supply of nutrients such as phosphorus and nitrogen), is situated in a high-risk range for the regional ecosystems (see Figs. 18.1 and 18.2).

According to the German Federal Ministry for the Environment, 70% of habitats in Germany are in a critical ecological condition, as are 60% of species (BMUV 2020). Currently, about 80% of biodiversity losses can be ascribed to effects resulting from intensive agriculture (Leopoldina 2020). Nothing has influenced nature like agriculture, and the current way of food production is harmful to the planet. It causes the loss of biodiversity as well as the net loss of individual wildlife (Bar-On et al. 2018), destroys ecosystems, and increasingly exacerbates water scarcity.

K. Frölich (✉)
Zoo, Ark Warder e.V., Warder, Germany, and Department of Biology, Institute of Biology and Chemistry, University of Hildesheim, Hildesheim, Germany
e-mail: kfroelich@arche-warder.de

B. Grabkowsky, T. Blaha (eds.), *Intensive Livestock Production in Transition*,
https://doi.org/10.1007/978-3-031-97872-2_18

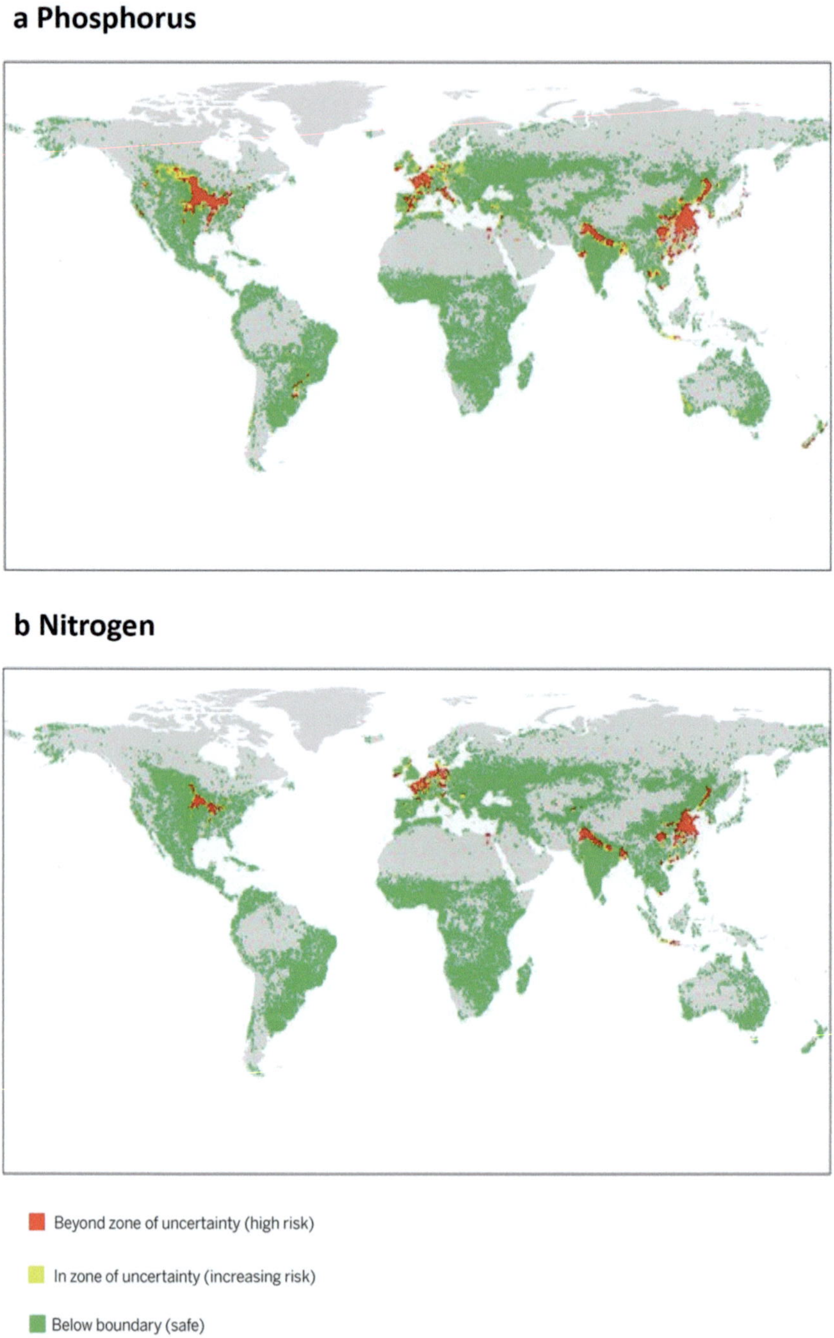

Fig. 18.1 The global distributions and current status of the control variables for (**a**) biogeochemical flows of phosphorus and (**b**) biogeochemical flows of nitrogen. (Modified according to Steffen et al. 2015)

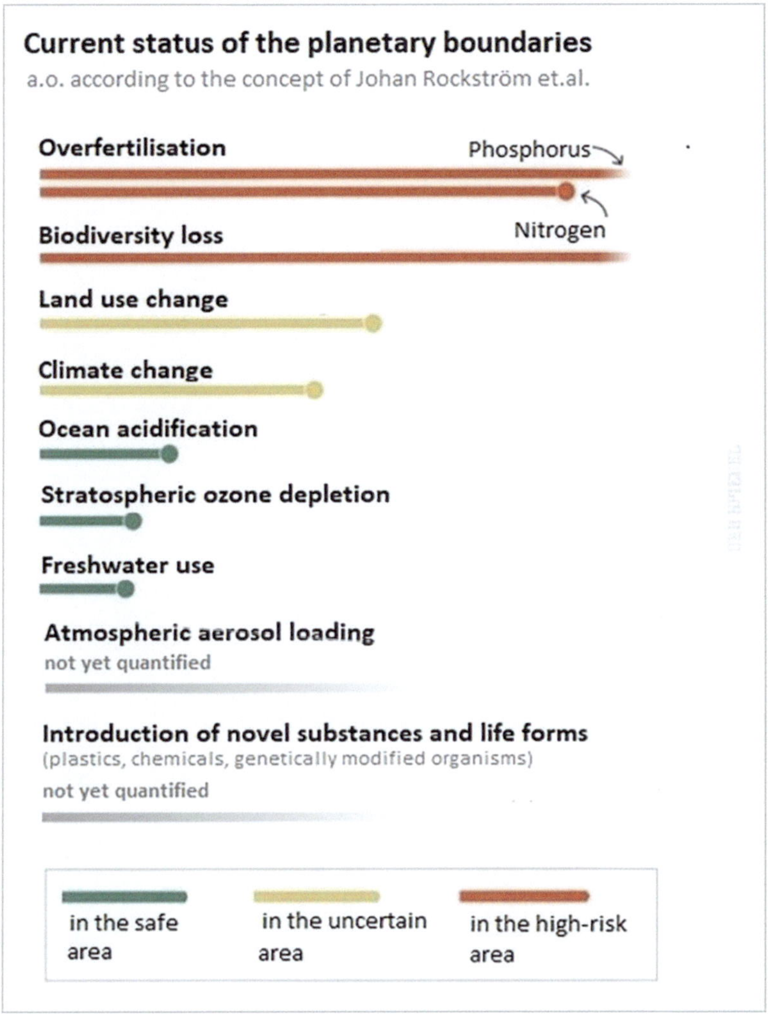

Fig. 18.2 Current status of the planetary boundaries modified according to Grolle (2021)

A distinction can be made between direct and indirect drivers of the dramatic ecosystem changes caused by human activities, as shown in Fig. 18.3 (IPBES 2019). The direct drivers, namely human interventions, and the active destruction of habitats, are the primary contributors (e.g., land–/sea-use change, direct exploitation of organisms, climate change, pollution, and invasive alien species) to ecosystem degradation. They are a result of an array of underlying societal causes. These causes can be demographic (e.g., human population growth), sociocultural (e.g., consumer behavior), economic (e.g., international trade), technological and political nature, or related to conflicts or epidemics (IPBES 2019). A list of the main indirect drivers is presented in Fig. 18.3.

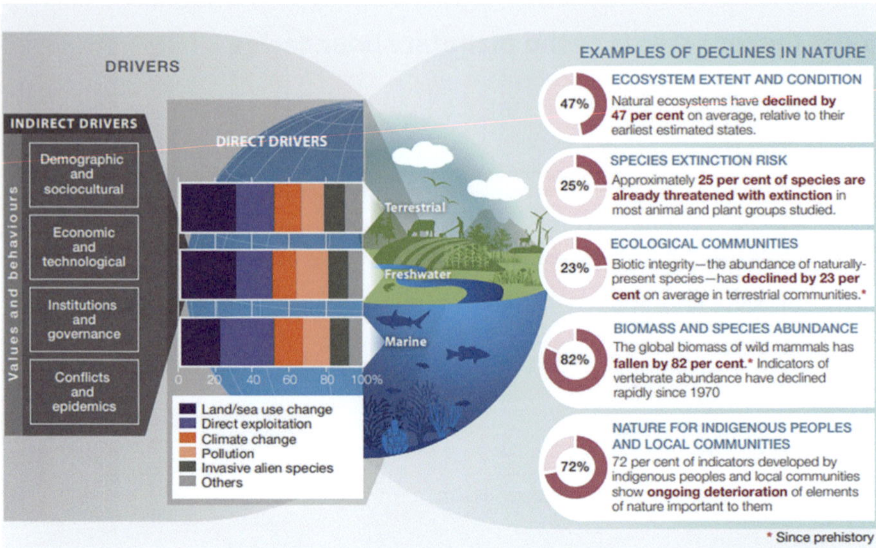

Fig. 18.3 Examples of global declines in nature, emphasising declines in biodiversity, that today and in the past have been caused by direct and indirect drivers of change. (Source: IPBES 2019)

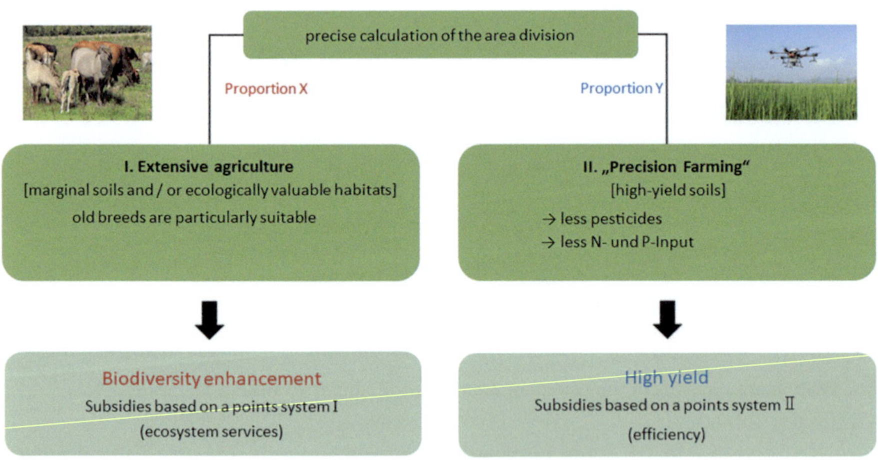

Fig. 18.4 Dichotomous land use concept that combines extensive and intensive agricultural production

Most researchers agree that agriculture must become more ecologically sustainable in the future and that new perspectives are needed to develop sustainable agricultural systems. The following approach, which is the essential objective of this article, is considered to provide such a possibility for sustainable agriculture:

Farming is based on a dichotomous concept (Fig. 18.4) that combines, in different areas, extensive land use with intensive agriculture ("Precision Farming"). Both approaches enable more ecologically sustainable and natural land management, as well as efficient agricultural production.

Of central importance is the identification and assignment of suitable areas that can be used either for intensive "Precision Farming" or for extensification of agricultural land use. A main factor for the allocation of areas to either category is the productivity of the respective soils expressed by soil quality points (arable land grade). Areas with soils that have low soil points and/or ecologically valuable habitats with already a relatively high biodiversity can be used for extensive agriculture, while those with high soil points can be used for intensive "Precision Farming." The following quality classes are distinguished: low 18–35 points, medium 35–55 points, high 55–75 points, and very high 75 and 100. Typically, soils with less than 50 points are considered marginal for agricultural production.

The use of extensive forms of land management in marginal areas can contribute to the preservation of biodiversity in these areas and thus provide various ecosystem services. This approach offers a promising path to reconcile food security as well as biodiversity preservation.

Old, robust livestock breeds would be particularly suitable for this form of landscape management. This management should also integrate modern wildlife management systems to get maximum conservation and ecological benefits.

On the other hand, in highly productive areas, modern ways of intensive but resource-conserving agriculture, such as "Precision Farming," could be carried out that enable high yields and efficiency with low pesticide and fertilizer use. High-performance livestock breeds and digitally based agricultural management practices can be used in these areas (Fig. 18.4).

State support instruments (e.g., subsidies) should no longer be linked primarily to the size of the cultivated areas but instead be awarded for management practices that enable ecological and sustainable forms of agriculture. In the case of "Precision Farming," subsidies would also be granted with the help of an adapted points system that primarily takes into account the degree of efficiency, including low pesticide input as well as low fertilizer (nitrogen and phosphorus) consumption.

Extensive agricultural management practices enable the conservation of open landscapes with high biodiversity and the simultaneous production of selected high-quality animal proteins. Different site-adapted livestock species can be used for this purpose (Fig. 18.5).

However, some high-performance breeds are currently also kept in extensive farming, although they are not always optimally suited for this purpose. This leads to specific problems because these breeds were bred for maximum performance under intensive husbandry conditions and are, for example, dependent on high-energy forage. In contrast, old breeds are particularly suitable for extensive land management and should in the future therefore be increasingly included in a management concept of near-natural large-scale pasture landscapes (Frölich and Kopte 2024).

A further advantage of the use of old breeds is the maintenance of genetic diversity within the respective species that enables them to cope with future challenges, such as changing market demands or production conditions, as well as resistance or

Fig. 18.5 Examples of site-adapted sheep breeds: (**a**) White polled heath (peaty pastures). (© Ark Warder_ Lisa Iwon), (**b**) Tiroler Steinschaf (alpine area). (© Ark Warder_ Lisa Iwon), and (**c**) North Ronaldsay sheep (salty seaweed). (© Ian Caldwell_https://en.wikipedia.org/wiki/North_Ronaldsay_sheep#/)

higher immunity to new diseases or emerging environmental stress factors (Simianer 2005).

In the context of the historical development of livestock farming, breeds have been selected that are adapted to different local conditions. These so-called autochthonous breeds are part of the cultural and historical heritage of a particular region and have contributed significantly to the development of Europe's typical regional landscapes. Breed names such as Rhön sheep (optimal adaptation to the climatic conditions of low mountain ranges), Heidschnucke (special suitability for lean fodder), Hinterwälder cattle (hard claws and very sure-footed) (Fig. 18.6), or European water buffalo (broad hooves and good suitability for wetlands) indicate the importance of these animals for the respective landscapes (Hoffmann 2013; Frölich and Kopte 2024).

Fig. 18.6 Examples of different autochthonous breeds: (**a**) Rhön sheep. (© Helge Busch-Paulick, Wikipedia, http://de.wikipedia.org/wiki/Benutzer:Grand-Duc) (**b**) German grey heath. (© Pixabay_Makalu) (**c**) Hinterwälder cattle. (© Ark Warder_ Lisa Iwon), and (**d**) European water buffalo. (© Kai Frölich)

These old livestock breeds would thus be able to occupy a key position in the context of extensive grazing systems, as they have the following characteristics (Frölich et al. 2018):

1. Optimal adaptation to local sites.
2. Efficient utilisation of nutrient-poor forage and independence from energy-rich forage.
3. Resistance to local weather conditions and climatic peculiarities (temperature adaptations).
4. Stress resistance.
5. Uncomplicated births and rearing of offspring.
6. High immunological competence against pathogens (Ballweg et al. 2016).

According to these criteria, robust, old breeds can be kept on marginal pastures potentially all year round and cope better with environmental influences than high-performance breeds (Fig. 18.7).

However, due to the lower stocking densities and the associated lower productivity, the economic efficiency of the respective areas is of course reduced, and the

Fig. 18.7 Extensive year-round grazing with Heck cattle and Koniks. (© Gaby Wiefel-Keysberg)

degree of self-sufficiency in food production decreases. This could result in an increase in imports and thus a decrease in international competitiveness. Upstream and downstream sectors of agriculture, such as stable construction technology or feed producers, are also influenced.

Nevertheless, extensive grazing, as a holistic management system, promotes the structural diversity of grazed areas through a high ratio of grazing, trampling, and recreation. The treading and feeding effects of the old livestock breeds resemble the stochastic impacts of former megaherbivores. This leads to an increase in the number of ecological niches and moreover has a positive effect on the establishment of complex biocenoses. The result is a significantly higher overall species diversity of extensively managed areas.

Furthermore, the principle of "agroforestry" offers an additional interesting approach to more sustainable land use. It describes a form of land management in which the production of arable crops (silvoarable systems) or animal husbandry (silvopastoral systems) is combined with the cultivation of trees and shrubs. Although this combination of agriculture and forestry requires increased labor and maneuvering requirements, it also offers numerous advantages: such as the creation of habitats for many animal species and thus greater biodiversity, improvement of soil structure and health through providing natural fertilizer and increased shading, as well as reduced erosion risk, and greater potential for carbon sequestration.

Another form of extensive land use is the so-called "energy pasture landscape." It combines renewable energy production, advancement of biodiversity, and climate protection. For this purpose, large, extensively cultivated year-round pasture landscapes are combined with elevated free-standing photovoltaic systems. The high-mounted solar panels allow livestock such as sheep, cattle, horses, and donkeys to be used as landscape managers (Fig. 18.8). In this way, besides the production of

Fig. 18.8 Agrophotovoltaics: (**a**) vertical solar modules. (© Next2sun AG), (**b**) ground-mounted solar panels with sheep. (© BayWa r.e. AG), and (**c**) high-mounted solar panels with livestock, e.g., donkeys as landscape managers. (AdobeStock ©Jenson)

renewable energy, high-quality meat can also be produced in an ethically appropriate way. For this purpose, the approval of renewable energies would have to be linked to the preservation of biodiversity, with the aim of providing 20% of the pasture landscapes (outside of adjacent protected areas) with photovoltaic systems.

However, agriculture is undeniably also faced with the challenge of not only preserving biodiversity but also feeding the rapidly growing world population. In order to make this possible to a sufficient extent while still being as environmentally friendly as possible, so-called "Precision Farming" respectively, "Precision Livestock Farming" should be used on particularly high-yield areas.

"Precision Farming" offers the possibility of realising precise and site-adapted management with a high level of mechanisation and automation. Site-specific sensing, sampling, and managing allow farmers to treat a field as a heterogeneous entity (Finger et al. 2019). There is no longer a general management of large areas, but the different soil zones and yield capabilities within a field are considered. With the help of various monitoring technologies, such as sensors, drones, and satellites, data are collected on plant and soil quality or weather conditions, among other things (Fig. 18.9). These are evaluated, and precise application maps are created. This enables site-specific and plant-specific input of fertilizers and crop protection products precisely in the quantity in which they are needed at a specific location.

A specific type of "Precision Farming," "Smart Livestock Farming," involves the management of individual animals through continuous real-time monitoring of health, welfare, production and reproduction, and environmental impact (24 h a day, 7 days a week). Real-time field data, known as biosignals, are collected from the animal through the use of sensors located on or around the animal (e.g., temperature measurements, GPS position, acceleration data, etc.), real-time image or sound analysis (Berckmans 2017). This allows one to react quickly in an individual-specific way to any problems as and when needed.

These types of "Precision Farming" can provide both a reduction of production costs and negative environmental effects of agricultural production, as well as an increase in yields. From an environmental perspective, "Precision Farming" can provide various positive effects. Through the targeted use of inputs, it reduces waste, thereby cutting both private and environmental costs such as those caused by agrochemical residuals (Finger et al. 2019).

However, introduction of new technologies is often accompanied by large investments. This capital-intensive approach to farming unfortunately excludes many farmers from technological progress and could widen the gap between small farmers and high-tech large-scale farms. Therefore, the use of "Precision Farming" methods should, in the beginning, be limited to large farms, while small farms should focus more on organic farming and organic products.

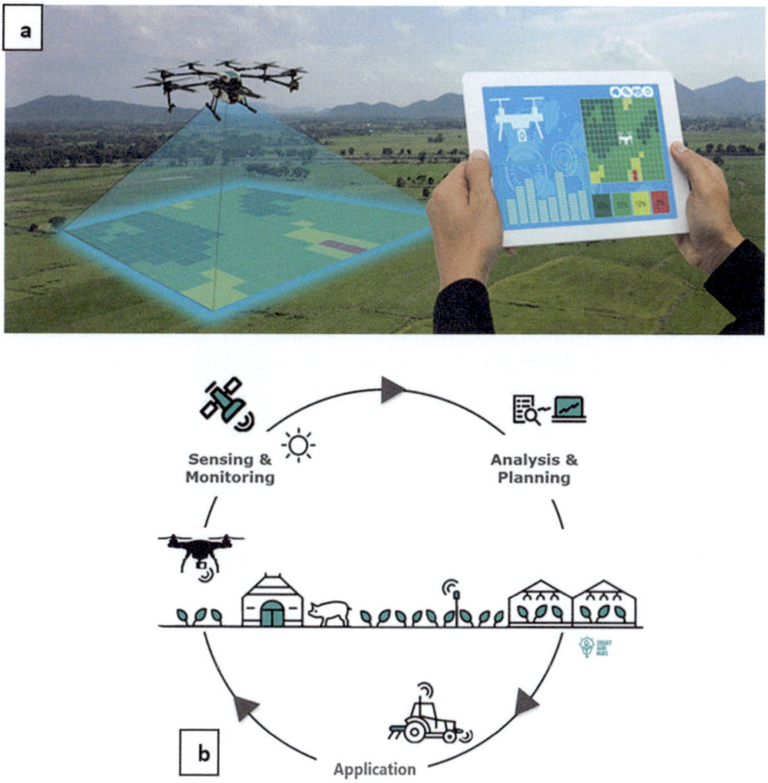

Fig. 18.9 Precision farming. (**a**) © AdobeStock © Monopoly919 and (**b**) Modified according to Smartagrihubs (2021)

1 Conclusions

While this suggested dichotomous approach provides a reasonable solution for the direct drivers of ecosystem change, the social, economic, and political framework conditions, i.e., the indirect drivers, must not be ignored.

A drastic change in consumer behavior is particularly crucial for the successful transformation of current production methods. This is primarily associated with a reduction in the production of animal-based foods. In the case of Germany, the annual per capita consumption of meat must be reduced from around 52 kg to a maximum of 30–40 kg, among other things through higher prices ("pricing-in the ecological footprint").

Direct payments (subsidies) from the European Common Agricultural Policy can also make a valuable contribution as an adjusting screw (according to the principle of "public funds for public services").

In addition, based on the EU biodiversity strategy, about 30% of Germany would have to be converted into fauna-flora-habitat (FFH) areas by 2030, and about 3–4% of the country's area would have to become national parks (currently 0.6%).

Although the ecological limits have already been reached in some areas, the situation is not hopeless. Theoretically, it could be possible to feed the ever-increasing population without endangering the stability of the planetary system. However, it requires the efforts of each one of us, above all politicians, in order to achieve this goal.

Interestingly, the current post-Brexit agriculture policy in England shows that farmers can also play a certain role as "land healers" in order to protect natural areas and are not only paid through agricultural products by the consumers. The new (post-Brexit) approach to subsidising farmers is that those farmers who produce high-quality agricultural food are paid by the consumers, and those farmers who identify themselves as "stewards of the lands" (Bittner 2024) are paid by public subsidies. This is a good example that a more complex, ecology-orientated agriculture subsidy policy may replace the current EU policy, which is mainly characterised by subsidies linked to the size of the cultivated areas. In addition, this also provides farmers with an important perspective for the future. Taken together, this post-Brexit agriculture policy in England represents practically the above-described dichotomous land management approach.

Competing Interests The author has no conflicts of interest to declare that are relevant to the content of this chapter.

References

Ballweg IC, Frölich K, Fandrey E et al (2016) Comparison of the immune competence of Turopolje, German landrace × Turopolje, and German landrace × Pietrain pigs after PRRSV vaccination. Vet Imm 174:35–44. https://doi.org/10.1016/j.vetimm.2016.04.003

Bar-On YM, Phillips R, Milo R (2018) The biomass distribution on earth. PNAS 115(25):6506–6511. https://doi.org/10.1073/pnas.1711842115

Berckmans D (2017) General introduction to precision livestock farming. Anim Front 7(1):6–11. https://doi.org/10.2527/af.2017.0102

Bittner J (2024) Ohne Krawall. DIE ZEIT, 06/2024

BMUV (2020) Bundesministerium für Umwelt, Naturschutz und nukleare Sicherheit (BMUV), Bundesamt für Naturschutz (BfN) (eds.) (2020) Die Lage der Natur in Deutschland. Berlin/Bonn. https://www.bfn.de/sites/default/files/BfN/natura2000/Dokumente/bericht_lage_natur_2020.pdf

Finger R, Swinton SM, El Benni N et al (2019) Precision farming at the nexus of agricultural production and the environment. Annu Rev Res Econ 11(1):313–335. https://doi.org/10.1146/annurev-resource-100518-093929

Frölich K, Kopte S (2024) Alte Nutztierrassen. Komplett neu überarbeitete Auflage, Cadmos Verlag GmbH

Frölich K, Ballweg I, Sorg D et al (2018) Sind alte Haustierrassen "robuster"?: Untersuchungen zur immunologischen Kompetenz. DTBL 66(12):1644–1648

Grolle J (2021) Spiegel-Gespräch: Erdsystemforscher warnt vor neuer Heißzeit. DER SPIEGEL, 20/2021, 102–106

Hoffmann I (2013) Adaptation to climate change – exploring the potential of locally adapted breeds. Animal 7(2):346–362. https://doi.org/10.1017/S1751731113000815

IPBES (2019) Summary for policymakers of the global assessment report on biodiversity and ecosystem services of the intergovernmental science-policy platform on biodiversity and ecosystem services. In: Díaz S, Settele J, Brondízio ES et al (eds) IPBES secretariat, Bonn. https://doi.org/10.5281/zenodo.3553579

National Academy of Sciences Leopoldina (ed) (2020) Global Biodiversity in Crisis – What can Germany and the EU do about it? Discussion Nr. 24, Halle (Saale). https://www.leopoldina.org/uploads/tx_leopublication/2020_Diskussionspapier_Biodiversitaetskrise_web.pdf

Rockström J, Steffen W, Noone K et al (2009) Planetary boundaries: exploring the safe operating space for humanity. Ecol Soc 14(2):1–33

Simianer H (2005) Decision making in livestock conservation. Ecol Econ 53(4):559–572. https://doi.org/10.1016/j.ecolecon.2004.11.016

Smartagrihubs (2021) The fundamentals for digital farming are being laid out. https://www.smartagrihubs.eu/latest-news/The-fundamentals-for-digital-farming-are-being-laid-out. Accessed 03 Nov 2023

Steffen W, Richardson K, Rockström J et al (2015) Planetary boundaries: guiding human development on a changing planet. Science 347(6223). https://doi.org/10.1126/science.1259855

Potential of Digital Technologies for Transforming Livestock Production

19

Henning Müller and Joachim Hertzberg

1 Introduction: Digital Technologies and Livestock Production

The United Nations identifies a triple planetary crisis: climate change, pollution, and biodiversity loss, including land degradation (United Nations 2022). A key challenge for humanity is to increase the sustainable production of biomass while at the same time taking steps to tackle this crisis and reduce the environmental footprint. Radically new processes and technologies need to be identified, developed, and optimized to meet the growing demand for biomass for food, feed, energy sources, and biomaterials while minimizing negative environmental impacts (FAO 2017).

However, the focus must not only be on crop production but also on livestock production, as farm animals play an essential role in the biomass cycle, in which far more non-edible biomass is produced than edible biomass. Sustainable food production cannot succeed without integrating animal production into the agricultural material cycle (Windisch 2022). In the context of transforming the agri-food system toward more sustainability, livestock farming will remain of great importance. The use of digital technologies, especially AI in livestock farming, supports this transformation (Fuentes et al. 2022). In view of the increasing global demand for animal products and the challenges associated with a growing population and limited

H. Müller (✉)
German Research Center for Artificial Intelligence (DFKI), Cooperative and Autonomous Systems, Osnabrück, Germany
e-mail: henning.mueller@dfki.de

J. Hertzberg
Institute of Computer Science, Osnabrück University, German Research Center for Artificial Intelligence (DFKI), Cooperative and Autonomous Systems Group, Osnabrück, Germany
e-mail: joachim.hertzberg@uos.de

B. Grabkowsky, T. Blaha (eds.), *Intensive Livestock Production in Transition*,
https://doi.org/10.1007/978-3-031-97872-2_19

resources, the industry is facing the need for transformation. In this context, digital technologies offer promising potential to improve efficiency, productivity, and sustainability in livestock farming.

2 Digital Transformation: Challenges

The introduction of digital technologies in livestock farming faces several challenges that may hinder their widespread use and, consequently, stand against the transformation. These challenges relate not only to technological aspects but also to infrastructural, social, and legal framework conditions.

2.1 Broadband Expansion

A fundamental obstacle to the widespread implementation of digital technologies in rural areas, at least in Germany, is the lack of broadband expansion and inadequate Internet connectivity. Many farms, especially remote stables, still only have limited access to fast broadband internet. One of the most important political measures from the perspective of farmers in Germany is the expansion of mobile phones and broadband in rural areas (Bitkom 2022). Without a reliable internet connection, data cannot be transmitted in real time, which limits the effectiveness of many digital applications. This is especially true for applications that require continuous monitoring of animal health and environmental conditions. This is not only of great importance for the agricultural and food industry but also for the future prosperity of rural areas in general.

2.2 Training and Expertise

Another obstacle lies in the education and training of farmers in the use of digital technologies. There is room for improvement when it comes to competencies in digital farming. Many farmers may not have the necessary technical expertise to implement and maintain complex systems. A high level of media expertise is too often confused with digital expertise. None of this is surprising, because:

- The education and training system in agriculture is a small part of the overall education system. According to the FIDA study ("Programme for International Digital Skills Assessment") from 2022 (Fujitsu 2022), relevant IT skills are still taught to a very limited extent in a European comparison, especially in Germany.
- The occupational profile for agriculture was last reorganized in 1995 (BGBI. Year 1995, Part I, p. 168 ff.). Aspects from the field of digital agriculture do not (yet) play a role in the framework curriculum. As a result, digital agriculture is an option in the curriculum, but not a must. It is important to note that trainers and teachers also need training: Complex systems explained incorrectly do not become correct just because you talk about them yourself.

2.3 IT and Cybersecurity

An important aspect in introducing and operating digital technologies in livestock farming is IT and cybersecurity. The use of networked, digital technologies, for example, in automated feeding and milking technology or ventilation control, creates dependencies. A failure caused by a cyber-attack at the wrong time can threaten the lives of the animals.

Moreover, the protection of data and ensuring the privacy of farmers is an important aspect. Sensor data, animal health data, and other sensitive information must be stored, transmitted, and processed securely to prevent misuse or data loss. At the same time, a clear legal framework must be created to regulate the handling of animal data and ensure data protection. To identify weaknesses and initiate targeted measures, it is necessary to carry out a comprehensive assessment of the IT and cybersecurity of agricultural companies and their process chains (Bitkom 2023).

2.4 Investments in Digital Technologies

Investment in the digital transformation of agriculture is undoubtedly crucial to meeting the challenges of modern agriculture (Bertschek et al. 2024). Nevertheless, the question arises as to what extent these investments could represent a hurdle. According to a study, many farmers see the high investment costs for digitalization as an obstacle (Bitkom Research 2022). Accordingly, most farmers would like to receive financial support when implementing digital technologies in agriculture. These findings suggest that barriers to investment are not only of a financial nature but are also based on uncertainties and information deficits.

Overall, overcoming these barriers requires a multidimensional approach that includes financial support, education, and advice.

3 State of the Art: Digital Technologies

Digital technologies offer a wide range of solutions to the transformation of livestock farming and are making rapid progress (Sturm et al. 2020). A key role is played by sensors and the Internet of Things (IoT), which make it possible to continuously monitor and analyze animal health, behavior, and environmental conditions. By integrating sensors into barns, watering systems, and grazing areas, for example, farmers can record important parameters such as temperature, humidity, air quality, water, and feed intake as well as animal movement patterns in real time.

This data not only provides insights into the individual well-being of each animal but also enables early detection of diseases or stress symptoms. For example, deviations in feeding behavior (Sivamani et al. 2016) or motion patterns (Fan et al. 2022) can indicate health problems that require immediate attention.

Especially in dairy farming, smart applications are becoming increasingly important (Nleya and Ndlovu 2021; Bernhardt 2023). For example, the milking robot makes it possible to milk cows efficiently and gently without the need for human intervention by using various sensors that enable the automation of the milking process and precise recording of important data, including milk quantity, milk quality, and cow behavior (Haidn and Harms 2016). This data is recorded and analyzed in real time to adapt the milking process to each individual cow. By integrating digital technologies such as image processing, the milking robot can identify the animals and monitor their health and milking speed. Of course, the focus of digital technologies in dairy farming is not just on the milking process. Sensor systems are finding their way into more and more areas and are increasingly networked with each other; the flow of information and data is increasingly viewed holistically and extends from the calf to the heifer to the dairy cow (Regler et al. 2022).

Newer approaches also go beyond the cowshed and increasingly address the grazing of dairy cows, so that the data technology catechism is becoming ever more closed. GPS tracking and geofencing systems are used to monitor animal movements on pastureland in livestock farming. By fitting animals with GPS collars or ear tags, farmers can track the location and movement patterns of their herds. This information can be used to develop optimized grazing strategies that maximize resource utilization while conserving grazing land (Aquilani et al. 2022). In new applications in the Netherlands, attempts are being made to use a virtual pasture fence and actively control grazing using a combination of different stimulus elements (Reimink 2024).

In addition, automatic feeding systems are revolutionizing livestock farming. Intelligent feeding robots can provide optimized feed rations based on specific needs and health (Oberschätzl-Kopp et al. 2016).

Another area in which digital technologies have made great strides toward practical usability is the automated recognition and sorting of livestock. Image processing algorithms make it possible to identify and classify individual animals based on their characteristics, such as size or weight (Pan et al. 2023).

Overall, existing digital technologies in livestock farming offer a variety of innovative solutions that can help improve the efficiency, productivity, and sustainability of the industry. From precise monitoring to control of production processes, digital technologies based on various forms of sensor data are already in use for "personalizing" animal welfare aspects. These technologies offer a promising way to meet the challenges of the future.

4 AI Technologies in Livestock Production

The scope of applications of AI in livestock production includes animal identification, behavior monitoring (feeding behavior, aggression behavior), health monitoring (disease detection, activity recognition), performance prediction, and production technology in different livestock species (e.g., cattle, pigs, poultry) (Bao and Xie 2022). The meta-study showed that the spectrum of AI methods used is wide and includes, for example, machine learning, machine vision, neural networks, and deep learning.

In pig farming, AI can be used to reliably detect upcoming births of piglets, to monitor the course of birth, and to significantly reduce the subsequent crushing of piglets by the sow. In the later fattening phase, tail biting should be detected, and the causes analyzed (Opderbeck et al. 2024). Among other things, computer vision-based activity recognition is used for this purpose (Hesse et al. 2023). Individual animal identification is particularly important to be able to reliably select culprit animals AI is also used to predict growth trends better and to select animals by weight before slaughter (Wang et al. 2022). Even if the use of AI differs in cattle and poultry farming due to the differences between the species, the above-mentioned central topics are identical. By integrating machine learning algorithms, data can be automatically analyzed to detect patterns and anomalies that may not be obvious to human observers. These systems can also monitor feed intake and identify irregularities that could indicate potential health problems (Fig. 19.1).

The AI application field of livestock farming is particularly interdisciplinary. More recent approaches not only incorporate expertise from the research fields of computer science, veterinary medicine, environmental sciences, mechatronics, etc., but also take ethical issues into account (Neethirajan 2023).

To summarize, AI, put on top of more classical digital technologies in livestock production, has the potential to enable far-reaching improvements in animal welfare and holistic monitoring, so that livestock products, such as meat, milk, or eggs, cannot simply be sold as physical goods. Instead, AI methods can be used to market future animal welfare and efficiency indicators. The research field of Explainable AI is likely to become increasingly important in the context of livestock production. Farmers have a high level of responsibility when dealing with animals, and the traceability of decision-making in AI approaches is not only important to increase acceptance. Rather, it enables humans to retain their responsibility towards farm animals.

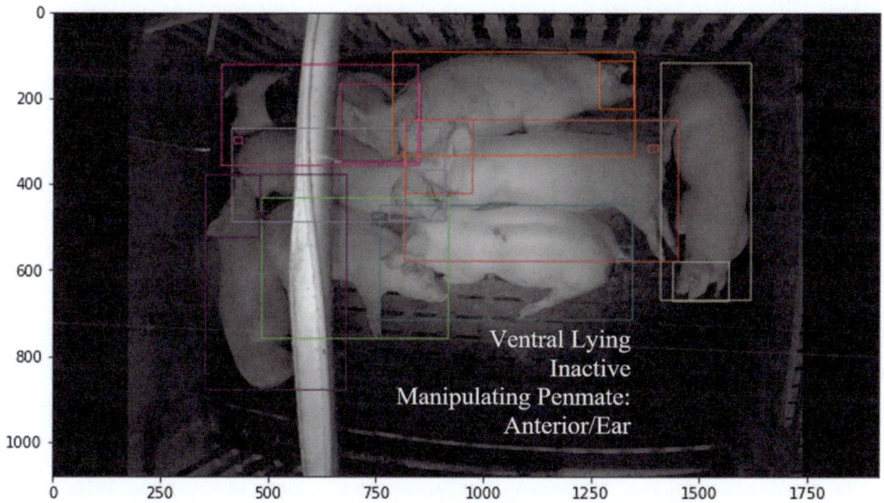

Fig. 19.1 Example image of the continuous recording with generated bounding boxes and exemplary activity labels for the blue bounding box. (Hesse et al. 2023)

5 Conclusion: The Role of Digital Technologies and AI in the Future of Transforming Livestock Farming

The integration of digital technologies into livestock farming has the potential to improve efficiency, productivity, and sustainability. However, certain challenges, including some of a non-technical character, currently stand in the way of the widespread use of these technologies. The lack of network expansion, the need for comprehensive education and training, and ensuring data protection and data security are key factors that need to be addressed to be able to realize the full potential of digital technologies in livestock farming. Through targeted measures and political support, these challenges can be overcome to enable sustainable and future-proof animal husbandry.

Acknowledgments The DFKI Niedersachsen is supported by the Ministry of Science and Culture of Lower Saxony from funds of the program zukunft.niedersachsen of the Volkswagen Foundation.

References

Aquilani C, Confessore A et al (2022) Review: precision livestock farming technologies in pasture-based livestock systems. Animal 16(1):100429. https://doi.org/10.1016/j.animal.2021.100429

Bao J, Xie Q (2022) Artificial intelligence in animal farming: a systematic literature review. J Clean Prod 331(129):956. https://doi.org/10.1016/j.jclepro.2021.129956

Bernhardt H (2023) Technik in der Rinderhaltung. In: Frerichs L (Hrsg.): Jahrbuch Agrartechnik 2022. Braunschweig: Institut für mobile Maschinen und Nutzfahrzeuge, S. 1–18. https://doi.org/10.24355/dbbs.084-202303081049-0

Bertschek I, Bünstorf G et al (2024) Transformative F&I-Politik am Beispiel neuer Technologien in der Landwirtschaft und sozialer Innovationen. Wirtschaftsdienst 104(4):225–229. https://doi.org/10.2478/wd-2024-0064

Bitkom (2022) High-Tech im Stall, KI auf dem Acker: Digitalisierung sichert Zukunft der Landwirtschaft. https://www.bitkom.org/Presse/Presseinformation/Digitalisierung-sichert-Zukunft-der-Landwirtschaft#_. Accessed 28 Apr 2024

Bitkom (2023) Cyberresilienz in der Landwirtschaft. https://www.bitkom.org/sites/main/files/2023-08/bitkom-whitepaper-cyberresilienz-in-der-landwirtschaft.pdf. Accessed 28 Apr 2024

Bitkom Research (2022) Die Digitalisierung der Landwirtschaft. https://www.bitkom.org/sites/main/files/2022-05/Bitkom-Charts%20Landwirtschaft.pdf. Accessed 28 Apr 2024

Fan B, Bryant R, Greer A (2022) Behavioral fingerprinting: acceleration sensors for identifying changes in livestock health. Multidiscip Sci J 5(4):435–454. https://doi.org/10.3390/j5040030

FAO (2017) The future of food and agriculture: drivers and triggers for transformation. https://doi.org/10.4060/cc1024en

Fuentes S, Gonzalez V et al (2022) The livestock farming digital transformation: implementation of new and emerging technologies using artificial intelligence. Anim Health Res Rev 23(1):59–71. https://doi.org/10.1017/S66252321000177

Fujitsu (2022) Program for International Digital Skill Assessment. https://www.fujitsu.com/de/about/resources/news/press-releases/2022/20220404.html/. Accessed 28 Apr 2024

Haidn B, Harms J (2016) Milchviehhaltung. In: Landwirtschaftliche Tierhaltung. Grundlagen zur landwirtschaftlichen Tierhaltung, -fütterung und -zucht. 14. Auflage 2016. ISBN 978-3-8186-0788-3

Hesse L, Fruhner M et al. (2023) Computer-Vision-basierte Aktivitätserkennung bei Schweinen. 43. GIL-Jahrestagung, Resiliente Agri-Food-Systeme. Bonn: Gesellschaft für Informatik e.V.. PISSN: 1617-5468. ISBN: 978-3-88579-724-1. pp 101–112. Osnabrück. 13.-14. Februar 2023

Neethirajan S (2023) The significance and ethics of digital livestock farming. Agric Eng 5(1):488–505. https://doi.org/10.3390/agriengineering5010032

Nleya S, Ndlovu S (2021) Smart dairy farming overview: innovation, algorithms and challenges. In: Smart agriculture automation using advanced technologies. Springer, Singapore, pp 35–59

Oberschätzl-Kopp R, Haidn B et al. (2016) Effects of an Automatic Feeding System with dynamic feed delivery times on the behaviour of dairy cows. CIGR-AgEng conference, Aarhus, Denmark. https://www.researchgate.net/publication/306002772_Effects_of_an_automatic_feeding_system_with_dynamic_feed_delivery_times_on_the_behaviour_of_dairy_cows. Accessed 05 Jan 2025

Opderbeck S, Grossklos-Bumbalo J, Gallmann E (2024) Technik in der Schweinehaltung. In: Frerichs L (Hrsg.): Jahrbuch Agrartechnik 2023. Braunschweig: Institut für mobile Maschinen und Nutzfahrzeuge S. 1–12. https://doi.org/10.24355/dbbs.084-202401171550-1

Pan Y, Zhang Y et al (2023) Low-cost livestock sorting information management system based on deep learning. Artif Intell Agric 9:110–126. https://doi.org/10.1016/j.aiia.2023.08.007

Regler F, Ziegler K et al (2022) Closing data-gaps between calves and cows – conceptualization of a specified sensor system for data acquisition in calf and heifer husbandry. International Cnference on Agricultural Engineering, 22.-23.11.2022, Berlin. In: VDI (Hrsg.): AgEng- Land. Technik 2022, Düsseldorf: VDI Verlag GmbH 2022, ISBN: 978-3-18092406-9, S. 387–396

Reimink A. (2024) Pilotprojekt in den Niederlanden: Virtueller Weidezaun für Milchkühe, In: top agrar online, https://www.topagrar.com/rind/news/virtuell-eingezaeunt-20002275.html. Accessed 28 Apr 2024

Sivamani S, Kim HG et al (2016) A system for early disease detection based on feed behaviors in livestock. Adv Sci Lett 22(9):2391–2395. https://doi.org/10.1166/asl.2016.7846

Sturm B, Nasirahmadi A et al (2020) Smart Livestock Farming – Eine Bestandsaufnahme. Züchtungskunde 92(6)

United Nations (2022) What is the triple planetary crisis? https://unfccc.int/news/what-is-the-triple-planetary-crisis. Accessed 28 Apr 2024

Wang S, Jiang H et al (2022) The research Progress of vision-based artificial intelligence in smart pig farming. Sensors 22(17):6541. https://doi.org/10.3390/s22176541

Windisch W (2022) Brauchen wir in Zukunft überhaupt noch Nutztiere? In: Nutztierschutztagung 2022, pp 13–20. https://raumberg-gumpenstein.at/jdownloads/Tagungen/Nutztierschutztagung/Nutztierschutztagung_2022/3n_2022_Windisch.pdf. Accessed 28 Apr 2024

Making Sustainability our Business: Entre- and Intrapreneurial Activities to Develop new Pathways for a Sustainable Agricultural Sector

20

Jantje Halberstadt and Sophia Kollhoff

1 Introduction: Agribusiness as Usual, or Rocking the Stage?

The agri-food sector is experiencing an unprecedented, urgent need for sustainability transformation. Research and practice are currently stressing the immense challenges agribusiness faces, describing the industry as fundamentally unsustainable (Wojtynia et al. 2021). Authors criticize the unsustainable, exploitative (over)use of soil, water, and the atmosphere (Gavrilescu 2021). Underestimating the value of functional biodiversity, agri-food systems are depicted as overemphasizing cost-price reduction, leading to lock-in effects that hinder transformation (Oliver et al. 2018; Plumecocq et al. 2018; Springmann et al. 2018). While this aspect at first glance appears to consider at least some level of economic sustainability, long-term success means it will require an intelligent integration of all pillars of sustainability. Neglecting or underestimating social and ecological sustainability might deliver positive short-term economic returns will lead to negative economic sustainability effects from a longer-term perspective. It is also important to note the potential agriculture has for improving sustainability: agricultural innovations can improve productivity and financial outcomes, as well as environmental and social sustainability. Agriculture also plays a significant role in, e.g., poverty reduction and improved health as it tackles global hunger, malnutrition, and food insecurity (Röling 2009; Sanga et al. 2023). It is therefore important to generate a common understanding of sustainability from different perspectives, as well as a holistic view of sustainability activities and their interconnections with reciprocal effects.

J. Halberstadt (✉) · S. Kollhoff
Economics of Sustainability, Institute of Sustainability Transformation in Rural Areas, University of Vechta, Vechta, Germany
e-mail: jantje.halberstadt@uni-vechta.de; sophia.kollhoff@uni-vechta.de

B. Grabkowsky, T. Blaha (eds.), *Intensive Livestock Production in Transition*,
https://doi.org/10.1007/978-3-031-97872-2_20

Understanding and valuing the overall effects of creating sustainable agri-food supply chains and agribusiness management will create the basis for innovations across various agricultural value chains. These are a precondition for the urgently needed transformation of agriculture, including the ways in which we produce, process, trade, and consume food (Gaitán-Cremaschi et al. 2019; Willett et al. 2019). Innovation is one of the most significant ways to increase business efficiency and strengthen sustainability. According to traditional thinking, innovation mainly plays an important role in light of how it can increase profit and strengthen economic development (Voeten et al. 2009; Pol and Ville 2009). Innovation is often associated with developing new products, services, processes, or technologies motivated by the maximization of profits (Bock 2012). However, social and ecological innovation is receiving increased academic and practical attention; they are considered especially promising when understanding both as having a positive impact on economic goals. Global trends in the development of key economic, social, and environmental factors in combination with increasing technological progress create both new opportunities and development threats for companies (Ahmad et al. 2024). For agribusiness, the influence of these factors is even more significant due to their peculiarities, as well as the changes occurring in this field (Sokolova and Litvinenko 2020).

On the one hand, due to their vast economic interests, size, and number of actors involved, transforming these kinds of complex systems needs to be a collective effort (European Commission 2020; Wojtynia et al. 2021). On the other hand, individual expertise and creativity are required to develop innovative solutions for sustainable change. These innovations are not only able to build the basis for transformative movements but can also create competitive advantages for the company, regional, and national levels with long-term effects on growth and stability. We, however, currently identify the risk of resting on former innovativeness and success laurels, leading to lock-in effects that make companies stick to established structures instead of reacting to global change. This is in essence a wake-up call from innovation fatigue that is poised to transform the agri-food industry into a desirable future system, but which currently neither sufficiently meets societal needs nor adheres to ecological boundaries or limits (Bui et al. 2016; Wanzenböck et al. 2020).

Our chapter will first focus on the importance of a common understanding of sustainability and how its specific aspects are relevant for and within agriculture. We will then carve out some challenges and potential for deriving innovative solutions to sustainability-related issues, examining potential changes in society (e.g., consumer behavior) and technology (e.g., digitalization and artificial intelligence), as well as their interrelations (e.g., changing work conditions). While considering innovative entrepreneurial activities, we also see untapped potential within existing institutions, which is why we add a brief introduction to the intrapreneurship concept that underlines the potential found, especially when including employees as experts in sustainable transformative innovative agriculture.

2 Sustainability in Agriculture: Challenges and Potentials

Agriculture and sustainability are interwoven, particularly in light of sustainability's roots in traditional agricultural practices that have aimed to maintain soil fertility and productive capacity for generations (Pretty 2008). Sustainable agriculture is today a cornerstone of global efforts to address pressing challenges such as climate change, food security, and rural development. It encompasses a holistic approach integrating ecological health with economic profitability and social well-being. This integration is essential for ensuring that agricultural systems not only sustainably produce food, materials, and fuel but also enhance ecosystem services, conserve natural resources, and promote rural livelihoods (FAO 2021). The adoption of sustainable agricultural practices is guided by principles that emphasize efficient resource use, biodiversity conservation, soil and water management, and resilience to environmental and economic shocks (FAO 2018). It represents a shift toward regenerative farming methods that mitigate environmental impacts while enhancing productivity and profitability for farmers (Pretty et al. 2011).

Sustainability as defined by the World Commission on Environment and Development can be seen as meeting "the needs of the present without compromising the ability of future generations to meet their own needs" (WCED 1978). This commission's work and its report have triggered growing interest in sustainable development in research and practice ever since. It is an essential part of an integrated perspective consisting of at least three key development dimensions: economic, environmental, and social (Mignaqui 2014), with some authors even stressing the relevance of additional pillars such as cultural development (Dessein et al. 2015). This integrated perspective counts as being extremely important: underestimating one perspective can potentially hinder the overall goal of sustainability. Neglecting the interconnected nature of economic, ecological, and social aspects not only leads to a flawed understanding of current agricultural systems but also reduces the chances of assessing and identifying successful innovation for transforming agribusiness. This makes it important to stick to an integrative view, specifically relevant in/for agriculture.

Realistically, carving out interconnections makes it necessary to first understand how agricultural activity works within various fields, and how it influences economic, social, and environmental areas while encompassing all of their positive and negative effects. Positive effects are often stressed when it comes to ensuring food security and battling hunger (Goodland 1997; Johnston et al. 2014). Progress in agricultural research over the last 100 years has led to vast (technological) innovations capable of achieving the productivity increases needed to feed the world's populations (Welch and Graham 1999). Current research, however, stresses how global and local food systems still fail to deliver adequate quantities of high-quality, healthy food, which holds specifically true for underprivileged communities (Woodhill et al. 2022). We also see rising concerns regarding how agricultural activity has negative effects on the environment, such as soil degradation and water pollution, with unforeseeable consequences for global (food) security. Agribusiness, as a result, has the duty of providing subsistence to society while at the same time minimizing the damage that this can cause to the environment (Fusco et al. 2023).

Social sustainability also plays an important, often underestimated role in agriculture. While some authors argue that social dimensions are often underestimated in the sustainability discourse (Binder and Feola 2013; De Olde et al. 2016), this may hold specifically true for agriculture, where its effects on the environment are the primary focus. Additional social effects, however, are in fact seen regarding, e.g., human and labor rights when it comes to conditions in the food production process, or farmers' ideas about life quality (Janker and Mann 2020). We also find gender aspects being integrated into research on sustainability in agriculture (Desiderio et al. 2022; Dutta and Thaker 2020). There are also consequences for society in general, in terms of, e.g., providing healthy/unhealthy food (Clark et al. 2019). As there are various key social aspects to consider in agriculture, this dimension urgently needs more attention in research and practice.

All sustainability effects, however, have to be considered in relation to the entirety of the value chain, and can differ from the different stages of production and related stakeholders. We see enormous differences when looking at different food production branches. For example, the topic of animal welfare and animal health comes into play when looking at (intensive) livestock farming (Hashem et al. 2020; Lovarelli et al. 2020). This is most often directly related to the breeding/production of animals, as well as the related effects on, e.g., wild animals caused by both livestock and arable or mixed farming. Here we stress the need to consider the effects of/on animals in a structured way, adding this as an additional pillar of sustainability in agriculture in light of how it remains unclear whether this should be categorized as a social or environmental impact.

In sum, the complexity of sustainability-related factors and effects in agribusiness makes it important to evaluate sustainability in agriculture. "Assessment of agricultural production sustainability is a multi-dimensional task that involves the quantification of various economic, social, and environmental indicators at different scales and uncertainty levels" (Bartzas and Komnitsas 2020: 223). Acknowledging this complexity helps structure and deliver a true picture of agribusiness's impacts on sustainability, even while it at the same time nevertheless complicates the measurements thereof due to its extensive variety of factors and interconnections. The numerous tools and techniques suggested by research and practice still fail to fully meet the needs of a realistic and practical method (Bathaei and Štreimikienė 2023). Most of this work only focuses on one aspect, primarily in the field of environmental sustainability (e.g., Sabiha et al. 2016), with more recent work dealing with considering and measuring social sustainability in agriculture (Janker and Mann 2020; Sannou et al. 2023). As criticized by Welsh and Graham (1999: 2) 25 years ago, we still fail to evaluate sustainability effects along value chains and do not properly "link farming-system sustainability to food-system sustainability as a whole." It is also important to note that there is no one-size-fits-all solution, meaning that an intensive individual evaluation process is required. Digital solutions may help provide more detailed information to customers and other stakeholders. With this being said, however, unsecured or uncontrolled transparency accompanies the risk of data misinterpretation and/or the risk of giving competitors access to data that can create competitive advantages for them.

As we agree with the growing importance of sustainability in agriculture while addressing the complex effects and current challenges being faced here, it is a great time to understand all of this as an opportunity for developing innovative solutions, from new processes and products to new farming techniques and methods, as well as advanced measurement and secured transparency measures. As a result, we focus on entrepreneurial thinking and acting as the basis for sustainability-driven change via innovation.

3 Entre- and Intrapreneurship: Innovation for Sustainable Transformation

3.1 Entrepreneurship and the Need for Sustainability Innovations

Twenty years ago, authors criticized entrepreneurial activity and innovation as not receiving sufficient academic attention, while at the same time stressing its practical importance: "While agricultural economists have not placed much emphasis on entrepreneurship and innovation, it has become a priority with policymakers and is a critical aspect of value-added agriculture" (Knudson et al. 2004: 1330). This holds especially true when it comes to sustainable transformation. As traditional agricultural systems are seen as having a negative impact on the environment due to their, e.g., carbon emissions, they need to be sustainably executed using the right kinds/amounts of production and resources (Balsalobre-Lorente et al. 2019; Ben Amara et al. 2020). The key to this lies in entrepreneurship, with entrepreneurial thinking and acting leading to innovative solutions—in the best case even leading to economic, social, and environmental improvements.

Agriculture's extensive environmental impact means that eco-innovation will play an increasing role in research and practice, connecting problem-solving and entrepreneurial opportunities to address environmental issues while at the same time supporting business performance (Haller et al. 2023; Mylan et al. 2015; Ociepa-Kubicka and Pachura 2017). Social and social-ecological innovation and transformation are also important topics in agriculture (Sanga et al. 2023). Social innovation can be applied to the enhancement of agricultural production processes, while agriculture contributes to social sustainability as it, e.g., improves the role of (disadvantaged) coproducers in the value chain and fair property use; achieves the fair treatment of employees, including fair wages; and promotes community-supported agriculture or social farming (Alberio and Moralli 2021; Egartner et al. 2020; Léger-Bosch et al. 2020; Mert-Cakal and Miele 2022). It is additionally important not to underestimate the interplay and dynamics of social-ecological systems. Olsson and Galaz (2012: 223) stress "the need for radical shifts to new approaches that can enhance the fit between human and biophysical systems and improve the capacity of ecosystems to generate services for human well-being."

Even as we find an increasing body of work on entrepreneurship and agriculture, the untapped potential remains for fostering entrepreneurial thinking and acting to produce innovative solutions for the current challenges found in agri-food chains (Jia 2021).

The time is ripe to fully recognize the potential of entrepreneurial orientation and action as important factors for transforming agribusiness. This includes increasing academic work that specifically elaborates upon the identification and seizing of entrepreneurial opportunities for sustainable agribusiness, as well as the factors motivating and empowering entrepreneurial action. And it also needs agribusiness to react. Stakeholders, including policymakers, owners, and company management, as well as those involved in agri-food value chains, must be aware of changes in innovation and the risks of ignoring the need to innovate and/or resting on past or current successes while being outperformed by more attentive, proactive competitors. From a practical perspective, this also makes it important to include entrepreneurial orientation and the teaching of sustainability and entrepreneurial skills and competencies to farmers and other agri-food commodity managers, perhaps even helping them transform into value-added entrepreneurs (Knudson et al. 2004). This is the basis for enhancing the agricultural sector's ability to overcome traditional, commodity-oriented agriculture toward an innovative, sustainability-oriented one that also takes into account changing consumer needs while building competitive advantages.

3.2 Intrapreneurship and the Potential of Employees for Innovation

Intrapreneurship, commonly defined as entrepreneurship within existing firms or organizations (Antoncic and Hisrich 2001), is gaining increasing attention in research and practice. Employees are an important factor in creating innovation within firms (Kraus et al. 2019), which will be specifically and increasingly slated to meet the current challenges in agribusiness (Fu et al. 2021). According to Renkema et al. (2021: 2), employee innovation is understood as the "bottom-up emergence of innovation based on employees' ideas." It can therefore be seen as a form of intrapreneurship. The concept of employee-driven innovation, however, shifts the focus beyond R&D departments to the initiation of ideas by individual employees who are not tasked with innovation, at least in terms of their job descriptions. This means that innovation includes employees at all levels, most notably in light of how they know the company via their personal insights and are aware of its strengths and challenges from an internal perspective. They can contribute with specific expert knowledge from various fields.

Fostering employees' innovative behavior will require entrepreneurial motivation and competencies. Studies show that this is achieved via a concept called *individual entrepreneurship orientation* that integrates various aspects such as risk-taking, proactivity, and innovativeness (Bolton and Lane 2011; Teles et al.

2021). The literature stresses the importance of individual entrepreneurship orientation for intrapreneurial activity. Kraus et al. (2019), for example, found that employees with a high level of IEO are more likely to work on explorative activities, even though their study could not detect an IEO influence on exploitative activities (two phases of intrapreneurship being explorative and exploitative activities). Although disputed, the overall literature suggests a positive correlation between the quality of human capital and innovation performance (Alpkan et al. 2010), human capital and productivity-enhancing entrepreneurship orientation, and the ability to become an entrepreneur or exploit entrepreneurial opportunities (Douglas and Fitzsimmons 2008; Parker 2009). Despite some research dealing with the antecedents of employees' innovative work behavior, it remains unclear which factors (such as overall work environment and satisfaction or individual incentives) influence this kind of behavior, as well as the entrepreneurial orientation serving as its basis (Khalili 2016; Le 2020; Lee et al. 2020).

This is why we call for acknowledging the so-far untapped potential of employees in innovation processes within agricultural practice that support individual entrepreneurship orientation and entrepreneurial competencies. Combining entrepreneurial motivation and expertise with further competencies relevant to sustainable development may lead to even more transformational innovations. This is why we also point to current work on training future agents of sustainable change (Rieckmann and Barth 2022), emphasizing the potential for fostering sustainable innovation from within; this will require both entrepreneurial and sustainability competencies as its transformation basis (Halberstadt et al. 2019). First studies even point at sustainability orientation as having a positive influence on innovative behavior (e.g., Dean and McMullen 2007; Kuckertz and Wagner 2010). Research should furthermore elaborate on the role of employees in transformative intrapreneurship, providing both theoretical and especially empirical insights into how to enable and motivate employees at all levels to adopt entrepreneurial behavior.

4 Summary and Outlook

The transformation of agriculture toward more sustainability is not only necessary but an enormous opportunity as well. Innovations that take into account economic, ecological, and social aspects can lead to long-term success and stability. Entrepreneurial and intra-entrepreneurial action is crucial to driving this transformation forward. Companies should strive to harness the innovative power of their employees while at the same time integrating sustainable practices. Instead of agricultural organizations evaluating employee performance by primarily measuring their productivity and quality of work (Vrabcová and Urbancová 2023), we hope to motivate practitioners and researchers to focus on the largely untapped potential of employees' contributions to transformative innovation development. Transforming agribusiness will require a rethink and a willingness to question existing structures to break new ground. Only in this way can agriculture meet the challenges of the future while simultaneously contributing to a more sustainable world. Players shouldn't wait until others innovate; they instead should seize the opportunity to cocreate an agri-food future as they develop successful sustainable businesses based on an intelligent, holistic sustainability perspective.

References

Ahmad H, Yaqub M, Lee SH (2024) Environmental-, social-, and governance-related factors for business investment and sustainability: a scientometric review of global trends. Environ Dev Sustain 26:2965–2987. https://doi.org/10.1007/s10668-023-02921-x

Alberio M, Moralli M (2021) Social innovation in alternative food networks. The role of co-producers in Campi Aperti. J Rural Stud 82:447–457

Alpkan L, Bulut C, Gunday G, Ulusoy G, Kilic K (2010) Organizational support for intrapreneurship and its interaction with human capital to enhance innovative performance. Manag Decis 48:732–755. https://doi.org/10.1108/00251741080000697

Antoncic B, Hisrich R (2001) Intrapreneurship: construct refinement and cross- cultural validation J Bus Ventur 16:495–527. https://doi.org/10.1016/S0883-9026(99)00054-3

Balsalobre-Lorente D, Driha OM, Bekun FV, Osundina OA (2019) Do agricultural activities induce carbon emissions? The BRICS experience. Environ Sci Pollut Res 26:25218–25234

Bartzas G, Komnitsas K (2020) An integrated multi-criteria analysis for assessing sustainability of agricultural production at regional level. Inf Process Agric 7(2):223–232

Bathaei A, Štreimikienė D (2023) A systematic review of agricultural sustainability indicators. Agriculture 13(2):241

Ben Amara D, Chen H, Hafeez M (2020) Role of entrepreneurial opportunity identification factors in the eco-innovation of agribusiness. Bus Strategy Dev 3(4):435–448

Binder CR, Feola G (2013) Normative, systemic and procedural aspects: a review of indicator-based sustainability assessments in agriculture. In: Marta-Costa AA, Soares da Silva ELDG (eds) Methods and procedures for building sustainable farming systems. Springer, Dordrecht, pp 33–46

Bock BB (2012) Social innovation and sustainability; how to disentangle the buzzword and its application in the field of agriculture and rural development. Stud Agric Econ 114(2):57–63

Bolton DL, Lane MD (2011) Individual entrepreneurial orientation: development of a measurement instrument. Education and Training 54(2/3): 219–233. https://doi.org/10.1108/00400911211210314

Bui S, Cardona A, Lamine C, Cerf M (2016) Sustainability transitions: insights on processes of niche-regime interaction and regime reconfiguration in agri-food systems. J Rural Stud 48:92–103

Clark MA, Springmann M, Hill J, Tilman D (2019) Multiple health and environmental impacts of foods. Proc Natl Acad Sci U.S.A. 116 (46):23357–23362

De Olde EM, Oudshoorn FW, Sørensen CAG, Bokkers EAM, de Boer IJM (2016) Assessing sustainability at farm-level: lessons learned from a comparison of tools in practice. Ecol Indic 66:391–404

Dean TJ, McMullen JS (2007) Toward a theory of sustainable entrepreneurship: reducing environmental degradation through entrepreneurial action. J Bus Ventur 22:50–76

Desiderio E, García-Herrero L, Hall D, Segrè A, Vittuari M (2022) Social sustainability tools and indicators for the food supply chain: a systematic literature review. Sustain Prod Consum 30:527–540

Dessein J, Soini K, Fairclough G, Horlings L (eds) (2015) Culture in, for and as sustainable development. Conclusions from the COST action IS1007 investigating cultural sustainability. University of Jyväskylä, Jyväskylä

Douglas E, Fitzsimmons J (2008) Individual intentions towards entrepreneurship vs intrapreneurship. Proceedings of the 5th AGSE International Entrepreneurship Research Exchange, 937–951
Dutta MJ, Thaker J (2020) Sustainability, ecology, and agriculture in women farmers' voices: culture-centering gender and development. Commun Theory 30(2):126–148
Egartner S, Ayrault J, Niedermayr J (2020) Development of community supported agriculture in Austria and France: a comparative analysis in the context of social innovation. Austrian J Agric Econ Rural Stud 29(28):243–250
European Commission (2020) Strategic Foresight Report 2020: Charting the course towards a more resilient Europe. Publications Office of the European Union
Food and Agriculture Organization (FAO) (2018) The state of food and agriculture 2018: migration, agriculture and rural development. FAO, Rome
Food and Agriculture Organization (FAO) (2021) The state of food and agriculture 2021: towards more efficient and inclusive food systems. FAO, Rome
Fu N, Zhang T, Nie X, Li M (2021) Research on the influence of leadership style on employee innovation behavior in agricultural science and technology enterprises. In: E3S web of conferences, vol 235. EDP Sciences, Les Ulis, p 01020
Fusco G, Campobasso F, Laureti L, Frittelli M, Valente D, Petrosillo I (2023) The environmental impact of agriculture: an instrument to support public policy. Ecol Indic 147:109961
Gaitán-Cremaschi D, Klerkx L, Duncan J, Trienekens JH, Huenchuleo C, Dogliotti S, Contesse ME, Rossing WAH. (2019) Characterizing diversity of food systems in view of sustainability transitions. A review. In Agronomy for sustainable development, vol. 39, issue 1. Springer, Paris
Gavrilescu M (2021) Water, soil, and plants interactions in a threatened environment. Water 13(19):2746. https://doi.org/10.3390/w13192746
Goodland R (1997) Environmental sustainability in agriculture: diet matters. Ecol Econ 23(3):189–200
Halberstadt J, Timm JM, Kraus S, Gundolf K (2019) Skills and knowledge management in higher education: how service learning can contribute to social entrepreneurial competence development. J Knowl Manag 23(10):1925–1948
Haller AP, Ştefănică M, Butnaru GI, Butnaru RC (2023) Climate neutrality through economic growth, digitalisation, eco-innovation and renewable energy in European countries. Kybernetes 53(4):1250–1280
Hashem NM, González-Bulnes A, Rodriguez-Morales AJ (2020) Animal welfare and livestock supply chain sustainability under the COVID-19 outbreak: an overview. Front Vet Sci 7:582528
Janker J, Mann S (2020) Understanding the social dimension of sustainability in agriculture: a critical review of sustainability assessment tools. Environ Dev Sustain 22(3):1671–1691
Jia X (2021) Agro-food innovation and sustainability transition: a conceptual synthesis. Sustain For 13(12):6897. https://doi.org/10.3390/su13126897
Johnston JL, Fanzo JC, Cogill B (2014) Understanding sustainable diets: a descriptive analysis of the determinants and processes that influence diets and their impact on health, food security, and environmental sustainability. Adv Nutr 5(4):418–429
Khalili A (2016) Linking transformational leadership, creativity, innovation, and innovation-supportive climate. Manag Decis 54:2277–2293
Knudson W, Wysocki A, Champagne J, Peterson HC (2004) Entrepreneurship and innovation in the agri-food system. Am J Agric Econ 86(5):1330–1336
Kraus S, Breier M, Jones P, Hughes M (2019) Individual entrepreneurial orientation and intrapreneurship in the public sector. Int Entrep Manag J 15(4):1247–1268. https://doi.org/10.1007/s11365-019-00593-6
Kuckertz A, Wagner M (2010) The influence of sustainability orientation on entrepreneurial intentions—investigating the role of business experience. J Bus Ventur 25(5):524–535. https://doi.org/10.1016/j.jbusvent.2009.09.001
Le PB (2020) How transformational leadership facilitates radical and incremental innovation: the mediating role of individual psychological capital. Asia-Pac J Bus Adm 12:205–222
Lee A, Legood A, Hughes D, Tian AW, Newman A, Knight C (2020) Leadership, creativity and innovation: a meta-analytic review. Eur J Work Organ Psychol 29:1–35

Léger-Bosch C, Houdart M, Loudiyi S, Le Bel PM (2020) Changes in property-use relationships on French farmland: a social innovation perspective. Land Use Policy 94:104545

Lovarelli D, Bacenetti J, Guarino M (2020) A review on dairy cattle farming: is precision livestock farming the compromise for an environmental, economic and social sustainable production? J Clean Prod 262:121409

Mert-Cakal T, Miele M (2022) 'Workable utopias' for social change through inclusion and empowerment? Community supported agriculture (CSA) in Wales as social innovation. In: Social innovation and sustainability transition. Springer, Cham, pp 307–326

Mignaqui V (2014) Sustainable development as a goal: social, environmental and economic dimensions. Int J Soc Qual 4(1), 57–77. Retrieved July 2, 2024. https://doi.org/10.3167/IJSQ.2014.040105

Mylan J, Geels FW, Gee S, McMeekin A, Foster C (2015) Eco-innovation and retailers in milk, beef and bread chains: enriching environmental supply chain management with insights from innovation studies. J Clean Prod 107:20–30

Ociepa-Kubicka A, Pachura P (2017) Eco-innovations in the functioning of companies. Environ Res 156:284–290

Oliver TH, Boyd E, Balcombe K, Benton TG, Bullock JM, Donovan D, Feola G, Heard M, Mace GM, Mortimer SR, Nunes RJ, Pywell RF, Zaum D (2018) Overcoming undesirable resilience in the global food system. Glob Sustain 1. https://doi.org/10.1017/sus.2018.9

Olsson P, Galaz V (2012) Social-ecological innovation and transformation. In: Social innovation: blurring boundaries to reconfigure markets. Palgrave Macmillan, London, pp 223–247

Parker S (2009) Intrapreneurship or entrepreneurship? IZA Discussion Papers, No. 4195. Institute for the Study of Labor (IZA), Bonn. https://nbn-resolving.de/urn:nbn:de:101:1-20090615104

Plumecocq G, Debril T, Duru M, Magrini M-B, Sarthou JP, Therond O (2018) The plurality of values in sustainable agriculture models: diverse lock-in and coevolution patterns. Ecol Soc 23(1):21

Pol E, Ville S (2009) Social innovation: buzz word or enduring term? J Socio-Econ 38:878–885. https://doi.org/10.1016/j.socec.2009.02.011

Pretty J (2008) Agricultural sustainability: concepts, principles and evidence. In: Hester RE, Harrison RM (eds) Sustainability in agriculture. Royal Society of Chemistry, London, pp 1–34

Pretty J, Benton TG, Bharucha ZP, Dicks LV, Flora CB, Godfray HCJ, Goulson D (2011) Global assessment of agricultural system redesign for sustainable intensification. Nat Sustain 1(8):441–446

Renkema M, Meijerink J, Bondarouk T (2021) Routes for employee-driven innovation: how HRM supports the emergence of innovation in a formalized context. Int J Hum Resour Manag:1–35. https://doi.org/10.1080/09585192.2021.1913625

Rieckmann M, Barth M (2022) Educators' competence frameworks in education for sustainable development. In: Competences in education for sustainable development: critical perspectives. Springer, Cham, pp 19–26

Röling N (2009) Pathways for impact: scientists' different perspectives on agricultural innovation. Int J Agric Sustain 7(2):83–94. https://doi.org/10.3763/ijas.2009.0043

Sabiha NE, Salim R, Rahman S, Rola-Rubzen MF (2016) Measuring environmental sustainability in agriculture: a composite environmental impact index approach. J Environ Manag 166:84–93

Sanga U, Schlüter M, Berrío-Martínez J (2023) Modelling agricultural innovations as a social-ecological phenomenon. Socio-Environ Syst Modell 5:18562–18562

Sannou RO, Kirschke S, Günther E (2023) Integrating the social perspective into the sustainability assessment of agri-food systems: a review of indicators. Sustain Prod Consum 39:175–190

Sokolova AP, Litvinenko GN (2020) Innovation as a source of agribusiness development. IOP Conf Ser Earth Environ Sci 421(2):022053

Springmann M, Clark M, Mason-D'Croz D, Wiebe K, Bodirsky BL, Lassaletta L, de Vries W, Vermeulen SJ, Herrero M, Carlson KM, Jonell M, Troell M, DeClerck F, Gordon LJ, Zurayk R, Scarborough P, Rayner M, Loken B, Fanzo J et al (2018) Options for keeping the food system within environmental limits. Nature 562(7728):519–525

Teles DDS, Nieuwenhuizen C, Schachtebeck C (2021) Entrepreneurial education and individual entrepreneurial orientation: An experts' perspective. An empirical Delphi study. EUREKA: Social and Humanities 4: 46–56. https://doi.org/10.21303/2504-5571.2021.001943

Voeten J, de Haan J, de Groot G (2009) Is that innovation? Assessing examples of revitalized economic dynamics among clusters of small producers in northern Vietnam, UNU-MERIT working paper 2009-055. United Nations University, Tokyo

Vrabcová P, Urbancová H (2023) Sustainable innovation in agriculture: building a strategic management system to ensure competitiveness and business sustainability. Agric Econ/Zemědělská Ekonomika 69(1). https://doi.org/10.17221/321/2022-AGRICECON

Wanzenböck I, Wesseling JH, Frenken K, Hekkert MP, Weber KM (2020) A framework for mission-oriented innovation policy: alternative pathways through the problem–solution space. Sci Public Policy 47:474–489

WCED (1978) Our common future. Oxford University Press, Oxford

Welch RM, Graham RD (1999) A new paradigm for world agriculture: meeting human needs: productive, sustainable, nutritious. Field Crop Res 60(1–2):1–10

Willett W, Rockström J, Loken B, Springmann M, Lang T, Vermeulen S, Garnett T, Tilman D, DeClerck F, Wood A, Jonell M, Clark M, Gordon L, Fanzo J, Hawkes C, Zurayk R, Rivera JA, De VW, Sibanda L et al (2019) Food in the Anthropocene: the EAT–Lancet Commission on healthy diets from sustainable food systems. Lancet 6736(18):3–4

Wojtynia N, van Dijk J, Derks M, Groot Koerkamp PW, Hekkert MP (2021) A new green revolution or agribusiness as usual? Uncovering alignment issues and potential transition complications in agri-food system transitions. Agron Sustain Dev 41:1–20

Woodhill J, Kishore A, Njuki J et al (2022) Food systems and rural wellbeing: challenges and opportunities. Food Sec. 14:1099–1121. https://doi.org/10.1007/s12571-021-01217-0

Transformation-Literate Citizens: Advancing Sustainable Development Through Education

21

Marco Rieckmann

1 Introduction

Education for Sustainable Development (ESD) seeks to cultivate competencies that empower individuals to critically assess their own actions in light of their immediate and long-term impacts on social, cultural, economic, and environmental dimensions, both locally and globally. It prepares individuals to navigate complex situations sustainably, which may necessitate pioneering approaches, and to engage in sociopolitical processes that advance their societies toward sustainable development. Viewed through this lens, ESD is designed to enable learners to take responsible actions that contribute to the construction of sustainable societies in the present and future.

ESD should be considered a fundamental component of quality education and, as such, is integral to the concept of lifelong learning. Educational institutions at all levels—from preschool, through tertiary education, to nonformal and informal learning environments—bear the responsibility to rigorously engage with issues of sustainable development and promote the development of sustainability competencies.

ESD plays a pivotal role in the collective endeavor to attain the Sustainable Development Goals (SDGs) by fostering cross-cutting sustainability competencies among learners. It not only provides individuals with the necessary understanding of the SDGs but also equips them with the competencies needed to actively participate, as informed citizens, in driving the transformation towards a more sustainable society (UNESCO 2017).

M. Rieckmann (✉)
Faculty I, Department of Education, Working Group Higher Education Development, University of Vechta, Vechta, Germany
e-mail: marco.rieckmann@uni-vechta.de

B. Grabkowsky, T. Blaha (eds.), *Intensive Livestock Production in Transition*,
https://doi.org/10.1007/978-3-031-97872-2_21

ESD represents a holistic and transformative approach to education, encompassing learning content and goals, methodologies, and the learning environment. Thus, it goes beyond merely integrating urgent topics like climate change, poverty, and sustainable consumption into the curriculum; ESD fosters interactive, learner-centric educational spaces. What ESD advocates for is a paradigm shift from traditional teaching to engaged learning. It demands an action-oriented transformative pedagogy, along with the fusion of formal and informal learning experiences. Only through such pedagogical strategies can the essential competencies necessary for advancing sustainable development be nurtured effectively.

This chapter introduces ESD as a transformative and competence-based form of (citizenship) education; outlines the sustainability competencies to be developed through ESD; expounds on the action-oriented, transformative pedagogy required for facilitating this development; and examines the competencies needed by educators to effectively deliver ESD.

2 ESD as Transformative and Competence-Based Education

As societies worldwide grapple with rapid advancements in technology and globalization, they face a multitude of new challenges. These challenges encompass escalating complexity and uncertainties, heightened individualism alongside burgeoning social diversity, increasing economic and cultural homogenization, deterioration of critical ecosystem services, and a growing susceptibility to natural and technological risks. Simultaneously, there is an exponential increase in accessible information. Addressing these conditions demands creative and self-organized responses, as the intricacy of the stakeholders, scenarios, and actions precludes strictly planned problem-solving processes.

To contribute to sustainable development, individuals need to grasp the complexities of the world they inhabit and navigate uncertainties, trade-offs, and risks, as well as the rapid pace of societal and global changes. They must possess the ability to collaborate, advocate, and take action for positive transformation in our world (UNESCO 2015). Such individuals may be referred to as "sustainability citizens" (Schank and Rieckmann 2019; Wals 2015; Wals and Lenglet 2016).

Since the late 1990s, the discourse about how to educate sustainability citizens has evolved from an input-focused orientation, which emphasized lists of essential educational content, to an outcome-based competence approach (Adomßent and Hoffmann 2013; Wiek et al. 2011). This shift poses the question: How can individuals competently navigate and engage with an increasingly complex world? How can they contribute to the transformation of its structures? The competence approach is grounded in identifying effective real-world applications (or instances where successful navigation of complex demands is evident) and explores how we can promote learning to achieve these outcomes.

Within the framework of sustainable development challenges, it is contended that education ought to equip individuals with the ability to reflect on their actions by considering both their immediate and long-term social and environmental impacts from a global standpoint. It should also empower them to productively engage in shaping these actions toward greater sustainability. Moreover, education should enable individuals to adeptly navigate complex situations, potentially prompting them to venture into new territories. A competency-based approach plays a crucial role in bridging the gap between knowledge and action.[1]

Rather than encouraging specific behaviors and ways of thinking ("instrumental approach"/"ESD 1"), an emancipatory concept of ESD ("ESD 2") focuses particularly on "building capacity to think critically about [and beyond] what experts say and to test sustainable development ideas" and "exploring the contradictions inherent in sustainable living" (Vare and Scott 2007: 194; cf. Wals 2015). In this context, ESD aims to develop competencies that enable individuals to engage in sociopolitical processes, thereby moving their societies toward sustainable development (Rieckmann 2012, 2020; Wiek et al. 2011). Viewing ESD from a competence perspective allows for the exploration of key areas deemed necessary for sustainability success.

2.1 Development of Sustainability Competencies

The emancipatory approach to ESD inquires into the key competencies required for learners to become agents of sustainability. Within the discourse on ESD, an expanding community of researchers is delving into the numerous interconnected aspects of ESD and the requisite competencies (e.g., de Haan 2010; Glasser and Hirsh 2016; Rieckmann 2012, 2018; Wiek et al. 2011, 2016). These scholars have identified a range of essential competencies for sustainable development, detailing the abilities individuals must possess to transition their lifestyles to more sustainable ones and to contribute to societal shifts toward sustainability. Notable among the proposed concepts of key sustainability competencies are:

- *OECD's key competencies* (Rychen 2003)*:* In its project "Definition and Selection of Competencies" (DeSeCo), OECD has identified key competencies in three categories: using tools interactively (the ability to use language, symbols, and texts interactively; the ability to use knowledge and information interactively; the ability to use technology interactively); interacting in heterogeneous groups (the ability to relate well to others; the ability to cooperate; the ability to manage and resolve conflicts); acting autonomously (the ability to act within the big picture; the ability to form and conduct life plans and personal projects; the ability to defend and assert rights, interests, limits and needs).

[1] Competencies, essential for navigating complex contexts, result from combining knowledge, skills, motives, and affective dispositions, emerging through action, experience, and reflection, rather than being directly taught (UNESCO 2015; Weinert 2001). Key competencies are vital cross-cutting competencies needed globally across all age groups, offering a broad foundation that complements but doesn't replace specific situational competencies (Rychen 2003; Weinert 2001).

- *Gestaltungskompetenz (shaping competence)* (de Haan 2010): This competence framework considers key competencies for shaping or transforming the future including the abilities to: gather knowledge in a spirit of openness to the world, integrating new perspectives; think and act in a forward-looking manner; acquire knowledge and acting in an interdisciplinary manner; deal with incomplete and overly complex information; cooperate in decision-making processes; cope with the individual dilemmatic situation of decision-making; participate in collective decision-making processes; motivate oneself as well as others to become active; reflect upon one's own principles and those of others; refer to the idea of equity in decision-making and planning actions; plan and act autonomously; show empathy for and solidarity with the disadvantaged.
- *Key competencies for sustainable development* (Rieckmann 2012): This list was developed through a Delphi study involving ESD experts from Chile, Ecuador, Germany, Mexico, and the United Kingdom: competency for systemic thinking and handling of complexity; competency for anticipatory thinking; competency for critical thinking; competency for acting fairly and ecologically; competency for cooperation in (heterogeneous) groups; competency for participation; competency for empathy and change of perspective; competency for interdisciplinary work; competency for communication and use of media; competency for planning and realizing innovative projects; competency for evaluation; competency for ambiguity and frustration tolerance.
- *Sustainability competence* (Wals 2015): Wals distinguishes the following dimensions of sustainability competence: dynamics and content of sustainability; critical dimension of sustainability; change and innovation dimension of sustainability; existential and normative dimension of sustainability.
- *Key competencies in sustainability* (Wiek et al. 2011, 2016): Wiek et al. expanded their original framework from five to six key competencies by including problem-solving competence in 2016. These competencies are systems thinking competence, anticipatory competence, normative competence, strategic competence, interpersonal competence, and integrated problem-solving competence. The contributions of Wiek et al. have been pivotal in synthesizing various concepts and lists, offering a structured foundation to foster discussions about the competencies deemed essential for sustainability. An international Delphi study validates the significance of these competencies and enhances them by including intrapersonal competence (Brundiers et al. 2021).
- *Sustainability core competencies* (Glasser and Hirsh 2016): Glasser and Hirsh identified five additional key competencies: affinity for life, state-of-the-planet knowledge, wise decision-making, modeling sustainable behavior, and transformative social change.
- *SD competencies* (Lozano et al. 2017): The following competencies are listed here: systems thinking; interdisciplinary work; anticipatory thinking; justice, responsibility, and ethics; critical thinking and analysis; interpersonal relations and collaboration; empathy and change of perspective; communication and use of media; strategic action; personal involvement; assessment and evaluation, and tolerance for ambiguity and uncertainty.

- *Transformative literacy* (Schneidewind 2013): It refers to the capacity to comprehend and apply information regarding societal transformation processes, interpret them, and actively participate in such processes. It involves understanding the multifaceted nature of transformation processes and actively contributing to them through one's actions. By categorizing transformative literacy into four dimensions, (1) technological, (2) economic, (3) institutional, and (4) cultural, a framework is established. This framework aids in identifying key areas and specific perspectives within the ongoing discourse on transformation processes.

It is evident that while these lists display variations, they also reveal commonalities regarding certain key sustainability competencies. Within the international ESD discourse, there exists a consensus on the particular importance of the following key competencies for fostering thought and action in support of sustainable development (Rieckmann 2018; UNESCO 2017):

- *Systems thinking competence:* The abilities to recognize and understand relationships, to analyze complex systems, to think of how systems are embedded within different domains and different scales, and to deal with uncertainty
- *Anticipatory competence:* The ability to understand and evaluate multiple futures—possible, probable, and desirable, to create own visions for the future, to apply the precautionary principle, to assess the consequences of actions, and to deal with risks and changes
- *Normative competence:* The ability to understand and reflect on the norms and values that underlie one's actions and to negotiate sustainability values, principles, goals, and targets—in a context of conflicts of interests and trade-offs, uncertain knowledge, and contradictions
- *Strategic competence:* The ability to collectively design and carry out innovative actions that promote sustainability at the local level and beyond.
- *Collaboration competence:* The ability to learn from others; to understand and respect the needs, perspectives, and actions of others (empathy); to understand, relate to, and be sensitive to others (empathic leadership); to deal with conflicts in a group; and to facilitate collaborative and participatory problem solving
- *Critical thinking competence:* The ability to question norms, practices, and opinions, to reflect on one's own values, perceptions, and actions to take a position in the sustainability discourse
- *Self-awareness competence:* The ability to reflect on one's own role in the local community and (global) society, to continually evaluate and further motivate one's actions, and to deal with one's feelings and desires
- *Integrated problem-solving competence:* The overarching ability to apply different problem-solving frameworks to complex sustainability problems and develop viable, inclusive, and equitable solution options that promote sustainable development—integrating the aforementioned competencies

Although each competence possesses distinct qualities and relevance, they are intrinsically interrelated. Consequently, integrated problem-solving competence emerges as particularly crucial. Additionally, basic competencies, like communication skills, play a significant role in addressing sustainable development issues. Nevertheless, this discussion accentuates those competencies, especially pivotal for sustainability that has traditionally received less attention in formal education. Still, it is unmistakable that these key sustainability competencies must be cultivated alongside basic competencies (Redman and Wiek 2021; Wiek et al. 2011).

The key competencies for sustainability highlight the specific skills needed by "sustainability citizens" to navigate today's complex challenges. However, although competencies represent the ability or inclination to act, they do not guarantee that individuals will behave in a certain manner in specific situations. Therefore, to convert these capacities into tangible sustainable actions, individuals require aligning values and motivational drivers.

Sustainability performance is also contingent on an individual's environment, which is seen as opportunities for action that are beyond personal control. From this perspective, opportunities are environmental and contextual factors that facilitate action, essentially serving as the conditions that provide the necessary support for sustainability-driven initiatives. Drawing on the capability approach, Nussbaum (2010) underscores the vital role of governance institutions in creating opportunity structures that empower individuals to act. Lozano et al. (2012: 4) define "capabilities" as the set of genuine opportunities to pursue what they value. Hence, sustainability performance is shaped by the interplay of knowledge and skills, values and motivational forces, and opportunities, with the interaction between these dimensions significantly influencing personal behavior.

2.2 Engaging with Sustainability-Relevant Values

ESD emphasizes intra- and intergenerational justice, aiming to sensitize individuals toward responsibility for sustainability and fostering critical value examination (Rieckmann 2020). This approach not only encourages reflection on personal values related to sustainability, such as in animal husbandry or genetic technology in agriculture, but also invites learners to participate in critical value discourses, potentially contributing to a sustainability-oriented value shift without overwhelming them (Balsiger et al. 2017).

Effective transformative learning within ESD necessitates environments where learners critically assess assumptions, facilitating a conceptual change in orientations, including attitudes and values toward sustainability (Scheunpflug 2019).

2.3 ESD as Citizenship Education

To foster transformation toward sustainability, structural and institutional shifts are essential (WBGU 2011), positioning ESD as a form of citizenship education that moves beyond individual responsibility to address systemic issues and promote collective action. ESD as citizenship education seeks not to reduce learners to mere consumers but to empower them as active participants in fostering sustainable development. This involves addressing not only individual behavior changes but also engaging with broader societal structures and promoting critical political agency (Fedorchenko 2021; Kenner and Nagel 2022). By challenging the social imaginary and encouraging social experiments (Muraca 2015), ESD as citizenship education aims to nurture "sustainability citizens" capable of contributing to the societal transformation towards sustainability (Schank and Rieckmann 2019).

3 Main Implications of ESD for the Practice of Education and Pedagogy

3.1 Whole-Institution Approach

ESD transcends the mere instruction on sustainable development or the mere addition of new content to courses and pieces of training. Educational institutions, including schools and universities, should perceive themselves as arenas for learning and experiencing sustainable development, aligning all their processes with sustainability principles. For enhanced efficacy, a transformative institutional change is necessary. This whole-institution approach incorporates sustainability into every facet of the institution, including curriculum design, campus operations, organizational culture, student involvement, leadership approaches, community engagement, and research undertakings (Holst 2023; UNESCO 2014), thereby setting a practical example for learners (Holst et al. 2024).

Hence, the adoption of whole-institution approaches is advocated across all educational levels and settings. All institutions, including those in the public and private sectors, are urged to implement sustainability strategies. Experiences with these approaches in higher education and secondary schooling should be scaled up and extended to other educational domains, such as early childhood education, Technical and Vocational Education and Training (TVET), and nonformal education for youth and adults.

While all aspects of the whole-institution approach hold significance, the essence of ESD delivery in classrooms and various learning environments lies in engaging in interactive, integrative, and critical learning methods—characterized by an action-oriented, transformative pedagogy.

3.2 Action-Oriented, Transformative Pedagogy

Competencies and value orientations cannot be imparted but must be developed by learners themselves (Weinert 2001). Therefore, ESD requires action-oriented, transformative pedagogy characterized by the following pedagogical principles (Rieckmann 2022; UNESCO 2017):

- Learner-centeredness and accessibility
- Action and reflection orientation

- Transformative and transgressive learning
- Focus on participation
- Inquiry-based learning
- Networking learning
- Focus on vision
- Integration of social, self-directed, and methodological learning with content-based learning

To truly foster processes of transformative education within ESD, learning environments are needed where learners can become aware of their own assumptions and those of others and critically reflect on them. In these environments, learners are encouraged to reflect on frameworks in a critical, deconstructive, and transgressive manner to stimulate genuinely transformative learning processes leading to conceptual change (Rieckmann 2022).

While the pedagogical principles outline the overarching approach to designing learning processes in ESD, tailored methods that align with these principles are essential for facilitating the learning experience. In ESD, methods that promote sustainability competencies through active learning are preferred. There are several methods that are especially recommended for ESD (UNESCO 2017):

- Collaborative real-world projects, such as a service-learning project and campaigns for different sustainability topics
- Vision-building exercises such as future workshops, scenario analyses, utopian/dystopian story-telling, science-fiction thinking, and fore- and back-casting
- Analyzing complex systems with community-based research projects, case studies, stakeholder analysis, actors' analysis, modeling, and systems games, for instance
- Critical and reflective thinking through fish-bowl discussions and reflective journals, for instance

These participatory teaching and learning methods enable learners to actively engage in sustainable development. In selecting teaching and learning methods for a specific scenario, it is crucial to tailor them to the learners' needs (such as age, previous knowledge, interests, and abilities) and the learning context (including curriculum space, pedagogical environment, and cultural norms). Additionally, the available resources and support (like teacher expertise, instructional materials, technology, and financial resources) must also be considered.

4 Teaching Competencies for ESD

Educators are powerful change agents for delivering the educational response needed in the context of sustainable development. Their knowledge and competencies are crucial for restructuring educational processes and educational institutions towards sustainability.

Working with the concept of ESD demands new pedagogical competencies from educators. Ideally, educators should possess sustainability competencies themselves and have the ability to support learners in developing similar competencies. This requires a critical understanding of sustainable development on one hand, and the educational approach of ESD on the other. They need knowledge of innovative teaching and learning methods, as well as skills in applying them. Additionally, they require the ability to guide learners, for example, in projects, which includes critically reflecting on them as educators and seeing themselves more as facilitators of learning. The ESD competence framework developed in the EU project "A Rounder Sense of Purpose" provides detailed guidance on the requirements for educators within the framework of ESD (Millican 2022). Rieckmann and Barth (2022) provide a comprehensive overview of the discourse concerning the competencies of educators in ESD.

Since many educators have not been exposed to ESD during their initial training, access to ESD-focused in-service training is essential. Such training would offer opportunities to acquire the requisite knowledge and competencies for engaging in sustainable development processes. Furthermore, professional development is critical for the reorientation of educational practices and institutions.

5 Conclusion

ESD plays a crucial role in promoting sustainable development. It enables learners to develop competencies relevant to sustainability and to critically engage with societal values, contributing also to their citizenship education. Learners are empowered to partake in shaping sustainable development. To facilitate such educational processes in line with ESD, action-oriented, transformative pedagogy is needed, and educators must be qualified to work with the ESD concept. This is also a key prerequisite for embedding ESD structurally across the entire education system.

In order to make it possible for all people worldwide to take action in favor of sustainable development, all educational institutions should consider it their responsibility to deal intensively with matters of sustainable development and to foster the development of sustainability competencies.

References

Adomßent M, Hoffmann T (2013) The concept of competencies in the context of education for sustainable development (ESD). http://esd-expert.net/assets/130314-Concept-Paper-ESD-Competencies.pdf

Balsiger J, Förster R, Mader C, Nagel U, Sironi H, Wilhelm S, Zimmermann AB (2017) Transformative learning and education for sustainable development. GAIA Ecol Perspect Sci Soc 26(4):357–359. https://doi.org/10.14512/gaia.26.4.15

Brundiers K, Barth M, Cebrián G, Cohen M, Diaz L, Doucette-Remington S et al (2021) Key competencies in sustainability in higher education—toward an agreed-upon reference framework. Sustain Sci 16(1):13–29. https://doi.org/10.1007/s11625-020-00838-2

de Haan G (2010) The development of ESD-related competencies in supportive institutional frameworks. Int Rev Educ 56(2):315–328

Fedorchenko L (2021) Kritisch, ganzheitlich und progressiv—zivilgesellschaftliche Perspektiven auf das neue UNESCO-Programm „BNE 2030". ZEP 44(2):35. https://doi.org/10.31244/zep.2021.02.06

Glasser H, Hirsh J (2016) Toward the development of robust learning for sustainability core competencies. Sustainability J Rec 9(3):121–134

Holst J (2023) Towards coherence on sustainability in education: a systematic review of whole institution approaches. Sustain Sci 18(2):1015–1030. https://doi.org/10.1007/s11625-022-01226-8

Holst J, Grund J, Brock A (2024) Whole institution approach: measurable and highly effective in empowering learners and educators for sustainability. Sustain Sci. https://doi.org/10.1007/s11625-024-01506-5

Kenner S, Nagel M (2022) Große Transformation mit jungen Change Agents? Partizipative politische Bildung für nachhaltige Entwicklung als Antwort auf multiple Krisen der Gegenwart. Zeitschrift für Didaktik der Gesellschaftswissenschaften (zdg) 13(2):99–116

Lozano JF, Boni A, Peris J, Hueso A (2012) Competencies in higher education: a critical analysis from the capabilities approach. J Philos Educ 46:132–147

Lozano R, Merrill M, Sammalisto K, Ceulemans K, Lozano F (2017) Connecting competences and pedagogical approaches for sustainable development in higher education. A literature review and framework proposal. Sustainability 9(11):1889. https://doi.org/10.3390/su9101889

Millican R (2022) A rounder sense of purpose: competences for educators in search of transformation. In: Vare P, Lausselet N, Rieckmann M (eds) Competences in education for sustainable development. Critical perspectives. Springer, Cham, pp 35–43

Muraca B (2015) Wider den Wachstumswahn: Degrowth als konkrete Utopie. Blätter für deutsche und internationale Politik 2:101–109

Nussbaum MC (2010) Creating capabilities: the human development approach. Belknap Press of Harvard University Press, Cambridge

Redman A, Wiek A (2021) Competencies for advancing transformations towards sustainability. Frontiers. Education 6:Article 785163. https://doi.org/10.3389/feduc.2021.785163

Rieckmann M (2012) Future-oriented higher education: which key competencies should be fostered through university teaching and learning? Futures 44(2):127–135

Rieckmann M (2018) Chapter 2—Learning to transform the world: key competencies in ESD. In: Leicht A, Heiss J, Jung Byun W (eds) Issues and trends in education for sustainable development. United Nations Educational, Scientific and Cultural Organization (Education on the move), Paris, pp 39–59

Rieckmann M (2020) Emancipatory and transformative global citizenship education in formal and informal settings—empowering learners to change structures. Tertium Comparationis: Journal für International und Interkulturell Vergleichende Erziehungswissenschaft 26(2):174–186

Rieckmann M (2022) Bildung für nachhaltige Entwicklung. Ziele, didaktische Prinzipien und Methoden. Religionsunterrichtheute 50(1):10–17. https://bistummainz.de/export/sites/bistum/schule/RU-heute-1-2022-web.pdf

Rieckmann M, Barth M (2022) Educators' competence frameworks in education for sustainable development. In: Vare P, Lausselet N, Rieckmann M (eds) Competences in education for sustainable development. Critical perspectives. Springer, Cham, pp 19–26

Rychen DS (2003) Key competencies: meeting important challenges in life. In: Rychen DS, Salganik LH (eds) Key competencies for a successful life and well-functioning society. Hogrefe and Huber, Cambridge, pp 63–107

Schank C, Rieckmann M (2019) Socio-economically substantiated education for sustainable development: development of competencies and value orientations between individual responsibility and structural transformation. J Educ Sustain Dev 13(1):67–91. https://doi.org/10.1177/0973408219844849

Scheunpflug A (2019) Transformatives Globales Lernen—eine Grundlegung in didaktischer Absicht. In: Lang-Wojtasik G (ed) Bildung für eine Welt in Transformation. Global Citizenship Education als Chance für die Weltgesellschaft. Verlag Barbara Budrich, Leverkusen, pp 63–74

Schneidewind U (2013) Transformative literacy. Gesellschaftliche Veränderungsprozesse verstehen und gestalten. GAIA 22(2):82–86

United Nations Educational, Scientific and Cultural Organization (UNESCO) (2014) Shaping the future we want. UN decade of education for sustainable development (2005–2014). Final report. http://unesdoc.unesco.org/images/0023/002301/230171e.pdf

United Nations Educational, Scientific and Cultural Organization (UNESCO) (2015) Rethinking education. Towards a global common good? http://unesdoc.unesco.org/images/0023/002325/232555e.pdf

United Nations Educational, Scientific and Cultural Organization (UNESCO) (2017) Education for sustainable development goals. Learning objectives. UNESCO, Paris. http://unesdoc.unesco.org/images/0024/002474/247444e.pdf

Vare P, Scott W (2007) Learning for a change: exploring the relationship between education and sustainable development. J Educ Sustain Dev 1(2):191–198

Wals AEJ (2015) Beyond unreasonable doubt. Education and learning for socio-ecological sustainability in the anthropocene. Wageningen University, Wageningen. https://arjenwals.files.wordpress.com/2016/02/8412100972_rvb_inauguratie-wals_oratieboekje_v02.pdf

Wals AEJ, Lenglet F (2016) Sustainability citizens: collaborative and disruptive social learning. In: Horne R, Fien J, Beza B, Nelson A (eds) Sustainability citizenship in cities: theory and practice. Routledge, London

WBGU—German Advisory Council on Global Change (2011) World in transition—a social contract for sustainability. WBGU, Berlin. https://www.wbgu.de/en/publications/publication/world-in-transition-a-social-contract-for-sustainability

Weinert FE (2001) Concept of competence: a conceptual clarification. In: Rychen DS, Salganik LH (eds) Defining and selecting key competencies. Göttingen, Hogrefe and Huber, Seattle, Toronto, Bern, pp 45–65

Wiek A, Withycombe L, Redman CL (2011) Key competencies in sustainability: a reference framework for academic program development. Sustain Sci 6(2):203–218

Wiek A, Bernstein MJ, Foley RW, Cohen M, Forrest N, Kuzdas C, Kay B, Withycombe Keeler L (2016) Operationalising competencies in higher education for sustainable development. In: Barth M, Michelsen G, Thomas I, Rieckmann M (eds) Routledge handbook of higher education for sustainable development. Routledge, London, pp 241–260

Consumer-Oriented Perspectives for Sustainable Animal Products and Consumption

22

Brianne A. Altmann, Sarah Kühl, Clara Mehlhose, Daniel Mörlein, and Annika Thies

1 "Less but Better" Animal Products as a Sustainability Transformation Goal

The increasing consumer awareness of the negative impact that intensive livestock farming has on climate, environment, and animal welfare has led to a shift in dietary patterns. On the one hand, consumer preferences for animal products are changing. Purchasing motives such as taste, quality, and appearance of the products are not always at the forefront for consumers. Sustainability characteristics related to

B. A. Altmann (✉)
Institute of Animal Hygiene, Animal Welfare and Farm Animal Behaviour, University of Veterinary Medicine Hannover, Hannover, Germany

Marketing for Food and Agricultural Products, Department of Agricultural Economics and Rural Development, University of Goettingen, Goettingen, Germany
e-mail: brianne.altmann@tiho-hannover.de

S. Kühl
Agricultural Marketing, Department of Sustainable Agriculture and Energy Systems, University of Applied Sciences Weihenstephan-Triesdorf, Freising, Germany
e-mail: sarah.kuehl@hswt.de

C. Mehlhose
Marketing for Food and Agricultural Products, Department of Agricultural Economics and Rural Development, University of Goettingen, Goettingen, Germany
e-mail: clara.mehlhose@uni-goettingen.de

D. Mörlein
Division for Quality of Animal Products, Department of Animal Sciences, University of Goettingen, Goettingen, Germany
e-mail: daniel.moerlein@uni-goettingen.de

A. Thies
Institute of Market Analysis, Johann Heinrich von Thuenen Institute, Braunschweig, Germany
e-mail: annika.thies@thuenen.de

B. Grabkowsky, T. Blaha (eds.), *Intensive Livestock Production in Transition*,
https://doi.org/10.1007/978-3-031-97872-2_22

animal, environmental, and societal impacts have come into focus (Alonso et al. 2020; Parlasca and Qaim 2022). In this context, labeling plays an important role, as sustainability properties are most frequently credence characteristics that consumers cannot recognize on the product sitting on a retailer shelf. On the other hand, sustainability considerations are also a leading reason for consumers to reduce the amount of animal proteins in their diet. Plant-based and animal-free food options have become accessible alternatives in many Western countries (Short et al. 2021). Thus, the percentage of self-stated flexitarians, vegetarians, and vegans is growing strongly, especially in Germany (Bryant et al. 2020), with consumers increasingly agreeing to consume fewer animal products and more plant-based alternatives. In addition to a dietary transformation, efforts to achieve more sustainable food systems must incorporate strategies to reduce food waste. As meat production is often linked to negative environmental impacts (e.g., air pollution, high nutrient run-off), reducing the amount of meat that is discarded is relevant for sustainable diets (Thies et al. 2021), with the amount of meat waste being highly influenced by purchasing and non-purchasing decisions of consumers (Ramanathan et al. 2022).

Therefore, livestock farming and the associated production of animal products are currently facing the challenge of meeting increasing consumer expectations. In this context, there are numerous developments to support the "less but better" approach to animal products. These can include, for example, the development of more animal-friendly husbandry systems, the use of old and local breeds, or the whole-body utilization of animals. However, many of these approaches remain niches in the food market, making a holistic transformation difficult or even impossible. The reasons for this are manifold and range from financial restrictions to cognitive dissonance, a lack of confidence in product information or labels, or the missing availability of corresponding products (Consumer-Citizen-Gap; Busch and Spiller 2020). Consumers are focused on intrinsic as well as extrinsic attributes when buying animal products, so ignoring a product's appearance and eating quality in turn for sustainability attributes does not make products marketable. Cardona et al. (2023) found that the most important intrinsic aspects of meat are taste/flavor, texture, appearance, color, and odor, whereas the most important extrinsic ones are price, origin, production method, and certification. This is more or less in line with the findings of Henchion et al. (2017) who summarized the most important aspects for consumers when choosing beef. They found that (1) price; (2) certification, labels, brands, and information; and (3) visible fat were ranked as the three most important search attributes. Flavor, freshness/shelf life, and tenderness are all important experience attributes. Origin, animal welfare, and production system/feeding are the three most important credence attributes. How relevant these aspects are for individual consumer groups depends on various aspects such as sociodemographics but also attitudes and values (Henchion et al. 2017; Cardona et al. 2023).

In summary, it is important to address consumers' concerns and adapt production processes accordingly in order to ensure the marketability of animal products in the future. This means taking a consumer-oriented approach. In the following, case studies will be used to describe why and how consumer perspectives need to be considered during the sustainability transformation of livestock farming.

2 Sustainability Transformations' and their Influence on Product Quality

2.1 Alternative Feedstuffs to Reduce Arable Land Requirements

To increase protein source availability as well as substitute current animal and fish feedstuffs linked to land use change and other environmental degradative practices, microalgae and insects have been investigated as substitutes for soybean meal and fishmeal. Like plants, microalgae are naturally able to convert light energy into caloric energy for animals through photosynthesis; unlike most plants, some species also can grow without sunlight (Daneshvar et al. 2021). Their cultivation can be detached from arable land and large-scale production takes place in tubular photobioreactors, ponds, or raceways (Daneshvar et al. 2021). Wastewater has been researched as a nutrient source for microalgae production with some success (Li et al. 2019). In addition, insects, specifically the black soldier fly, can be reared without arable land using waste streams and offer nutritious protein (Maroušek et al. 2023) for livestock, poultry, and fish. Both alternative feedstuffs have the potential for increasing resource use efficiency by including waste as inputs and therefore contributing to a circular economy.

Alternative feedstuffs marginally affect physicochemical and sensory product quality in pork and poultry meat. Because swine finishing diets only include small proportions of protein feeds, the inclusion of either *Spirulina platensis* (a microalga) or black soldier fly meal, as alternative protein feedstuffs, did not result in relevant meat quality differences compared to a soybean meal reared product (Altmann et al. 2019). Poultry farming is much more reliant on soybean meal for production in Europe. Protein feeds make up a larger proportion of poultry rations. For this reason, alternative feedstuffs have a greater impact on overall meat quality. Microalgae with a high proportion of carotenoids, such as *Spirulina platensis*, produce an orange meat color that is obvious to the naked eye (Altmann et al. 2020, 2022). Black soldier fly meal has also been shown to increase yellowness in chicken breast meat across three different breeds (Altmann et al. 2020, 2023a). Both alternative feedstuffs also increased flavor attributes in breast meat (Altmann et al. 2020), which could be linked to differences in flavor-related compounds (Gkarane et al. 2020). Feeding black soldier fly meal also led to increased levels of saturated fatty acids, primarily lauric (C12:0) and myristic acid (C14:0) in intramuscular fat from poultry meat (Altmann et al. 2020) and backfat in from pork (Altmann et al. 2019). The marginal increase in the proportion of saturated fatty acid is unlikely to be perceived by consumers but may increase the processing quality attributes, such as reducing rancidity for cured and dried meat products and for products stored in highly oxygenated modified atmosphere packaging. Overall, research shows that alternative feedstuffs could be integrated into current swine and poultry farming systems as substitutes for soybean meal in order to increase process quality aspects without degrading product quality.

Microalgae and insects have also been investigated as a substitute for fishmeal in aquaculture. Various species of microalgae can be included in aquaculture diets at low to moderate levels without negatively impacting animal weight gain (Nagappan et al. 2021); however, inclusion at a high level (e.g., complete substitution of fishmeal) of *Spirulina platensis* has been shown to reduce final body weight in brook and rainbow trout (Rosenau et al. 2022) as well as (Rosenau et al. 2021) catfish. Microalgae compared to fishmeal decreases the proportion of polyunsaturated fatty acids in fish and seafood products (Nagappan et al. 2021; Rosenau et al. 2022), while *Spirulina platensis* influences raw and cooked filet color (Rosenau et al. 2021, 2022). Insects in aquaculture diets have been shown to negatively affect growth (Tran et al. 2022), likely due to chitin content. Therefore, insect fat as a component of aquaculture feed may be more promising (Hossain et al. 2023). Yet, insect feed quality varies immensely based on insect species (Tran et al. 2022) and rearing-substrate (Spranghers et al. 2017), and improving and stabilizing feed quality could go a long way in improving fish quality. For example, insect meal can be produced on fish offal, which increases the overall proportion of polyunsaturated fatty acids in feed (St-Hilaire et al. 2007), likely to subsequently increase such levels in the end product (Maroušek et al. 2023).

2.2 Transition from Castration to Immunocastration in Swine Fattening

Surgical castration of male piglets intended for fattening is one of the so-called non-curative interventions that are under debate mainly for animal welfare reasons. Historically, it was practiced to prevent unwanted mating behavior as well as to control breeding (Mörlein 2020). Nowadays, the main reason is to prevent "boar taint," i.e., a malodor occurring mainly in fatty parts of the carcass of sexually mature male pigs, which often decreases sensory liking of pork and pork products among sensitive consumers (e.g., Bonneau and Weiler 2019). The case of piglet castration and its alternatives illustrates the dilemmas and conflicting aims associated with pork production sustainability. Kress et al. (2019) outline a framework that describes the sustainability conundrum of swine fattening; societal aspects of sustainability (expected product quality and animal welfare) conflict often with environmental and economic goals, where economic and environmental goals may be synergistic as animal performance brings more economic benefits with decreasing environmental burdens pro kg/meat.

Animal welfare critiques are one reason to move from castration toward immunocastration of male pigs. It was shown that surgical castration, when practiced without anesthesia and/or analgesia causes intra- and/or postoperative pain, and may result in post-operation infections that ultimately can depress growth (Prunier et al. 2006). For these reasons, surgical castration is to be phased out in the EU; in

Germany, castration without anesthesia and pain relief has been forbidden by law since 2021. In addition, raising intact (i.e., uncastrated) male pigs can be associated with reduced animal welfare, due to mounting and agonistic behavior in male pigs resulting in injuries. Immunocastration, which uses active immunization against gonadotropin-releasing hormone is an efficient alternative to surgical castration—it can also reduce such sex-specific behavior after the second vaccination (von Borell et al. 2020).

Economically speaking, immunocastrated pigs show a more favorable feed conversion ratio, a higher carcass leanness but lower dressing percentage compared to surgically castrated pigs. Overall, conventionally raised immunocastrates yielded a higher share of valuable meat compared to surgically castrated (+0.628 kg) and entire male pigs (+1.385 kg). Compared to entire male pigs, immunocastrates show greater daily gains (+65.04 g/day), such as they reach their market weight earlier, as well as greater feed efficiency (+0.075 kg/kg) while achieving a similar dressing percentage (Poulsen Nautrup et al. 2018). Entire males are widely considered superior in terms of lean meat content compared to immunocastrates, which have thicker backfat. However, when compared to surgically castrated pigs, immunocastrates are leaner (Kress et al. 2019). That is, in terms of carcass quality, immunocastrated male pigs take an intermediate position between entire males and surgically castrated pigs (e.g., Batorek et al. 2012). The timing of the second vaccination can be used to shift the balance between entire male-like or more castrated-like growth performance and carcass traits (Kress et al. 2019).

Global warming potential per kg carcass weight is lower for immunocastrates compared to surgical castrates, as long as all animals are slaughtered at a similar weight. The main reason is this sustainability improvement links back to immunocastrates' improved feed efficiency. Also, nitrogen efficiency was reportedly higher for immunocastrates than surgical castrates, yet slightly lower than entire males (Kress et al. 2019). A recent project conducted by one of the co-authors showed a reduced (approximately 7%) global warming potential of pork production when using entire males or immunocastrated males as compared with castrated pigs (Mörlein, unpublished data). It remains to be noted that although ceasing the physical castration of male piglets may mitigate the global warming potential of pork production, principles such as precision feeding, the use of enzymes such as phytase to increase nutrient digestibility, and the application of free amino acids to reduce total crude protein content may have a more pronounced effect than abandoning castration (Kress et al. 2019).

The application of immunocastration can improve animal welfare and production performance while reducing the environmental burden of pig production. Yet, concerns about product quality and consumer acceptance are widespread. Certainly, the greatest barrier against a widespread implementation is the risk of boar taint in pork from (sexually mature) entire males or incomplete immunocastrates. However, procedures to analytically or sensorially detect tainted carcasses at slaughter have been developed (Burgeon et al. 2021). Also, breeding strategies can be used to effectively reduce the risk of boar taint (Souza Duarte et al. 2021). Immunocastrated male pigs are in terms of meat and fat quality often range in between entire male pigs and surgically castrated males and present a quality improvement over entire males (Batorek et al. 2012; Škrlep et al. 2020).

However, the widespread use of immunocastration is still impaired in Europe. Some actors along the value chains discredit immunocastration as a sustainable alternative to surgical castration based on an assumed rejection of the practice and negative association by consumers. There are myths and concerns that human fertility may be negatively affected by ingesting immunocastrated pork (Bonneau and Weiler 2019), despite the European food safety authority EFSA approving the vaccine as safe. Knowledge dissemination, however, will also need to play a role in the future in order to transform male swine fattening. Despite societal debates on surgical piglet castration, studies show that pig producers prefer to continue surgical castration, as they consider it the most reliable way to produce a high product quality (Kress et al. 2019). Finally, communication to consumers' needs to be tailored; a quality-oriented approach was appreciated over technical details about the immunocastration as the latter raises negative associations (Mörlein and Schübeler 2017).

3 Sustainability Transformations as Marketing Tools

3.1 Feedstuff Labeling to Communicate Sustainability Transformations in Livestock Farming

Labeling feedstuffs as a product credence attribute is not a new concept in animal farming. In North America, examples of credence attributes linked to feedstuffs are "grass-fed beef" or "grain-finished" beef, as well as "grain-fed" chicken. In Germany, "Weidemilch" translating to "meadow-milk" is also a commonly found product indicating that the livestock is primarily kept on pasture. Internationally, consumers commonly associate animal welfare and reduced environmental impact with pasture-raised products, of which "Weidemilch" is included (Stampa et al. 2020). Although pasture-raised products may resemble "grass-fed beef," it is important to keep in mind that they communicate a housing form that corresponds with one type of available feed (but often not the only feed), whereas "grass-fed beef" depicts the feedstuff itself (e.g., fresh pasture grass, or pasture hay) and does not need to be fed on pasture. Likely the most common feedstuff-related and one of the most preferred labeling (Profeta and Hamm 2018) in Western markets is "GMO-free." Despite the common usage of feedstuff claims in retail settings, there is limited research on consumer preferences for explicit feedstuffs labels, and their overall importance as a search attribute for consumers at the point of purchase.

When prompted, consumers have expressed preferences for feedstuff attributes, such as "free of veterinary additives or drugs," "GMO-free," pertaining to a specific origin, and "local" (Profeta and Hamm 2018). Focusing on characteristics like "origin" and "local," Profeta and Hamm (2019) showed that consumers exhibit preferences for feedstuff sourcing (e.g., "90% local feed share in diet) on local animal products such as eggs, milk, pork cutlets, and beef steak. Yet, feedstuffs themselves can influence process and product intrinsic quality. Therefore, using a discrete choice experiment with a split-sample design and information treatments, Altmann et al. (2022) tested consumer preferences for "spirulina" and "insect meal" as feedstuffs used in broiler production. During a discrete choice experiment, half of the consumers were shown a chicken breast product either produced using spirulina, insect meal, or soybean meal as the main protein feed ingredient, but the feedstuffs used were not explicitly communicated or labeled. In this scenario, consumers preferred chicken breast raised on insect meal due to its moderately yellow color, consumers rejected the unfamiliar orange-colored chicken breast raised using *Spirulina platensis*. When informed about the effects that each feedstuff can have on product quality, e.g., altering meat color, and with a label of either "spirulina-fed," "insect-fed," or "soy-fed," consumers preferred the insect-fed chicken breast more intensely than without information, especially environmentally conscious consumers. "Spirulina-fed" products were accepted but not preferred above "soy-fed" chicken breast once the unfamiliar color was explained to consumers (Altmann et al. 2022). In trout fillet, consumers informed about the use of *Spirulina platensis* as a fish feed preferred the appearance of spirulina-fed fish fillet, i.e., orange color compared to a conventional fish fillet. Rosenau et al. (2023) investigated the influence of feedstuff claims, such as "fed using sustainable resources," "fed without fish meal," or "fed with spirulina," with limited effect on consumer choice. Consumers in the study slightly preferred the most ambiguous claim "fed using sustainable resources, but the more specific claims regarding feedstuffs were neither preferred nor rejected (Rosenau et al. 2023).

Although not exhaustive, these studies point out the potential benefits and limitations to including feedstuff characteristics as marketing tools. Information regarding feedstuffs is likely necessary to communicate unfamiliar intrinsic attributes, such as an unfamiliar color. If feedstuff is a trending topic and familiarly associated with a positively perceived extrinsic or intrinsic attribute, such as environmental sustainability, then it could be beneficial to use specific labeling to cater to consumers. Overall, the labeling of feedstuffs and associated labeling claims do not appear to be important criteria for consumer choice of animal products. This does not mean that such sustainability transformations are not relevant for livestock farming, but rather that product labeling is not the appropriate avenue to drive sustainability transformations related to animal feedstuffs.

3.2 Multimedia Technologies to Satisfy Consumer Interests in Animal Welfare

Public concerns have emerged regarding the balance between increased productivity in livestock farming systems and the welfare of farm animals (Alonso et al. 2020; Cao et al. 2021). Consumers demand more ethical systems and are inclined to purchase products that align with their animal welfare concerns. Consumers value various aspects of animal welfare, such as larger living spaces, less invasive procedures, and shorter transportation times (Gracia et al. 2011; Clark et al. 2017; Latacz-Lohmann and Schreiner 2018). However, positive attitudes toward farm animal welfare do not always translate into corresponding purchasing behavior, indicating a gap between consumer sentiment and action (Verbeke 2009). Reasons for this can be seen in consumers' limited understanding of modern animal husbandry systems (Camilleri et al. 2019), coupled with a lack of direct interaction with livestock farming, due to urbanized consumer bases (Vanhonacker et al. 2008).

To address citizens' interest in agricultural issues and overcome their lack of direct contact, one possible approach to increase production transparency is to provide visual media of housing conditions to consumers. Virtual farm tours have emerged as a promising alternative to on-site farm tours that are often hindered by hygiene regulations and geographical distance from consumers. In a recent study by Schütz et al. (2022), 17 participants were presented with a 360-degree video showcasing a conventional pig fattening pen via both a tablet and virtual reality (VR) glasses. Participants' perceptions of the pig fattening pen varied based on the recording perspective and the medium used. Nonetheless, participants generally viewed the housing conditions more favorably than they initially anticipated, considering the housing conditions as mostly positive. Therefore, virtual farm tours can be an effective tool for enhancing transparency, disseminating information, and providing consumers with insights into livestock farming practices. Participants appreciated the convenience and entertainment value offered by both mediums. They expressed a preference for VR glasses due to their heightened realism and entertainment value, yet acknowledged the ease-of-use advantages associated with a tablet. It should be noted that presenting video sequences without additional contextual information about the farm or housing conditions was described as insufficient for facilitating a comprehensive understanding of the content.

At point of sale, consumers are often reliant on credence attributes such as labels to make informed conditions regarding housing conditions. The effectiveness of labels hinges on the provision of adequate information and how consumers interpret such information cues. The German "Haltungsform" label has been introduced by German retailers for fresh meat and dairy products. The label is widely available on animal food products in German supermarkets and discounters and is intended to provide a quick overview of housing conditions by classifying existing animal welfare programs and standards into four label levels. For all animal species, it comprises four levels ranging from low (i.e., = 1) to high (i.e., = 4) animal welfare levels in relation to the housing system. Another study by Schütz et al. (2023) investigated the impact of different media portraying information on housing conditions on consumer understanding, evaluation, willingness to buy (WTB), and willingness to pay (WTP) for an animal product, labeled with level 3 of the German "Haltungsform" label. Four experimental groups were presented: (1) text form only; (2) text and pictures of usual housing condition; (3) text and a 360° video of housing conditions via tablet; and (4) text and a 360° housing conditions video via virtual reality (VR) glasses. Irrespective of the medium, participants demonstrated enhanced comprehension of the label along with improved perceptions of animal welfare and acceptance of the exemplified housing conditions. Moreover, the provision of information positively impacted participants' WTB and WTP. Participants particularly valued

VR glasses as a technology to communicate housing conditions, despite expressing a preference for simpler formats for use at the point of sale.

Therefore, German consumers were approached at two supermarkets to observe how the different media are used by consumers on-site at the point of sale (Kühl et al. 2024). Of 182 consumers looking to purchase pork, the largest proportion opted for information in the form of pictures (40%). The second most frequently chosen medium was informational texts (25%); however, nearly 40% of these consumers also opted for the text in combination with another medium. Eighteen percent chose the video, with approximately half of these choosing to watch it via the tablet and the others via the VR glasses. The main reasons for choosing pictures and/or text included the ease of understanding and reduced time needed. Detailed information was also a motivation for choosing the text. Consumers chose the video for its portrayal of realistic and detailed information. In the case of the VR glasses, interest in the technology and "fun" were cited as motivations. However, throughout the experiment, it became clear that the time needed at the point of sale played a decisive role in the choice of medium. Many consumers want to do their grocery shopping quickly, and this is why pictures were chosen as the most common medium and rated as the most suitable. In addition, pictures can be quickly understood and portray realism. Nevertheless, many consumers wanted additional information in text form to be able to better comprehend what they were seeing. Overall, all media led to a better understanding and a higher willingness to buy and pay for products associated with the housing conditions. Overall, a variety of media are suitable in conveying information on housing conditions and associated animal welfare efforts, which can lead to more informed consumer decisions at the point of sale.

3.3 Habitual Consumer Behavior and Reducing Meat Waste

Consumers' perceptions of food quality at the time of purchase are often frequently ingrained, habitual, and linked to color preferences (Lee et al. 2013). This is especially true for meat, with meat color strongly influencing consumer purchase decisions (Altmann et al. 2022). As beef shoppers often prefer a cherry red beef color and reject discoloration (Altmann et al. 2023b), beef turning brownish red to brown is typically discounted or even discarded by retailers, resulting in negative environmental effects as well as economic losses (Ramanathan et al. 2022).

An evaluation of US consumers' beef color preferences using digital photos of packaged beef steaks indicates that beef shoppers prefer the appearance of fresher meat; preference intensity is nearly linear across days of retail display. Consequently, consumers expect large price discounts for what they perceive as discolored beef steaks. In a study conducted by Thies et al. (2024), consumers preferred a bright cherry red-colored beef steak corresponding to 0 days of retail display, followed by 4, 7, and 9 days of retail display. Thus, consumers' willingness to pay (WTP) for these beef steak products decreased nearly proportionately with an increase in display time. Beef steaks in retail display for up to 4 days could be marketed at a price discount of $5.08/lbs., while consumers expect a price discount of $9.89/lbs. for steaks in retail display for up to 7 days. Beef retailers need to consider that beef steaks with 9 days of retail display are difficult to sell even with heavy price discounts, especially with large populations of female shoppers as these buyers exhibited stronger preferences for color compared to their male counterparts.

Discount labeling was included in the study to understand its impact on consumer choice. The presence of a discount label did not reduce the utility derived from a product purchase, which likely means that consumers are not using discount labels to convey product quality information, but rather product appearance is used to judge quality. However, discounts of 30%, as communicated through the discount label in the study, were not sufficient to sway consumer choices toward 7- and 9-day displayed steak products. In follow-up focus group sessions (Thies et al. 2025), participants indicated that they expect significant price reductions for discolored beef products. Beef shoppers would consider buying a discolored beef steak, with the guarantee that discoloration has no impact on flavor, tenderness, juiciness, and meat quality overall. Participating consumers anticipated that their purchasing behavior would change with repeated positive experiences of cooking and eating discolored beef and stressed the importance of confirmation by an official and independent authority regarding the food safety of these products. Price discounts alone are not enough, but in combination with information measures, such as display signs at the point of sale, they could combat food waste of raw beef products at the retail level. Male beef shoppers should be a target segment for promoting discolored beef products and ensuring timely consumption prior to spoilage.

4 Discussion and Conclusion

4.1 How to Communicate "Less but Better" Animal Products

Animal welfare and environmental burden are two of the many reasons justifying the need for sustainability transformation in livestock farming. Consumers in Western countries are increasingly aware of the negative effects of current-day livestock farming practices, which underlie reduced consumption and higher expectations regarding the sustainability of animal products. This awareness on the consumer side is what drives much of the research on sustainable alternatives in livestock farming: for example, alternative feedstuffs, castration alternatives, and different housing forms. From a production technology and animal science perspective, the approaches indicate great potential for more sustainable production of animal foodstuffs. However, there are also challenges. For example, perceptible differences in food quality, such as color or concerns regarding technology safety may impede the acceptance of such technological innovations. Targeted marketing measures addressing individual groups of buyers could be used to communicate product color while implementing consumer education initiatives on why product appearance may differ from habituated expectations. Informative measures at the point of sale hold significant promise in educating meat consumers about product quality, thereby reshaping consumer preferences and effectively combating meat waste. Information on product quality when applying technological innovation, such as immunocastration, should be disseminated instead of information on the technology itself. Consumers at the end of the supply chain may not be able to evaluate technological information correctly, resulting in negative acceptance of sustainability transformations, such as immunocastration. Information dissemination should not become too one-sided so as to lead to consumer deception; livestock farming needs to maintain transparency in order to win and keep trust with its consumer bases. Finally, the provision of information to consumers is necessary to communicate production processes that lead to higher prices, such as animal-friendly husbandry systems. Price plays a very important role at the point of sale. In order to be able to sell products at higher prices, good and credible communication of the underlying sustainable production processes is necessary to convince consumers. This is particularly relevant as consumers cannot check the authenticity of credence attributes at the point of sale. Simplified labels are often not enough to convey complex information. More comprehensive information measures are required, such as texts, pictures, or videos at the point of sale, to convince concerned consumers to buy animal products from transformed systems.

4.2 Possibilities and Limitations of Sustainability Transformation in Livestock Farming

Transformations have always been a part of livestock farming, starting with the tipping point from hunting societies toward the domestication of livestock. Today, the sector finds itself in a state of transition, leading to transformation while looking for solutions to ensure accessible and safe animal foodstuffs. Therefore, sustainability transformations are a continuation of a historically dynamic system and are rightfully needed to ensure the provision of animal products for future generations. The above case studies highlight some of the potential sustainability transformations from a research perspective; some of which have transformed livestock farming at the local or national levels. However, none have pushed the livestock sector past its current tipping point, globally. The case studies highlight that not all transformations will be driven by consumer demand, and labeling products with credence attributes is not suitable for moving all sustainability transformations forward. Providing specific and technical information is not always better, even for credence attributes with a high level of consumer acceptance; at some point, consumers may become overwhelmed with labels and information. Furthermore, not all consumer segments may value sustainability transformations. As such, it should be noted that consumers, although in some cases can even drive sustainability transformations, do not simply accept animal products because they are derived from sustainability transformations. Understanding the consumer perspective of how transformations influence intrinsic and extrinsic quality and how to best communicate this (if at all) through the supply chain will be key to the success of sustainability transformations in livestock farming. Avoiding systems transformation is no longer an option, and consumers offer both possibilities and limitations in their success.

4.3 Outlook: Transformation or Disruption?

To overcome the complex dilemma of modern livestock farming to meet growing global demand with environmental, economic, and ethical solutions, the emergence of cellular agriculture presents itself as a possible solution. Cellular agriculture encompasses techniques such as precision fermentation and cultured meat production and offers a glimpse into a future where animal products are produced without housing large numbers of animals; rather a minimal number of donor animals are kept. Contrary to popular belief, cellular agriculture is unlikely to entirely replace conventional livestock farming. Instead, it is poised to complement existing systems, offering alternative sources of animal-derived products for mass consumption. The scalability and efficiency of precision fermentation and cultured meat production make them viable candidates for supplying standardized products to meet the demands of a burgeoning global population. They allow for the production of homogenous products, which could reduce food loss and waste. For example, cellular agriculture can specialize in producing dairy products without having to rear cows (Kühl et al. 2024). Yet, it is essential to recognize that this technological advancement is in its infancy and does not signify the end of conventional animal agriculture but heralds a new era of coexistence.

As seen in the case studies described above, there remains a steadfast commitment to the preservation of conventional farming practices and the production of high-quality, artisanal animal products. For specialty and premium food items, such as artisanal cheeses, heritage breed meats, and ethically sourced dairy products, the role of animals in agriculture and their surrounding natural environment remains irreplaceable. The unique flavors, textures, and cultural significance associated with these products cannot be replicated through cellular agriculture alone. Therefore, it is imperative to recognize that while technological advancements may in the future alter the assortment of animal products on grocery store shelves, they do not diminish the value of livestock farming. The coexistence of conventional and cellular agriculture presents an opportunity for synergy and innovation within the food industry. By embracing cellular agriculture alongside livestock farming, a diverse and resilient food system can develop to meet the evolving needs and preferences of all consumers. Moreover, collaboration between stakeholders across the agricultural spectrum—from farmers and scientists to policymakers and consumers—will be essential in navigating the transition toward a more sustainable and ethical future within livestock farming systems.

References

Alonso ME, González-Montaña JR, Lomillos JM (2020) Consumers' concerns and perceptions of farm animal welfare. Animals 10. https://doi.org/10.3390/ani10030385

Altmann BA, Neumann C, Rothstein S, Liebert F, Mörlein D (2019) Do dietary soy alternatives lead to pork quality improvements or drawbacks? A look into micro-alga and insect protein in swine diets. Meat Sci 153:26–34. https://doi.org/10.1016/j.meatsci.2019.03.001

Altmann BA, Wigger R, Ciulu M, Mörlein D (2020) The effect of insect or microalga alternative protein feeds on broiler meat quality. J Sci Food Agric 100:4292–4302. https://doi.org/10.1002/jsfa.10473

Altmann BA, Anders S, Risius A, Mörlein D (2022) Information effects on consumer preferences for alternative animal feedstuffs. Food Policy 106:102192. https://doi.org/10.1016/j.foodpol.2021.102192

Altmann BA, Geisler S, Morthorst F, Angeli S, Bortolini S, Gauly M, Hummel J, Sünder A, Mörlein D, Traulsen I, Ammer S (2023a) Animal performance and meat quality of two slow-growing chicken genotypes fed insects reared on municipal organic waste. J Insects Food Feed 9:1445–1459. https://doi.org/10.1163/23524588-20230035

Altmann BA, Trinks A, Mörlein D (2023b) Consumer preferences for the color of unprocessed animal foods. J Food Sci 88:909–925. https://doi.org/10.1111/1750-3841.16485

Batorek N, Čandek-Potokar M, Bonneau M, Van Milgen J (2012) Meta-analysis of the effect of immunocastration on production performance, reproductive organs and boar taint compounds in pigs. Animal 6:1330–1338. https://doi.org/10.1017/S1751731112000146

Bonneau M, Weiler U (2019) Pros and cons of alternatives to piglet castration: welfare, boar taint, and other meat quality traits. Animals 9. https://doi.org/10.3390/ani9110884

Bryant C, van Nek L, Rolland NCM (2020) European markets for cultured meat: a comparison of Germany and France. Food Secur 9. https://doi.org/10.3390/foods9091152

Burgeon C, Debliquy M, Lahem D, Rodriguez J, Ly A, Fauconnier ML (2021) Past, present, and future trends in boar taint detection. Trends Food Sci Technol 112:283–297. https://doi.org/10.1016/j.tifs.2021.04.007

Busch G, Spiller A (2020) Warum wir eine Tierschutzsteuer brauchen-Die Bürger-Konsumenten-Lücke. Available at https://www.uni-goettingen.de/de/document/download/7e3ee3661cf99df26d7796608a11d817.pdf/2001_Diskussionsbeitrag_Busch und_Spiller_2020.pdf. Verified 16 June 2024

Camilleri AR, Larrick RP, Hossain S, Patino-Echeverri D (2019) Consumers underestimate the emissions associated with food but are aided by labels. Nat Clim Chang 9:53–58. https://doi.org/10.1038/s41558-018-0354-z

Cao Y (Jessica), Cranfield J, Chen C, Widowski T (2021) Heterogeneous informational and attitudinal impacts on consumer preferences for eggs from welfare enhanced cage systems. Food Policy 99. https://doi.org/10.1016/j.foodpol.2020.101979

Cardona M, Izquierdo D, Barat JM, Fernández-Segovia I (2023) Intrinsic and extrinsic attributes that influence choice of meat and meat products: techniques used in their identification. Eur Food Res Technol 249:2485–2514. https://doi.org/10.1007/s00217-023-04301-1

Clark B, Stewart GB, Panzone LA, Kyriazakis I, Frewer LJ (2017) Citizens, consumers and farm animal welfare: a meta-analysis of willingness-to-pay studies. Food Policy 68:112–127. https://doi.org/10.1016/j.foodpol.2017.01.006

Daneshvar E, Sik Ok Y, Tavakoli S, Sarkar B, Shaheen SM, Hong H, Luo Y, Rinklebe J, Song H, Bhatnagar A (2021) Insights into upstream processing of microalgae: a review. Bioresour Technol 329. https://doi.org/10.1016/j.biortech.2021.124870

Gkarane V, Ciulu M, Altmann BA, Mörlein D (2020) Effect of alternative protein feeds on the content of selected endogenous bioactive and flavour-related compounds in chicken breast meat. Food Secur 9:392. https://doi.org/10.3390/foods9040392

Gracia A, Loureiro M, Nayga R (2011) Valuing an EU animal welfare label using experimental auctions. Agric Econ 42. https://doi.org/10.1111/j.1574-0862.2011.00543.x

Henchion MM, McCarthy M, Resconi VC (2017) Beef quality attributes: a systematic review of consumer perspectives. Meat Sci 128:1–7. https://doi.org/10.1016/j.meatsci.2017.01.006

Hossain MS, Small BC, Hardy R (2023) Insect lipid in fish nutrition: recent knowledge and future application in aquaculture. Rev Aquac 15:1664–1685. https://doi.org/10.1111/raq.12810

Kress K, Millet S, Labussière É, Weiler U, Stefanski V (2019) Sustainability of pork production with immunocastration in Europe. Sustainability (Switzerland) 11. https://doi.org/10.3390/SU11123335

Kühl S, Schäfer A, Kircher C, Mehlhose C (2024) Beyond the cow: consumer perceptions and information impact on acceptance of precision fermentation-produced cheese in Germany. Future Foods 10:100411. https://doi.org/10.1016/j.fufo.2024.100411

Latacz-Lohmann U, Schreiner JA (2018) Assessing consumer and producer preferences for animal welfare using a common elicitation format. J Agric Econ 70. https://doi.org/10.1111/1477-9552.12297

Lee SM, Lee KT, Lee SH, Song JK (2013) Origin of human colour preference for food. J Food Eng 119:508–515. https://doi.org/10.1016/j.jfoodeng.2013.06.021

Li K, Liu Q, Fang F, Luo R, Lu Q, Zhou W, Huo S, Cheng P, Liu J, Addy M, Chen P, Chen D, Ruan R (2019) Microalgae-based wastewater treatment for nutrients recovery: a review. Bioresour Technol 291. https://doi.org/10.1016/j.biortech.2019.121934

Kühl S, Löbbert T, Schütz A (2024) How does information impact meat purchase decisions? A consumer survey on German husbandry labeling (Haltungsformkennzeichnung) at the point of sale. Fleischwirschaft International 2:30–33

Maroušek J, Strunecký O, Maroušková A (2023) Insect rearing on biowaste represents a competitive advantage for fish farming. Rev Aquac 15:965–975. https://doi.org/10.1111/raq.12772

Mörlein J (2020) History of pig castration. Digest J Foodways Cult 7:57–61

Mörlein D, Schübeler AS (2017) This is how the dialogue with the customers succeeds: vaccination against boar taint-how can the procedure be best communicated? Fleischwirtschaft 97:36–40

Nagappan S, Das P, AbdulQuadir M, Thaher M, Khan S, Mahata C, Al-Jabri H, Vatland AK, Kumar G (2021) Potential of microalgae as a sustainable feed ingredient for aquaculture. J Biotechnol 341:1–20. https://doi.org/10.1016/j.jbiotec.2021.09.003

Parlasca MC, Qaim M (2022) Meat consumption and sustainability. https://doi.org/10.1146/annurev-resource-111820

Poulsen Nautrup B, Van Vlaenderen I, Aldaz A, Mah CK (2018) The effect of immunization against gonadotropin-releasing factor on growth performance, carcass characteristics and boar taint relevant to pig producers and the pork packing industry: a meta-analysis. Res Vet Sci 119:182–195. https://doi.org/10.1016/j.rvsc.2018.06.002

Profeta A, Hamm U (2018) Consumers' expectations and willingness-to-pay for local animal products produced with local feed. Int J Food Sci Technol:1–9. https://doi.org/10.1111/ijfs.13933

Profeta A, Hamm U (2019) Do consumers prefer local animal products produced with local feed? Results from a discrete-choice experiment. Food Qual Prefer 71:217–227. https://doi.org/10.1016/j.foodqual.2018.07.007

Prunier A, Bonneau M, von Borell EH, Cinotti S, Gunn M, Fredriksen B, Giersing M, Morton DB, Tuyttens FAM, Velarde A (2006) A review of the welfare consequences of surgical castration in piglets and the evaluation of non-surgical methods. Anim Welf 15:277–289. https://doi.org/10.1017/S0962728600030487

Ramanathan R, Lambert LH, Nair MN, Morgan B, Feuz R, Mafi G, Pfeiffer M (2022) Economic loss, amount of beef discarded, natural resources wastage, and environmental impact due to beef discoloration. Meat Muscle Biol 6. https://doi.org/10.22175/mmb.13218

Rosenau S, Oertel E, Dietz C, Wessels S, Tetens J, Mörlein D, Ciulu M (2021) Total replacement of fishmeal by spirulina (Arthrospira platensis) and its effect on growth performance and product quality of african catfish (clarias gariepinus). Sustainability (Switzerland) 13. https://doi.org/10.3390/su13168726

Rosenau S, Ciulu M, Reimer C, Mott AC, Tetens J, Mörlein D (2022) Feeding green: spirulina (Arthrospira platensis) induced changes in production performance and quality of salmonid species. Aquac Res 53:4276–4287. https://doi.org/10.1111/are.15925

Rosenau S, Wolgast T, Altmann B, Risius A (2023) Consumer preference for altered color of rainbow trout (Oncorhynchus mykiss) fillet induced by spirulina (Arthrospira platensis). Aquaculture 572. https://doi.org/10.1016/j.aquaculture.2023.739522

Schütz A, Kurz K, Busch G (2022) Virtual farm tours—virtual reality glasses and tablets are suitable tools to provide insights into pig husbandry. PLoS One 17. https://doi.org/10.1371/journal.pone.0261248

Schütz A, Kühl S, Busch G, Mehlhose C (2023) Making animal welfare labelling more transparent—the potential of different information types from simple text to highly immersive stable tours via VR glasses. J Agric Food Res 14. https://doi.org/10.1016/j.jafr.2023.100712

Short EC, Kinchla AJ, Nolden AA (2021) Plant-based cheeses: a systematic review of sensory evaluation studies and strategies to increase consumer acceptance. Food Secur 10. https://doi.org/10.3390/foods10040725

Škrlep M, Tomašević I, Mörlein D, Novaković S, Egea M, Garrido MD, Linares MB, Peñaranda I, Aluwé M, Font-I-furnols M (2020) The use of pork from entire male and immunocastrated pigs for meat products—an overview with recommendations. Animals 10:1–26. https://doi.org/10.3390/ani10101754

Souza Duarte DA, Schroyen M, Reis Mota R, Vanderick S, Gengler N (2021) Recent genetic advances on boar taint reduction as an alternative to castration: a review. J Appl Genet 62:137–150. https://doi.org/10.1007/s13353-020-00598-w/Published

Spranghers T, Ottoboni M, Klootwijk C, Ovyn A, Deboosere S, De Meulenaer B, Michiels J, Eeckhout M, De Clercq P, De Smet S (2017) Nutritional composition of black soldier fly (Hermetia illucens) prepupae reared on different organic waste substrates. J Sci Food Agric 97:2594–2600. https://doi.org/10.1002/jsfa.8081

Stampa E, Schipmann-Schwarze C, Hamm U (2020) Consumer perceptions, preferences, and behavior regarding pasture-raised livestock products: a review. Food Qual Prefer 82. https://doi.org/10.1016/j.foodqual.2020.103872

St-Hilaire S, Cranfill K, McGuire MA, Mosley EE, Tomberlin JK, Newton L, Sealey W, Sheppard C, Irving S (2007) Fish offal recycling by the black soldier fly produces a foodstuff high in omega-3 fatty acids. J World Aquacult Soc 38:309–313

Thies AJ, Schneider F, Efken J (2021) The meat we do not eat. A survey of meat waste in German hospitality and food service businesses. Sustainability (Switzerland) 13. https://doi.org/10.3390/su13095059

Thies AJ, Altmann BA, Countryman AM, Smith C, Nair MN (2024) Consumer willingness to pay (WTP) for beef based on color and price discounts. Meat Sci 217:109597

Thies A, Holloway M, Altmann BA, Countryman AM, Smith C, Nair MN (2025) Addressing retail meat waste: A focus group study on discolored beef. Food and Humanity 4:100639

Tran HQ, Nguyen TT, Prokešová M, Gebauer T, Van DH, Stejskal V (2022) Systematic review and meta-analysis of production performance of aquaculture species fed dietary insect meals. Rev Aquac 14:1637–1655. https://doi.org/10.1111/raq.12666

Vanhonacker F, Verbeke W, Van Poucke E, Tuyttens FAM (2008) Do citizens and farmers interpret the concept of farm animal welfare differently? Livest Sci 116:126–136. https://doi.org/10.1016/j.livsci.2007.09.017

Verbeke W (2009) Stakeholder, citizen and consumer interests in farm animal welfare. Anim Welf 18:325–333. https://doi.org/10.1017/S0962728600000725

von Borell E, Bonneau M, Holinger M, Prunier A, Stefanski V, Zöls S, Weiler U (2020) Welfare aspects of raising entire male pigs and immunocastrates. Animals 10:1–12. https://doi.org/10.3390/ani10112140

The Consumer as a Change Agent in the Green Transition of the Food Sector

23

Klaus G. Grunert

1 Introduction

The transition of the food system towards a greater degree of sustainability in food production and a healthier diet requires changes from all actors in the food chain. All actors in the food chain contribute to greenhouse gas emissions and other environmental impacts. Farmers, food manufacturers and retailers form the supply of food products among which consumers can choose and therefore set boundary conditions for the healthiness of people's diets. Authorities, through regulations of the food supply, set standards for which products can enter the market and how they can be marketed.

However, consumers have a special role in this transition process (Grunert 2011). Consumers, by their choices, can speed up or delay the green transition. Healthier and more sustainable products can be brought to the market, but if consumers do not adopt them, these products will not contribute to the green transition. As is well known, a large share of the new products that are introduced to the market fail (Dijksterhuis 2016). There are many reasons for this—consumers do not like the taste, lack the convenience, do not trust the information about health effects, cannot find out how to integrate the product into their diet and think that the product is too expensive in relation to what it offers. The standard answer in marketing on how to deal with high failure rates in new product development is to become better in understanding consumer preferences and develop products that are aligned with them (Grunert et al. 2008). But the green transition involves changing some of these preferences—like reducing the amount of meat consumption, eating different kinds of meat and increasing the consumption of plant-based food. In addition, consumers may reject new products not only because of their characteristics but also because

K. G. Grunert (✉)
Aarhus University, Aarhus, Denmark
e-mail: klg@mgmt.au.dk

B. Grabkowsky, T. Blaha (eds.), *Intensive Livestock Production in Transition*,
https://doi.org/10.1007/978-3-031-97872-2_23

of the way they have been produced. Gene technology and irradiation are standard examples of how consumers can delay or even prohibit the adoption of new technologies in food production (Grunert et al. 2008).

From the consumer perspective, the green transition requires behaviour change—adopting new products, new meal patterns and new ways of cooking and handling waste. With regard to meat production, the discussion about necessary behaviour changes has been dominated by the call for less meat consumption, in line with nutritional recommendations. However, meat can be produced in more and less sustainable ways, and differences exist both between different types of farmed animals and also within those categories. Sustainable consumer choices when buying meat are an important part of the green transition.

Enacting this behaviour change is a task for all food chain actors. In order to tackle it, one needs a basic understanding of factors that affect behaviour change. This chapter aims to provide such a basic understanding with special emphasis on issues related to meat production.

2 The COM-B Model

People don't change their behaviour easily, especially not behaviours that are performed regularly and have become routine. Whether people change their behaviour is affected by a host of factors and there is a wide range of behaviour change models that try to group and systematize them. Many of them share a common core, though, distinguishing three groups of factors: those dealing with whether people want to change, whether they are able to change and whether the environment gives them the opportunity to change. This is the essence of (among others) the COM-B model (West and Michie 2020), which we will use in this chapter. The model is shown in Fig. 23.1.

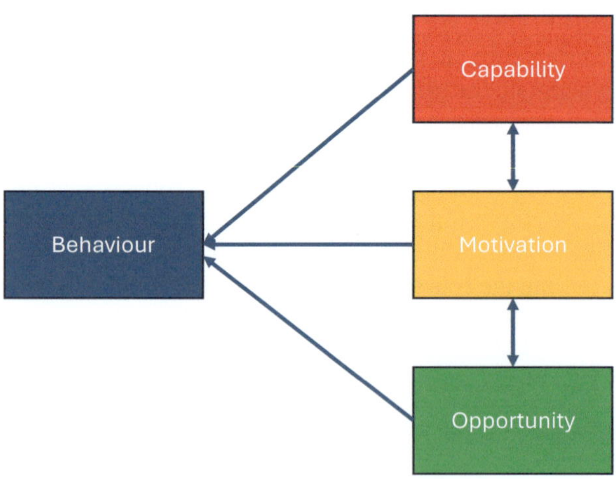

Fig. 23.1 The COM-B model of behavioural change

The basic premise of the COM-B model is that behaviour is affected by capability, opportunity and motivation. Capability refers to whether people think that they could perform the new behaviour if they wanted. People may think that they lack the cooking skills to prepare plant-based dishes or lack the insight that would allow them to distinguish between more and less sustainably produced meat. Opportunity refers to whether the environment supports the new behaviour. If we want consumers to choose healthier and more sustainable food products, we need to give them information on which products are healthier and which are more sustainable. Nutritional labelling addresses this issue for the healthiness aspect, whereas distinguishing the more from the less sustainable is still difficult. Motivation is whether people think that the behaviour change is desirable. If people think that climate change is a hoax, they will not be motivated for more sustainable eating.

As a general rule, all three factors need to be in place to bring about behaviour change. The COM-B model has been applied to a broad range of issues, including healthy and sustainable eating. It is important to realize, though, that it does not provide solutions—it merely attracts our attention to which problems need to be solved.

In the remainder of this chapter, we will use the COM-B model to structure our discussion of consumer behaviour change in the green transition of the food sector.

3 Motivation for Sustainable Eating

EIT Food carries out a yearly survey of consumers in 18 European countries and measures, among other matters, motivation for sustainable living by asking for agreement with three items: *I am concerned about the development of the global environment, I feel it is a moral obligation to use environmentally friendly products* and *It concerns me that people do not care enough for the environment.* Figure 23.2 shows that the majority of consumers agree with these statements—but also that the level of agreement actually has fallen from 2020 to 2024. This is a measure of motivation for sustainable living in general—not a measure of motivation for sustainable food choices.

When it comes to food, making more sustainable choices is not top of mind when it comes to changing one's diet. As Fig. 23.3 shows, the by far most frequently mentioned way in which people want to change their diet is towards a healthier diet.

Figure 23.3 shows the predominance of health concerns when it comes to changing one's diet, but it also shows that food choice is governed by multiple motives. This goes not only for desired changes with regard to one's diet but also for everyday food choices. A major study by Eurobarometer conducted in all EU countries asked people to select the three most important criteria when selecting food (EU 2022). The three most frequently mentioned items were taste, cost and safety. Of

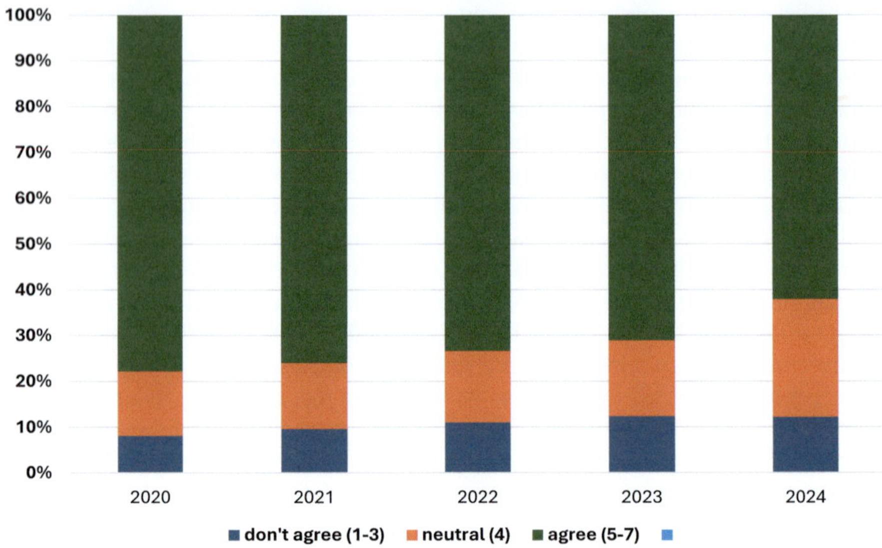

Fig. 23.2 Motivation for sustainable living in 18 European countries. Based on data from the EIT Food TrustTracker, 7-point scale with 1 = totally disagree and 7 = totally agree, mean of items *I am concerned about the development of the global environment, I feel it is a moral obligation to use environmentally friendly products,* and *It concerns me that people do not care enough for the environment.*

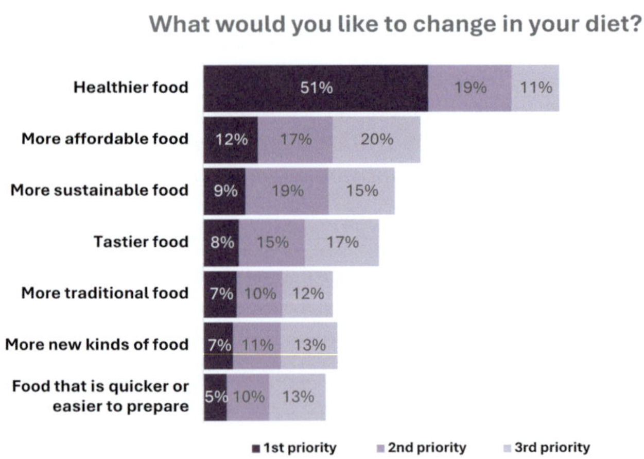

Fig. 23.3 Desired changes to one's diet in 18 European countries. Based on data from the EIT Food Trust Tracker, respondents indicated first, second and third priority from a list

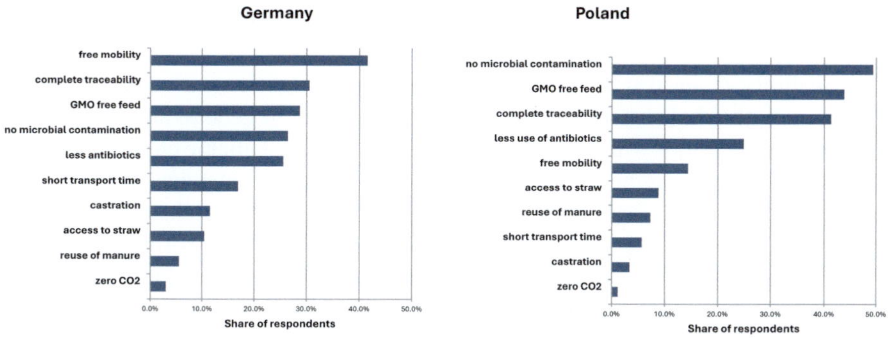

Fig. 23.4 Most important ethical attributes of pork meat in Germany and Poland. Respondents had to pick the two most important attributes from a list. (Grunert et al. 2018)

course, this is neither new nor surprising—taste has always been the major driver of food choice, weighed up against the price of the food (Grunert 2005). Still, the way in which consumers view food quality has expanded, now including—in addition to taste—convenience, healthiness and to some extent even sustainability, but changes are slow and aspects of sustainability are most likely to influence food choice when they do not conflict with the other criteria and when the price premium for added sustainability is limited.

Several studies have by means of choice experiments analysed the impact of sensory attributes, price and various attributes related to sustainability in the choice of meat products (e.g. Eldesouky et al. 2020; Grunert et al. 2018; Peschel et al. 2024). These studies show that while there is a minority of consumers that is not interested in sustainability at all or at least are not willing to pay any premium for it, the majority do. When looking at various aspects of sustainability, those related to animal welfare range clearly at the top. For example, in a study conducted in Germany and Poland, out of ten attributes related to ethical aspects of meat production, CO_2 emissions were ranked last in terms of importance to consumers, with all animal welfare-related aspects rated higher (Fig. 23.4).

Taken together, the various results mentioned show a differentiated picture in terms of consumer motivation to make sustainable food choices, both in general and with regard to meat. A majority of consumers are motivated, but the motive for sustainable living can be in conflict with other motives that are related to aspects like pleasure, health, convenience and thriftiness. Those aspects that consumers view as important when choosing meat products are not necessarily those that are most important with regard to the environmental impact of meat production.

4 Ability of Consumers to Make Sustainable Choices

Sustainability is an abstract and complex concept. If consumers have difficulties grappling with what sustainability actually means, both in general and with regard to meat production, it will impair their ability to make sustainable choices.

The term itself may be a source of confusion. Different languages have used or coined different terms for it, both newly formed terms ('Nachhaltigkeit' in German), adaptations of existing terms ('sustainability' in English, 'sostenibilidad' in Spanish) and terms that also have another, everyday language meaning ('hållbarhet' in Swedish, 'durabilité' in French). Several studies have investigated the way in which consumers understand the term (Grunert et al. 2014; Hanss and Böhm 2012; Sánchez-Bravo et al. 2021). In a study in six European countries (Grunert et al. 2014), when asked to define the term in their own words, respondents in Germany, France, Spain and the United Kingdom linked it mostly to protection of the environment, whereas respondents in Poland linked it mostly to the more general issue of maintaining a standard of living. In Sweden, many respondents came up with the other, everyday language meaning of 'hållbarhet', namely the sell-by-date of a product. When asked to pick from a list of 18 potential sustainability aspects of food production, the items picked most often were related to the environmental dimension of sustainability (e.g. the *environmental impact of the use of land and water, the environmental impact of food production*), whereas items related to the ethical dimension of sustainability (e.g. *working conditions in food, child labour in food, world food supply*) were picked less often.

A recent literature review of studies on consumer perception of food sustainability concluded that consumers still lack knowledge about this domain (Van Bussel et al., 2022). It is clear that when people are fumbling with regard to what exactly sustainability means when buying food products, this impairs their ability to do so, even when the motivation is present.

Another possible impairment to the perceived ability to make sustainable food choices is related to the trust people have in food choice actors. Sustainability characteristics, no matter whether they are related to CO_2 emissions, animal welfare, land, and water use or fair-trade conditions, cannot be seen or smelled when inspecting the product, i.e. they are what in economic theory is called credence attributes (Fernqvist and Ekelund 2014). People need to be informed about them, and people will be confident to use this information only when they trust the information sources.

Consumer trust in the food chain, or rather the lack of it, is a major concern in the green transition. The green transition requires that consumers have trust in the information they receive about food products and feel confident to try and adopt new products that are produced in a more sustainable way. The EIT Food TrustTracker (Macready et al. 2020) has monitored consumer trust in food chain actors for the past years. As Fig. 23.5 shows, trust has been stable over the years, with about half of consumers exhibiting some or even a lot of trust and the other half exhibiting no or little trust. Consumers generally trust farmers more than the other actors in the food chain. Trust levels also differ between countries and are generally higher in

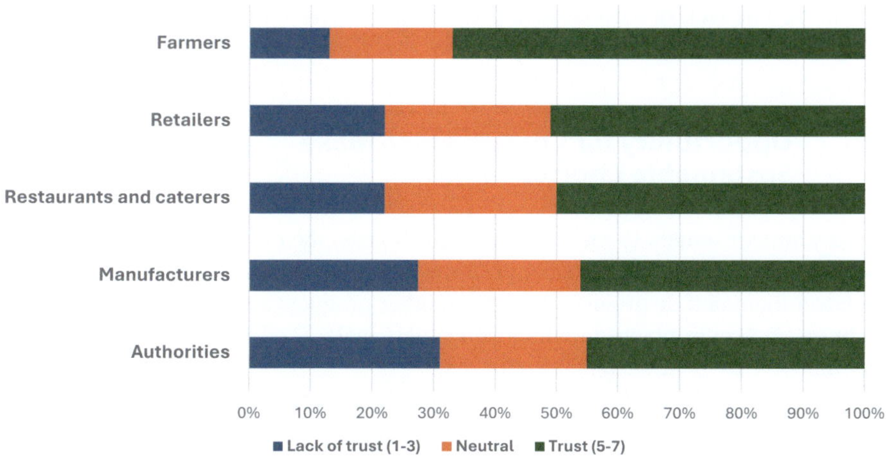

Fig. 23.5 Trust in food chain actors in 18 countries. Scale 1–7 with 1 = very little trust and 7 = very high level of trust. (Data from the EIT Food TrustTracker 2025)

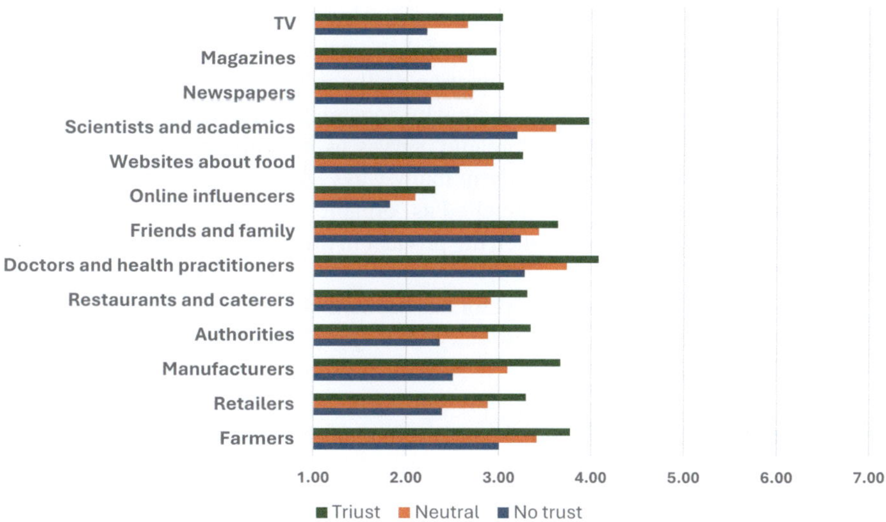

Fig. 23.6 Trust of consumers in different information sources about food. 7-point scale with 1 = very little trust and 7 = very high level of trust. Split by respondent groups according to Fig. 23.5. (Data from EIT Food TrustTracker 2025)

Northern compared to Southern Europe. Differences between countries are especially prevalent with regard to trust in authorities.

As Fig. 23.6 shows, the degree of trust that people have carries over to the trust they have in different information sources. Importantly, we can see that consumers with low levels of trust have lower trust levels also with regard to those information sources that are generally regarded as the most trustworthy, like doctors and scientists. The lack of trust is therefore a major barrier for consumers to become a positive agent in the green transition.

5 Opportunity for Consumers to Make Sustainable Choices

If we want consumers to make sustainable choices, the environment needs to provide two things: More sustainable products and information that allows consumers to detect them and distinguish them from other products already on the market. In addition, the environment in which consumers make food choices can be more or less inducive to make sustainable choices.

New product development is a main pillar in the green transition. This involves not only developing products that have been produced in a more sustainable way, but also products that avoid or reduce trade-offs, real or perceived, between the degree of sustainability and other product characteristics of relevance for consumers, not least taste and price. Many consumers associate sustainable food with organic production. In spite of some recent slumps, the organic sector has been a huge success, achieving in some product categories and in some countries market shares of 20% and more, and this in spite of the fact that many consumers have only a vague idea about what organic production really entails (Thøgersen et al. 2012). There has been considerable research on why consumers buy organic food, and one of the major conclusions is that 'organic' serves as a summary quality indicator, used to infer that the product is better in many ways (Aertsens et al. 2009). Indeed, research has shown that products labelled as 'organic' are perceived to taste better, even in a real taste test and even when the product in question in fact was not organic (Sörqvist et al. 2015).

Organic production in the meat sector has been much less successful than organic production in fruits, vegetables and dairy. A major element here is that the price premium for organic products in the meat sector is much higher than in these other categories, and this in turn is related to requirements like more space for the animals and slower growing times (for chicken). Attempts to position meat products somewhere in between conventional and organic have not been overly successful and are mainly related to animal welfare considerations. There is clearly room in livestock production for products that combine some of the less costly elements of animal welfare (e.g. space enrichment) with sustainability elements linked to factors like CO_2 emissions and land and water use.

As already noted, the success of any such products hinges on the ability of consumers to recognize them. Their advantages in terms of sustainability need to be communicated credibly to consumers, and here labelling is the major tool. Eco-labels are still in their infancy, though (Geldres-Weiss et al. in press). Those that exist on the market mostly inform consumers only about one particular aspect of sustainability, like animal welfare, fair trade or CO_2 footprint. Metalabels and eco-scores have been discussed (Torma and Thøgersen 2021), but their comprehensive introduction to the market has still not happened. A potential credibility problem can be mitigated by third-party certification. A more difficult question is what such a label should indeed cover. The EU product environmental footprint concept covers 17 different dimensions, and this does not even include animal welfare. A summary eco-score is easy to use for consumers but will need communicative backup in order to explain what it actually stands for—and what not.

While attempts to prompt consumers to make more sustainable food choices are relatively new, attempts to make them choose healthier food have a long history, and there is broad agreement that the provision of information alone in the choice environment is not enough to change choices (Capacci et al. 2012). The concept of 'nudging' has since been coined to describe a broader class of changes in the choice environment that can influence choice behaviour without necessarily requiring a lot of conscious action on the side of the consumer (Thaler and Sunstein 2008). The term 'nudging' covers a range of quite different instruments, like making the product to be promoted more accessible on the shelf, using social norms to influence behaviour and priming the motive to be bolstered by in-store stimuli. Much empirical research has been devoted to developing, testing and classifying nudges that produce positive results toward healthier food choices and more recently more sustainable food choices, although the effects reported are usually small (Cadario and Chandon 2020). The number of field studies is still limited, and those that exist suggest that behaviour changes in food service settings like canteens are easier to achieve than influencing in-store behaviour. Choosing a meal in a canteen or similar food service happens in a relatively closed environment with few available options. Choosing food in a store occurs in a vastly more complex environment, and when people come to the store, many decisions on what to buy and eat will already have been made, either explicitly or implicitly.

6 Implications for the Green Transition

We have in this chapter looked at the prospects of making the consumer an agent of change in the green transition, somebody who by his/her food choices speed up the purchase and consumption of more sustainably product food products. We have looked at this by analysing consumers' motivation for sustainable choices, their ability and the opportunities provided by the choice environment.

The analysis showed that a lack of motivation for sustainable living is not the major bottleneck, although it should not be taken for granted and may even have slightly declined over the years. However, food choice is governed by multiple motives, among which sustainability is only one and, for most consumers, not the most important. In addition, sustainability is a complex concept and consumer understanding of it is currently limited, meaning that consumers' ability to make sustainable choices is limited. Changing choice environments by better labelling and by other means of encouraging sustainable choices will help, but experience shows that the effects of such measures are small and need time to develop. Also, all food choices that people make in the store are subject to a post-purchase assessment once people get home and start incorporating the product in a meal, and if this assessment is not positive, there will be no long-term change in people's diets (Grunert and Hieke in press). Indeed, the choices that consumers make in the store are only a subset of the many choices that form people's diets—like deciding when to eat, how often to eat, with whom to eat, and designing meals based on iconic major meal components. Research on promoting the use of plant- or insect-based products suggests that while people are quite willing to try such products and, to some extent, incorporate them into their diets, they do not necessarily substitute meat (Maya et al. 2024). More sustainable meat production is a major element in the green transition and has the potential to result in products that consumers perceive as being both pleasurable and sustainable. This requires that such products are developed taking consumer preferences into account and that the members of the food chain work together to provide a choice environment where consumers will find credible and easily understandable information.

References

Aertsens J, Verbeke W, Mondelaers K, Van Huylenbroeck G (2009) Personal determinants of organic food consumption: a review. Br Food J 111:1140–1167

Cadario R, Chandon P (2020) Which healthy eating nudges work best? A meta-analysis of field experiments. Mark Sci 39:465–486

Capacci S, Mazzocchi M, Shankar B, Brambila Macias J, Verbeke W, Pérez-Cueto FJ, Kozioł-Kozakowska A, Piórecka B, Niedzwiedzka B, D'Addesa D (2012) Policies to promote healthy eating in Europe: a structured review of policies and their effectiveness. Nutr Rev 70:188–200

Dijksterhuis G (2016) New product failure: five potential sources discussed. Trends Food Sci Technol 50:243–248

Eldesouky A, Mesias FJ, Escribano M (2020) Consumer assessment of sustainability traits in meat production. A choice experiment study in Spain. Sustain For 12:4093

EIT Food Consumer Obsevatory (2025). Trust report 2024. Leuven: EIT Food. Can be downloaded from https://www.eitfood.eu/reports/trust-report-2024

EU. (2022) Food safety in the EU. European Union, Briussels

Fernqvist F, Ekelund L (2014) Credence and the effect on consumer liking of food–a review. Food Qual Prefer 32:340–353

Geldres-Weiss VV, Nicolas C, Massa NP (2024) Toward a sustainable research agenda on food eco-labelling in the business and management research domain. Eur J Manag Bus Econ 33:429–444.

Grunert K (2005) Food quality and safety: consumer perception and demand. Eur Rev Agric Econ 32:369–391

Grunert K (2011) Sustainability in the food sector: a consumer behaviour perspective. Int J Food Syst Dyn 2:207–218

Grunert K, Hieke S (2025) Nudging for more sustainable and healthy food choices: a dual processing approach with a time perspective. In: Reisch L, Sunstein C (eds) The Elgar companion to consumer behaviour and the UN sustainable development goals (pp. 191-204). Edward Elgar

Grunert K, Jensen BB, Sonne A-M, Brunsø K, Byrne DV, Clausen C, Friis A, Holm L, Hyldig G, Kristensen NH (2008) User-oriented innovation in the food sector: relevant streams of research and an agenda for future work. Trends Food Sci Technol 19:590–602

Grunert K, Hieke S, Wills J (2014) Sustainability labels on food products: consumer motivation, understanding and use. Food Policy 44:177–189

Grunert K, Sonntag W, Glanz-Chanos V, Forum S (2018) Consumer interest in environmental impact, safety, health and animal welfare aspects of modern pig production: results of a cross-national choice experiment. Meat Sci 137:123–129
Hanss D, Böhm G (2012) Sustainability seen from the perspective of consumers. Int J Consum Stud 36:678–687
Macready AL, Hieke S, Klimczuk-Kochańska M, Szumiał S, Vranken L, Grunert KG (2020) Consumer trust in the food value chain and its impact on consumer confidence: a model for assessing consumer trust and evidence from a 5-country study in Europe. Food Policy 92:101880
Maya C, Wilderspin D, Costa A, Cunha L, Roos N (2024) Introducing menus of three weekly insect-or plant-based dinner meals slightly reduced meat consumption in Danish families: results of a randomized intervention study. Appetite 203:107689
Peschel AO, Thomsen K, Tsalis G, Grunert KG (2024) The role of ethical orientation in animal welfare choice behaviour: a segmentation study. Food Qual Prefer 123:105334
Sánchez-Bravo P, Chambers E, Noguera-Artiaga L, Sendra E, Chambers E IV, Carbonell-Barrachina ÁA (2021) Consumer understanding of sustainability concept in agricultural products. Food Qual Prefer 89:104136
Sörqvist P, Haga A, Langeborg L, Holmgren M, Wallinder M, Nöstl A, Seager PB, Marsh JE (2015) The green halo: mechanisms and limits of the eco-label effect. Food Qual Prefer 43:1–9
Thaler RH, Sunstein m CR (2008) Nudge: improving decisions about health, wealth, and happiness. Yale University Press, New Haven
Thøgersen J, Jørgensen AK, Sandager S (2012) Consumer decision making regarding a "green" everyday product. Psychol Mark 29:187–197
Torma G, Thøgersen J (2021) A systematic literature review on meta sustainability labeling–what do we (not) know? J Clean Prod 293:126194
Van Bussel L, Kuijsten A, Mars M, Van't Veer, P. (2022) Consumers' perceptions on food-related sustainability: a systematic review. J Clean Prod 341:130904
West R, Michie S (2020) A brief introduction to the COM-B model of behaviour and the PRIME theory of motivation [v1]. Qeios

The Role of Sustainable Food Labels in Transforming the Food System

24

Achim Spiller

1 Consumer-Citizen Gap: The Contested Role of Consumers in the Transformation Process

Labels are product markings in the form of text and often images that neutrally assess certain properties of a food product. Labels are available not only to a single company but to all or at least a larger group of companies. The following chapter examines the role of such labels with a focus on Germany, considering key sustainability attributes.

The transformation process towards climate and environmental protection, social sustainability, animal welfare, and healthier nutrition presents significant challenges for agricultural and food industries. The "farmers' protests" in Germany and many other EU countries in 2024 clearly demonstrated that the sector is struggling with these demands. One explanation for these difficulties lies in the contradictions between societal demands on the one hand and consumer purchasing behaviour in the market on the other. The gap between citizens' expectations for sustainability and consumer behaviour at the checkout, the so-called Consumer-Citizen Gap, is relatively large. For example, 94% of citizens in Germany expect animals to have had a good life before being slaughtered (Hölker et al. 2019). However, the market share of animal welfare products in Germany has long been in a niche sector and has only recently started to slowly increase (Spiller et al. 2015; Kompetenznetzwerk Nutztierhaltung 2020). Therefore, the role of animal welfare labels is viewed with some scepticism in the German agricultural sector today.

Recently, there have been increasing suggestions to focus on changes on the supply side in transformation policy. For example, in Germany, the Kompetenznetzwerk

A. Spiller (✉)
Marketing for Food and Agricultural Products, Department of Agricultural Economics and Rural Development, Georg-August-Universität Göttingen, Göttingen, Germany
e-mail: a.spiller@agr.uni-goettingen.de

B. Grabkowsky, T. Blaha (eds.), *Intensive Livestock Production in Transition*,
https://doi.org/10.1007/978-3-031-97872-2_24

Table 24.1 Development stages of animal welfare labelling in the food market

Stage (and period in Germany)	Positions on animal welfare labelling
1. No labelling (prior to the early 2000s)	Animal welfare was not labelled; the topic was deliberately excluded by the industry
2. Label euphoria (from the early 2000s to the end of the 2010s)	Initial labels by NGOs, then the hesitant entry of the industry, also based on positive experiences in the Netherlands
3. Label scepticism (beginning in the mid-2010s).	Label jungle, niche labels, consumer-citizen gap, stakeholders distance themselves from labelling
4. Realistic expectations (necessary in the future)	Improvement of labelling policy, deeper understanding of the opportunities and limits of the labelling instrument

Source: Own presentation

Nutztierhaltung (2020), a high-profile and broadly composed stakeholder committee, directs its recommendations to the government. By raising the VAT from 7% to 19%, the state should generate a budget that it makes available to agricultural enterprises on the basis of long-term contracts to finance the additional costs of animal-friendly husbandry. Due to this subsidy, animal-friendly husbandry would not be more expensive. A change in consumer behaviour would not be necessary.

Regarding animal welfare labels, various stages of development can be distinguished (see Table 24.1 and Spiller & Zühlsdorf 2018). Until the early 2000s, there was only a small niche label (Neuland) and organic meat, both with very low market shares, in Germany. The main market players were opposed to animal welfare labelling. In the early 2000s, important animal welfare labels were developed with the animal welfare label of the German Animal Welfare Federation and later with the "Initiative Tierwohl" (i.e. Initiative for Animal Wellbeing). The first is organised by the largest Animal Welfare Association of Germany (Deutscher Tierschutzbund), the latter is organised by the voluntary national quality assurance system (QS) for the value chain for food of animal origin run by the five industry sectors: feed industry, farming community, slaughter industry, processors and retailers. Subsequently, several other labels were introduced, some by individual companies, some by regional initiatives. Since the mid-2010s, a veritable label jungle has developed (Sonntag et al. 2023). However, market success has remained limited. Due to the confusion, the first political initiatives for a state label emerged but initially failed. In the early 2020s, the leading companies in the German food retail sector introduced the "Haltungsform-label", which increasingly gained market traction. At the political level, a mandatory national animal husbandry label was adopted in 2023 to be introduced in 2025. Meanwhile, doubts about the contribution of labelling to solving the animal welfare problem have grown as demand remains subdued.

In Germany, the situation is currently characterised by the coexistence of an increasing number of animal welfare labels on the one hand and growing scepticism about the labelling instrument on the other. The proposal of the Kompetenznetzwerk Nutztierhaltung (2020), which has dominated the political and economic debate in the German animal husbandry community over the last 5 years, relies entirely on subsidising the supply side and explicitly justifies this with the Consumer-Citizen

Gap and the limits of labelling. In the following, I will argue why this very sceptical assessment is "throwing the baby out with the bathwater".

2 Limits of Supply-Side Regulation

There are two reasons to critically view a sole focus on animal welfare and overall sustainability policies on the supply side and subsidies. First, financing reaches its limits because, in an era of multiple crises, the state simply lacks the necessary funds, and tax increases are unpopular. For example, the proposed increase in VAT on animal products to enhance animal welfare has so far failed in Germany due to the liberal party's fundamental stance and the concern of other parties about a public outcry over rising meat prices.

Second, a significant portion of people want to make a contribution themselves, want to know how animal-friendly a product is and how they can shop more sustainably. Many consumers want to be part of the solution and demand transparency. Surveys show that they are well aware of their limitations and do not want to bear sole responsibility for solving the problem (Zühlsdorf et al. 2016). They see the main responsibility with businesses and politics. Nonetheless, most people want to be involved, to understand market contexts and to avoid being patronized. Labels are an important instrument of consumer empowerment.

However, this tension does not only concern animal welfare labels. Other sustainability labels are also very controversial. For many years, there has been a debate in Germany about a comprehensive health-related labelling system, modelled after the British "traffic light" food labelling system. After long lobbying battles, the Nutri-Score developed in France seemed to be gaining acceptance in Germany as well. An EU-wide introduction of the Nutri-Score, possibly even mandatory, seemed imminent in 2023. However, the EU Commission backtracked due to massive resistance from Italy and Hungary (EU Scientists 2023). There are also increasingly critical contributions in the German public discourse regarding the usefulness of the Nutri-Score. The spread of the Nutri-Score is stalling.

The debate over the importance of labels also extends to the issue of health-promoting foods. Here too, other instruments such as mandatory reformulation (changing recipes) regarding sugar, salt and fat content, advertising bans or binding guidelines for communal catering could be possible. Politics also struggles with corresponding proposals, such as those formulated by the Scientific Advisory Board of the Federal Ministry of Food and Agriculture in Germany (WBAE 2020; Martinez et al. 2021).

The examples of animal welfare labels and the Nutri-Score illustrate the dilemma of current labelling policies. Existing labels are often inadequately present in the market and get lost in the unregulated labelling jungle. At the same time, alternative policy instruments are hardly implemented in Germany.

3 The Four Central Dimensions of a More Sustainable Diet

It is widely accepted in scientific circles that agriculture and food can and should make a significant contribution to achieving key sustainability goals (Poore & Nemecek 2022). The Planetary Health Diet has provided an important impetus in the debate on a healthy and environmentally friendly diet (Willett et al. 2019). A core component of the Planetary Health Diet is a significant reduction in the consumption of animal products, which would lead to a reduction in animal husbandry, albeit not necessarily to the same extent. Over the past 10 years, meat consumption in Germany has indeed shifted slightly in this direction, decreasing by about 20% (BLE 2024). However, the consumption of dairy products has remained stable, with less fresh milk consumption being offset by increased cheese consumption. Further important steps towards a more sustainable diet would include increased consumption of legumes and vegetables, which are consumed in lesser amounts in Germany.

The Scientific Advisory Board on Agricultural Policy, Food and Consumer Health Protection (WBAE) at the Federal Ministry of Food and Agriculture in Germany (BMEL) has outlined a broader, four-pillar policy recommendation for sustainable nutrition in its report (2020) (see Fig. 24.1).

Regarding environmentally friendly foods, the organic label is by far the most widespread mark and is also the most well-known label in the German food market. For organic farming, a systematic and accredited control procedure to ensure the credibility of the labelling was first developed with EU certification in the early 1990s. Moreover, in the early 2000s, a national label was also introduced and promoted through a state information campaign. Organic farming represents environmentally friendly agriculture but is not a comprehensive label for all essential environmental dimensions of a food product along the entire value chain. Instead, it primarily signifies a specific form of agriculture that completely avoids synthetic chemical pesticides and synthetic fertilisers while guaranteeing higher animal husbandry standards. Other criteria, such as the environmental friendliness of packaging or transportation, are not evaluated. Additionally, organic farming is not the only

Fig. 24.1 The big four of sustainable food consumption. (Source: Own graphic based on WBAE 2020)

form of sustainable land management. For this reason, new labels have recently been developed to indicate the environmental friendliness of food products, following the concept of Product Life Cycle Assessment (LCA), also referred to by the EU as the Product Environmental Footprint (PEF). The most advanced debate in this regard concerns climate protection labelling (Lemken et al. 2021; Edenbrandt and Nordström 2023), leading to a rapidly emerging "carbon footprint landscape" (Deconinck et al. 2023).

However, climate protection labelling faces similar problems to those described above for animal welfare. There are numerous labelling initiatives. One of them is the particularly popular, simple, cost-effective and marketing-friendly label "climate-neutral" claim. It is based on offsetting greenhouse gas emissions through more or less reliable, verified offset projects, while the actual greenhouse gas emissions remain unclear. This has led to intense legal disputes between parts of the industry and consumer protection agencies over greenwashing in recent years. Only recently, there have been indications of a ban on this greenwashing and extended discussions about a reliable environmental and climate label on food products (Lemken et al. 2021; Zühlsdorf et al. 2024).

The fourth dimension of sustainable food consumption involves social aspects of production, with the Fairtrade label playing the most significant role in promoting fair producer prices in the Global South. In addition to the label issued by Fairtrade International, there are also other labels such as Rainforest Alliance or GGN (a system run by GLOBALG.A.P.). The prevalence of these labels varies. Fairtrade is considered more demanding and expensive, but it also occupies smaller market segments. Rainforest Alliance and GGN aim to address mass markets with lower standards (Kühl et al. 2017; Gassler and Spiller 2018).

Since there are trade-offs between the mentioned sustainability goals, all four dimensions of a more sustainable diet must be considered together, which significantly increases the complexity for both consumers and companies (Schütz et al. 2024).

4 Effectiveness of Labelling as a Policy Instrument for Credence Attributes

The described four central sustainability attributes are, in economic terms, credence and process features (Jahn et al. 2005), meaning attributes that cannot be reliably determined by consumers or other external parties at the final product stage. For example, evaluating animal welfare or determining greenhouse gas emissions is not possible by simply looking at a steak on the supermarket shelf. Therefore, there must be labelling based on process-oriented verification. Without such information, market-based governance fails, leading to adverse selection, where consumers focus solely on price due to a lack of reliable quality signals (Akerlof 1970).

Given the limited information-processing capacity of consumers, it is impractical for each provider to develop its own approach. Such a "label jungle", where consumers neither recognise the signs nor can assess their reliability, does not help overcome market failures. A manageable number of reliable labels is needed. This requires agreements on important methodological standards, such as problem weighting, evaluation algorithms and verification standards (Lemken et al. 2021; Ramos et al. 2022).

Moreover, consumer information involves not just labels or quality marks. Advertising is the primary communication tool, particularly advertising claims and key marketing messages. In reality, consumers do not distinctly differentiate between labels and claims. Advertising statements like "boosts the immune system" or "climate-neutral" are also perceived as sustainability indicators and can have a stronger impact than labels (Zühlsdorf et al. 2023). To prevent misleading influence through advertising claims, the EU has initially set guidelines in the food sector with the Health Claims Regulation and the Food Information to Consumers Regulation. Most importantly, health-related advertising claims must be scientifically proven and approved; the same applies to nutritional labels like the Nutri-Score.

Similarly, the European Union is now aiming to regulate "green claims", which will need to be substantiated according to precise rules in the future (vzbv 2023). However, the details of the implementation are still open. The EU has not yet presented specific proposals for the design of the envisaged Sustainability Labelling Framework, contrary to its initial plans. Therefore, many market observers are currently looking to France, which has taken the lead in developing a state environmental label. A climate or environmental label is expected to be legally anchored there shortly (Zühlsdorf et al. 2024).

Summarising the developments in the four central labelling areas over recent years in Germany and many other countries, three problem areas can be distinguished (see Table 24.2): there are too many labels on the market, the existing labels are largely unknown and not very successful, and sustainability progress remains limited because the labels often have qualitative deficiencies.

Without delving into all the issues outlined in Table 24.2, it is clear that there is an implementation deficit in sustainability labelling. The reasons are multifaceted. Companies in competition are often not interested in transparency through a neutral label but favour their own marketing approaches that promise a unique selling proposition. For some NGOs with a well-known trademark (like WWF), labels are instruments for generating revenue. Politics often contributes to the problem through regulatory deficits and a limited understanding of the market. Additionally, the challenges of labelling are sometimes underestimated. Labels are mistakenly considered simple and cost-effective instruments, although experiences, such as with the state animal welfare label in Germany, clearly show the challenges. On the consumer side, knowledge about sustainability is sometimes low, and preferences are correspondingly unstable. Increasingly polarised societal debates, for example, about climate protection and nutrition policy, also contribute to labels being attacked and discredited in media debates (EU Scientists 2023). Overall, it is a delicate mix of market, policy and discourse failures.

Table 24.2 Central problem areas of food labelling

Problem area	Description
Label jungle	
Too many labels	Companies, associations, and NGOs introduce a multitude of signs into the market
Fragmented single-topic labels	Companies, associations, and NGOs develop labels for individual sub-dimensions of the four sustainability dimensions, leading to halo effects
Fake labels	Companies use signs that resemble neutral labels but are not subject to external control
Competing claims	Companies use terms in advertising that are perceived similarly to a label by the audience
Ineffective labels	
Poor label design	Labels, including state labels, are sometimes unprofessionally designed and not intuitively understandable
Lack of promotion	Many labels are hardly advertised. The low communication pressure limits their effectiveness in marketing
Public discreditation	Labels are sometimes the subject of political controversy and lose credibility with the audience
Qualitatively problematic labels	
Deficient criteria and indicators	Some criteria and measurement indicators do not sufficiently correspond to the state of science
Inadequate verification of participating companies	Reliable, neutral certification is sometimes lacking

Source: Own compilation

5 Labelling as Part of a Coordinated Policy Mix

A sustainability transformation of food consumption can be promoted with various policy instruments (WBAE 2020). For example, steering taxes can make the consumption of products with low animal welfare more expensive. Nudging can bring climate-friendly options to the forefront in communal catering or supermarkets. Reformulation requirements can make processed food recipes healthier and more climate-friendly, etc. In the scientific discussion, a broad mix of instruments from various policy measures is increasingly recommended to effectively influence consumer behaviour. Drawing on the relatively successful policy to reduce tobacco consumption, it seems sensible to use a range of measures from information and education campaigns, mandatory labelling, advertising bans, as well as voluntary and mandatory nudging, to product bans (Spiller et al. 2017). However, nutrition policy is controversial in many countries, and major political initiatives are not currently expected in Germany.

Improving consumer information and education, on the other hand, appears politically much easier to implement. Such soft measures are widely accepted in society and only cautiously interfere with market processes. Therefore, they are also less controversial economically. Regarding education, more action-oriented approaches are being discussed, which in schools could, for example, link nutritional education with school catering. However, most discussions revolve around the area of sustainability labelling of products.

Market failure occurs not only when there is no information about a product attribute but also when there are too many, including false or misleading, seals and claims on food packaging. The "right to be informed" (Kennedy 1962) is undermined for credence attributes if the state does not set an appropriate framework. In such cases, market and policy failures complement each other in a fatal way.

To prevent this, the EU has set guidelines in the health sector with the Health Claims Regulation and the Food Information to Consumers Regulation. The Nutritional Declaration is mandatory on food products. Health-related advertising claims must be scientifically proven and approved; the same applies to nutritional labels like the Nutri-Score. The European Union is now aiming for similar regulations for so-called "green claims" and environmental labels. There is still much room for professionalising label policies, but its elements are increasingly emerging (Weinrich and Spiller 2016; Torma and Thøgersen 2021; Zühlsdorf et al. 2024):

- Only a few preferably mandatory national labels for key sustainability attributes instead of a multitude of single-item labels
- Multi-level labels with a colour-coded, interpretative implementation similar to the Nutri-Score (multi-level labels instead of binary signs)
- Scientifically sound evaluation algorithms based on defined indicators, reducing lobby influence
- Intensive communicative support to publicise the trustworthy signs

The extreme scepticism towards sustainability labels that is sometimes expressed today is nevertheless exaggerated. The example of the organic label, with market shares of more than 10% in Germany and significantly higher shares in Denmark, Austria and Switzerland, shows that a credible label combined with supply-side instruments can indeed make an important contribution to transformation processes (ZKL 2021). This requires a realistic assessment of the potential of sustainability labelling and stricter, well-designed regulation. This also includes a stronger orientation towards science and a limitation of lobby influence.

References

Akerlof GA (1970) The market for "lemons": quality uncertainty and the market mechanism. Q J Econ 84(3):488–500

BLE (2024) Bericht zur Markt- und Versorgungslage mit Fleisch 2024, Bonn. https://www.ble.de/SharedDocs/Downloads/DE/BZL/Daten-Berichte/Fleisch/2024BerichtFleisch.pdf?__blob=publicationFile&v=2#:~:text=Durchschnittlich%20wurden%20in%20Deutschland%20pro,19%2C9%20kg%20Geflügelfleisch%20verbraucht

Deconinck K, Jansen M, Barisone M (2023) Fast and furious: the rise of environmental impact reporting in food systems. Eur Rev Agric Econ 50(4):1310–1337. https://doi.org/10.1093/erae/jbad018

Edenbrandt AK, Nordström J (2023) The future of carbon labeling – factors to consider. Agric Resour Econ Rev 52(1):151–167. https://doi.org/10.1017/age.2022.29

EU Scientists & Health Professionals for Nutri-Score (2023) Why the European Commission must choose the Nutri-Score Nutrition Label, Report. https://nutriscore-europe.com/wp-content/uploads/2023/01/NS_rapport-EU-V10_230202.pdf

Gassler B, Spiller A (2018) Is it all in the Mix? Consumer preferences for segregated and mass balance certified sustainable palm oil. J Cleaner Prod 195:21–31. https://doi.org/10.1016/j.jclepro.2018.05.039

Hölker S, Steinfath H, Meyer-Höfer Mv, Spiller A (2019) Tierethische Intuitionen in Deutschland: Entwicklung eines Messinstrumentes zur Erfassung bereichsspezifischer Werte im Kontext der Mensch-Tier-Beziehung. German J Agric Econ 68(4):299–315

Jahn G, Schramm M, Spiller A (2005) The reliability of certification: quality labels as a consumer policy tool. J Consum Policy 28(1):53–73

Kennedy JF (1962) Consumer bill of rights. https://www.mass.gov/info-details/consumer-bill-of-rights

Kompetenznetzwerk Nutztierhaltung (2020) Empfehlungen des Kompetenznetzwerks Nutztierhaltung, Stakeholder-Empfehlung, Berlin. https://www.bmel.de/SharedDocs/Downloads/DE/_Tiere/Nutztiere/200211-empfehlung-kompetenznetzwerk-nutztierhaltung.pdf?__blob=publicationFile&v=3

Kühl S, Gassler B, Spiller A (2017) Labeling strategies to overcome the problem of niche markets for sustainable milk products: the example of pasture-raised milk. J Dairy Sci 100(6):5082–5096

Lemken D, Zühlsdorf A, Spiller A (2021) Improving consumers' understanding and use of carbon footprint labels on food: proposal for a climate score label. EuroChoices 20(2):23–28. https://doi.org/10.1111/1746-692X.12321

Martinez J, Spiller A, Renner B et al (2021) Faire Ernährungsumgebungen gestalten. Zugleich eine Replik auf die Stellungnahme von Di Fabio zum WBAE-Gutachten „Politik für eine nachhaltigere Ernährung". ZLR (Zeitschrift für das gesamte Lebensmittelrecht) 5:589–614

Poore J, Nemecek T (2022) Reducing food's environmental impacts through producers and consumers. Science 360(6392):987–992. https://doi.org/10.1126/science.aaq0216

Ramos S, Segovia L, Melado-Herreros A et al (2022) Enviroscore: normalization, weighting, and categorization algorithm to evaluate the relative environmental impact of food and drink products. npj Sci Food 6:54. https://doi.org/10.1038/s41538-022-00165-z

Schütz A, Faletar I, Sonntag WI, Christoph-Schulz I, Spiller A (2024) Tierschutz first? Wie wichtig sind unterschiedliche Nachhaltigkeitsaspekte in der Nutztierhaltung: Ein systematischer Vergleich. J Consum Prot Food Saf 19:21–27. https://doi.org/10.1007/s00003-024-01480-1

Sonntag WI, Lemken D, Spiller A, Schulze M (2023) Welcome to the (label) jungle? Analyzing how consumers deal with intra-sustainability label trade-offs on food. Food Qual Prefer 104:1–13. https://doi.org/10.1016/j.foodqual.2022.104746

Spiller A, Zühlsdorf A (2018) Haltungskennzeichnung und Tierschutzlabel in Deutschland: Anforderungen und Entwicklungsperspektiven. Wissenschaftliches Gutachten erstellt im Auftrag von Greenpeace e.V. https://www.greenpeace.de/sites/www.greenpeace.de/files/publications/haltungskennzeichnung-tierschutzlabel-gutachten-15.10.2018.pdf

Spiller A et al (2015) Wege zu einer gesellschaftlich akzeptierten Nutztierhaltung, Gutachten des Wissenschaftlichen Beirats für Agrarpolitik beim Bundesministerium für Ernährung und Landwirtschaft, Berichte über Landwirtschaft, Sonderheft 221

Spiller A, Zühlsdorf A, Nitzko S (2017) Instrumente der Ernährungspolitik. Ein Forschungsüberblick – Teil 1. Ernährungs. Umschau 64(3):M146–M153. https://doi.org/10.4455/eu.2017.012

Torma G, Thøgersen J (2021) A systematic literature review on meta sustainability labeling – what do we (not) know? J Cleaner Prod 293. https://doi.org/10.1016/j.jclepro.2021.126194

vzbv (2023) Green Claims stoppen, Green Claims regulieren, Positionspapier des Verbraucherzentrale Bundesverbandes Februar 2023. https://www.vzbv.de/sites/default/files/2023-03/23-02-22_vzbv_Positionspapier%20Green%20Claims.pdf

WBAE (2020) Politik für eine nachhaltige Ernährung. Eine integrierte Ernährungspolitik entwickeln und eine faire Ernährungsumgebung gestalten. Gutachten. Berlin. https://www.bmel.de/SharedDocs/Downloads/DE/_Ministerium/Beiraete/agrarpolitik/wbae-gutachten-nachhaltige-ernaehrung.html

Weinrich R, Spiller A (2016) Developing food labelling strategies: multi-level labelling. J Clean Prod 137(2016):1138–1148

Willett W et al (2019) Food in the anthropocene: the EAT–Lancet Commission on healthy diets from sustainable food systems. Lancet., Published online January 16, 2019. https://doi.org/10.1016/S0140-6736(18)33179-9

Zühlsdorf A, Spiller A, Kühl S, Gauly S (2016) Wie wichtig ist Verbrauchern das Thema Tierschutz? Präferenzen, Verantwortlichkeiten, Handlungskompetenzen, Politikoptionen. Zusammenfassender Ergebnisbericht zur repräsentativen Umfrage im Auftrag des Verbraucherzentrale Bundesverband e. V. (vzbv), Göttingen. https://www.uni-goettingen.de/de/document/download/682825f9a6c855772775e3e196d37b6d.pdf/Verbraucherbefragung%20Tierschutz_Ergebnisbericht%20mit%20Kurzzusammenfassung_08.01.16.pdf

Zühlsdorf A, Kühl S, Radda D, Spiller A (2023) Grüne Marketing-Claims auf Lebensmitteln. Verbraucherstudie zum Verständnis von umwelt- und klimabezogenen Werbeaussagen. Zusammenfassender Ergebnisbericht zur repräsentativen Umfrage im Auftrag des Verbraucherzentrale Bundesverband e. V. (vzbv), Göttingen. https://jimdo-storage.global.ssl.fastly.net/file/9268e025-6dc3-4ab6-91b6-607b0dfa8792/Verbraucherstudie%20Grüne%20Marketingclaims.pdf

Zühlsdorf A, Schulze-Ehlers B, Spiller A (2024) Wegweiser dringend gesucht! Hintergrund, Entwicklungsstand und Herausforderungen von Klima- und Umweltlabeln auf Lebensmitteln. Ernährungs Umschau, 3/24: M 142-M 153

Zukunftskommission Landwirtschaft (ZKL) (2021) Zukunft. Landwirtschaft. Eine gesamtgesellschaftliche Aufgabe. Empfehlungen der Zukunftskommission Landwirtschaft, Berlin. https://www.bmel.de/SharedDocs/Downloads/DE/Broschueren/abschlussbericht-zukunftskommission-landwirtschaft.pdf?__blob=publicationFile&v=10

Future Requirements for Animal Husbandry Structures

25

Alfons Balmann

1 Introduction

Over the past decades, global production of and demand for animal-based products have substantially increased and are supposed to further increase due to an increasing world population and changing consumption patterns driven by increasing incomes in developing and emerging economies (OECD/FAO 2021). Production growth occurs through increasing stock sizes as well as through intensification and productivity increases. However, these processes differ substantially across as well as within countries and sectors. These differences include production structures characterised by farms or herd size distributions, the organisation of farms, the used technologies and breeds, as well as the organisation of the supply chains. While the poultry sector is highly integrated, including contract farming, the dairy and beef sectors show rather low levels of integration. The organisation of the pork sector is very heterogeneous across countries: On the one hand, highly integrated systems exist, such as Smithfield Foods in the USA, while on the other hand, the German pork sector shows a low level of integration.

Despite the differences regarding structures of farms and supply chains across sectors, certain similarities, characteristic drivers and mechanisms exist. Structural changes in animal husbandry can be characterised by increasing farm sizes, increasing knowledge and capital intensity as well as standardisation to address the needs of modern global food supply chains (Boehlje 1999; Balmann et al. 2006). Due to the globalisation of food chains and the increasing role of international trade, this process has been fostered additionally by the so-called supermarket revolution in the former socialist countries in the 1990s and in developing countries since the

A. Balmann (✉)
Leibniz Institute of Agricultural Development in Transition Economies (IAMO), Halle (Saale), Germany
e-mail: balmann@iamo.de

B. Grabkowsky, T. Blaha (eds.), *Intensive Livestock Production in Transition*,
https://doi.org/10.1007/978-3-031-97872-2_25

early 2000s (Reardon and Timmer 2012). Key elements of the supermarket revolution were high food safety and quality standards to gain and secure consumer trust, particularly through certification systems, such as GlobalG.A.P., as well as vertical integration and coordination (e.g. contract farming). These processes require adequate governance through, e.g. focal companies such as large processors or retailers (Hanf and Gagalyuk 2009). According to Boehlje (1999), a key challenge for such chains is adequate risk management, particularly if there is high capital intensity. Moreover, participation in these globalised supply chains requires the ability to adapt to the required standards.

Against this background, and in order to identify the future requirements for animal husbandry structures, the remainder of this section presents current and future drivers of change. In addition, it will be discussed how the diverse animal husbandry sectors in Western countries, such as Germany, can cope with these drivers as well as what this means for future production structures.

2 Current and Upcoming Drivers of Change

2.1 Societal Concerns

In recent years, animal husbandry in most Western economies, and more recently also in emerging and transition countries, has been and continues to be increasingly confronted with societal concerns on environmental impacts, greenhouse gas emissions and animal welfare. Concerns and debates about surface and groundwater pollution from intensive animal husbandry with phosphorus and nitrogen, in, e.g. Western Germany, date back to the 1980s (e.g. Rat der Sachverständigen für Umweltfragen 1985; Conrad 1990). Though there were substantial improvements, such as modification of feedstuff composition in order to reduce the nitrogen and phosphorus content of manure and increasing storage capacity for manure, enormous problems remain (BMU and BMEL 2020). Moreover, animal husbandry contributes substantially to agricultural greenhouse gas emissions through methane from ruminants as well as carbon dioxide and nitrous oxide from feedstuff production (e.g. WBAE 2020). Animal welfare issues have been a matter of public concern in rather rich developed countries, such as Germany, since the mid-1970s. An early example is the industrial use of cages for laying hens. Despite significant improvements, such as the replacement of cages for laying hens by floor and free-range systems, substantial animal welfare problems remain, partly driven by the enormous focus of breeding and housing on productivity (e.g. WBA 2015).

Resolving these well-documented problems will require fundamental and challenging adjustments. One reason for these needs results from trade-offs. Higher animal welfare standards usually require more feed for the same amount of product, which means more land for feed production and/or higher production intensities. Another reason is related to the economic costs of fixing the existing problems. The Scientific Advisory Board on Agricultural Policy of the German Federal Ministry of Food and Agriculture estimated that a substantial improvement in animal welfare of

German animal husbandry would cost around three to five billion Euros annually (WBA 2015). This corresponds to an increase in meat production costs by some 20–30%. Though many opinion polls have found that consumers claim to be willing to pay higher prices for improved animal welfare, real buyers' behaviour seems to contradict these claims (Clark et al. 2017). One reason for this attitude-intention gap (Vermeir and Verbeke 2006) or citizen-consumer gap (de Bakker and Dagevos 2012) is that consumers face a social dilemma because animal welfare is a public good (Vermeir and Verbeke 2006). Given that animal welfare products are comparatively more expensive, it is individually rational for consumers to buy products based on low animal welfare standards. Partially, retailers responded to this dilemma by taking the lead. Examples are the de-listing of cage eggs from German supermarkets in 2009, or specific programs such as the Dutch initiative "Beter Leven" or the German "Initiative Tierwohl" that establishes labels or funds to support farmers producing at higher animal welfare standards (WBA 2015; WBAE 2020). These initiatives are likely driven by the interest of large retailers in avoiding scandals and being blamed for not being responsible. The existence of certain market power can ease such initiatives. As an alternative to market solutions, agricultural policies can address these problems either through regulations, subsidies for farmers supporting preferred practices, or by taxation. However, policymaking is facing restrictions which may prevent political solutions. Examples include budget limitations, international trade agreements, lobbying activities as well as public beliefs about agricultural practices.

2.2 Changing Consumption Patterns

Globally, meat consumption is expected to continue to increase, with particularly strong growth in poultry (OECD/FAO 2021). This is different in the EU and Germany. EU per capita meat consumption tends to stagnate at around 67–69 kg annually but shows a steady substitution of pork, beef and sheep meat by poultry (European Commission 2021). Per capita, poultry consumption in the EU increased during the decade 2011–2021 from 19.6 kg to 23.5 kg (i.e. by 20%) and is expected to increase further.

In recent years, Germany showed a substantial decline in meat consumption (Fig. 25.1). Since 2018, total meat consumption per capita declined from 60.9 kg to 51.6 kg in 2023, which means a decline by 15% within 5 years. Compared to 2010, it declined even by 18%. Particularly strong was the decline of pork consumption, which fell between 2018 and 2023 from 34.1 kg to 27.5 kg per capita, i.e. by minus 19%. Compared to 2010, pork consumption declined even by almost 30%. The consumption of beef is also declining in Germany, though at a lower speed and with some fluctuations. While it fell between 2018 and 2023 by 17%, the decline since 2010 is only 7%. German poultry consumption is currently rather stable and fluctuates around 13 kg per capita after an increase of some 15% since 2010.

Whether and to what extent the recent demand shifts in Germany will continue is difficult to assess. Demographics, such as the aging of the population and

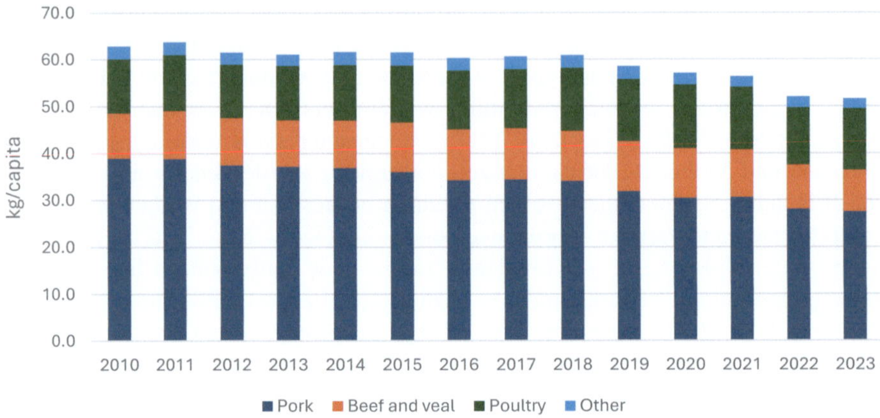

Fig. 25.1 The development of meat consumption in Germany. (Source: BZL 2024)

immigration of people who reject pork consumption, are one important driver. However, demographic change can hardly explain the huge recent shifts. Other drivers are to be seen in the increasing awareness of animal welfare and environmental as well as health issues, which may also explain the increasing share of vegetarian and vegan consumers in the population (Forsa 2023). Moreover, the market for alternatives to animal-based products is growing. The dynamics will depend not just on consumer attitudes but also on the progress in fields such as precision fermentation or cell-based substitutes (GFI 2024).

2.3 Demographic Change Effects on Production

In addition to its impacts on consumption, demographic change will also have significant implications for future production. Figure 25.2 shows a substantial share of higher age groups among farm managers in the EU in 2020. About one-third of all farm managers in the EU are older than 65. Another quarter of the farmers are between 55 and 65 years old. Only about 20% are younger than 45. In part, these figures result from specific regional effects, such as a large share of small and semi-subsistence farms in the south-east of the EU, which are often managed by elderly, often retired people. Moreover, there are specific sample effects resulting from rather small farms without a successor, which continue farming most likely just until retirement. These groups of farms are large in number, but their share in production is rather limited. A much bigger production share applies to intensive animal husbandry, which is dominated by larger and commercially oriented businesses. These farms are particularly dependent on hired labour. The results of the EU Farm Survey 2023 show for the north-west of Germany, with its very high livestock intensity that about 60% of the persons working in agriculture are permanent and seasonal hired employees (Destatis 2024). Overall, some 55% of persons working in German agriculture are permanent and seasonal hired workers with an increasing

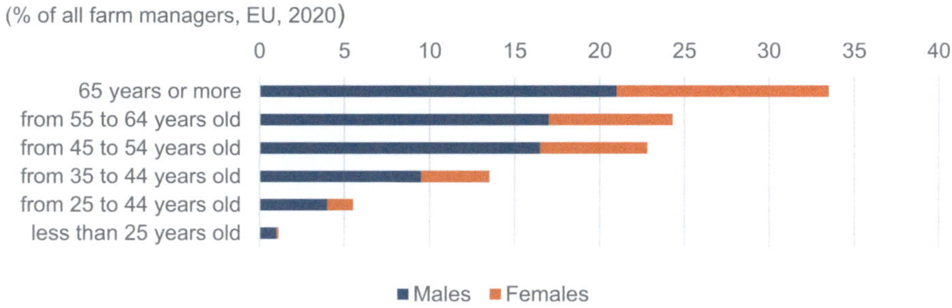

Fig. 25.2 Age classes of farm managers, by gender. (Source: Eurostat 2022)

share of migrants. The salary level is relatively low compared to other sectors of the economy. While the amount of family labour decreased, the number of permanent hired farm workers increased. This tendency has to be seen in contrast to the fact that the Boomer generation will retire by the early 2030s, and the following Z and Alpha generations are much smaller in number, resulting in increased competition for qualified labour across sectors and regions. These trends will not just affect the number of farm managers but particularly hired labour, which is quite important for modern intensive animal husbandry. Particularly qualified labour will become scarce and increasingly expensive.

2.4 Climate Change Issues

Animal husbandry contributes directly and indirectly to greenhouse gas (GHG) emissions through, e.g. digestion (CH_4), manure management (N_2O) and feed production (CO_2, N_2O). Reducing GHG emissions of animal husbandry can occur by reductions of animal-based food consumption (i.e. of meat, dairy, eggs) or by reducing GHG emissions per unit of production. While the first approach, i.e. reducing the production of animal-based foods, would be quite straightforward if consumption patterns would change in parallel, the latter approach, i.e. increasing production efficiency, has to be viewed in a more nuanced way. Several studies show substantial differences in carbon footprints across countries and production systems. For dairy production, e.g. Mazzetto et al. (2022) and Sorley et al. (2024) find that carbon footprints depend on, e.g. intensity, housing or grazing, fodder base and management. For the pork and poultry sector, there is considerable variation, depending on, e.g. conventional versus organic production, feeding diets, genetics and manure management (Kool et al. 2009; van Grinsven et al. 2019). Given this heterogeneity, both private and public initiatives should consider reducing carbon footprints by optimising production systems. However, there are trade-offs, as certain strategies to reduce carbon footprints may have a negative impact on animal welfare or biodiversity.

A particular problem for approaches to reducing the carbon footprint of animal husbandry results from the fact that GHG emissions cause global damage. The public good nature of mitigating carbon emissions creates low political and economic incentives. A related problem is the leakage effects of policies aimed at reducing the national carbon footprints of animal-based products (WBAE/WBW 2016). If measures to reduce the carbon footprint in a given country lead to a reduced domestic supply of animal-based products, while consumers do not respond in the same way, prices for these products are likely to increase. This, in turn, may create incentives to import more and thus to increase production abroad, potentially resulting in even higher carbon emissions. In this regard, the international accounting practice of determining country-specific carbon emissions based on production alone and not considering consumption equally may create adverse incentives if there are, e.g. no border adjustment taxes or other measures that internalise the social costs of GHG emissions. As long as such measures are not applied, policies aiming to reduce animal husbandry need to consider potential leakage effects, e.g. by applying the concept of comparative advantages not only in economic terms but also in terms of emissions.

In addition to the need to reduce the carbon footprint, animal husbandry will also be affected by climate change through, e.g. an increased competition of food versus feed, heat stress for animals and diseases (Rojas-Downing et al. 2017), and requires also adaptation to climate change through, e.g. adjustments of production and management systems, breeding strategies and farmers' perception and adaptive capacities.

2.5 Technological Drivers

Modern digitalisation is driven in particular by the exponential increase in computational power combined with the declining costs of modern computers and the growth of the Internet, which connects not only people but also things. Traditional automatisation allowed for the substitution of labour, such as automated feeding or milking robots. Modern technologies allow us to collect enormous amounts of data during their use. Connecting these databases enables the generation of "big data", which can be exploited by statistical tools and, in particular, by machine learning. The automated collection of data from the monitoring of individual animals, as well as of herds and flocks, will enable to develop unknown opportunities for an improved and automated identification of technological, management and health problems as well as autonomous and optimised responses regarding, e.g. feeding and environmental conditions (Neethirajan 2020). These new technologies will further increase the already enormous capital intensity of modern livestock systems. Based on existing technologies, the creation of one job in hog feeding in Germany requires an initial investment of more than one million Euros (Deblitz et al. 2024). This means not only that capital costs (i.e. depreciation and interests) are already now a multiple of labour costs in intensive animal husbandry, but also that access to capital and the ability to cope with associated financial risks are decisive for the future development of the sector (Boehlje 1999). Digitalisation is likely to raise the capital-labour ratio significantly.

A second technological driver must be seen in biotechnological advancements such as genome sequencing, molecular marker technology and genome editing, which will allow faster and more specific breeding (e.g. Oliver and Bill 2022). Moreover, modern biotechnology provides new opportunities for veterinary medicine and the treatment of feedstuff for improved digestibility and addition of micronutrients, e.g. through precision fermentation.

Given the globalisation of modern food systems, the future development of animal husbandry will decisively depend on the accessibility of technological innovations and the ability to implement them. This depends on and will be linked to the public and private institutional frameworks that can hinder or promote implementation. Recent examples of frictions for German pork producers can be seen in the failure of the pork supply chain to allow for immunocastration as an alternative to surgical castration of pigs (Kress et al. 2019) or the ban of GMO-based feed for higher animal welfare labels. In globalized food systems, such frictions cause comparative disadvantages and risks for producers, which may lead to leakage effects.

2.6 Sectoral Drivers and Frictions

The previously mentioned drivers refer to external push and pull factors of structural change in animal husbandry. However, there are internal drivers of the animal husbandry sectors as well. In particular, there are huge differences among the farms' performance indicators. Table 25.1 shows structural and performance indicators

Table 25.1 Key figures of German specialised full-time animal husbandry farms according to profitability groups (average of financial years 2020/21, 2021/22, 2022/23)

Farm type	**Units**	**Dairy and cattle**		
		–	**middle**	+
Farm size standard-output	1000 €	193	228	329
Farmed land	ha	77	75	96
Labor units	Units	1. 1	1.2	1.7
Cows	Number	38	56	90
Milk yield	kg/cow	7590	8218	8809
Profit	€/ha	237	989	1657
Profit	€/farm	17,802	74,067	161,821
Farm type	**Units**	**Pigs and poultry**		
		–	**middle**	+
Farm size standard-output	1000 €	403	388	607
Farmed land	Ha	64	69	92
Labor units	Units	1.2	1.1	1.7
Milk yield	Kg/cow	5464	8126	9377
Piglets per sow (born)	Piglets/sow	25. 8	28. 3	31. 1
Profit per ha	€/ha	−162	988	1771
Profit	€/farm	−10,536	69,550	163,377

BMLEH (2025)

from the German Farm Accountancy Network for full-time family farms, which are specialised in dairy and cattle or in pigs and poultry. They are classified into three profitability groups (i.e. the upper, middle lower third) for the financial years 2020/21, 2021/22 and 2022/23. Although not all farms are in the same groups every year, several conclusions can be drawn. A key point is that there are enormous differences in profitability. While the top third of farms achieved an average profit of more than 160,000 €, the bottom third had an average profit close to zero or even negative. Comparing the lower groups of poorly performing farms and the middle group, there are rather small size differences while the enterprises in a group of very profitable farms are on average substantially larger. There are also substantial productivity differences. The poorly performing dairy farms have milk yields that are about 7% lower than the middle group, while the more profitable farms have yields that are about 7% higher than the middle group. Compared to the middle group, the number of piglets per sow is almost 10% lower for the less profitable farms, while it is about 10% higher for the more profitable farms. These differences in productivity and profitability will have substantial effects on structural change within the sector. The less-performing farms are likely to leave the sector.

These differences in profitability provoke the question of why they emerged, and in particular, why non-profitability can persist at all given the high competition among farmers. Several explanations for these phenomena relate to sunk costs. If farms invest in modern and expensive stables for livestock production, a large share of the initial investment costs is sunk and cannot be recovered by stopping production (Balmann et al. 1996). Continuing production may allow to recover at least a part of the investment outlays. Such sunk costs do not just result from investments in physical capital but also in human capital, i.e. farmers may continue to produce even if their labour income is lower than that of average workers. Specific reasons are concerns of becoming unemployed or mobility costs. Considering that (a) investments and disinvestments in animal husbandry are usually irreversible, there is (b) an ongoing and substantial uncertainty on returns, and that (c) investment decisions can be postponed, has specific implications for investment decisions. Under such conditions, the real options approach to investments should be applied for decision-making instead of traditional approaches such as the net present value approach (Odening et al. 2005). Pieralli et al. (2017) find empirically a substantial reluctance of dairy farmers to stop production even if farming becomes unprofitable, particularly in times of high volatility—which is in accordance with the real options theory. According to the real options approach, postponing an irreversible disinvestment of an expected unprofitable activity can be rational in the presence of persistent uncertainty because it accounts for the value of waiting for new information. Musshoff et al. (2013) and Sandri et al. (2010) find empirical and experimental evidence that farmers wait even too long to disinvest compared to the suggestions of the real options approach. Reasons may, e.g. result from emotional frictions.

3 Consequences and Challenges

3.1 Policy Level

Figure 25.3 illustrates the structural problem of identifying a social optimum given the existing trade-offs between production and environmental objectives. The area below the blue curve illustrates the space of possibilities. The blue curve represents the set of technologically efficient points, i.e. the Pareto set. The social optimum would be represented by the point of tangency of the Pareto set and the social indifference curve of preferences. Due to existing inefficiencies, such as outdated technologies, suboptimal farm sizes, and management deficiencies, the reality or status quo is under usual conditions far below the Pareto set and located somewhere in the possibility space. Under such conditions, a rational policy focusing on win-win opportunities would be to try shifting reality towards the upper right in order to exploit efficiency potentials which would allow for higher production and a better environment. Analysing real policy processes and discourses leads however to the impression that, e.g. environmental and farmers' lobby groups try to shift reality either to the upper left or lower right. One reason could be that the simplified illustration of the environment versus production trade-off ignores further relevant objectives of the actors, such as social or animal welfare dimensions. This could include the objective of avoiding structural change because certain farmers may have no ability to adjust and would lose.

Another friction that hinders moving towards Pareto superior or Pareto efficient solutions may be found in a discourse failure. One particular reason for discourse failure is public misconception, such as believing in the universal superiority of certain types of farming, such as organic, local, small and peasant farming. Irrespective of the specific benefits of such types of farming, Rausser et al. (2019) understand such beliefs as a *Naturalist Paradigm* which ignores scientific evidence. They argue that "current findings demonstrate that restrictions on industrial

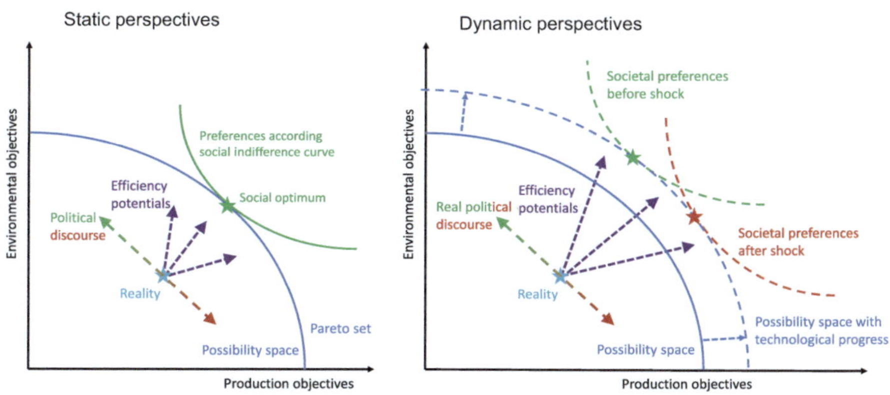

Fig. 25.3 Coping with trade-offs between production and societal objectives

practices imposed by policy or consumer preferences sacrifice productivity, thereby imposing environmental costs and raising food prices" and that the "intended benefits of such practices for human, animal, or environmental well-being must be weighed against the costs they impose". Based on Thompson's (2010) distinction between an *agrarian vision* and an *industrial vision* of agriculture, Jauernig et al. (2020) argue that the ongoing public debate between both sides is driven by political rent-seeking, although there is common ground in the underlying values. Despite an ongoing institutionalisation of corporate social responsibility of business actors to address societal concerns, De Olde and Valentinov (2019) understand moral complexity as a further reason preventing the transformation of ethical responsibilities into legal and economic responsibilities. Practically, this means that business actors, as well as political institutions, may not be able to address societal concerns in an economically and socially efficient way and may even fail to find effective solutions.

Managing trade-offs and inefficiencies may even be more important as well as challenging in dynamic environments in which due to technological progress the Pareto frontier is shifted outwards or when societal preferences change due to certain shocks due to scandals or due to new situations such as before and after a war. In such cases, negative effects for potential losers may be even larger. Moreover, there may be even bigger uncertainties and disputes in win-win situations.

In recent years, several attempts have been made to overcome discourse failures in the agricultural sector. Examples for Germany are the Competence Network Animal Husbandry (Kompetenznetzwerk Nutztierhaltung), established in 2019, and the Commission for the Future of Agriculture (Zukunftskommission Landwirtschaft), established in 2020. Both aimed to identify pathways for the sustainable development of the livestock sector and the entire agricultural sector in Germany. Both commissions provided concepts and visions on how to link societal and business perspectives (Kompetenznetzwerk Landwirtschaft 2020; Zukunftskommission Landwirtschaft 2021). However, both proposals are not yet realised in 2025. The main problem is that the requested financial means for implementation have not been provided. Indeed, the required financial means are considerable and amounting to several billion Euros per year. The German Competence Network Animal Husbandry proposed to establish public funding of some 80–90% of the additional costs of higher animal welfare standards and to guarantee this support for up to 20 years. The proposed annual budget just for the pork sector is estimated to be initially some 1.2 billion Euro in 2025, rising to 2.4 billion € in 2030 and 3.6 billion Euro in 2040. However, apart from the question of how to establish the necessary funding, the provided visions and concepts may not be sufficiently convincing. With reference to Fig. 25.3, it is questionable whether the proposals really provide a substantial step towards the upper right to achieve improvements in environmental, animal welfare, productivity and economic dimensions, or whether they address, e.g. the societal objectives in a very or even too costly way.

3.2 Business Level

From a business perspective, political frictions cause frustrations and can be seen as one of the reasons that provoked the fierce farmer protests in Germany and across the EU in 2024. Many farmers, particularly those engaged in animal husbandry, feel insecure. In the absence of clear rules and perspectives, many farmers are not able or willing to invest in the construction or reconstruction of expensive stables. Other farmers may even struggle with the decision to continue or exit animal husbandry.

However, given the list of drivers of change presented in the previous section (e.g. consumption patterns, demographics, climate change, technologies), it should be obvious that there are currently many uncertainties on how the animal husbandry sectors will develop in the future. Political uncertainties and risks are just one uncertainty among others. Thus, even a clear political framework will not provide guarantees for livestock farmers that ensure, e.g. profitability or competitiveness of certain technologies or standards in the future. Guaranteed compensation of costs for higher animal welfare standards for the next 20 years, as proposed by the Competence Network Animal Husbandry, may be particularly attractive for those farmers who underestimate market and technological dynamics. Even the political use of a time horizon of 20 years of promised compensations may already result in a normative framing that farmers "should" base their decisions on investment planning horizons of 20 years. Such expectations and norms may lead to frustrations for investing farmers if, e.g. current technologies are becoming quickly outdated or if prices are not developing as expected. Given the opportunity to postpone an irreversible investment and given an ongoing uncertainty, Odening et al. (2005) suggest to apply the real options approach to investments. According to the real options approach, farmers should consider the opportunity costs of postponing slightly profitable investments and wait for new information which allows to expect a significantly positive net return.

Despite farmers' requests for political planning reliability may be justified for certain reasons, investment decisions of farmers should be based on entrepreneurship and reflections about local, national and international comparative advantages. Political uncertainties should be addressed by entrepreneurial assessments. If farmers do not invest due to unresolved political uncertainties, these non-investments should also be considered as an entrepreneurial decision. Vice versa, investments stimulated mainly by political promises may be considered as fatalistic dependence on future political support.

3.3 Societal Level

Despite the hybrid nature of people to be citizens and consumers at the same time, the roles and perspectives of consumers differ from those of citizens. While citizens have good reasons to demand more sustainable animal husbandry and higher animal welfare standards, consumers have good reasons to spend their money according to their budget, individual preferences and supply. However, this does not mean that there is no good argument for consumers to buy products according to their sustainability and animal welfare standards (de Bakker and Dagevos 2012). This is particularly true if higher standards are hardly more expensive than conventional products. Moreover, citizens may have good reasons to reflect critically on their own beliefs. Possibly, cheap products are cheap due to efficient production rather than due to low environmental or animal welfare standards. The need for critical reflection also applies to the promises of the suppliers who advertise their products with nostalgic framing, such as cows grazing in touristic landscapes or produced by small farmers. A need for critical self-reflection on the part of citizens may also be derived from the existing trade-offs. Thus, consumers and citizens may reflect on their awareness of, e.g. the environmental implications of higher animal welfare.

References

Balmann A, Odening M, Weikard HP, Brandes W (1996) Path-dependence without increasing returns to scale and network externalities. J Econ Behav Organ 29(1):159–172. https://doi.org/10.1016/0167-2681(95)00055-0

Balmann A, Dautzenberg K, Happe K, Kellermann K (2006) On the dynamics of structural change in agriculture: internal frictions, policy threats and vertical integration. Outlook Agric 35(2):115–121. https://doi.org/10.5367/000000006777641543

BMU, BMEL (2020) Nitratbericht 2020 - Gemeinsamer Bericht der Bundesministerien für Umwelt, Naturschutz und nukleare Sicherheit sowie für Ernährung und Landwirtschaft. https://www.bmuv.de/fileadmin/Daten_BMU/Download_PDF/Binnengewaesser/nitratbericht_2020_bf.pdf

Boehlje M (1999) Structural changes in the agricultural industries: how do we measure, analyze and understand them? Am J Agric Econ 81(5):1028–1041. https://doi.org/10.2307/1244080

BZL – Bundesinformationszentrum Landwirtschaft (2024) Fleisch und Geflügel—Versorgungsbilanzen. https://www.ble.de/DE/BZL/Daten-Berichte/Fleisch/fleisch_node.html

BMLEH (2025) Archiv Buchführungsergebnisse Landwirtschaft. https://www.bmelstatistik.de/landwirtschaft/testbetriebsnetz/testbetriebsnetz-landwirtschaft-buchfuehrungsergebnisse/archivbuchfuehrungsergebnisse-landwirtschaft/

Clark B, Stewart GB, Panzone LA, Kyriazakis I, Frewer LJ (2017) Citizens, consumers and farm animal welfare: a meta-analysis of willingness-to-pay studies. Food Policy 68:112–127. https://doi.org/10.1016/j.foodpol.2017.01.006

Conrad J (1990) Nitratdiskussion und Nitratpolitik in der Bundesrepublik Deutschland, ISBN 3-89404-101-3, Edition Sigma, Berlin. https://www.econstor.eu/bitstream/10419/122902/1/209724.pdf

de Bakker E, Dagevos H (2012) Reducing meat consumption in today's consumer society: questioning the citizen-consumer gap. J Agric Environ Ethics 25:877–894. https://doi.org/10.1007/s10806-011-9345-z

de Olde EM, Valentinov V (2019) The moral complexity of agriculture: a challenge for corporate social responsibility. J Agric Environ Ethics 32:413–430. https://doi.org/10.1007/s10806-019-09782-3

Deblitz C, Rohlmann C, Verhaagh M, Krause M, Hartmann W (2024) Berechnung der einzelbetrieblichen Kosten einer Teilnahme am "Bundesprogramm zur Förderung des Umbaus der Tierhaltung" in der Schweinehaltung. Braunschweig: Johann Heinrich von Thünen-Institut, 36 p, Thünen Working Paper 233, https://doi.org/10.3220/WP1708069947000

Destatis (2024) 7% weniger Arbeitskräfte in der Landwirtschaft von 2020 bis 2023. Pressemitteilung Nr 176 vom 3. Mai 2024. https://www.destatis.de/DE/Presse/Pressemitteilungen/2024/05/PD24_176_41.html

European Commission (2021) EU agricultural outlook for markets, income and environment 2021–2031, Luxembourg. https://agriculture.ec.europa.eu/document/download/6ffe6666-676f-4d54-adb7-a36f4d15449a_en?filename=agricultural-outlook-2021-report_en_0.pdf

Eurostat (2022) Farmers and the agricultural labour force—statistics. https://ec.europa.eu/eurostat/statistics-explained/index.php?oldid=431368

Forsa (2023) Pflanzenbetonte Ernährung – Ergebnisse einer repräsentativen Bevölkerungsbefragung. https://www.bvlh.net/fileadmin/redaktion/downloads/pdf/2023/forsa-Umfrage_Pflanzenbetonte_Ern%C3%A4hrung.pdf

GFI – Good Food Institute (2024) Plant-based: Meat, seafood, eggs, and dairy. 2023 State of the Industry report. https://gfi.org/wp-content/uploads/2024/04/2023_State-of-the-Industry-Report-Plant-based-meat-seafood-eggs-and-dairy.pdf

Hanf J, Gagalyuk T (2009) Supply chain quality and its managerial challenges: insights from Ukrainian agri-food business. J East Eur Manage Stud 14(4):332–356

Jauernig J, Pies I, Thompson PB, Valentinov V (2020) Agrarian vision, industrial vision, and rent-seeking: a viewpoint. J Agric Environ Ethics 33:391–400. https://doi.org/10.1007/s10806-020-09830-3

Kompetenznetzwerk Landwirtschaft (2020) Empfehlungen des Kompetenznetzwerks Nutztierhaltung. https://www.bmel.de/SharedDocs/Downloads/DE/_Tiere/Nutztiere/200211-empfehlung-kompetenznetzwerk-nutztierhaltung.pdf

Kool A, Blonk H, Ponsioen T, Sukkel W, Vermeer HM, de Vries JW et al (2009) Carbon footprints of conventional and organic pork: assessments of typical production systems in The Netherlands, Denmark, England and Germany. Blonk Milieu Advies BV, Gouda. https://library.wur.nl/WebQuery/wurpubs/fulltext/50314

Kress K, Millet S, Labussière É, Weiler U, Stefanski V (2019) Sustainability of pork production with immunocastration. Europe Sustain 11(12):3335. https://doi.org/10.3390/su11123335

Mazzetto AM, Falconer S, Ledgard S (2022) Mapping the carbon footprint of milk production from cattle: a systematic review. J Dairy Sci 105:9713–9725. https://doi.org/10.3168/JDS.2022-22117

Musshoff O, Odening M, Schade C, Maart-Noelck SC, Serena S (2013) Inertia in disinvestment decisions: experimental evidence. Eur Rev Agric Econ 4(3):463–485. https://doi.org/10.1093/erae/jbs032

Neethirajan S (2020) The role of sensors, big data and machine learning in modern animal farming. Sens Bio-Sens Res 29:Article 100367. https://doi.org/10.1016/J.SBSR.2020.100367

Odening M, Mußhoff O, Balmann A (2005) Investment decisions in hog finishing: an application of the real options approach. Agric Econ 32:47–60. https://doi.org/10.1111/j.0169-5150.2005.00004.x

OECD/FAO (2021) Agricultural outlook 2021–2030. Chapter 6. https://openknowledge.fao.org/server/api/core/bitstreams/af0d6d72-b15b-46d6-af82-4949ddc0d004/content

Oliver A, Bill T (2022) Animal biotechnology—past, present and future. J Appl Biotechnol Bioeng 9(1):8–10. https://doi.org/10.15406/jabb.2022.09.00278

Pieralli S, Hüttel S, Odening M (2017) Abandonment of milk production under uncertainty and inefficiency: the case of western German farms get access arrow. Eur Rev Agric Econ 44(3):425–454. https://doi.org/10.1093/erae/jbx001

Rat der Sachverständigen für Umweltfragen (1985) Umweltprobleme der Landwirtschaft, Sondergutachten, März 1985. Verlag W. Kohlhammer, Stuttgart und Mainz. https://multimedia.gsb.bund.de/SRU/Dokumente/1985_SG_Umweltprobleme_der_Landwirtschaft.pdf

Rausser, Gordon C. and Sexton, Steven and Zilberman, David, (2019). The economics of the naturalist food paradigm Ann Rev Resour Econ, 11(1), 217–236, https://doi.org/10.1146/annurev-resource-100516-053623

Reardon T, Timmer CP (2012) The economics of the food system revolution. Annu Rev Resour Econ 4(1):225–264

Rojas-Downing MM, Nejadhashemi AP, Harrigan T, Woznicki SA (2017) Climate change and livestock: impacts, adaptation, and mitigation. Clim Risk Manag 16:145–163. https://doi.org/10.1016/j.crm.2017.02.001

Sandri S, Schade C, Mußhoff O, Odening M (2010) Holding on for too long? An experimental study on inertia in entrepreneurs' and non-entrepreneurs' disinvestment choices. J Econ Behav Organ 76(1):30–44. https://doi.org/10.1016/j.jebo.2010.02.011

Sorley M, Casey I, Styles D, Merino P, Trindade H, Mulholland M, Resch Zafra C, Keatinge R, Le Gall A, O'Brien D, Humphreys J (2024) Factors influencing the carbon footprint of milk production on dairy farms with different feeding strategies in western Europe. J Clean Prod 435:140104. https://doi.org/10.1016/j.jclepro.2023.140104

Thompson PB (2010) The agrarian vision: sustainability and environmental ethics. University Press of Kentucky, Kentucky

van Grinsven HJM, van Eerdt MM, Westhoek H, Kruitwagen S (2019) Benchmarking eco-efficiency and footprints of Dutch agriculture in European context and implications for policies for climate and environment. Front Sustain Food Syst 3:13. https://doi.org/10.3389/fsufs.2019.00013

Vermeir I, Verbeke W (2006) Sustainable food consumption: exploring the consumer "attitude—behavioral intention" gap. J Agric Environ Ethics 19:169–174. https://doi.org/10.1007/s10806-005-5485-3

WBA – Wissenschaftlicher Beirat für Agrarpolitik beim Bundesministerium für Ernährung und Landwirtschaft (2015) Wege zu einer gesellschaftlich akzeptierten Nutztierhaltung. Gutachten, Berlin. https://www.bmel.de/SharedDocs/Downloads/DE/_Ministerium/Beiraete/agrarpolitik/GutachtenNutztierhaltung.pdf

WBAE – Scientific Advisory Board on Agricultural Policy, Food and Consumer Health Protection of the BMEL (2020) Promoting sustainability in food consumption—developing an integrated food policy and creating fair food environments. Expertise, Berlin. https://www.bmel.de/SharedDocs/Downloads/EN/_Ministry/Scientific_Advisory_Board-Promoting-sustainability-food-consumption.pdf

WBAE/WBW – Scientific Advisory Board on Agricultural Policy, Food and Consumer Health Protection and Scientific Advisory Board on Forest Policy (2016) Climate change mitigation in agriculture and forestry and in the downstream sectors of food and timber use. Executive Summary, Berlin. https://www.bmel.de/SharedDocs/Downloads/EN/_Ministry/ClimateChangeMitigation.pdf

Zukunftskommission Landwirtschaft (2021) Zukunft Landwirtschaft. Eine gesamtgesellschaftliche Aufgabe – Empfehlungen der Zukunftskommission Landwirtschaft, Berlin. https://www.bmel.de/SharedDocs/Downloads/DE/Broschueren/abschlussbericht-zukunftskommission-landwirtschaft.pdf

26 Areas of Responsibilities (Who Has to Take Responsibility for What) for the Well-Being of Food Animals

Sebastian Hartstang and Peter Kunzmann

1 On the Term of Responsibility

An operational definition of responsibility is the "attribution of actions and consequences of actions in light of value aspects" (Kunzmann 2010). Responsibility emerges when individuals steer events based on their own decisions, which can only be attributed to them if they have the opportunity to decide. Claiming responsibility for someone unable to realise it is nonsensical. Decisions relevant to responsibility can be viewed as "value preference decisions", where individuals select options that prioritise the realisation of their most important values or goods. Establishing this is not straightforward, particularly because animal welfare often lacks overriding priority among these goods.

Animal welfare, defined as the well-being and welfare of animals, holds immediate moral relevance under widely shared contemporary premises (Zühlsdorf et al. 2016). However, as a moral issue, animal welfare is inherently altruistic and challenging to enforce against competing goods like income and food safety.

Responsibility does not exist in isolation but rather emerges through discourse, conversation, debate and argumentation. It is crucial to discern whether responsibility is attributed by others (passive) or self-attributed and embraced (active). In the livestock farming debate, responsibility is wielded in two directions: self-attribution emphasises good intentions while acknowledging limitations, then shifts responsibility to others. This confirms the initial suspicion that responsibility

The remarks in this chapter are primarily based on the results of the research project "Systematics of responsibilities in the livestock sector (SysVerNu)". See the relevant publications (Kunzmann et al. 2020; Hartstang et al. 2022).

S. Hartstang (✉) · P. Kunzmann
University of Veterinary Medicine Hannover, Hannover, Germany
e-mail: sebastian.hartstang@tiho-hannover.de; peter.kunzmann@tiho-hannover.de

B. Grabkowsky, T. Blaha (eds.), *Intensive Livestock Production in Transition*,
https://doi.org/10.1007/978-3-031-97872-2_26

is being shifted; the rhetoric employed reveals it. Recognising this rhetoric facilitates its dismissal as a discourse accessory. Consequently, trust in players' language games may diminish aimed at changes that should realise goods like animal welfare.

2 A Question of Heuristics: Identifying Players

To exhibit a more realistic perspective, a matrix can be distilled from existing general models of responsibility (Bayertz 1995; Heidbrink et al. 2017; Lenk 2017; Loh 2017; Lübbe 1998; Isaacs 2017; Maring 2001; Ropohl 2002), enabling the identification of responsibility holders. It is noteworthy that typically only a select few entities are identified and debated as accountable for animal welfare and/or animal protection in livestock farming. The outcome is a comprehensive table facilitating the identification of potential recipients for appeals or demands. Alongside the usual suspects (cf. Hoischen-Taubner et al. 2014), the list encompasses individuals and groups seldom discussed in public discourse. While the inclusion of entities like "research funding agencies" or "trade press" may not be surprising, it broadens the spectrum to encompass subjects less frequently mentioned in advocating for animal welfare, potentially holding them accountable if necessary. The example of piglet castration highlights the significant influence wielded by the trade press. Thus, the objective of this list is to delineate necessary distinctions, such as not categorising "the media" broadly but distinguishing between the specialised/trade press and public media. Similarly, it is crucial to underscore the oft-neglected differentiation between "consumer" and "citizen".

Furthermore, the list can be utilised to identify and name the so-called hidden players—entities seldom addressed in mainstream discourse and sometimes not even recognised as responsibility bearers in specialised discussions. For instance, the critical domain of extension and advisors, pivotal in actualising animal welfare, is seldom externally acknowledged. Transitioning such entities from their "virtual responsibility" to tangible obligation can emerge as a vital policy sphere. Such a list is seldom exhaustive; nonetheless, it significantly expands the realm of possible recipients (cf. Kunzmann et al. 2020).

A brief insight into such a possible list is given here as an example. For the livestock sector, the actors involved can be categorised into the following main groups: consultants, animal transport, veterinary medicine, slaughtering, animal husbandry, equipment suppliers, food industry, veterinary pharmaceutical production, animal nutrition, general public, trade, press/media, education/research, chambers, associations, government, certification bodies, (informal) alliances. Within each of these main groups, a number of players are subsumed. For example, within the group "slaughter", the following actors can be found: official veterinarian for antemortem/post-mortem inspection, slaughterhouse operator, animal welfare officer, slaughterer, lairage supervisor and slaughterhouse technician. Within the main group "government", European, national, state and communal players can be identified (cf. Kunzmann et al. 2020).

A consequential revelation in the discourse on responsibility recipients is the identification of "bogus players". These collective entities, such as "consumers", "food retailers" and "farmers", are often attributed as responsibility in surveys and responses, yet they lack unified intentionality or group agency (cf. Albertzart 2015). Consequently, they cannot be held accountable. Conversely, contemplating "alliances"—orchestrated units of players, like food retailers within their often-oligopolistic structure—proves immensely valuable and pertinent.

Central categories can be delineated for sorting the myriad of players and groups. These classifications depart from action chains (breeder, fattener, slaughterer, etc.) and instead categorise players based on responsibility attributes. This includes discerning between those structurally weaker in power and those exercising influence, setting norms and thereby broadening their responsibility sphere. Notably, the interplay among players is not solely economic; information exchange, advisory roles, quality demands and mutual monitoring, among others, forge significant connections and intertwine responsibilities within this system.

3 Criteria for Attributing Responsibilities

In order to attribute responsibility in a meaningful manner, some criteria for doing so have to be named. These criteria allow us to characterise the spaces of action and power for each individual player, and consequently, the sense and extent of their responsibility.

We distilled it into four categories:

1. Type of contact with animals: non-existent, direct or intermediate
2. Means of establishing norms: legal or non-legal (e.g. via technical norms)
3. Power of influence: compelling or facultative (e.g. via consultation)
4. Sphere of influence: within their own realm or extending into others' realms of action

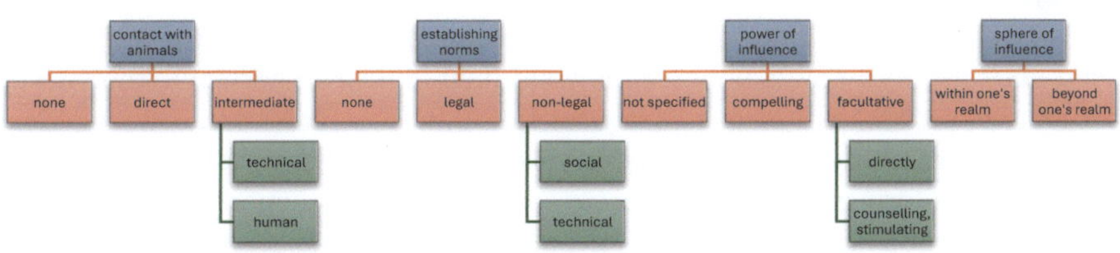

Chart: overview of the categories

Chart: overview of the categories

Applying these categories to the list of players yields a comprehensive catalogue of responsibilities in farm animal husbandry. We crafted brief descriptions for all players, focusing not on where in the system a player holds responsibility, but rather on the sense and extent to which we can reasonably attribute responsibility to them. This aids in making informed judgements in relevant discussions regarding whether a player is indeed a pertinent addressee.

Facilitating the application of the report's results is the transfer into an easy-to-use database program. The categories can be combined to enable targeted searches for players with specific characteristics. A pivotal category is whether the respective players have direct contact with animals. Such players are typically subject to media scrutiny, and their behaviour is regulated by animal protection laws. However, our tool reveals the limited scope of responsibility these players usually possess. Further criteria classifications highlight their lack of norm-setting power and the localised impact of their actions.

Conversely, other combinations yield a list of players, such as bottlenecks or so-called big players, who wield considerable power and assertiveness. This underscores the assumption of special power nuclei existing in these places. While it is unsurprising that corporations, system gastronomy and alliances are mentioned, our model provides plausible reasons for their heightened responsibility. Simply naming big players is not the primary output here; it is the rationale behind why they serve as responsibility bottlenecks and are therefore particularly obligated.

4 Levels of Responsibility: Room for Manoeuvre and Power to Act

From this, a scheme of levels of responsibility is derived: The central concept revolves around a dual attribution of responsibility. We distinguish between responsibility for action, pertaining to the specific realm of activity and framework responsibility: (co)responsibility for the contextual conditions surrounding one's actions. When someone is held accountable for their actions, it is essential to discern whether they're being criticised for shortcomings within their sphere of activity or for inadequately shaping their domain of action within their capacity. For instance, a driver of an animal transport may face reproach for handling the animals too roughly or recklessly, but not for route planning or the appropriateness of the transport's subtleties. To sensibly constrain responsibility, we apply the principle of *ultra posse nemo obligatur*: No one is obligated beyond their ability. Importantly, this means that individuals bear full responsibility for fulfilling their duties and cannot deflect them to higher authorities. A notable example illustrating responsibility in this sense is the self-monitoring of animal owners as stipulated in § 11 section 8 of the German Animal Protection Act. The requirements outlined therein are achievable at the animal owner's level of responsibility, thereby obligating them to fulfil it.

In essence, a crucial aspect in the prudent exercise of responsibility is to precisely ascertain the *posse*, or ability, at the respective level, and demand its fulfilment rather than succumb to the practice of shifting responsibility to others, as is often observed. Nonetheless, the responsibility of individual actors does not solely entail fulfilling their role-specific responsibilities: If prevailing conditions hinder a satisfactory outcome altogether, players are obligated to strive for an improvement in the contextual conditions as well. It is seldom evident to what extent players access a higher level to enhance the conditions. For instance, food retailers frequently assert their inability to educate customers (or similar) and claim to merely trade in accordance with consumer preferences. However, food retailing significantly influences customer attitudes and consequently shapes their participation in the system to a significant extent.

Upholding this responsibility and ensuring its realisation at all levels would serve as an effective means of significantly curbing the shifting of responsibility (Te Velde et al. 2002). Individual groups of players are appropriately addressed based on their capacity to act.

5 Subsidiarity and Reliability: Individual and Supra-individual Responsibilities

If dissatisfaction arises with someone's performance, and if someone is not doing a good job, it may indicate a deficiency in their execution of duties. This implies an accusation of enforcement shortfall. However, if someone performs poorly despite executing their duties adequately, it prompts examination of the nature of the jobs themselves, necessitating changes initiated by those responsible for designing these tasks. Such a model enables systematic attributions of responsibility on a case-by-case basis: Players with limited capacity to act (often those with direct animal contact) are situated at lower levels, while those with substantial capacity to act are positioned at higher levels, culminating in the so-called players at the top level, endowed with the greatest authority in the livestock farming system. At this level, the focus shifts to orchestrating players responsible for the prevailing conditions within a network (Beckmann and Pies 2006).

The presence of state entities and large alliances among these top-level players may not come as a surprise. However, demonstrating why certain tasks within livestock systems can only be effectively regulated at this level underscores a politically significant demand: subsidiarity. If it can be demonstrated that other players or groups lack the requisite efficacy, it logically follows that the top-level players have a direct obligation and justification for intervention.

An approach rooted in the attribution of responsibility offers another advantage: it fosters reliability within the system. A consistent theme throughout the discourse on responsibility in the farm animal system has been the recurring call for "reliability". We have elucidated the concept of reliability (Meijboom et al. 2006) by distinguishing it from "trust": trust presupposes that the trusted party acts in one's interest (Hartmann 2010; Herzog 2013). Trust assumes goodwill towards the trusted individual and deems them trustworthy. Additionally, we have differentiated it from predictability, as even someone known to be unreliable is at least predictable in that aspect (Hartmann 2011; Herzog 2013; Weigel et al. 2017).

The framework for long-term change cannot be established by individual players within the system alone. It relies on all components of the mechanism working cohesively. This leaves only the top-level players and government action as guarantors of reliability.

6 Outlook: Possible Futures

Given the ongoing political and social discourse surrounding animal welfare and protection in livestock farming, it is challenging to anticipate the trajectory of demands for animal-friendly husbandry and the extent to which they will be prioritised in the future. The trajectory, speed and direction of animal welfare initiatives remain largely unpredictable.

Government interventions offer a degree of planning security, as evidenced by initiatives like the livestock strategy (BMEL 2019, p. 7). However, if state support for animal welfare measures fails to cover the additional costs or ensure higher purchase prices, the endeavour remains financially unsustainable. Favourable loans provide little relief, as the costs for extensive animal welfare measures must still be recouped in the future. Moreover, safeguarding existing facilities perpetuates outdated forms of animal husbandry, hindering progress in animal welfare efforts.

Legal regulations mandating higher animal welfare standards present a potential solution, as they apply universally (assuming appropriate structural and institutional safeguards). However, given the current context of global market orientation, such regulations must adequately offset potential competitive disadvantages to prevent adverse effects on farms and businesses. Moreover, there is uncertainty regarding the permanence of legal regulations, as subsequent governments may repeal them. Consequently, the timeframe for legislative changes aligns with electoral cycles unless there is bipartisan consensus.

Longer-term models, such as the European Common Agricultural Policy (CAP), offer considerable potential but are subject to similar uncertainties.

The broader landscape remains rife with uncertainties: the influence of animal rights movements, the prominence of animal welfare concerns, shifts in consumer behaviour and the reliability of citizens' political will are all unpredictable. This uncertainty complicates long-term planning and implementation of animal welfare measures along the product chain. Consequently, economically rational behaviour may entail awaiting concrete requirements before taking action.

Conversely, manageable measures that entail minimal investments, offer reliable compensation and yield returns within a reasonable timeframe may garner immediate attention and implementation.

At the crux of this analysis is the observation that the livestock farming system is characterised by a pervasive shifting of responsibility. To counteract this trend, there must be consistent accountability for actions and their consequences. Establishing mechanisms to ensure reliability is a fundamental aspect of governance responsibility, particularly for major players and governmental bodies. Consequently, promoting reliability can foster proactive assumption of responsibility, which, in turn, bolsters trust in both individual actors and the system as a whole.

References

Albertzart M (2015) Der Vorrang des Pflichtbegriffs in kollektiven Kontexten. Zeitschrift für Praktische Philosophie 2:87–120

Bayertz K (1995) Eine kurze Geschichte der Herkunft der Verantwortung. In: Bayertz K (ed) Verantwortung. Prinzip oder Problem? Wissenschaftliche Buchgesellschaft, Darmstadt, pp 3–71

Beckmann M, Pies I (2006) Ordnungsverantwortung – Konzeptionelle Überlegungen zugunsten einer semantischen Innovation., Diskussionspapier des Lehrstuhls für Wirtschaftsethik an der Martin-Luther-Universität Halle-Wittenberg Nr. 2006-10., Halle-Wittenberg, Germany

BMEL (2019) Bundesministerium für Ernährung und Landwirtschaft: Nutztierhaltungsstrategie. Zukunftsfähige Tierhaltung in Deutschland. Stand Juni 2019. Available at: heeps://www.bmel.de/SharedDocs/Downloads/DE/Broschueren/Nutztierhaltungsstrategie.pdf. Accessed 10 Mar 2022

Hartmann M (2010) Die Komplexität des Vertrauens. In: Maring M (ed) Vertrauen — Zwischen sozialem Kitt und der Senkung von Transaktionskosten. KIT, Karlsruhe, pp 15–25

Hartmann M (2011) Die Praxis des Vertrauens. Suhrkamp, Berlin

Hartstang S, Preuss D, Giersberg M, Kunzmann P (2022) Tracing responsibilities in food production with animals. In: Bruce D, Bruce A (eds) Transforming food systems: ethics, innovation, and responsibility, EurSafe 2022. Wageningen Academic Publishers, pp 288–294

Heidbrink L, Langbehn C, Loh J (2017) Handbuch Verantwortung. Springer, Wiesbaden

Herzog L (2013) Persönliches Vertrauen, Rechtsvertrauen, Systemvertrauen. Deutsche Zeitschrift für Philosophie 61:529–548

Hoischen-Taubner S, Bielecke A, Sundrum A (2014) Reflexionen zu den Voraussetzungen und möglichen Hemmnissen eines zielführenden Wissenstransfers am Beispiel der Tiergesundheit in der ökologischen Nutztierhaltung. Available at: https://www.orgprints.org/id/eprint/32077/. Accessed 10 Mar 2022

Isaacs T (2017) Kollektive Verantwortung. In: Heidbrink L, Langbehn C, Loh J (eds) Handbuch Verantwortung. Springer, Wiesbaden, pp 453–475

Kunzmann P (2010) Die Verantwortung des Verbrauchers und einige ihrer Grenzen. Journal für Verbraucherschutz und Lebensmittelsicherheit 5:1–5

Kunzmann P, Giersberg M, Hartstang S, Preuß D, Rieke L (2020) Systematik der Verantwortung für den Tierschutz in der Nutztierhaltung (SysVerNu). Abschlussbericht (Förderkennzeichen 2816HS011), Hannover: 2020. Available at: https://service.ble.de/ptdb/index2.php?detail_id=440850&ssk=673b20eafa&site_key=141&lej=2020&einrtg=hannover&stichw=Systematik+der+Verantwortung+f%C3%BCr+den+Tierschutz+in+der+Nutztierhaltung&zeilenzahl_zaehler=79&NextRow=30&pId=440850&dId=43196351. Accessed 31 July 2022

Lenk H (2017) Verantwortlichkeit und Verantwortungstypen: Arten und Polaritäten. In: Heidbrink L (ed) Handbuch Verantwortung. Springer, Wiesbaden, pp 57–84

Loh J (2017) Strukturen und Relata der Verantwortung. In: Heidbrink L, Langbehn C, Loh J (eds) Handbuch Verantwortung. Springer, Wiesbaden, pp 35–56

Lübbe W (1998) Verantwortung in komplexen kulturellen Prozessen. Alber, Freiburg/München

Maring M (2001) Kollektive und korporative Verantwortung. Begriffs- und Fallstudien aus Wirtschaft, Technik und Alltag. Münster, zugl.: Habil.-Schr. Karlsruhe 2000

Meijboom FLB et al (2006) From trust to trustworthiness: why information is not enough in the food sector. J Agric Environ Ethics 19:427–442

Ropohl G (2002) Welche Schwierigkeiten die Technik mit der Ethik hat. In: Arnswald U, Kertscher J (eds) Herausforderungen der angewandten Ethik, Paderborn, pp 97–108

Te Velde H et al (2002) Dealing with ambivalence: Farmers' and consumers' perceptions of animal welfare in livestock breeding. J Agric Environ Ethics 15:203–219

Weigel TF et al (2017) Vertrauen und Verlässlichkeit in der Chirurgie. Der Chirurg 88:219–225

Zühlsdorf A et al (2016) 'Wie wichtig ist Verbrauchern das Thema Tierschutz? Präferenzen, Verantwortlichkeiten, Handlungskompetenzen und Politikoptionen'. https://www.vzbv.de/sites/default/files/downloads/Tierschutz-Umfrage-Ergebnisbericht-vzbv-2016-01.pdf. Accessed 24 Feb 2021

How to Govern the Transition to Sustainable Animal Husbandry in High-Income Open Economies?

27

Harald Grethe and Jasmin Zöllmer

1 Introduction

Animal husbandry constitutes a large part of the agricultural sector in most regions of the world. Livestock constitutes about 48% of production value in the United States (USDA 2022) and 43% in the European Union (FAO 2024). About 36% of the farms in the EU keep animals (Eurostat 2022).

Especially in high-income countries with high consumption levels of animal products and intensive production systems, animal husbandry is criticised for predominantly two reasons: ethical concerns about low animal welfare levels and the environmental footprint of animal husbandry. This chapter discusses the challenges of governing animal welfare and the environmental effects of animal husbandry in high-income countries, with a focus on the integrated markets of the European Union and the United States.

While increasing animal welfare is recommended by scientists (Broom 2017; EFSA 2023) and supported by international organisations (United Nations 2016) and the general public (for the EU see Eurobarometer 2023), factual progress remains slow in most countries (European Commission 2022; EFSA 2023). Whether substantial improvements in animal welfare beyond niche development are feasible is politically contested. At the core of the issue is the fact that substantially improving animal welfare in intensive husbandry systems is often costly: Providing more space per animal, more variation in the equipment of the husbandry systems and access to different climate conditions and floorings involve higher fixed costs for construction and also often higher management cost because of more intensive animal management practices. For example, Spiller et al. (2015) estimate that increasing animal welfare substantially across the entire sector in Germany would cost

H. Grethe (✉) · J. Zöllmer
Humboldt-Universität zu, Berlin, Germany
e-mail: grethe@hu-berlin.de; jasmin.zoellmer@hu-berlin.de

B. Grabkowsky, T. Blaha (eds.), *Intensive Livestock Production in Transition*,
https://doi.org/10.1007/978-3-031-97872-2_27

approximately €3–5 billion annually. Similarly, Agora Agriculture (2024) estimates that achieving a comparable animal welfare improvement in the EU would cost €10–20 billion annually if animal production were to be halved at the same time. In countries integrated into international markets, where producers face international competition, increasing public minimum animal welfare standards reduces competitiveness and may therefore result in more imports and fewer exports (Zöllmer and Grethe 2024; Grethe 2017). Partially, this may be counteracted by raising domestic willingness to pay for animal welfare through information campaigns and effective labelling systems. However, as Spiller (2025) discusses in this book, Chapter 24, such labelling systems must be carefully designed to function effectively. Even well-designed labels provide no "silver bullet" for transforming the entire market but instead tend to influence only specific market segments.

If aiming at the transformation of a complete sector towards a higher animal welfare level without relocation of production to countries with lower standards, governments may compensate animal welfare costs with public animal welfare premiums or alternatively limit imports below certain animal welfare levels. Section 2 of this chapter describes these policy challenges in detail and suggests options to overcome them.

The second criticism of intensive animal husbandry concerns its environmental footprint in various dimensions, for example, on biodiversity, GHG emissions and air and water quality. Partially, this footprint can be reduced through improved technologies such as methane-reducing feed additives (Kebreab et al. 2023) or more efficient feed rations and partially through improved management such as better storage and distribution of manure (de Vries et al. 2015). But part of the problem stems from the fact that animal products are more land intensive compared to plant products per nutrient unit (Cassidy et al. 2013; Berners-Lee et al. 2018; Parlasca and Qaim 2022). It is therefore the sheer quantities of animals kept in high-income countries to satisfy the high demand for meat, dairy products and eggs, which constitute a problem that cannot be solved through technology and management alone. These quantities result in a high share of animal production in total GHG emissions from agriculture (e.g. about 60% in the EU (Agora Agriculture 2024)), a significant contribution to agriculture expansion into carbon-rich and biodiverse natural or semi-natural habitats such as forests and savannas (Watson et al. 2018; Foucher et al. 2023) and regionally high nutrient surpluses and respective effects on air and water quality. Finally, animal product-rich diets cannot be internationally copied: consider that current average EU and US levels of animal product consumption would prevail globally. This would imply that global beef as well as dairy production would have to increase by about 160%, while global pork and poultry production would have to increase by about 130% and 110%, respectively (own calculations based on FAO (2024)). Global land demand for feeding animals would thus more than double—that land is not available.

While the current level of animal products consumed and produced in high-income countries constitutes a problem, animals can have an important role in sustainable agricultural systems. Ruminants can digest fodder from grassland, which can constitute positive externalities for biodiversity (Wróbel et al. 2023) and carbon storage (De Rosa et al. 2024). Pigs and poultry can be fed to some extent based on residues, waste and side streams of the food industry and convert low-value plant material into high-value animal products. However, the quantity of animal products consumed results in a situation where a substantial share of animal feed is based on crops competing for land that could otherwise be used for food production (Cassidy et al. 2013; Windisch 2021).

Therefore, next to improving production technologies and management, reducing animal product consumption and production is a major lever to reduce its environmental footprint (see also von Cramon-Taubadel and Arcia (2025) in this book, Chapter 5). Section 3 of this chapter describes the challenge to reduce the environmental footprint of animal husbandry, especially through reducing consumption based on a policy mix including labelling, price incentives or improved public canteens. Section 3 also discusses whether reducing the consumption of animal products in high-income countries would necessarily imply reducing production as well.

Finally, Section 4 concludes. We argue that substantially increasing animal welfare across the sector is both technically feasible and politically manageable. It aligns effectively with a broader strategy reducing the overall consumption and production of animal products, resulting in animal husbandry being more sustainable and better integrated in sustainable agricultural and food systems.

2 How to Govern Animal Welfare in an Open Market?

2.1 Challenges

Animal welfare is becoming an increasingly relevant public concern, especially in high-income countries: 84% of Europeans favour better protection of farm animals (Eurobarometer 2023), and in the United States, every state-level public ballot initiative aimed at improving farm animal welfare has passed since the 2000s (Ufer 2022; Zöllmer 2025). Germany is a country with particularly high preferences for animal welfare, as repeatedly shown in a yearly survey by the Federal Ministry of Agriculture (BMEL 2023). However, despite this strong public support, tangible progress towards higher animal welfare standards has been slow. Animal welfare problems in livestock farming include insufficient space, lack of opportunities for expressing natural behaviours and high-stress levels caused by intensive confinement systems. Already in 2015, the Scientific Advisory Board on Agricultural Policy at the Federal Ministry of Agriculture in Germany identified significant deficits in farm animal welfare and recommended a major transformation of the livestock sector to increase societal acceptance of animal husbandry (Spiller et al. 2015). Ten years later, however, little progress has been made to improve farm animal welfare in Germany.

A central issue in advancing welfare standards lies in the substantial costs associated with improving conditions within highly intensified livestock systems. Enhanced welfare measures—such as reducing stocking densities, providing enriched housing or offering outdoor access—can easily drive up the cost of production by 30% or 40% (Grethe 2017) in the sectors of cattle, pork and poultry fattening, often without enhancing the tangible quality of the product.

Since most consumers are somewhat price-sensitive when purchasing food (Andreyeva et al. 2011; Busch and Spiller 2020), many would opt for cheaper imported animal products produced under lower standards over more costly domestic products that follow stricter welfare standards. As a result, regulations that apply solely to domestic producers can significantly weaken their competitiveness against imports from countries with less ambitious legislation. These risks are particularly prevalent within highly integrated markets, such as the European Union, where all goods are to move freely across borders, and each Member State must accept any product lawfully produced in any other Member State (Zöllmer and Grethe 2024). In Sweden, for instance, a Member State with comparatively high animal welfare standards, a large share of the cost-intensive domestic pig production was crowded out by imports after the country joined the EU in 1995. Sweden's self-sufficiency in pork production fell from around 100% to just 70% over 15 years (Bräunig 2020).

The Swedish example illustrates how domestic policies can be undermined by imported products produced at lower costs. Real risks of competitiveness loss and resulting leakage effects make it in turn challenging for policymakers to justify stricter standards in the first place. Yet, the actual degree of regulatory competition and potential for leakage vary case by case and depend on factors such as the product's exposure to international trade and differences in animal protection levels among countries. A particularly decisive factor is the cost of compliance: the larger the cost difference between producing domestically under new welfare-enhancing regulations and producing in regions with lower standards, the higher the potential risk of relocation (Grethe 2007).

Another important factor is the degree to which production is linked to land. For example, poultry and pork production is often highly concentrated and largely independent of land. Large-scale pig or poultry farms, which rely primarily on purchased feed, are much more mobile and thus prone to relocation risks than, for example, land-based dairy husbandry. This shows that while the risks of competitiveness loss and relocation can be substantial, they do not apply uniformly to every welfare measure.

Notwithstanding their actual degree, these potential risks have laid the groundwork for a dominant discourse that significantly shapes animal welfare policymaking. For example, the "competition argument" has become so predominant in the animal welfare policy discourse in Germany that it shapes the political process whenever raising animal welfare standards is under discussion. Zöllmer and Grethe (2024, p. 21) show how, over the past two decades, this competitiveness frame—and the associated risks of replacing parts of the domestic production with imports from countries with less rigid legislation—has served as the primary rationale against raising standards. Their study reveals both a visible lowering of standards, such as lifting an already decided ban on battery cages, and more invisible forms of policy stagnation, such as rejecting stricter standards and postponing deadlines for implementing new regulations. As an example, the German government justified insufficient legislation, in an official response to a 2016 inquiry by the Green Party, by stating that increasing national animal welfare standards beyond EU regulations would bear high risks of competition losses for domestic producers. When specifically questioned about the absence of legislation for dairy cows, fattening bulls, turkeys, ducks, broiler parents, laying hens and ducks, the government repeatedly referred to this justification, highlighting concerns over the economic impact of exceeding EU-wide legal requirements. This response underscores the powerful influence of the 'competitiveness frame' on animal welfare policymaking in Germany.

A similar debate over animal welfare and economic competitiveness emerged in the United States with California's Proposition 2 in 2008. Sumner et al. (2010) analysed the law's impact, which imposed minimum space requirements for laying hens, effectively banning traditional battery cages. Their analysis estimated a cost increase of at least 20% for California producers compared to those in other states. This was predicted to lead to a near-total replacement of local egg production with imports from states with less stringent regulations, as California producers would struggle to remain cost-competitive under the new standards. In response to these debates and subsequent political pressures, California amended the law two years later to apply its standards to out-of-state eggs as well. These examples highlight the challenges of unilaterally enforcing animal welfare improvements within open or even fully integrated markets.

2.1.1 The Role of Labelling Systems to Improve Market-Based Governance

In discussions about raising farm animal welfare standards, it is often suggested that consumers themselves should bear the higher costs of welfare improvements by voluntarily paying more for ethically produced, high-welfare products. However, this approach has clear limitations. Despite widespread public support for animal welfare improvements, many consumers do not opt for higher welfare products when making purchasing decisions. This disparity between citizens' stated preferences and their actual consumption choices is referred to as the "consumer-citizen" or "vote-buy" gap (see Spiller (2025) in this book, Chapter 24).

Price sensitivity is a key barrier to a broader market acceptance of high-welfare goods. While consumer demand for animal welfare-friendly products, such as organic meat, has steadily grown over the past decades, it is driven primarily by higher-income groups and individuals with a strong commitment to ethical consumption. For example, organic animal production remains small within the EU (about 3.6% for poultry, 1% for pigs and 6% for cattle in 2020, see European Commission (2023a).

One reason for the prevailing price sensitivity is the information asymmetry in the market for animal products: only if consumers know about "poor" welfare conditions associated with certain products, they may be willing to pay a higher price for products produced under higher animal welfare standards. Thus, for consumer-driven demand for high-welfare animal products to function effectively as a market solution, consumers must have full, clear and easily understandable information about the animal welfare practices behind each product. If such information is not available, consumers cannot make informed choices genuinely reflecting their welfare preferences and are more likely to opt for lower-priced, lower-welfare goods (Akerlof 1970). This perpetuates a cycle where the market provides little incentive for producers to invest in improved welfare standards, ultimately stalling progress in expanding higher welfare conditions into larger market segments.

As a result, relying solely on consumer-driven market solutions is insufficient to achieve widespread improvements in farm animal welfare. However, a clear, trustworthy and comprehensive labelling system can help to reduce information asymmetry regarding animal welfare practices and unlock the potential of market-based governance. Such a system should enable consumers to understand how animals were raised, transported and slaughtered, empowering them to make informed purchasing decisions. At the same time, labels must be simple and free from ambiguity to affect purchasing decisions effectively, such as through a colour-coded design. The different categories should be based on scientific evidence, rather than stakeholder interests, accompanied by intensive communication to increase awareness, and the total number of labels should be limited to avoid confusion (see Spiller (2025) in this book, Chapter 24).

Beyond labelling, other policies can help expand the market for high-welfare products. Increasing their share in public canteens, for instance, can drive demand, while supporting high-welfare production can help reduce prices. By combining these measures with improved labelling systems, policymakers can expand the niche for high-welfare products and drive a broader transformation in the food sector.

Nevertheless, although niche markets for higher animal welfare standards offer significant potential, which should be fully exploited, the impact of such niche markets on the broader sector remains limited. While instruments to unlock the full potential of market-based solutions are crucial, achieving a transformation of the whole sector requires more substantial measures, which are outlined in the next section.

2.2 Policy Options

Three main policy options exist to mitigate deregulatory pressures from international competition in open economies and allow for increasing animal welfare across the entire sector, rather than just within specific market segments: (1) international harmonisation of standards (positive integration), (2) public animal welfare payments to offset the cost gap resulting from stricter standards in the domestic market and (3) applying the domestic standard equally to imports (national treatment).

First, policymakers can increase harmonisation efforts, i.e. work to establish common standards across countries to ensure a level playing field, for instance, via multilateral agreements. However, high levels of common animal welfare standards are very difficult to achieve, not only on the international level but even in already fully integrated markets, such as the United States (Ufer 2022; Vogeler 2020) or between EU Member States (Zöllmer and Grethe 2024). Where preferences diverge and higher levels of protection would significantly increase the cost of production, countries typically meet at the lowest common denominator when negotiating a common standard—if arriving at any agreement at all.

As a second option, public payments could be used to offset the "cost gap" caused by stricter farm animal welfare standards compared to neighbouring countries. Under this approach, domestic producers would receive financial compensation to offset the additional costs associated with higher welfare standards, thus helping to level the playing field with countries that have less stringent requirements. Such payments have been proposed for Germany by scientific advisory boards (Spiller et al. 2015) as well as multistakeholder commissions (Kompetenznetzwerk Nutztierhaltung 2020; Zukunftskommission Landwirtschaft 2021) and for the EU as a whole (Agora Agriculture 2024). During the last CAP funding period (2014–2020), such payments were granted under voluntary animal welfare programmes implemented by 15 European regions and 14 member states, amounting to €360 million per year. For the 2021–2027 funding period, this budget is expected to rise to €900 million annually (European Commission 2023b). However, this amount contrasts sharply with the estimated €10–20 billion required to elevate animal welfare standards across the entire EU agricultural sector (Agora Agriculture 2024). This highlights the significant gap between current funding levels and the financial needs for a sector-wide transformation.

Allocating public funds for such a transformation requires both societal and political acceptance, which is often tied to the source of the funds. Next to financing animal welfare payments through general national budgets, they may also be supplemented by public taxes or levies being charged on animal product consumption. These taxes or levies could be justified not only by generating revenue to support welfare improvements but also by incentivising reduced animal product consumption (see Sect. 3). This dual approach could enhance public acceptance, as farm animal husbandry would simultaneously contribute to government revenue and require government outlays for animal welfare measures. In Germany, for instance, specific consumer taxes or an increase in the value-added tax (VAT) rate for animal products have been identified as legally feasible options (Grethe et al. 2021; see also Sect. 3 below).

Where harmonisation at a sufficiently high level is not achievable and sufficient public funding for animal welfare payments does not gain the necessary societal and political support, the introduction of border measures may serve as an alternative. Such measures could require imported products to either comply with domestic welfare standards ("national treatment") or be subject to compensatory tariffs to offset the cost advantage of producing at lower animal welfare levels.

A notable example of "national treatment" in animal welfare regulation can be found in the United States: In 2018, California passed a major animal welfare law by voter referendum with 63% approval—banning both the production and the sale of animal products stemming from confined animals. Proposition 12 sets minimum space requirements for certain farm animals and uniformly applies these to farms in and outside California. The law effectively mandates that any pork, eggs or veal sold in California meet its specific welfare standards, regardless of where the animals were raised (Ballotpedia 2018). Notably, by imposing these requirements on imports, California established a de facto trade barrier within the highly integrated market of the United States. For illustration, the biggest pork-producing state, Iowa, must now comply with the new Californian housing conditions for sows, prohibiting the confinement of sows in crates during gestation.

Consequently, the regulation prompted protests and legal challenges from industry groups and farmers' associations in other states, who argued that these regulations restricted interstate commerce by imposing California's standards on out-of-state producers. Despite these challenges, the US Supreme Court upheld the law in 2023, affirming California's right to enforce its standards on imported products to protect the welfare of animals sold within the state (Supreme Court of the United States 2023). Since the beginning of 2024, the regulation is fully implemented and in force.

In addition to its stated impact on producers aiming to sell their pork, eggs or veal to the big Californian market, the law might create a so-called California effect, i.e. induce other states to increase their own regulatory standards (Zöllmer 2025). This "race to the top" of standards has been observed by Vogel (1995) for the environmental policy field. He showed how California set the leading tone in environmental regulation over the past decades in the USA and was followed by other states.

Thus, if policymakers aim to raise animal welfare standards without risking replacing their domestic production with imports produced at lower standards—which in turn, would render their policy ineffective—national treatment seems to provide an effective option. However, there are also important drawbacks to this instrument, as it comes with a considerable political cost: Autonomous import measures can create significant tensions with trading partners and can potentially lead to retaliatory measures. While California's actions are legally justified within the context of the US legal framework, international trading partners may view these regulations as a form of protectionism. Countries that export animal products to California could argue that the law discriminates against their products, especially if the costs of compliance disproportionately affect foreign producers. Surprisingly, so far, only a few international voices have raised quite concerns about Proposition 12. For example, while Canada's Minister of Agriculture highlighted the potential for these rules to "restrict trade and disrupt supply chains" (Wichers 2024), he did not suggest any retaliatory measures, such as restrictions on US exports.

Can California's example of import requirements be easily transferred to other contexts? For the EU, an equivalent regulation would face legal and institutional challenges, as such measures would generally be considered to violate the free movement of goods. However, the possibility of derogating from the free movement provisions on the basis of Article 36 of the Treaty on the Functioning of the European Union (TFEU) could provide sufficient policy space to explore this option (Zöllmer and Grethe 2024).

In summary, in the interest of an open trading system and since trade policy often is a building stone of general foreign policy, national treatment may pose challenges, while public funding for animal welfare appears technically a more adequate solution. Should public funding be politically unfeasible, however, national treatment warrants serious consideration, given its demonstrated success in the USA

3 How to Govern the Environmental Effects of Animal Husbandry in an Open Market?

3.1 The Challenges

The second major criticism of intensive animal husbandry lies in its significant environmental footprint, impacting biodiversity, greenhouse gas (GHG) emissions, as well as air and water quality. In this section, the issues of land use for animal production, regional nutrient accumulation and GHG emissions from enteric fermentation and manure management are briefly described.

3.1.1 Land Use

The consumption of animal products is land-intensive and requires significantly more land per nutrient unit than plant-based food production (Cassidy et al. 2013; Berners-Lee et al. 2018; Parlasca and Qaim 2022). This is largely due to the need for large-scale production of feed crops such as soy and corn. With the rapid increase in global meat production, which more than tripled over the past 50 years (OECD/FAO 2023), global demand for land to produce animal feed has also significantly increased. This has led to large-scale land-use changes, which are particularly problematic if biodiverse and carbon-rich regions like forests and savannahs are converted for agricultural use, accelerating biodiversity loss and contributing to greenhouse gas (GHG) emissions through carbon release. The reduction of the Amazon rainforest is a typical example of a critical area affected by such land-use changes for soy production and cattle ranching.

Moreover, land is increasingly scarce, and food production competes globally with other critical land uses, such as biomass production for the decarbonisation of other sectors or conservation needs. The bioeconomy increasingly relies on land for cultivating biogenic raw materials for non-food purposes, such as bioenergy and renewable material uses as alternatives to fossil sources (Material Economics 2021). At the same time, land is needed for protecting biodiversity (IPBES 2019) and as an essential carbon sink (Arneth et al. 2021), as forests, savannas or peatlands store significant amounts of carbon.

3.1.2 Nutrient Accumulation

In regions of high animal density, such as West Germany and South Netherlands in Europe (Eurostat 2024) or Iowa in the United States (USDA 2024, intensive livestock production results in nutrient accumulation, especially nitrogen (N) and phosphorus (P). These nutrients, if not managed properly, become pollutants through emissions into the environment. Nitrogen-based emissions, such as ammonia (NH_3) and nitrous oxide (N_2O), contribute to air pollution and GHG emissions, while nitrate (NO_3) runoff contaminates water systems. Phosphorus accumulation contributes to the eutrophication of surface waters, a process that disrupts aquatic ecosystems by promoting excessive algal growth and depleting oxygen levels. The accumulation of these nutrients has far-reaching environmental and health impacts (Cheng et al. 2022; Luo et al. 2024; Rojas-Downing et al. 2017).

3.1.3 GHG Emissions from Fermentation and Manure Management

Animal husbandry is also a significant contributor to global greenhouse gas emissions, largely through methane emissions from ruminants, such as cattle. Methane, a potent greenhouse gas, has a much higher warming potential than carbon dioxide, if considered for a period of 100 years, making it a critical target for climate mitigation efforts. Ruminants are of particular concern because their digestive processes produce large amounts of methane: methane from cattle, sheep and goats constitutes more than half of the emissions from the EU's agricultural sector (EEA 2023).

On the other hand, pasture-based systems can also have substantial positive environmental impacts by sequestering carbon (De Rosa et al. 2024) and enhancing biodiversity (Wróbel et al. 2023). In addition, ruminants can digest grass and other fibrous plants that humans cannot consume, effectively converting otherwise unusable land resources into valuable protein (Milazzo et al. 2023). Thus, when raised extensively on grassland without being fed large quantities of human-edible crops and maintained at ecologically sustainable population levels, ruminants can increase biodiversity while their grazing lands serve as valuable carbon sinks.

3.2 Policy Options

As the environmental impact of animal agriculture includes many dimensions, a policy mix of various measures is needed to address the diverse problems.

One of the most impactful strategies to address the described challenges and reduce the environmental footprint of food production is to decrease overall animal product consumption in high-income countries. According to a recent study, halving animal product consumption within the EU, combined with a somewhat smaller decline in production and a somewhat higher share of poultry in total meat consumption, could reduce emissions from animal husbandry by about 54% (Agora Agriculture 2024). In addition, this scenario would reduce arable land used for feeding animals inside and outside the EU from 77 million hectares to 29 million hectares annually.

Importantly, this reduction must primarily address demand: simply reducing production within the EU while importing animal products from third countries would merely shift, rather than alleviate, emissions and other environmental burdens. This could even increase the global environmental footprint due to varying standards and production methods across the world. Moreover, reduction strategies that specifically target animal products with high environmental footprints, such as beef and dairy—especially when mostly fed with crops which could also be used for human consumption—can yield the most substantial environmental benefits.

To effectively lower animal product consumption, a broad policy mix is required, aiming to create fair food environments (WBAE 2020) that make sustainable and healthy food choices easier. Such food environments would support informed consumer choices, e.g. through information and education, while also acknowledging the habitual nature of many food decisions. Core elements of such fair food environments include public price incentives, improved public canteens and transparent labelling.

3.2.1 Price Incentives

The environmental costs of food production and consumption are largely externalised. Policy measures aiming at internalising these external costs can shift consumption patterns towards more sustainable diets by making them more affordable. Subsidies, taxes and levies are traditional policy tools used to internalise the external costs and benefits of production or consumption. This way, external costs of animal production (such as environmental impacts) or consumption (such as health impacts) can be better reflected in market prices, thereby creating direct incentives for consumers and producers. As taxing animal product production in an open economy would reduce the competitiveness of domestic production, leading to higher imports without much effect on consumption, raising the relative consumer prices of animal-based foods compared to plant-based products through taxes would be a more effective approach to shift consumption patterns.

To reduce demand for animal products, either a per-unit or an ad valorem consumer tax may be applied. A per-unit tax applies a fixed monetary amount for each unit of a good sold, such as cents per kilogram of meat, and thus remains proportional to the particular quantity of a good sold, regardless of its price. The advantage of a specific tax is that it can be directly linked to the externalities associated with each product category. For example, a higher tax could be charged on products which have a higher climate impact per product unit. By contrast, an ad valorem tax charges a fixed percentage of the product price; its absolute value thus increases proportionally to the product price. The disadvantage of an ad valorem tax is that it may not reflect the actual externalities of production well. High-value, more expensive animal products, for example, organic meat, would incur a larger absolute tax. On the other hand, an ad valorem tax is easier to implement, as it could be integrated into existing value-added tax systems. For instance, in Germany, the Scientific Advisory Board on Agricultural and Food Policy recommended abolishing the reduced VAT rate of 7% on meat, dairy and other animal products and applying the standard rate of 19% instead (WBAE 2020).

To enhance public acceptance of consumption taxes, it is important to clearly communicate their objectives, including their environmental or health benefits and potentially relieve consumers in other fields, such as reducing taxes on plant-based products with a low environmental footprint. Due to the regressive nature of taxes on food items, such taxes should be accompanied by compensation measures for low-income households (Eykelenboom et al. 2019).

3.2.2 Pricing Greenhouse Gas Emissions in Agriculture

Pricing greenhouse gas emissions in agriculture has the potential to indirectly reduce the consumption of animal products, particularly those with a high environmental footprint. This depends on whether such a pricing system would drive up animal product prices significantly, which in turn depends on a potential Carbon Border Adjustment Mechanism (CBAM).

At the European Union level, agriculture-related emissions could for example be included in an Emissions Trading System (ETS). Such an approach may impact prices for food products with high greenhouse gas emissions, such as meat and dairy (Pérez et al. 2016; Stepanyan et al. 2023). However, in open markets, pricing GHG emissions in agricultural production would substantially raise the production costs of some animal products, such as milk and beef, both of which generate significant emissions per nutrient unit. This would potentially undermine the competitiveness of domestic producers against those from regions with fewer or no emission-related costs. This, in turn, potentially leads to a rise in imports of these products from regions without similar emissions pricing, an effect known as "carbon leakage". Carbon leakage contradicts the overarching policy goal of global GHG reduction, as emissions merely shift rather than decrease. To counteract this, a CBAM for selected high-emission products such as beef, lamb, milk powder and butter could be considered to level the playing field for domestic producers and contribute to consumption reduction incentivised by higher prices while maintaining a commitment to reducing GHG emissions at the global level (Agora Agriculture 2024).

3.2.3 Improved Public Canteens

Another key lever to shift consumption patterns is to make plant-based options more prominent and affordable in public canteens. Policymakers can encourage dietary shifts by introducing attractive, nutritious plant-based meals in institutions where people regularly eat, such as kindergartens, schools, retirement homes and hospitals. These public spaces shape habitual dietary choices for many people and thereby influence food consumption norms and production practices via the purchasing power of the public sector (WBAE 2020; Swensson et al. 2021). Establishing mandatory minimum health and sustainability criteria for public procurement across the EU would support leveraging this potential (Agora Agriculture 2024). Such standards could build on the European Commission's voluntary Green Public Procurement guidelines (European Commission 2019), particularly by prioritising plant-rich menus.

3.2.4 Transparent Labelling

A typical market-based instrument to shift consumption patterns is transparent labelling, which empowers consumers to make informed purchasing decisions. Sustainability labelling is currently fragmented, and there are almost no labels for key sustainability dimensions such as environmental and climate change effects and social standards. Many quite unknown, unintuitive, partially misleading and too detailed labels hinder market transparency. The WBAE (2020), therefore, recommends a public interpretative label for key sustainability dimensions. Such a label would preferably be mandatory and involve dimensions such as animal welfare, GHG footprint, environmental sustainability and human health.

3.2.5 Could Production Levels in High-Income Countries Be Sustained Despite Reduced Consumption?

The argument is often made that reducing animal product consumption in high-income countries would not necessarily need to translate into reducing production in these countries, as the excess supply could simply be exported. This may be beneficial from a global climate perspective, as production processes in high-income countries are typically efficient and often have lower GHG emissions than production systems in lower-income regions (examples for this argument include Searchinger et al. (2021) and Von Cramon-Taubadel and Arcia (2025) in this book, Chapter 5).

While this argument would technically be correct when environmental externalities in production are handled well, we argue that sustaining production levels despite the reduction in consumption is not a likely outcome of the policy process in high-income countries as it may conflict with other policy goals.

First, maintaining production levels in high-income countries would make it more expensive to address the local environmental problems associated with high livestock densities. Without a reduction in production, air and water pollution through nutrient surpluses would remain a challenge in regions where animal husbandry is concentrated.

Second, if the production remains in high-income countries, the GHG emissions would have to be reported there and, due to the political objective of climate neutrality, would need to be compensated by expensive negative emissions.

Third, if high animal welfare levels are funded through public funds (see above), exporting large quantities of animal products would imply exporting publicly funded animal welfare to other countries. The societal acceptability of such a model seems low.

Fourth, from a development policy perspective, the development of domestic production in countries which are strong importers today may be a desirable goal. In the long run, efficient production systems of pork and poultry meat can be built up where animal feed is available, such as close to ports of feed-importing or exporting countries. Supporting value creation domestically in low-income countries, rather than importing excess production from high-income countries, supports domestic agricultural development and strengthens regional supply chains. Dairy products may be an exception to some extent as many high-income regions such as the United States, Canada, New Zealand and the EU have favourable production conditions in terms of permanent grassland and climate. Moreover, the cost implied in increasing the animal welfare level is low compared to pig and poultry production systems. The political will to maintain grassland for reasons of carbon storage and the protection of certain habitats may therefore result in the maintenance or even increased exports of dairy products from high-income countries if consumption declines.

Next to reducing animal product consumption and production, other policies enhancing the efficiency of production can reduce the environmental impact of animal husbandry. As examples, the governance of nutrient surpluses and the uptake of GHG-reducing technologies in animal husbandry are mentioned here.

3.2.6 The Central Role of Fertilisation Policy

To mitigate nitrogen and phosphorus surpluses, which are often particularly high in farm animal-dense regions, optimising nutrient management in agricultural systems is crucial. This implies aligning fertiliser use with the actual nutrient requirements of crops, thereby reducing nutrient surplus and related emissions to the environment. Improving nitrogen use efficiency (NUE), which is the ratio of crop nitrogen uptake to available soil nitrogen, contributes to this objective (de Vries et al. 2022). High NUE reduces nitrogen pollution, which is responsible for exceeding multiple planetary boundaries (Sutton et al. 2021).

To address the issue of nutrient surpluses in agricultural systems, a robust fertilisation policy that requires nutrient balances and restricts nutrient surpluses for all farming operations is essential. By mandating that farmers track and balance the nutrients applied to their soils against what is actually required for optimal crop production, governments can contribute to increasing NUE and reducing nutrient surplus. A recent study by Agora Agriculture (2024) shows that nitrous oxide emissions from agricultural soils may be reduced by 40% through improved NUE.

Implementing clear, enforceable limits on nutrient application can help reduce the overall environmental footprint of agriculture (de Vries et al. 2022). Alternatively, nitrous oxide emissions may be included in an emission trading system for agriculture, and nitrogen surplus may be used as a proxy for such emissions and thus be priced.

3.2.7 Advancing GHG-Reducing Technologies and Practices in Animal Husbandry

A wide range of GHG-reducing technologies and management practices, such as improved manure management and optimised livestock feeding, are available. Among the most promising approaches are methane-inhibiting feed additives, anaerobic digestion of manure, and the use of manure additives (Agora Agriculture 2024). In addition, reducing livestock numbers would allow for a higher share of grass in the feed ration of ruminants and for a greater use of food industry by-products in the feeding of pigs and poultry. These measures would not only result in lower greenhouse gas emissions but also reduce competition for cropland.

4 The Transition Towards Sustainable Animal Husbandry Is Achievable

This chapter explored ways to successfully transition to sustainable animal husbandry in high-income, open economies, focusing on the European Union and the United States. It outlined the challenges policymakers face in open economies when aiming to increase animal welfare to socially acceptable levels while significantly reducing the environmental footprint of animal husbandry.

For a successful transition, a comprehensive policy mix of various measures is required. Regulatory and market-based instruments, such as price incentives and transparent labelling, alongside public funding for public goods, play essential roles in transforming the livestock sector effectively.

Substantially raising animal welfare standards for the sector as a whole in open economies is achievable, but it necessitates either public compensatory payments to offset increased production costs or border measures that require imports to comply with domestic animal welfare standards as well ("national treatment"). In the case of public compensatory payments, the cost of animal welfare may be borne either by taxpayers in general, or by consumers of animal products, if such payments are accompanied by consumption taxes on these products. Under national treatment, consumers of animal products would bear most of the costs/most of the cost would be borne by consumers of animal products, as prices would rise due to higher standards and the prevention of international competition. While national treatment may appear unviable in open economies at first glance, the example of California shows that it is feasible.

We argue that it is essential to communicate the costs of improving animal welfare transparently and openly debate the policy options to cover these costs. Each of these options comes at a price: national treatment may cause conflicts with trading partners, while compensatory payments require political consensus to spend taxpayers' money on animal welfare. If neither option is chosen, substantially raising animal welfare standards across the entire sector without relocation of production to regions with lower standards is practically infeasible.

Key to reducing the environmental footprint of agriculture in high-income countries is a significant decline in animal product consumption. Policymakers can apply various instruments to shift consumption patterns towards more sustainable diets. Price incentives, such as consumption taxes or greenhouse gas emissions pricing in production (potentially cushioned by a Carbon Border Adjustment Mechanism for selected products), provide effective instruments for driving behavioural change. These measures should be complemented by initiatives that foster healthier food environments, such as improving meal offerings in public canteens and improving information on the sustainability impacts of food choices.

Given the challenges ahead for animal farmers as well as related up- and downstream sectors, a successful transition to sustainable animal husbandry requires a long-term political strategy. Such a strategy should be operationalised through targets, milestones and a timeline, to monitor progress and provide the sector with the planning security necessary to adapt to changes in the governance framework.

References

Agora Agriculture (2024) Agriculture, forestry and food in a climate-neutral EU: The land use sectors as part of a sustainable food system and bioeconomy. https://www.agora-agriculture.org/publications/agriculture-forestry-and-food-in-a-climate-neutral-eu

Akerlof GA (1970) The market for "lemons": quality uncertainty and the market mechanism. Q J Econ 84(3):488–500

Andreyeva T, Long MW, Brownell KD (2011) The impact of food prices on consumption: A systematic review of research on the price elasticity of demand for food. Am J Public Health 100(2):216–222. https://doi.org/10.2105/AJPH.2008.151415

Arneth A, Olsson L, Cowie A, Erb KH, Hurlbert M, Kurz WA, Mirzabaev A, Rounsevell MDA (2021) Restoring degraded lands. Annu Rev Environ Resour 46(1):569–599. https://doi.org/10.1146/annurev-environ-012320-054809

Ballotpedia (2018) California Proposition 12, Farm Animal Confinement Initiative (2018). https://ballotpedia.org/California_Proposition_12,_Farm_Animal_Confinement_Initiative_(2018)

Berners-Lee M, Kennelly C, Watson R, Hewitt CN (2018) Current global food production is sufficient to meet human nutritional needs in 2050, provided there is radical societal adaptation. Elementa: Sci Anthropocene 6:52. https://doi.org/10.1525/elementa.310

BMEL (Bundesministerium für Ernährung und Landwirtschaft (2023). Deutschland, wie es isst – der BMEL-Ernährungsreport 2023

Busch G, Spiller A (2020) Warum wir eine Tierschutzsteuer brauchen: Die Bürger-Konsumenten-Lücke. Diskussionsbeitrag No. 2001. Georg-August-Universität Göttingen, Department für Agrarökonomie und Rurale Entwicklung (DARE), Göttingen

Bräunig U (2020) Schweden: Vorreiter in Sachen Tierwohl. Agrarheute. https://www.agrarheute.com/tier/schwein/schweden-vorreiter-sachen-tierwohl-565336

Broom DM (2017) Animal welfare in the European Union. In: European Parliament Directorate General for Internal Policies, Policy Department C: Citizens' Rights and Constitutional Affairs, Petitions. http://www.europarl.europa.eu/supporting-analyses

Cassidy ES, West PC, Gerber JS, Foley JA (2013) Redefining agricultural yields: from tonnes to people nourished per hectare. Environ Res Lett 8(3):034015. https://doi.org/10.1088/1748-9326/8/3/034015

Cheng M, McCarl B, Fei C (2022) Climate change and livestock production: A literature review. Atmos 13(1). https://doi.org/10.3390/atmos13010140

De Rosa D, Ballabio C, Lugato E, Fasiolo M, Jones A, Panagos P (2024) Soil organic carbon stocks in European croplands and grasslands: how much have we lost in the past decade? Glob Chang Biol 30(1):e16992. https://doi.org/10.1111/gcb.16992

de Vries J, Hoogmoed W, Groenestein C, Schröder J, Sukkel W, De Boer I, Koerkamp PG (2015) Integrated manure management to reduce environmental impact: I Structured design of strategies. Agric Sys 139:29–37. https://doi.org/10.1016/j.agsy.2015.05.010

de Vries W, Römkens P, Kros J, Voogd JC, Schulte-Uebbing LF (2022) Impacts of nutrients and heavy metals in European agriculture: current and critical inputs in relation to air, soil, and water quality. In: ETC DI report 2022/01. https://www.eionet.europa.eu/etcs/etc-di/products/impacts-of-nutrients-and-heavy-metals-in-european-agriculturecurrent-and-critical-inputs-in-relation-to-airsoil-and-water-quality

EEA (European Environment Agency) (2023) Greenhouse gas emissions from land use, land use change, and forestry in Europe. https://www.eea.europa.eu/en/analysis/indicators/greenhouse-gas-emissions-from-land

EFSA (European Food Safety Authority) (2023) Animal welfare. https://www.efsa.europa.eu/en/topics/topic/animal-welfare

Eurobarometer (2023) Attitudes of Europeans towards animal welfare. https://europa.eu/eurobarometer/surveys/detail/2996

European Commission (2019) Commission Staff Working Document: EU green public procurement criteria for food, catering services, and vending machines (SWD(2019) 366 final). https://data.consilium.europa.eu/doc/document/ST-12672-2019-INIT/en/pdf

European Commission (2022) Commission staff working document: fitness check of the EU animal welfare legislation (SWD(2022) 328 final). https://food.ec.europa.eu/system/files/2022-10/aw_eval_revision_swd_2022-328_en.pdf

European Commission (2023a) Organic farming in the EU: *A decade of organic growth.* A market brief. https://agriculture.ec.europa.eu/system/files/2023-04/agri-market-brief-20-organic-farming-eu_en.pdf

European Commission (2023b) Summary of CAP Strategic Plans for 2023–2027: Joint effort and collective ambition. https://agriculture.ec.europa.eu/cap-my-country/cap-strategic-plans_en

Eurostat (2024) Agri-environmental indicator – livestock patterns: livestock density at regional level in 2020. European Commission. https://ec.europa.eu/eurostat/statistics-explained/index.php?title=Agri-environmental_indicator_-_livestock_patterns#Livestock_density_at_regional_level_in_2020

Eurostat (2022) Farms and farmland in the European Union - statistics. European Commission. https://ec.europa.eu/eurostat/statistics-explained/index.php?title=Farms_and_farmland_in_the_European_Union_-_statistics

Eykelenboom M, van Stralen MM, Olthof MR, Schoonmade LJ, Steenhuis IHM, Renders CM (2019) Political and public acceptability of a sugar-sweetened beverages tax: A mixed-method systematic review and meta-analysis. Int J Behav Nutr Phys Act 16:78. https://doi.org/10.1186/s12966-019-0843-0

FAO (Food and Agriculture Organization of the United Nations) (2024) FAOSTAT: Agriculture data. https://www.fao.org/faostat/en/#data/Q

Foucher A, Tassano M, Chaboche PA, Chalar G, Cabrera M, Gonzalez J, Cabral P, Simon AC, Agelou M, Ramon R, Tiecher T, Evrard O (2023) Inexorable land degradation due to agriculture expansion in south American Pampa. Nature Sustain 6:662–670. https://doi.org/10.1038/s41893-023-01074-z

Grethe H (2007) High animal welfare standards in the EU and international trade—how to deal with potential 'low animal welfare havens'? Food Policy 32(3):315–333. https://doi.org/10.1016/j.foodpol.2006.06.001

Grethe H (2017) The economics of farm animal welfare. Ann Rev Resour Econ 9(1):75–94. https://doi.org/10.1146/annurev-resource-100516-053419

Grethe H, Martinez J, Osterburg B, Taube F, Thom F (2021) Klimaschutz im Agrar- und Ernährungssystem Deutschlands: Die drei zentralen Handlungsfelder auf dem Weg zur Klimaneutralität. In: Stiftung Klimaneutralität. https://www.stiftung-klima.de/app/uploads/2021/06/2021-06-01-Klimaneutralitaet_Landwirtschaft.pdf

IPBES (Intergovernmental science-policy platform on biodiversity and ecosystem services) (2019) Summary for policymakers of the global assessment report on biodiversity and ecosystem services doi:https://doi.org/10.5281/zenodo.3553579

Kebreab E, Bannink A, Pressman EM, Walker N, Karagiannis A, van Gastelen S, Dijkstra J (2023) A meta-analysis of effects of 3-nitrooxypropanol on methane production, yield, and intensity in dairy cattle. J Dairy Sci 106(2):927–936. https://doi.org/10.3168/jds.2022-22211

Kompetenznetzwerk Nutztierhaltung (2020) Empfehlungen des Kompetenznetzwerks Nutztierhaltung. https://www.bmel.de/DE/themen/tiere/nutztiere/umbau-nutztierhaltung.html

Luo L, Cohan DS, Gurung RB, Venterea RT, Ran L, Benson V, Yuan Y (2024) Impacts assessment of nitrification inhibitors on U.S. agricultural emissions of reactive nitrogen gases. J Environ Manag 359:121043. https://doi.org/10.1016/j.jenvman.2024.121043

Material Economics (2021) EU biomass use in a net-zero economy—A course correction for EU biomass and bioresources within a net-zero emissions economy: making a sustainable approach possible. Assumptions and methodology underlying the sectoral analysis Technical Annex, Version 1.0. https://materialeconomics.com/sites/default/files/2024-06/2021-019-etc-bio-economy-technical-annex-web-aw.pdf

Milazzo F, Francksen RM, Abdalla M, Ravetto Enri S, Zavattaro L, Pittarello M, Hejduk S, Newell-Price P, Schils RLM, Smith P, Vanwalleghem T (2023) An overview of permanent grassland grazing management practices and the impacts on principal soil quality indicators. Agronomy 13(5):1366. https://doi.org/10.3390/agronomy13051366
OECD/FAO (2023) OECD-FAO Agricultural Outlook 2023–2032. OECD Publishing, Paris. https://doi.org/10.1787/08801ab7-en
Parlasca MC, Qaim M (2022) Meat consumption and sustainability. Ann Rev Resour Econ 14:17–41. https://doi.org/10.1146/annurev-resource-111820-032340
Pérez DI, Fellmann T, Weiss F, Witzke HP, Barreiro HJ, Himics M, Jansson T, Salputra G, Leip A (2016) An economic assessment of GHG mitigation policy options for EU agriculture (EcAMPA 2). https://doi.org/10.2791/843461
Rojas-Downing MM, Nejadhashemi AP, Harrigan T, Woznicki SA (2017) Climate change and livestock: impacts, adaptation, and mitigation. Clim Risk Manag 16:145. https://doi.org/10.1016/j.crm.2017.02.001
Searchinger TD, Zionts J, Wirsenius S, Peng L, Beringer T, Dumaj P (2021) Carbon-neutral agriculture: A vision for Denmark. World Resources Institute. https://doi.org/10.46830/wrirpt.20.00006
Spiller A, Gauly M, Balmann A, Bauhus J, Birner R, Bokelmann W, Christen O, Entenmann S, Grethe H, Knierim U, Latacz-Lohmann U, Martinez J, Nieberg H, Qaim M, Taube F, Tenhagen BA, Weingarten P (2015) Wege zu einer gesellschaftlich akzeptierten Nutztierhaltung [Ways to societally acceptable livestock farming]. Berichte über Landwirtschaft 221:1–172. https://doi.org/10.12767/buel.v0i221.82
Spiller A (2025) The role of sustainable food labels in transforming the food system. In: Grabkowsky and Blaha (ed) Livestock production in transition, pp 422–428
Stepanyan D, Heidecke C, Osterburg B, Gocht A (2023) Impacts of national vs European carbon pricing on agriculture. Environ Res Lett 18(7):074016. https://doi.org/10.1088/1748-9326/acdcac
Sumner DA, Matthews WA, Mench JA, Rosen-Molina JT (2010) The economics of regulation on hen housing in California. J Agric Appl Econ 42(3):429–438. https://cail.ucdavis.edu/publications/eggs/egginitiative.pdf
Supreme Court of the United States (2023) National pork producers council v. Ross, 598. ___ U.S. https://www.supremecourt.gov/opinions/22pdf/21-468_5if6.pdf
Sutton MA, Howard CM, Kanter DR, Lassaletta L, Móring A, Raghuram N, Read N (2021) The nitrogen decade: mobilizing global action on nitrogen to 2030 and beyond. One Earth 4(1):10–14. https://doi.org/10.1016/j.oneear.2020.12.016
Swensson LFJ, Hunter D, Schneider S, Tartanac F (2021) Public food procurement as a game changer for food system transformation. Lancet Planetary Health 5(8):e495–e496. https://doi.org/10.1016/S2542-5196(21)00176-5
Ufer DJ (2022) State policies for farm animal welfare in production practices of U.S. Livestock and poultry industries: an overview (EIB-245). U.S. Department of Agriculture, Economic Research Service. https://www.ers.usda.gov/publications/pub-details/?pubid=105480
United Nations (2016) High level panel on food security and nutrition, sustainable agricultural development for food security and nutrition, including the role of livestock. FAO, Rome, Italy
USDA (United States Department of Agriculture) (2022) Census of agriculture: Farm economics. https://www.nass.usda.gov/Publications/Highlights/2024/Census22_HL_FarmEconomics_FINAL.pdf
USDA (United States Department of Agriculture) (2024) NASS QuickStats Ad-hoc Query Tool. https://quickstats.nass.usda.gov/#5E7FF53A-8629-37FA-A140-FF06DB255E6B
Vogel D (1995) Trading up: consumer and environmental regulation in a global economy. Harvard University Press
Vogeler CS (2020) Politicizing farm animal welfare: A comparative study of policy change in The United States of America. J Comparat Policy Analysis: Res Practice 23(5–6):526–543. https://doi.org/10.1080/13876988.2020.1834859

Von Cramon-Taubadel S, Arcia K (2025) The transformation of livestock production in a globalised world – barriers and opportunities. In: Grabkowsky, Blaha (ed) Livestock production in transition, pp 139–160

Watson JE, Evans T, Venter O, Williams B, Tulloch A, Stewart C, Thompson I, Ray JC, Murray K, Salazar A, McAlpine C (2018) The exceptional value of intact forest ecosystems. Nature Ecology and Evolution 2(4):599–610. https://doi.org/10.1038/s41559-018-0490-x

WBAE (Wissenschaftlicher Beirat für Agrarpolitik, Ernährung und gesundheitlichen Verbraucherschutz) (2020) Politik für eine nachhaltige Ernährung. Eine integrierte Ernährungspolitik entwickeln und eine faire Ernährungsumgebung gestalten. Gutachten, Berlin. Retrieved from https://www.bmel.de/SharedDocs/Downloads/DE/_Ministerium/Beiraete/agrarpolitik/wbae-gutachten-nachhaltige-ernaehrung.html

Wichers G (2024) Ag minister talks Proposition 12, VCOOL with the U.S.'s Vilsack. The trip marks MacAulay's first ministerial visit to the U.S. since regaining the role. AgCanada, January 19. https://www.agcanada.com/daily/ag-minister-talks-proposition-12-vcool-with-the-u-s-s-vilsack

Windisch W (2021) Can we do without livestock? J Consum Prot Food Saf 16(3):193–194. https://doi.org/10.1007/s00003-021-01338-w

Wróbel B, Zielewicz W, Staniak M (2023) Challenges of pasture feeding systems: opportunities and constraints. Agriculture 13(5):974. https://doi.org/10.3390/agriculture13050974

Zöllmer J, Grethe H (2024) Enabling free movement but restricting domestic policy space? The price of mutual recognition. European Policy Anal 10:380–411. https://doi.org/10.1002/epa2.1208

Zöllmer J (2025) Race to the top of farm animal welfare policies in US states: What can explain the new development? A Qualitative Comparative Analysis. Journal of Comparative Policy Analysis: Research and Practice, 1–19. https://doi.org/10.1080/13876988.2025.2493820

Zukunftskommission Landwirtschaft (2021) Zukunft Landwirtschaft. Eine gesamtgesellschaftliche Aufgabe. Empfehlungen der Zukunftskommission Landwirtschaft. https://www.bmel.de/SharedDocs/Downloads/DE/_Landwirtschaft/abschlussbericht-zukunftskommission-landwirtschaft.html

Operationalising Sustainability: A Structured Approach to Systemic Transformation

28

Barbara Grabkowsky

1 Barriers to Sustainability Transformation – Structural, Systemic, and Geopolitical Dimensions

Early attempts at sustainability transformation provide valuable lessons, revealing common pitfalls that have hindered implementation. Despite well-intended policies and scientifically sound strategies, recurrent challenges have hindered effective implementation and led to public resistance, economic inefficiencies, and political backlash. Three key obstacles emerge consistently: an overreliance on expert-driven policies, the moralisation of sustainability narratives, and state-centric governance approaches that fail to engage market and civil society actors. To better understand why past transformation efforts have failed, the three major barriers—technocratic overreliance, moralising narratives, and excessive dependence on state-driven solutions—are visualised as drivers of resistance, polarisation, and inefficiency, ultimately leading to transformation failure (Fig. 28.1).

1.1 How the Barriers Reinforce Each Other

Overreliance on Scientific-Only Approaches → Policies developed solely on scientific expertise, without public acceptance, often face resistance.

- **Moralizing Narratives →** Framing sustainability as a *moral duty rather than an opportunity* provokes *backlash and polarisation.*

B. Grabkowsky (✉)
Center for Sustainability Transformation Research in AgriFood Systems, Vechta Institute of Sustainability Transformation in Rural Areas, University of Vechta, Vechta, Germany
e-mail: barbara.grabkowsky@uni-vechta.de

B. Grabkowsky, T. Blaha (eds.), *Intensive Livestock Production in Transition*,
https://doi.org/10.1007/978-3-031-97872-2_28

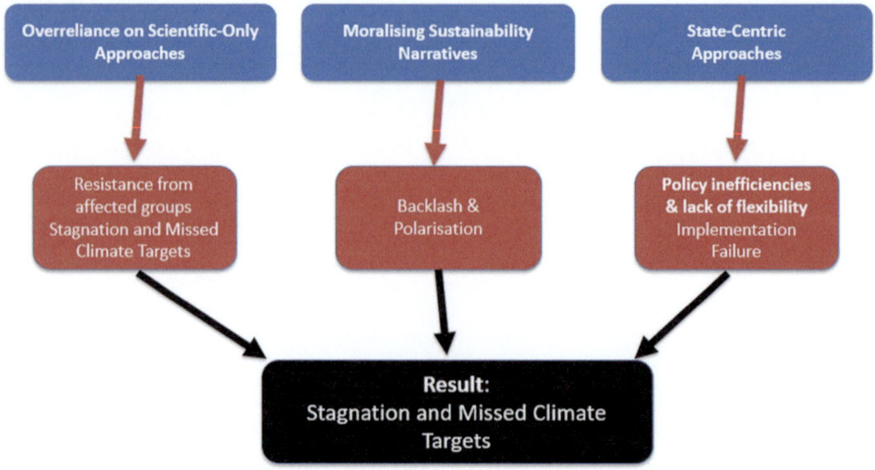

Fig. 28.1 Causal barriers to T-Transformation: a cause-effect structure (© Grabkowsky). Legend: blue boxes = root causes; red boxes = consequences; arrows indicate cause-effect relationships

- **State-Centric Approaches** → Overdependence on *centralised regulation*, without economic incentives or private-sector engagement, causes *policy inefficiencies.*

Each of these barriers triggers *negative socio-political reactions*, making transformation difficult to implement and sustain. If unaddressed, these failures accumulate, resulting in *stagnation, policy reversals, and missed climate targets*.

1.2 The Limits of Scientific-Only Approaches: Why Technical Solutions Alone Do Not Drive Transformation

Scientific expertise is fundamental for evidence-based policymaking, but *expert-driven transformation efforts* that neglect public engagement often fail (Stirling 2008). Policies that rely exclusively on scientific recommendations, without considering social and economic realities, risk generating scepticism, resistance, and implementation failure (Jasanoff 2018).

For instance, the European Union's *Nitrate Directive*, aimed at reducing groundwater contamination, encountered strong resistance from farmers in Germany and the Netherlands. Although scientifically justified, the policy failed to integrate farmers' perspectives and *ignored economic feasibility*, leading to widespread protests and difficulties in enforcement. This case demonstrates that *even well-justified environmental regulations require careful alignment with economic realities and stakeholder engagement* to ensure successful implementation.

Similarly, during the COVID-19 pandemic, early climate policy proposals linking recovery funds to sustainability measures faced opposition from economic sectors prioritising short-term survival over long-term environmental goals. The lesson here is clear: Even scientifically justified policies require societal acceptance and economic feasibility to be successfully implemented.

Thus, sustainability transitions cannot be purely expert-driven. Public participation and co-design processes are crucial to ensure that policies are both effective and socially embraced (Kuhlmann and Rip 2018).

1.3 Moralising Sustainability Narratives: The Risk of Public Backlash

Many sustainability movements present transformation as a **moral obligation**, urging people to act in the interest of future generations. However, such narratives and normative approaches can *alienate stakeholders*, particularly when they fail to provide *realistic, actionable pathways for change. Overly prescriptive sustainability discourses* that appeal to guilt rather than offering *pragmatic solutions* risk triggering *public resistance rather than fostering societal buy-in.*

A prominent example is the Fridays for Future movement, which has significantly raised awareness about climate change. However, its demand for a 50% reduction in meat consumption by 2035 was perceived by farmers as a direct attack on their livelihoods, leading to tensions between activists and agricultural stakeholders (Fridays for Future 2023). This reaction illustrates how moral appeals that overlook or even neglect economic and social realities can provoke resistance instead of fostering cooperation.

Another example is the public debate on meat consumption in Germany. Surveys indicate that consumers support stronger animal welfare regulations, yet actual meat consumption remains relatively high, particularly for low-cost meat products (Heinemann 2021). This gap between stated preferences and actual consumer behaviour underscores the limitations of moral arguments without practical alternatives.

Further examples can be found in policy contexts. The European Union's early attempts to reform the Common Agricultural Policy (CAP) failed to effectively tie subsidies to measurable sustainability outcomes, leading to continued environmental degradation. Similarly, Germany's transition to renewable energy (Energiewende) initially overlooked the social and economic impacts on coal-dependent regions, resulting in resistance and subsequent policy reversals. In the United States, repeated efforts to tighten environmental regulations in agriculture have been met with strong opposition from industrial farming lobbies, delaying the shift towards more sustainable practices.

These examples illustrate that moralising sustainability narratives alone are often insufficient to drive systemic change. Instead, more inclusive approaches are needed—ones that acknowledge economic realities, provide practical alternatives, and encourage broad participation to foster societal acceptance and active, creative engagement.

1.4 Overreliance on State-centric Approaches: The Limits of Centralised Governance

A common misconception in past transformation efforts has been the assumption that governments alone can drive systemic change through top-down regulations. However, experience has shown that multi-actor governance—which combines state regulations, market incentives, and participatory decision-making—is far more effective in achieving long-term sustainability (Ostrom 2010).

A prominent example of the limitations of state-centric governance is the European Union's Common Agricultural Policy (CAP). Originally designed to ensure food security and rural development, the CAP has undergone multiple reforms aimed at integrating sustainability objectives. However, despite decades of adjustments and alignment with the European Green Deal, structural misalignments persist. Recent analyses by the European Court of Auditors (2024) reveal that direct payments continue to disproportionately benefit large-scale farms rather than incentivise sustainability. Eco-schemes introduced to promote biodiversity and carbon sequestration often lack enforceable ecological outcomes, resulting in minimal impact on farming practices (Pe'er et al. 2020; Matthews 2021). Moreover, the increased flexibility granted to member states has led to fragmented and inconsistent national strategies, undermining collective climate ambitions.

This case illustrates the inherent limitations of top-down governance approaches: without robust economic incentives, clearly defined and measurable sustainability criteria, mechanisms for regional adaptation, and meaningful stakeholder participation, policy implementation risks devolving into administrative compliance without substantive change—ultimately leading to inefficiency and stagnation. The transition towards sustainable agri-food systems therefore necessitates multi-level governance frameworks that not only integrate regulatory instruments with market-based incentives but also foster dynamic collaboration between policymakers, industry stakeholders, and civil society. Only through such adaptive and participatory structures can systemic transformation be both effective and resilient.

1.5 Structural Barriers to Transformation

Sustainability transitions are rarely linear; they are shaped by institutional inertia, economic constraints, and societal resistance (Geels 2002; Köhler et al. 2019). Transformation processes often trigger counterforces, as stakeholders and established industries attempt to preserve their existing structures, investments, and economic models. The IÖW Report (2025) identifies several major barriers that must be addressed to ensure progress in sustainability transitions:

- **Transformation generates its own resistance:** Initial enthusiasm often gives way to scepticism as people experience real-world economic and social impacts. Public resistance tends to increase when sustainability policies are perceived as costly, disruptive, or unequally distributed (Messner 2025).

- **Economic uncertainty breeds hesitation:** Uncertainty about who bears the costs of transformation and which societal groups are perceived as "losers" creates reluctance among policymakers and businesses (Ostrom 2010).
- **Path dependencies slow down progress:** Long-established industries—such as fossil fuel extraction, industrialised agriculture, and resource-intensive manufacturing—actively resist transformation to maintain market dominance and protect sunk investments. In particular, the agri-food sector demonstrates strong inertia, with deeply entrenched technologies, institutional lock-ins, and market structures designed around conventional large-scale production systems. This resistance is reinforced by the protection of capital-intensive infrastructure and regulatory frameworks favouring the status quo, making shifts towards sustainable practices both politically and economically challenging (Conti et al. 2021; Béné 2022).

One of the *greatest challenges in sustainability transformation* is *bridging the gap between initial enthusiasm and long-term resilience.* As Messner (2025) highlights, sustainability transitions often follow a three-stage pattern:

1. *Initial momentum*, driven by public awareness, policy announcements, and media attention
2. *Resistance and scepticism*, as stakeholders experience *economic costs, social disruptions, and political pushback*
3. *Adaptation or stagnation*, depending on how well governance mechanisms manage systemic resistance

Historically, many sustainability initiatives have stalled in the second phase—facing political reversals, funding withdrawals, and industry lobbying efforts. This was evident in the rollback of environmental policies in the United States under the first Trump administration, where climate regulations were dismantled due to perceived economic costs (Dryzek et al. 2019). Alarming signs indicate that a second Trump presidency could trigger an even more profound backlash, with discussions already emerging around dismantling key institutions such as the Environmental Protection Agency (EPA) and withdrawing from international climate commitments. This highlights the vulnerability of sustainability progress to political shifts. Similarly, Germany's agrarian transformation has faced strong resistance from farmers and industrial actors who fear losing competitiveness due to increasingly strict environmental regulations (Pe'er et al. 2020).

1.6 Geopolitical Uncertainties as a Barrier to Transformation

Sustainability transformations do not occur in isolation; they are deeply shaped by global economic dependencies, geopolitical tensions, and security concerns. The IÖW Report (2025) highlights three key geopolitical barriers to transformation that interfere with sustainability transitions:

1. **Climate action competes with geopolitical priorities:** National security, economic growth, and energy stability often take *precedence over long-term sustainability goals.*
2. **Global trade dependencies create uncertainty:** Western democracies remain *highly dependent on international supply chains*, making *sustainable economic planning vulnerable* to global trade disruptions.
3. **Rising political polarisation hampers coordinated action:** Populist movements often *capitalise on transformation anxieties*, using sustainability policies as *a political wedge issue* (Dryzek et al. 2019).

A prime example of geopolitical uncertainty disrupting sustainability efforts is the global energy crisis following the *war in Ukraine*. The sudden *increase in fossil fuel prices* in 2022 led to a temporary revival of coal-fired power plants, as European governments prioritised energy security over climate targets. Similarly, in the agricultural sector, the war disrupted fertiliser and grain supply chains, leading to increased costs for farmers and delayed sustainability investments (OECD 2023). In the United States, a second Trump administration could significantly impact agriculture by deregulating environmental laws, thereby undermining global climate commitments, while expanding subsidies for industrial agriculture, which would slow the transition to more sustainable farming models. This illustrates how geopolitical risks can override sustainability priorities, forcing governments to make short-term economic and security-driven decisions that derail long-term transformation efforts.

1.7 Interdependencies of Transformation Barriers

Figure 28.2 shows how structural challenges, systemic barriers, and geopolitical uncertainties interact and reinforce each other, requiring a multi-level, integrated governance response. While Fig. 28.1 illustrates the immediate cognitive and governance-related errors that have historically undermined transformation processes, Fig. 28.2 expands the perspective to the structural, systemic, and geopolitical dimensions that interact to shape—and often impede—sustainability transitions. Figure 28.2 illustrates the complex interrelationships between three core categories of barriers that hinder sustainability transformation:

- **Structural Challenges** are represented by institutional rigidity, fragmented policy frameworks, and lack of adaptive governance. These structural weaknesses create inertia and slow down transformative processes.
- **Systemic Barriers** encompass factors such as resistance to change, economic uncertainty, and entrenched path dependencies, which reinforce existing regimes and obstruct innovation.
- **Geopolitical Uncertainties** highlight external disruptions caused by supply chain vulnerabilities, political polarisation, and resource conflicts, which can derail even well-planned transformation initiatives.

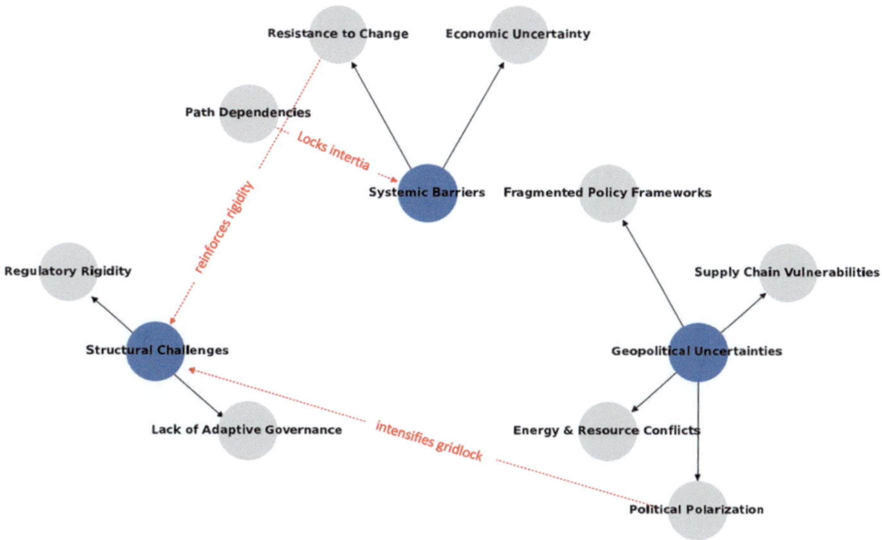

Fig. 28.2 Interdependencies of Transformation barriers. (adapted from IÖW 2025; illustration by Grabkowsky)

Arrows between nodes indicate causal relationships and feedback loops:

- Systemic barriers exacerbate structural challenges, as economic uncertainty and path dependencies lead to regulatory rigidity and inflexible institutions.
- Structural challenges, in turn, amplify geopolitical risks by weakening the international coordination needed for coherent sustainability policies.
- Geopolitical uncertainties further intensify systemic resistance by generating external shocks, which trigger conservative responses and policy reversals.

Figure 28.2 demonstrates that these barriers do not act in isolation but form a mutually reinforcing network, which can lock systems into unsustainable trajectories. Effective transformation governance, therefore, requires an integrated approach that simultaneously addresses structural rigidity, systemic inertia, and geopolitical vulnerabilities.

To prevent transformation failures and ensure the long-term success of sustainability transitions, governance structures must address structural, systemic, and geopolitical challenges proactively. Based on current research and empirical evidence, the following key recommendations emerge.

First, resistance to change should be anticipated early on, with flexible and adaptive policies that address societal and sector-specific concerns before they escalate into political opposition (Markard et al. 2012). Financial and economic feasibility must be ensured by embedding clear and transparent incentive structures rather than relying solely on regulatory mandates (Köhler et al. 2019). Participatory governance should be strengthened by engaging civil society, businesses, farmers, and local actors in collaborative decision-making processes (Ostrom 2010).

Moreover, sustainability policies need to be closely aligned with market mechanisms, for example by increasing support for regenerative agriculture and alternative farming systems. This alignment helps to ensure that climate and environmental policies remain robust and politically stable, even during times of economic or geopolitical crises (Porter and Van der Linde 1995). Increasing regional self-sufficiency in both agriculture and energy will reduce dependency on volatile international markets and enhance long-term resilience.

It is equally essential to foster international cooperation and policy coherence, avoiding fragmented landscapes of conflicting regulations. Subsidy structures, particularly within frameworks like the Common Agricultural Policy, need to be reformed by shifting from volume-based payments towards performance-based models that reward measurable sustainability outcomes (Pe'er et al. 2020). Finally, systemic resilience should be enhanced by supporting diversified and regionalised food production systems and resilient supply chains capable of withstanding external shocks (OECD 2023).

This comprehensive governance approach lays the groundwork for successful and lasting transformation. It also demonstrates why purely top-down or technocratic solutions are insufficient. Instead, dynamic multi-level governance frameworks are required that integrate scientific evidence, policy design, economic incentives, and cultural acceptance. The following section introduces the Burger Model—a holistic framework that builds on these insights and provides structured guidance for integrating ecological limits, systemic enablers, and economic viability into transformation pathways.

This *generalised governance approach* lays the foundation for models like the *Burger Model*, which will be introduced in the following section.

2 A Structured Framework for Transformation: A Multi-Layered Model for Sustainability Transformation

Building on the lessons learned from past transformation efforts and the barriers identified, it becomes clear that successful sustainability transitions require an integrated and dynamic framework. Such a framework must balance non-negotiable ecological constraints, systemic enablers, and long-term economic viability while remaining responsive to geopolitical uncertainties and societal dynamics.

The Burger Model, introduced in this section, provides exactly such a framework. It builds upon and extends established concepts, including the Planetary Boundaries framework (Rockström et al. 2009; Steffen et al. 2015) and Raworth's Doughnut Economics (Raworth 2017). While these models define ecological ceilings and social foundations for a safe operating space, the Burger Model adds a critical third dimension: the operationalisation of transformation through systemic enablers and the embedding of economic viability as a bridging condition between theory and practice.

2.1 The Structure of the "Burger" Model

The Burger Model visualises sustainability transformation as a multi-layered, hierarchical architecture:

1. **Ecological Foundation:** The non-negotiable planetary boundaries that define the safe operating space for human activity and agricultural production
2. **Systemic Enablers:** Seven structural pillars surrounding the central catalyst of Culture, acting as accelerators and framework conditions for transformation
3. **Strategic Principles:** The core ethical and operational guidelines of Efficiency, Consistency, and Sufficiency, guiding decision-making and societal behaviour
4. **Economic Viability:** The capstone, symbolising the necessity of aligning sustainability objectives with long-term economic feasibility and resilience

This structure reflects the complexity of transformation and emphasises that sustainability is not achieved through isolated interventions but through the synergy of ecological, systemic, and economic dimensions (Fig. 28.3).

2.2 The Ecological Foundation: Non-negotiable Boundaries and Economic Rationality

At the very base of the Burger Model lies the ecological foundation—the set of non-negotiable planetary boundaries that define the safe operating space for humanity (Rockström et al. 2009; Steffen et al. 2015; Richardson et al. 2023). These boundaries delineate the biophysical limits within which societies, economies, and agri-food systems must evolve. They constrain not only ecological but also economic and political action, providing the fundamental framework for any sustainability transformation. In alignment with the latest scientific understanding, the ecological foundation of the Burger Model encompasses seven core dimensions:

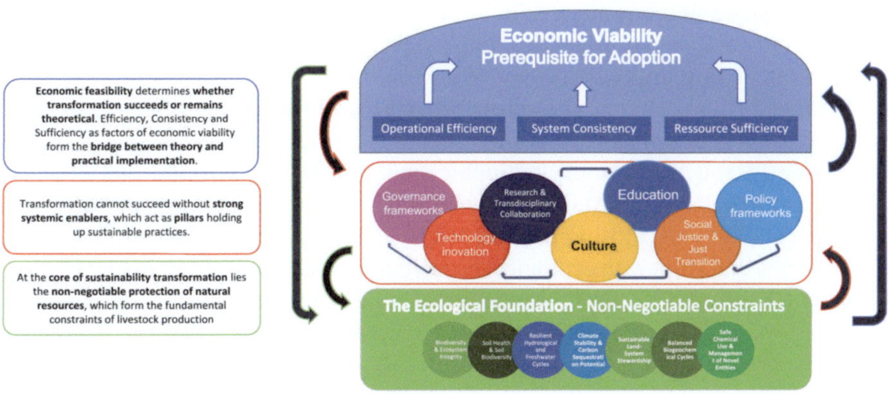

Fig. 28.3 The Burger Model for Sustainability Transformation © Grabkowsky

1. **Biodiversity and Ecosystem Iintegrity:** Biodiversity underpins essential ecosystem services such as pollination, pest control, nutrient cycling, and resilience to climatic extremes. Intact ecological networks are critical for agricultural stability and long-term productivity. Their degradation erodes adaptive capacity and increases vulnerability to shocks (Cozim-Melges et al. 2024). Protecting genetic diversity, species richness, and ecosystem functionality is therefore a non-negotiable foundation for sustainable agri-food systems.
2. **Soil Health and Soil Biodiversity:** Healthy soils are dynamic living systems that support carbon sequestration, nutrient retention, water regulation, and overall ecosystem resilience. Soil biodiversity is crucial for maintaining fertility and productivity while contributing significantly to climate regulation and adaptation (Delgado-Baquerizo et al. 2019). The loss of soil health undermines both food security and climate mitigation efforts.
3. **Resilient Hydrological and Freshwater Cycles:** Freshwater availability, groundwater quality, and stable hydrological patterns are essential for both ecosystems and agricultural productivity. Agricultural practices influence and depend on these water systems. Disruptions—through overuse, pollution, or climate-induced variability—threaten yields and ecological stability. Maintaining the integrity of hydrological cycles is therefore central to long-term food security and landscape resilience.
4. **Climate Stability and Carbon Sequestration Potential:** Agriculture holds a unique position as both a source of greenhouse gas emissions and a key driver of carbon sequestration. Sustainable soil management, regenerative practices, agroforestry, and peatland restoration contribute to carbon capture and long-term climate stability. These measures are not only environmental necessities but also present economic opportunities through emerging carbon markets (Roe et al. 2021; IPCC 2022).
5. **Sustainable Land-System Stewardship:** The expansion of croplands, deforestation, and habitat fragmentation are major drivers of biodiversity loss, soil degradation, and declining ecosystem resilience. Sustainable land-system stewardship requires regenerative land-use practices, multifunctional landscapes, and the protection of critical habitats, ensuring that land use respects ecological thresholds and avoids triggering tipping points that threaten food security and environmental stability.
6. **Balanced Biogeochemical Cycles (Nitrogen and Phosphorus Management):** Excess nutrient application disrupts natural nitrogen and phosphorus cycles, leading to soil degradation, eutrophication, aquatic dead zones, and climate-relevant imbalances. Sustainable nutrient management must focus on balancing inputs and outputs, closing nutrient loops, and employing precision application strategies that align with ecological capacities and minimise environmental harm.
7. **Safe Chemical Use and Management of Novel Entities**: The widespread release of pesticides, plastics, pharmaceuticals, and other synthetic chemicals poses escalating risks to ecosystems, soil and water quality, and human health. A future-oriented approach requires safe chemical use, reduction of hazardous substances, and promotion of green chemistry innovations that replace persistent pollutants with ecologically compatible alternatives.

These ecological dimensions in the ecological foundation layer of the Burger Model are not arbitrary; they reflect both scientific consensus and economic foresight. Increasingly, they are recognised by global economic leaders as key determinants of long-term economic resilience and market stability. The (World Economic Forum's Global Risks Report 2025) underlines this reality. Among the top ten global risks identified for the coming decade, five are directly linked to ecological and environmental challenges:

- Extreme weather events
- Critical Earth system changes
- Biodiversity loss and ecosystem collapse
- Natural resource shortages
- Pollution and waste-related risks

This convergence of ecological science and corporate risk assessment signals a profound paradigm shift. Ecological boundaries are no longer environmental concerns alone; they define the future operating space for global markets, trade, and geopolitical stability.

By placing the ecological foundation at the core of the Burger Model, we acknowledge that these constraints are not negotiable. Rather, they are absolute conditions that define the boundaries of all political, economic, and societal action. The question for transformative governance is no longer whether to respect these boundaries—but how to enable societies and economies to thrive within them while remaining competitive, resilient, and socially inclusive.

This figure from the WEF Global Risks Report 2025 illustrates the top ten global risks identified by global business leaders, policymakers, and experts. Strikingly, five out of the top six risks—extreme weather events, critical Earth system changes, biodiversity loss, natural resource shortages, and pollution-related risks—are ecological in nature. These results underscore that environmental degradation is not only a planetary crisis but also a direct threat to global economic stability, security, and prosperity. Figure 28.4 further confirms that ecological boundaries are not only theoretical concepts but also recognised determinants for global economic resilience. These findings illustrate that environmental and ecological risks are not merely theoretical considerations but are increasingly recognised as key determinants of global economic stability and competitiveness. Against this backdrop, anchoring the ecological foundation as the base layer in the Burger Model reflects this growing acknowledgement and ensures that sustainability strategies are grounded in both scientific evidence and global economic realities. The ecological boundaries shape the rules of tomorrow's global economy. However, recognising boundaries is only the first step. The question that follows is: *How can societies, economies, and agricultural systems operate within these constraints while remaining competitive, resilient, and inclusive?*

This brings us to the second layer of the Burger Model—the systemic enablers—the structural conditions and accelerators that make transformative change possible.

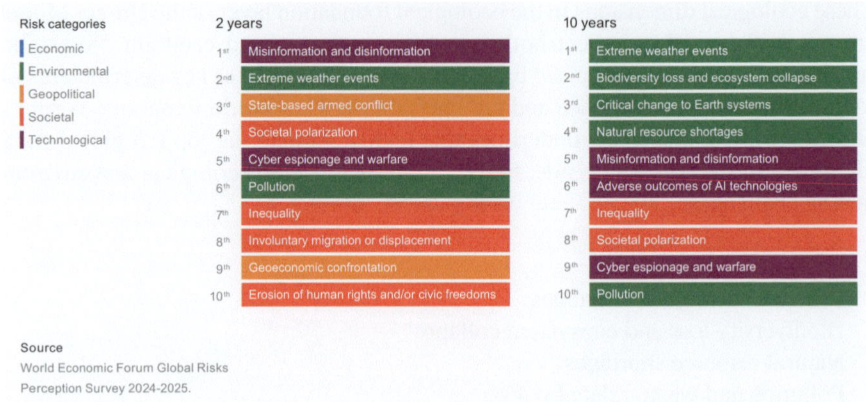

Fig. 28.4 WEF Global Risks Report 2025—Top Global Risks by Severity Over the Next 10 Years

2.3 The Systemic Enablers: The Pillars of Transformation

Resting on this foundation are seven systemic enablers, represented by vertical pillars and surrounding the core catalyst of Culture. These enablers are dynamic conditions that accelerate change, foster resilience, and provide the structural and institutional capacities for sustainability transformation:

- **Culture** (the central catalyst): The most decisive factor shaping behaviour, norms, and societal narratives. Cultural change is the prerequisite for acceptance of sufficiency, consistency, and efficiency as guiding principles.
- **Policy Frameworks**: Stable, long-term regulatory structures that provide clear goals and adaptable instruments, fostering predictable environments for sustainable investments.
- **Technological Innovation and Digitalisation**: Integration of AI, Big Data, precision agriculture, and digital tools for transparency, resource optimisation, and value chain monitoring.
- **Research and Transdisciplinary Collaboration**: Interdisciplinary and transdisciplinary research approaches, supported by real-world laboratories and living labs that connect science with policy and practice (Kuhlmann and Rip 2018).
- **Education**: Lifelong learning opportunities and broad sustainability literacy, enabling societal readiness for systemic transformation.
- **Social Equity and Justice/Just Transition**: Fair distribution of costs and benefits, protection of vulnerable groups, and cooperative models for resilience and societal buy-in (Bennett et al. 2019; EEA 2024).
- **Governance Frameworks**: The combination of regulatory mechanisms, market incentives, participatory governance, and data-driven monitoring structures (Köhler et al. 2019).

These enablers are not isolated; they function interdependently and evolve in concert with one another to facilitate transformative dynamics.

2.4 Strategic Principles (Efficiency, Consistency, and Sufficiency) as Ethical Foundations of Transformation

The Burger Model incorporates three strategic principles that represent not only operational approaches but also core ethical guidelines—a code of conduct for sustainable transformation. These principles align with the broader idea of a societal contract, as introduced in Chap. 29, emphasising the collective responsibility to act within ecological boundaries while ensuring economic and social resilience.

Efficiency The principle of efficiency focuses on maximising resource productivity while minimising environmental and social harm. Examples include precision farming, optimised input use, and advanced breeding strategies. However, efficiency gains alone cannot lead to sustainability if they trigger rebound effects—where increased efficiency leads to increased overall consumption. This phenomenon highlights the need for systemic integration and careful policy design (Santarius and Soland 2018).

Consistency Consistency refers to aligning production and consumption patterns with natural systems and regenerative processes. This includes circular economy approaches, regenerative agriculture, and closed-loop systems that restore rather than deplete ecosystems. Consistency strategies are crucial for ensuring that production cycles mimic natural flows and do not generate persistent waste or degradation (Kirchherr et al. 2017).

Sufficiency Sufficiency addresses the most challenging dimension of transformation: limiting resource use and consumption to what is necessary for well-being, thus confronting growth paradigms and consumerist habits. It requires individuals and societies to reduce excess, avoid waste, and embrace 'enough' as a guiding principle. Because this principle challenges comfort zones and established socio-economic norms, it has the closest link to Culture, which serves as the central enabler in the Burger Model. Sufficiency, therefore, calls for continuous societal dialogue, behavioural change, and strong governance frameworks that support degrowth-compatible consumption models (Lorek and Fuchs 2013; von Weizsäcker et al. 2009).

2.5 Economic Viability: The Bridging Condition Between Theory and Practice

At the top of the Burger Model lies economic viability—not as a focus on short-term profit maximisation, but as the capacity of agri-food systems to remain competitive, resilient, and sustainable over the long term. While ecological boundaries define the limits of action, and systemic enablers provide the conditions for change, economic viability represents the critical bridge that connects sustainability theory with practical, scalable implementation. Moreover, it is the key bottleneck for the acceptance and adoption of transformative measures within the business sector; only if sustainability measures are economically feasible and compatible with market realities will they be implemented at scale. Unlike the ecological foundation, economic parameters are adaptable. They can and must be shaped through intelligent governance, financial instruments, and market structures to facilitate—rather than obstruct—sustainable transformation (Wiek et al. 2012). Without this alignment, even the most ambitious transformation efforts will stall. Economic viability in this context includes several essential components:

- **Fair pricing models** for sustainable agricultural and livestock products that internalise environmental and social costs and help farmers cover transition expenses.
- **Access to sustainable financing mechanisms**, including green bonds, climate adaptation funds, and blended finance solutions to catalyse investment in innovative, climate-positive practices.
- **Risk diversification and adaptive business models**, which enable farmers and food producers to respond to climate volatility, market fluctuations, and shifting consumer demands.
- **Stable, long-term policy frameworks** that provide investment security, reduce uncertainty and foster market predictability for all actors along the agri-food value chain.
- **Market structures and incentive systems** that reward sustainability outcomes—such as carbon sequestration, biodiversity protection, and animal welfare—rather than penalise them through bureaucratic complexity or short-term cost competition.

Thus, economic viability in the Burger Model is not a passive outcome but an active design element. It enables transformational change to take root and flourish by making sustainable actions economically feasible, socially acceptable, and politically resilient.

3 Comparative Perspective and Integration into the Broader Sustainability Discourse

To fully appreciate the conceptual lineage and advancement of the Burger Model, it is helpful to visually juxtapose it alongside the two most influential frameworks in sustainability science: the Planetary Boundaries framework (Rockström et al. 2009; Richardson et al. 2023) and Doughnut Economics (Raworth 2017).

- **The Planetary Boundaries** framework establishes scientifically defined ecological thresholds that humanity must not transgress to preserve Earth system stability. It provides the foundational concept of environmental ceilings and delineates the limits within which human development can occur without risking systemic collapse.

- **Doughnut Economics** builds upon this ecological framing by integrating a social foundation beneath the environmental ceiling, defining a "safe and just space for humanity". While this model elegantly links ecological and social imperatives, it remains predominantly conceptual in nature.
- **The Burger Model** advances both approaches by providing a hierarchical and actionable framework that anchors ecological boundaries as a non-negotiable foundation, integrates systemic enablers (such as governance, culture, technology, and market incentives), and culminates in economic viability as a dynamic, adaptable capstone. This model emphasises operational implementation pathways and highlights the ethical principles of efficiency, consistency, and sufficiency as guiding transformation strategies.

Figure 28.5 presents these three models side-by-side, highlighting their conceptual connections, distinctions, and evolution. This visual comparison underscores how the Burger Model synthesizes and extends the established frameworks into an integrative structure designed for practical, large-scale application in sustainability transformations.

As illustrated in the comparative visualisation of key frameworks (Fig. 28.5), the Burger Model builds upon the conceptual foundations of the Planetary Boundaries framework (Rockström et al. 2009; Richardson et al. 2023) and the Doughnut Economics model (Raworth 2017), while extending them into actionable structures for real-world implementation. By integrating non-negotiable ecological boundaries, systemic enablers, and economic viability into a dynamic, multi-layered architecture, the Burger Model transforms abstract conceptual thinking into a pragmatic orientation framework for sustainability transitions.

While the Planetary Boundaries framework has successfully defined the environmental ceilings that delineate a safe operating space for humanity, and Doughnut Economics has framed these boundaries in conjunction with social foundations, both models remain predominantly conceptual. The Burger Model responds to this gap by introducing systemic enablers—including governance frameworks, technological innovation, scientific research, education, social justice, market incentives,

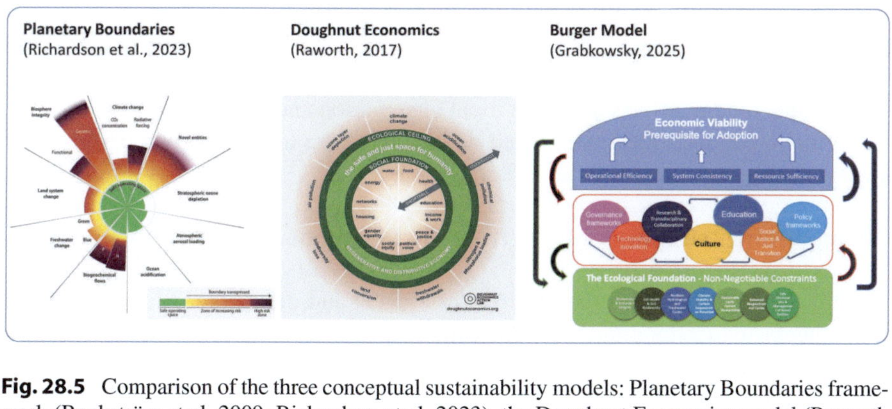

Fig. 28.5 Comparison of the three conceptual sustainability models: Planetary Boundaries framework (Rockström et al. 2009; Richardson et al. 2023), the Doughnut Economics model (Raworth 2017), and the Burger model (Grabkowsky 2025) - (**c**) Grabkowsky

and culture—as well as by positioning economic viability as the bridging condition between ecological imperatives and practical application. In doing so, the model not only advances theoretical discourse but also offers concrete guidance for designing and managing complex transformation processes in agri-food systems and beyond.

These actionable transformation pathways are grounded in the three strategic sustainability principles of efficiency, consistency, and sufficiency. Efficiency refers to the optimisation of resource use, reducing environmental impact without compromising productivity. Consistency emphasises the alignment of production and consumption with natural cycles, highlighting regenerative agriculture and circular economy approaches as core strategies. Sufficiency, arguably the most challenging principle, calls for a fundamental shift in societal behaviour and cultural norms by limiting resource use to what is socially and ecologically justifiable. This principle directly links to the model's cultural pillar, recognising that lasting transformation requires deep-rooted shifts in individual behaviour, collective values, and societal priorities.

Beyond extending existing conceptual frameworks, the Burger Model also transcends the traditional three-pillar sustainability model by embedding culture and governance as dynamic enablers rather than static categories. It emphasises the need for adaptive structures that integrate scientific evidence, policy innovation, economic mechanisms, and participatory governance into coherent and context-sensitive approaches.

Aligned with the growing call for a new societal contract on how to produce and to consume abundant food including sustainability, as discussed by Th. Blaha in Chap. 29 and reflected in the recommendations of the German Future Commission on Agriculture (ZKL 2021) (compare to Chap. 30), the Burger Model illustrates that successful transformation cannot rely on incremental adjustments or isolated, fragmented efforts. It requires a coherent, cross-sectoral framework that unites policy design, scientific insight, market structures and mechanisms, technological progress, and cultural change in order to navigate the complexities of agri-food transformation.

The Burger Model is thus more than an academic construct; it is the result of years of practical engagement with complex transformation processes, enriched by transdisciplinary collaboration between science, policy, industry, and civil society. Its strength lies in its integrative architecture, linking ecological imperatives, systemic enablers, and economic realities as interdependent and mutually reinforcing forces. This model reflects both empirical observation and strategic design, offering a pragmatic compass for those tasked with shaping the transformation of agri-food systems at regional, national, and international levels.

Distinguishing the Burger Model from the Doughnut Economics

The Doughnut Economics model (Raworth 2017) and the Burger Model share the fundamental understanding that economic activity must operate within planetary boundaries while ensuring social equity. Yet, the Burger Model extends and sharpens this approach by offering a structured and operational framework. While the Doughnut Economics approach defines a safe operating space between social foundations and ecological ceilings, the Burger Model establishes a hierarchical structure in which ecological integrity forms the foundation, supporting systemic enablers and sustainability principles and culminating in long-term economic viability.

Where the Doughnut Economics remains conceptual, the Burger Model proposes concrete governance mechanisms—such as decoupling agricultural policy from short-term legislative cycles to enable long-term sustainability planning. It advocates for a Social Contract for Sustainable Agriculture, defining how food is produced, valued, and consumed, drawing on the work of Blaha (compare to Chap. 29) and the ZKL Report (2021). In addition, the Burger Model recommends reducing regulatory rigidity by replacing prescriptive rules with goal-oriented benchmarks, allowing sector-specific flexibility and fostering innovation.

While Doughnut Economics acknowledges the relevance of technology, the Burger Model explicitly elevates digital transformation as a systemic enabler. It emphasises the necessity of full digital traceability in food production—encompassing animal health documentation, land use, and consumer behaviour data—and highlights artificial intelligence and real-time monitoring as key tools for managing sustainability across complex value chains. Moreover, it promotes cross-sectoral integration, linking agricultural innovation with developments in energy, mobility, and urban systems to advance circular economy solutions.

Cultural and social transformation are addressed with particular nuance in the Burger Model. Whereas Doughnut Economics emphasises social justice as a principle, the Burger Model delineates how cultural transformation must manifest across consumption patterns, production practices, and policymaking. Participatory governance structures, regional self-sufficiency, and cooperative models in food production are identified as essential levers for fostering resilience and reducing dependency on volatile global markets.

4 Implementation Strategies for the Burger Model

The Burger Model thus provides a comprehensive, action-oriented framework for guiding the sustainability transformation of agri-food systems. It moves beyond conceptual thinking and delivers practical orientation for policymakers, business leaders, and civil society actors alike. It is not a static blueprint, but a dynamic, adaptable tool for navigating the complex and interconnected landscape of agricultural transformation

To translate the Burger Model into real-world impact, the following implementation strategies are recommended:

- **Anchor long-term planning beyond political cycles:** Agricultural policies should be decoupled from short-term legislative terms and aligned with multi-decade sustainability goals to ensure consistency and investment security.
- **Adopt goal-oriented regulatory frameworks:** Replace rigid, prescriptive rules with outcome-based sustainability benchmarks, allowing sector-specific flexibility and fostering innovation.

- **Strengthen cooperative and regional structures:** Support cooperative models and regional food systems to enhance resilience, reduce geopolitical dependencies, and stabilise local economies (Benton and Harwatt 2022).
- **Leverage digitalisation and AI:** Mandate full digital traceability across production chains and implement AI-driven tools for resource optimisation and proactive sustainability management (Finger 2023).
- **Promote a societal contract for sustainable agriculture:** Foster dialogue between consumers, producers, and policymakers to define shared goals regarding food production, valuation, and responsible consumption (please compare to Chap. 29 (Blaha); ZKL Report 2021).
- **Integrate cross-sectoral synergies:** Connect agricultural sustainability strategies with innovations in energy systems, mobility, and urban development for holistic transformation.
- **Position sustainability as a competitive advantage:** Shift perception from regulatory burden to strategic opportunity. By investing in green innovation and climate-smart practices, agricultural sectors can enhance long-term competitiveness (Porter and Kramer 2011).
- **Implement adaptive governance models:** Enable flexibility in policy design to tailor sustainability measures to regional contexts. Experimental approaches, such as real-world laboratories, are essential for testing and refining transformative solutions (Kuhlmann and Rip 2018).
- **Boost investment in sustainable technologies:** Encourage public and private investments in precision agriculture, regenerative farming, and digital solutions to accelerate sustainability transitions (FAO 2024).

References

Béné C (2022) Why the great food transformation may not happen—a deep-dive into our food systems' political economy, controversies and politics of evidence. World Dev 154:10588. https://doi.org/10.1016/j.worlddev.2022.105881

Bennett NJ, Blythe J, Cisneros-Montemayor AM, Singh GG, Sumaila UR (2019) Just transformations to sustainability. Sustainability 11(14):3881. https://doi.org/10.3390/su11143881

Benton, T. G., & Harwatt, H. (2022). Sustainable agriculture and food systems. https://future-food4climate.eu/wp-content/uploads/2022/05/2022-05-24-sustainable-agriculture-benton-harwatt_3.pdf

Conti C, Zanello G, Hall A (2021) Why are agri-food systems resistant to new directions of change? A systematic review. Global Food Secur 31:100576., ISSN 2211-9124. https://doi.org/10.1016/j.gfs.2021.100576

Cozim-Melges F, Ripoll-Bosch R, Veen GF et al (2024) Farming practices to enhance biodiversity across biomes: a systematic review. NPJ Biodivers 3:1. https://doi.org/10.1038/s44185-023-00034-2

Delgado-Baquerizo M, Maestre FT, Reich PB, Jeffries TC, Gaitan JJ, Encinar D et al (2019) Multiple elements of soil biodiversity drive ecosystem functions across biomes. Nat Ecol Evol 3:1–9. https://doi.org/10.1038/s41559-019-1084-y

Dryzek JS, Norgaard RB, Schlosberg D (eds) (2019) The Oxford handbook of climate change and society. Oxford University Press. https://doi.org/10.1093/oxfordhb/9780199566600.001.0001

European Court of Auditors (ECA) (2024) Special report 20/2024 – common agricultural policy plans. Luxembourg: Publications Office of the European Union. Available at: https://www.eca.europa.eu/en/publications?ref=SR-2024-20

European Environment Agency (EEA) (2024) Just sustainability transitions — from concept to practice. ISBN: 978-92-9480-684-0. https://doi.org/10.2800/6238023. Retrieved at: https://www.eea.europa.eu/en/analysis/publications/just-sustainability-transitions

FAO (2024) The state of food and agriculture 2024—value-driven transformation of agrifood systems. Rome. https://doi.org/10.4060/cd2616en

Finger R (2023) Digital innovations for sustainable and resilient agricultural systems. Eur Rev Agric Econ 50(4):1277–1309. https://doi.org/10.1093/erae/jbad021

Fridays for Future (2023) Positionspapier „Ernährung und Klima: Unsere Forderungen bis 2035". Retrieved 23/03/2025 from https://fridaysforfuture.de/agrarwende-jetzt/

Geels FW (2002) Technological transitions as evolutionary reconfiguration processes: a multi-level perspective and a case study. Res Policy 31(8–9):1257–1274. https://doi.org/10.1016/S0048-7333(02)00062-8

Heinemann A (2021, February 11) Konsum und Nachhaltigkeit: Ständig lockt das Fleisch. Frankfurter Allgemeine Zeitung. Retrieved [23/03/2025], von https://www.faz.net/aktuell/wirtschaft/klima-nachhaltigkeit/konsum-und-nachhaltigkeit-staendig-lockt-das-fleisch-17358957.html

Institut für ökologische Wirtschaftsforschung (IÖW) (2025). Zwischen Zumutung und Zuversicht: Wie die sozial-ökologische Transformation gelingen kann. Berlin: Bundeskanzleramt. Retrived from: https://www.bundesregierung.de/bkamt-data

IPCC (2022) Climate change 2022: mitigation of climate change. In: Contribution of working group III to the sixth assessment report of the intergovernmental panel on climate change. Cambridge University Press. https://www.ipcc.ch/report/ar6/wg3/

Jasanoff S (2018) Just transitions: a humble approach to global energy futures. Energy Res Soc Sci 35:11–14. https://doi.org/10.1016/j.erss.2017.11.025

Kirchherr J, Reike D, Hekkert M (2017) Conceptualizing the circular economy: an analysis of 114 definitions. Resour Conserv Recycl 127:221–232. https://doi.org/10.1016/j.resconrec.2017.09.005

Köhler J, Geels FW, Kern F, Markard J, Onsongo E, Wieczorek A et al (2019) An agenda for sustainability transitions research: state of the art and future directions. Environ Innov Soc Trans 31:1–32. https://doi.org/10.1016/j.eist.2019.01.004

Kuhlmann S, Rip A (2018) Next-generation innovation policy and grand challenges. Sci Public Policy 45(4):448–454. https://doi.org/10.1093/scipol/scy011

Lorek S, Fuchs D (2013) Strong sustainable consumption governance—precondition for a degrowth path? J Clean Prod 38:36–43. https://doi.org/10.1016/j.jclepro.2011.08.008

Markard J, Raven R, Truffer B (2012) Sustainability transitions: an emerging field of research and its prospects. Res Policy 41(6):955–967. https://doi.org/10.1016/j.respol.2012.02.013

Matthews A (2021) Why the CAP still struggles with environmental challenges. CAP Reform Blog. Retrieved from https://capreform.eu

Messner D (2025) Transformationen gestalten: Die Kunst, Wandel zu ermöglichen. In: Institut für ökologische Wirtschaftsforschung (ed) Zwischen Zumutung und Zuversicht—Wie die sozial-ökologische Transformation gelingen kann (S. 7–13). Bundeskanzleramt, Berlin. Retrived from: https://www.bundesregierung.de/bkamt-data

OECD (2023) Global energy crisis: the impact on agriculture and food security. OECD Publishing. https://doi.org/10.1787/d0836b6e-en

Ostrom E (2010) Polycentric systems for coping with collective action and global environmental change. Glob Environ Chang 20(4):550–557. https://doi.org/10.1016/j.gloenvcha.2010.07.004

Pe'er G, Bonn A, Bruelheide H, Dieker P, Eisenhauer N, Feindt PH et al (2020) Action needed for the EU common agricultural policy to address sustainability challenges. People Nat 2(2):305–316. https://doi.org/10.1002/pan3.10080

Porter ME, Kramer MR (2011) Creating shared value: how to reinvent capitalism — and unleash a wave of innovation and growth. Harv Bus Rev 89(1/2):62–77. Available at: https://www.communitylivingbc.ca/wp-content/uploads/2018/05/Creating-Shared-Value.pdf

Porter ME, Van der Linde C (1995) Toward a new conception of the environment-competitiveness relationship. J Econ Perspect 9(4):97–118. https://doi.org/10.1257/jep.9.4.97

Raworth K (2017) Doughnut economics: seven ways to think like a 21st-century economist. Chelsea Green Publishing. https://doi.org/10.33568/rbs.2409

Richardson K, Steffen W et al (2023) Earth beyond six of nine planetary boundaries. Science. Advances 9(37):eadh2458. https://doi.org/10.1126/sciadv.adh2458

Rockström J, Steffen W, Noone K, Persson Å, Chapin FS, Lambin E et al (2009) A safe operating space for humanity. Nature 461(7263):472–475. https://doi.org/10.1038/461472a

Roe S, Streck C, Beach R, Busch J, Chapman M, Daioglou V, Deppermann A, Doelman J, Emmet-Booth J, Engelmann J, Fricko O, Frischmann C, Funk J, Grassi G, Griscom B, Havlik P, Hanssen S, Humpenöder F, Landholm D et al (2021) Land-based measures to mitigate climate change: potential and feasibility by country. Glob Chang Biol 27:6025–6058. https://doi.org/10.1111/gcb.15873

Santarius T, Soland M (2018) How technological efficiency improvements change consumer preferences: towards a psychological theory of rebound effects. Ecol Econ 146:414–424. https://doi.org/10.1016/j.ecolecon.2017.12.009

Steffen W, Richardson K, Rockström J, Cornell SE, Fetzer I, Bennett EM et al (2015) Planetary boundaries: guiding human development on a changing planet. Science 347(6223):1259855. https://doi.org/10.1126/science.1259855

Stirling A (2008) 'Opening up' and 'closing down': power, participation, and pluralism in the social appraisal of technology. Sci Technol Hum Values 33:262–294. https://doi.org/10.1177/0162243907311265

von Weizsäcker EU, Hargroves C, Smith MH, Desha C, Stasinopoulos P (2009) Factor five: transforming the global economy through 80% improvements in resource productivity, 1st edn, Routledge. https://doi.org/10.4324/9781849774475

Wiek A, Fariolo F, Fukushi K, Yarime M (2012) Sustainability science: bridging the gap between scienceand society. Sustain Sci 7(supplement 1):1–4. https://doi.org/10.1007/s11625-011-0154-0

World Economic Forum (WEF) (2025) Global risks report 2025. Retrieved from https://www.weforum.org/reports/global-risks-report-2025, World Economic Forum. Available at: https://reports.weforum.org/docs/WEF_Global_Risks_Report_2025.pdf

Zukunftskommission Landwirtschaft (ZKL) (2021) The future of agriculture: a societal task — recommendations of the German Commission on the Future of Agriculture. Federal Ministry of Food and Agriculture (BMEL), Berlin. Retrieved from https://www.bmel.de/SharedDocs/Downloads/DE/_Landwirtschaft/ZKL-Abschlussbericht.pdf (23/03/2025)

The Transition of Agri-food Systems Is a Task for Society as a Whole

29

Thomas Blaha

1 Animal Welfare Improvements and the Transition Towards Sustainability Stay Far Behind Societal Expectations

As well documented in this book, conventional and intensive agriculture despite growing criticisms is still harming the environment and the climate, reduces biodiversity, and most animals are still kept in confinement without sufficient enrichment; lack of choice of climatic zones; no access to open areas; and few opportunities to express their behavioural repertoires; chicken beaks are still trimmed; pigtails are still docked, and not all herds and flocks are healthy and get sufficiently competent care by the humans that have the responsibility for them.

One reason for this is that there seems to be a general expectation that the farming community is responsible for changing the way they produce our food with and from their livestock. However, this is unrealistic since our farmers are captured in the lowest-price ideology of the food value chain dictated by the few retail chains that compete for buyers by attracting them with the lowest possible prices. The consequence is that in the internationally open value chain, where suppliers can easily be replaced by others, farmers must accept the low prices they get offered by the retail sector. Those farmers that are not in one of the still rare, closed value chains for producing more sustainably (e.g. organic) or "animal-friendly" meat products, for which they get a higher price, cannot decide on their own that "from tomorrow on I will use less fertilizer or reduce the number of my animals and provide them with the very best living conditions". The reality is that the retail prices in the open value chain will not cover their additional costs.

T. Blaha (✉)
Former Head of the Field Station for Epidemiology (Bakum), University of Veterinary Medicine Hannover, Hannover, Germany
e-mail: thomas.blaha.ir@tiho-hannover.de

B. Grabkowsky, T. Blaha (eds.), *Intensive Livestock Production in Transition*,
https://doi.org/10.1007/978-3-031-97872-2_29

To understand the reasons for the sustainability shortcomings and the suboptimal housing and management for our food animals, it is necessary to explain that improving animal welfare is, like acting against climate change, not a "tame" (one-dimensional and solvable), but one of the typical "wicked" problems (multi-dimensional and not solvable). The phenomenon of "wicked" problems was defined in the late 1960s by the German Wilhelm Horst Jakob Rittel, who was a design and sociology professor in Ulm, Germany and Berkeley, USA (Rittel and Webber 1973). According to him, wicked problems are very complex, and different stakeholders with different expectations are involved. This means that changes in the right direction require many people to change their mindsets, habits, and behaviours. Due to their social complexity, wicked problems have no stopping point. They are characterised by very complex interdependencies, which means that the effort to solve only one aspect of a wicked problem creates several other problems, which are especially complicated if genuine conflicts of societal goals are involved. In the case of animal welfare, competing societal goals are food security and food safety, a decent income for farmers, and the competitiveness of farmers in the global free market. Maximising one of these societal goals will automatically lead to a reduction of the other interdependent goals. Maximising, for example, the focus on animal welfare is that those that are strongly affected by the decrease in the appreciation and support for "their" values will oppose any animal welfare improvement efforts. The aim of corrective actions within wicked problems needs to be shifted from "solutions" to "continuous intervention to achieve socially accepted balances". The major consequence of the fact that the transition of the current agri-food system is a typical "wicked" problem is that there are no simple, one-dimensional answers.

2 Not "Old McDonald's Farm" and Not Veganism

The public discourse about the need for change is often dominated by outspoken groups that are convinced that the avoidance of the adverse effects of our current agri-food system is quite simple. One of the ideas is that a move back to the idyllic small-scale family farming of our great-grandparents and grandparents would be the solution to all problems. This idea is mainly supported by the younger, technology-affine urban people who have no connections to agriculture and buy their food in urban mega-retail stores. Unfortunately, the idea of "back to the cosy little farm" is heavily supported by books for children and schoolbooks that use illustrations of a nice farm family together with chickens, a rooster, a milking cow milked by grandma, and happily playing piglets running around for explaining agriculture to urban kids. People who have never seen a current real livestock farm and have the books of their childhood in mind are prone to the Old-McDonald's-Farm idea as an ideal solution for the adverse effects of the existing highly intensive livestock production. We need no published scientific investigations of the possibility of feeding the world with "Old McDonald's farms" since history tells us that in the 1940s and 1950s, people could not be sufficiently supplied with food due to the steeply rising industrialisation and urbanisation. Apart from jeopardising food security, this concept of moving back to small-scale, low-input (minimal technology) and exclusively regional agricultural production will fail since the young, urban generation thinks the idea attractive but only a few people will be willing to do the necessary hard work 8 to 12 h on 365 days a year. It must be, however, said that local projects of alternative agriculture approaches such as cooperatives for solidary agriculture, family farms with farm shops, or vertical agriculture (Al-Kodmany 2018) are very welcome, but they will only be able to supply niche markets for a small proportion of affluent consumers, and they are not able to considerably contribute to food security.

Another idea for solving all problems with just one "silver bullet" measure is the demand for veganism for all people. It is thought to be "the new ethics of eating" (Marcus 2000) since no animals are being exploited and killed for human nutrition. This ethical knockdown argument (stop killing animals to enjoy meat as food for people) seems to be so convincing that it is astonishing that only smaller parts of the population worldwide refrain from eating meat and meat products. The reason for this is not only the (unethical?) desire for meat as a delicious foodstuff, but three-quarters of global agricultural areas consist of permanent grasslands that cannot be used as arable land for mainly topographical and climatic reasons. These grasslands solely generate non-edible biomass. In addition, arable lands provide more non-edible than edible biomass. The maintenance of soil fertility requires crop rotation that includes non-edible plants and organic manure. Without organic manure, the worldwide production of chemical fertilisers would have to be increased considerably (Windisch 2021). The latter alone is proof that the positive impact on climate change, biodiversity, and the environment would be greatly overestimated if the entire world went vegan (Dorgbetor et al. 2022).

However, there is no doubt that in most high-income countries, a shift of the human diet from a very high meat consumption per capita to a more plant-based diet has a measurable ecological benefit (Grethe et al. 2021) without disrupting traditional production cycles.

None of the single-dimensional "simple" solutions will be feasible and guarantee food security; that is why the researchers of the renowned Potsdam Institute for Climate Impact Research (Rockström et al. 2017) claim that for human prosperity and global sustainability not a change back to an extensive agricultural production system but a development towards "sustainable intensification" is indispensable.

3 We Need a New Social Contract for a Sustainable and Animal-Friendly Agri-food System

The traditional, unwritten social contract on how to produce and consume food has been in place for millennia and is still working today. In a nutshell, the *old social contract on how to produce and consume food* can be explained as: steered by demand and supply rules, farmers produce raw materials for food and foodstuffs at the lowest possible costs, food processors convert raw materials into food products at the lowest possible costs, and consumers buy the foodstuffs and food products from traders that are offered for the lowest possible prices. The competition in this field leads as a rule to food prices that everybody can afford to sustain their lives and those of their families. So far, so good. This traditional agri-food system has led to a stable food supply, even for the growing urban centres. However, the existing system has in itself no regulation mechanisms that curb the negative effects of exploiting soils, the environment, and animals. The resulting superabundance of affordable food in the countries of the Global North is one reason for the too-high post-harvest food losses. In most high-income countries, some hunger and malnutrition exist, however, not due to food shortages, but to social inequalities and dysfunctional social security systems.

This system, explained here in a simplified way, has never been questioned until recently. With the rise of the industrialisation of the agri-food system leading to intensive animal holdings, there were single critics of the then-rising "factory farming", criticising the cruelty to animals in industrialised intensive husbandry systems, but these first animal welfare pioneers had been for several centuries voices in the wilderness. However, these early animal welfarists (e.g. Harrison 1964; Singer 1975), as "lonely" as they were, have triggered a more generally scrutinising process of the seemingly functioning food system. Today, there is little doubt for most people that the "ideal world" of producing abundant food for all is far from being this ideal. The animals kept for food production lack most of the "Five Freedoms" (defining animal well-being quite accurate) that, from an ethical point of view, they deserve; the climate is at risk due to an ever-increasing emission of GHGs that contribute to the global warming, the biodiversity is shrinking due to agricultural monocultures, and there is a steady loss of natural reserves.

In other words, for too long, we have, with our old social contract, neglected the increasingly adverse effects of our highly productive system of providing mostly high-quality food for all. The current intensive husbandry systems are seen as systematic violations of the right of animals to lead a decent life. They are also seen as contributing to the exploitation of the planet, which is a threat to the well-being of future generations. Consequently, this means that we have a deep moral responsibility to "write" a new social contract on how to permanently maintain the capability to feed all people with nutritious, healthy, safe, and affordable food, and simultaneously avoid or minimise the detrimental consequences of our current agri-food system (WBGU—Germany Council on Global Change 2011 and 2021). The *new social contract on how to produce and consume food* can be described as: farmers produce high-quality food using intensive but sustainable production methods and get decent prices for their products, and all capital and labour that farmers additionally invest in measures that provide their animals a decent quality of life, maintain the biodiversity, protect the environment, and slow down the climate change will be compensated by public funding (tax money), since these investments are not paid for by selling the farmers' products.

This need for a new social contract for a future-proof agri-food system makes clear that pushing only the farmers to change their way of producing food or pushing the consumers to only buy plant-derived food or food from animal-welfare-optimised husbandry systems are both unjust to the addressed stakeholder group and inefficient to achieve considerable moves towards animal-friendliness and sustainability.

Only *multi-stakeholder approaches* that aim at societal compromises take the high complexity of the existing and evolving agri-food systems into consideration. It is very important to make sure that all stakeholders are borne in mind, including those that cannot speak for themselves (children, the mentally ill, animals, and Nature).

To moderate such multi-stakeholder commissions, clear goals must be the guardrails of any dialogues, such as the 17 UN SDG's and the European Commission's "Vision of the Future of Agriculture" issued on the EU Commission's homepage on 19 February 2025. Nations need to develop their specific multi-stakeholder approaches, such as the Final Report of the "Future Commission Agriculture" in Germany, issued in 2022. However, governments need to implement and enforce the suggested measures, as the Netherlands has done by buying farms from farmers to reduce the national agricultural production, or Denmark by passing a law on putting a CO_2 tax on farm products at the end of 2024. The transition does not occur without governments taking on the tasks the commissions suggest.

References

Al-Kodmany K (2018) The vertical farm: a review of developments and implications for the Vertical City. Buildings 8(2):24. https://doi.org/10.3390/buildings8020024. Berlin. ISBN 978-3-946830-35-1

Dorgbetor IK, Ondrasek G, Kutnjak H, Miku O (2022) What if the world went vegan? A review of the impact on natural resources, climate change, and economies. Agriculture 12:1518. https://doi.org/10.3390/agriculture12101518

Grethe H, Martinez J, Osterburg B, Taube F, Thom F (2021) Klimaschutz im Agrar-und Ernährungssystem Deutschlands: Die drei zentralen Handlungsfelder auf dem Weg zur Klimaneutralität. Stiftung Klimaneutralität. https://www.stiftung/klima.de/app/uploads/2021/06/2021-06-01-Klimaneutralitaet_Landwirtschaft.pdf

Harrison R (1964) Animal machines: the new factory farming industry. Vincent Stuart Publishers Ltd., London

Marcus E (2000) Vegan: the new ethics of eating. Rowman & Littlefield

Rittel HWJ, Webber MM (1973) Dilemmas in a general theory of planning. Pol Sci 4(2):155–169

Rockström J, Williams J, Daily G, Noble A, Gordon L, Wetterastrand H, DeClerck F, Shah M, Sreduto P et al (2017) Sustainable intensification of agriculture for human prosperity and global sustainability. Ambio 46:4–17. https://doi.org/10.1007/s13820-016-0793-6

Singer P (1975) Animal liberation, 2nd edn. Pimlico, London

WBGU – German Advisory Council on Global Change (2011) World in transition—a social contract for sustainability. WBGU, Berlin. ISBN 978-3-936191-47-9

WBGU – German Advisory Council on Global Change (2021) Beyond climate neutrality, Policy paper 12. WBGU, Berlin

Windisch W (2021) Can we do without livestock? J Consum Prot Food Saf 16(3):193–194. https://doi.org/10.1007/s00003-021-01338-w

30 Multi-Stakeholder Governance in Agricultural Policy: Insights from the Commission on the Future of Agriculture in Germany

Nana Zubek and Barbara Grabkowsky

1 Introduction

Society's focus and expectations regarding agriculture and food have significantly evolved in recent decades. This shift is driven not only by the sector's environmental impact but also by changing consumer behavior, increasing demands for sustainability, and heightened public awareness of ethical and climate-related issues.

Despite the heightened attention, a coherent approach has been lacking, resulting in unmet environmental goals. The debates are often marked by controversial statements and sweeping judgments, underscoring the need for solutions that are based on a solid technical foundation (Zukunftskommission Landwirtschaft 2021). These solutions must also gain social acceptance and be economically, ecologically, and socially viable. To address these needs, the Federal Government of Germany established the Commission on the Future of Agriculture (ZKL) on July 8, 2020.

The ZKL includes representatives from all agricultural policy groups, ensuring that economic aspects are considered alongside environmental and social factors. The commission comprises 31 members, convenes thematic working groups, and invites external experts from various fields. The members represent agriculture, business, environment, nature, animal welfare, and consumer protection sectors (BMEL 2024). Decisions are ideally made by consensus, with the chairman of the commission having the deciding vote in case of a tie. The chairman of the ZKL was

N. Zubek (✉)
Department of Agricultural Economics, Faculty of Agricultural Sciences and Landscape Architecture, Hochschule Osnabrück – University of Applied Sciences, Osnabrück, Germany
e-mail: n.zubek@hs-osnabrueck.de

B. Grabkowsky
Center for Sustainability Transformation Research in AgriFood Systems, Vechta Institute of Sustainability Transformation in Rural Areas, University of Vechta, Vechta, Germany
e-mail: barbara.grabkowsky@uni-vechta.de

B. Grabkowsky, T. Blaha (eds.), *Intensive Livestock Production in Transition*,
https://doi.org/10.1007/978-3-031-97872-2_30

Prof. em. Dr. Peter Strohschneider, who is also tasked with external representation and coordination of the commission's activities. Peter Strohschneider is a noted academic specializing in German Medieval Studies. He held professorships at the Dresden University of Technology and LMU Munich. His key roles include chairing the German Council of Science and Humanities (2006–2011) and serving as President of the German Research Foundation (2013–2019). Since the beginning of 2024, he has been involved in the Strategic Dialogue on the Future of EU-Agriculture. The federal government supports the commission through an office at the Federal Ministry of Food and Agriculture.

The commission's work primarily focuses on future agricultural policy priorities, agricultural funding objectives, and the role of the European Common Agricultural Policy. Additional topics include the economic viability of agriculture, its environmental and climate impacts, and the future of animal husbandry and animal health. The ZKL was established as a politically mandated commission by the German Federal Government in July 2020. Its secretariat was embedded within the Federal Ministry of Food and Agriculture (BMEL), ensuring close coordination with federal policymaking processes. This institutional anchoring highlights the strategic relevance and governmental ownership of the commission. The composition of the ZKL was designed to reflect a broad societal spectrum. Members were drawn from the following actor groups:

- Agriculture (e.g., farming associations)
- Food industry (processing and trade)
- Environmental, animal welfare, and consumer protection organizations
- Scientific community (agricultural, ecological, and social sciences)

This inclusive multi-stakeholder structure aimed to ensure that economic, environmental, and societal perspectives were integrated into the development of shared policy recommendations. It was also intended to help rebuild trust and foster dialogue between previously opposed societal groups following years of conflict surrounding the farmers' protests (Münch and Schulze Pals 2025).

2 Subject and Aim of the Study

To develop viable recommendations, the commission pursued a multi-stakeholder approach involving all relevant parties, i.e., farmers, environmental associations, consumer organizations, scientists, and policymakers. This methodological approach aims to foster constructive dialogue and consider diverse perspectives to develop joint solutions. The focus is on achieving consensus and formulating practical recommendations for a sustainable agricultural sector.

Numerous scientific sources support the efficacy of multi-stakeholder approaches in addressing complex challenges. Freeman (1984) was one of the first to explain how involving different stakeholder groups can lead to better decision-making and more sustainable management. Bäckstrand (2006) further examined the legitimacy, accountability, and effectiveness of multi-stakeholder partnerships in sustainable development, highlighting their role in addressing global challenges. Dentoni et al. (2018) explored the central role of these partnerships in tackling complex and "wicked" problems, emphasizing the benefits of participatory processes in developing innovative solutions. Recent studies, such as those by Jäger et al. (2023), Dentoni et al. (2018), Innes and Booher (2020), Eweje et al. (2021), Hajer and Versteeg (2018), and Austin and Seitanidi (2012), provide contemporary perspectives and empirical examples of applying the multi-stakeholder approach in various contexts, including sustainable development, climate adaptation, decarbonization pathways, "living labs" as contemporary techniques of futuring, and organizational transformation. Pahl-Wostl (2002) had already formulated approaches in linking analytical methods and participatory approaches in order to improve stakeholder-based policy design and modeling processes innovation and research. Kuenkel (2019) promotes a strategic design of collaboration to create tangible results and enhance actors' collaboration literacy in navigating complex change. However, in the specific context of agricultural transformation, existing literature lacks an in-depth examination of sector-specific dynamics. These include long investment cycles, complex value chains, and the integration of environmental and economic interests across scales—from farm level to EU policy.

With regard to the agricultural transformation process, however, the literature so far leaves open an in-depth analysis of the strengths and limitations of a multi-stakeholder approach, i.e., what lessons can be learned? What potential for improvement has emerged? Moreover, in 2021, the Commission on the Future of Agriculture published comprehensive recommendations for the creation of a sustainable agricultural and food system. However, these have not yet been implemented. This raises the question of which opportunities, barriers, and prospects for implementing these recommendations have arisen over the last 3 years from the perspective of those who have contributed to them. Against this background, this chapter aims to achieve two objectives. First, it examines the strengths and limitations of the multi-stakeholder approach in the transformation process using the example of the ZKL. Second, it explores the opportunities, barriers, and prospects for implementing the recommendations of the Commission on the Future of Agriculture published in 2021. For this reason, three qualitative expert interviews were conducted and evaluated. Qualitative expert interviews are widely recognized for their ability to explore complex issues in depth. They are particularly useful in understanding decision-making processes and investigating causal mechanisms (Kvale and Brinkmann 2009; Patton 2015; Bryman 2016). In the context of this chapter, expert interviews enable us to uncover individual attitudes, motives, experiences, and perceived challenges among members of the ZKL during their participation. Notable references underscore the methodological rigor and flexibility of semi-structured interviews, making them well-suited for investigating the complexities of agricultural transformation.

The advantages of qualitative expert interviews are particularly pertinent to this study for several reasons (von Soest 2023):

1. Practical Decision Insights: Expert interviews reveal how decisions are made in practice, adding context-specific information that is often not publicly available.

2. Information Aggregation and Weighting: Experts can synthesize and evaluate various pieces of information, crucial for addressing complex questions that do not lend themselves easily to experimental or statistical analysis.
3. Linking Macro and Micro Levels: These interviews provide data to connect macro-level phenomena with micro-level actions. For the ZKL, this approach helps understand the impact of its recommendations on all players along the agricultural value chain, as well as institutions and regions within and outside the EU.

3 Interview Process

In spring 2024, qualitative expert interviews were conducted with three key figures from the ZKL:

Prof. Dr. Kai Niebert: President of the German League for Nature Conservation and Environmental Protection (DNR), an umbrella organization for German nature, animal, and environmental protection organizations.

Hubertus Paetow: President of the German Agricultural Society (DLG).

Prof. Dr. Achim Spiller: Professor of Marketing for Food and Agricultural Products at the Department of Agricultural Economics and Rural Development, Georg-August University of Göttingen.

The three interviewees were selected due to their prominent and active roles and expertise as long-standing members within the ZKL, representing a broad spectrum of perspectives. This makes them ideal representatives to provide comprehensive and informed views on the commission's workings and outcomes.

- *Kai Niebert* represents environment and conservation interests, providing insights into how environmental and nature conservation issues are integrated into agricultural policy.
- *Hubertus Paetow* represents the agricultural industry, offering practical agricultural insights and economic considerations.
- *Achim Spiller contributes* academic and market-oriented perspectives to the agricultural sector, bridging the gap between theory and practice in agricultural marketing and economics.

3.1 Development of the Interview Guide

A questionnaire was developed based on existing literature on multi-stakeholder approaches and networks to conduct semi-structured interviews. This guide comprised ten questions aimed at discussing the strengths and limitations of a multi-stakeholder approach in the transformation process using the ZKL as a case study. Additionally, the questions focused on the opportunities, barriers, and prospects for implementing the ZKL's recommendations.

3.2 Conducting and Analyzing Interviews

The interviews were conducted online, recorded, transcribed, and edited for analysis. To identify and interpret patterns, similarities, and differences in the statements, the material was analyzed using qualitative content analysis (Kuckartz and Rädiker 2024). The analysis was supported by MAXQDA software and involved systematic coding using inductive category formation. Key steps in the analysis included *Initial Coding*, i.e., categories were formulated directly from the material for relevant text passages based on predefined category definitions, and *Category Refinement*, i.e., categories were generalized into main categories through a process of summarization, developing super- and subcategories to structure the results. The form of open coding was used to record the statements without distortions. In total, 180 minutes of transcribed interview material were analyzed, providing a comprehensive understanding of the experts' perspectives. Moreover, feedback mechanisms for returning results to stakeholders should be considered a vital part of transparent research processes. Although this study presents anonymized findings, results may be shared with participants through scientific communication formats such as conference presentations, policy briefs, or publications.

4 Results

Based on the developed categories, this section describes the key statements on the perceived strengths, weaknesses, and potential for improvement of the multi-stakeholder process.

4.1 Strengths of the ZKL Process

A perceived general advantage was that the ZKL's work coincided with the COVID-19 pandemic, which resulted in many meetings being held online. This allowed the participants to attend with fewer scheduling conflicts. At the beginning of the ZKL process, the respondents' expectations varied from low to optimistic regarding the potential for developing innovative solutions. However, the start of the Commission's work was perceived as particularly well structured and productive. Here, each participant was asked by the chair to write down their position from the perspective that the problems could be solved and that it would also be possible to reach consensus/common ground among all stakeholders involved. This exercise helped set a collaborative tone for further discussions. During the commission's work, some of the initially low expectations were partly far exceeded, leading to the development of and innovative approaches. The fact that there were no substitution rules for the participants in the ZKL was identified as a decisive factor. This meant that the presidents of the participating institutions and interest groups were required to take part in the meetings personally and were not allowed to send representatives. Another emphasized aspect was that the members trusted in the results of the intensive work in diverse working groups. This ensured continuous engagement at the highest level of decision-making.

Another key strength was the trust built within the working groups. Respondents emphasized that decisions made in the working groups were generally respected in the plenary sessions, reducing the need for redundant debates. In this context, the role of the "critical friends of the chair" was mentioned as an instrumental mechanism. These individuals, drawn from various stakeholder groups, provided structured feedback to the moderator—both on methodological aspects and in negotiating positions, ensuring a balanced and effective process.

The importance of facilitation and leadership was emphasized unanimously by all three interviewees. Beyond organizational competence, the moderator's objectivity and neutrality were seen as essential. The role of Peter Strohschneider as chair of the commission was considered particularly influential in ensuring impartiality and effective crisis management. One expert stated:

> *"[…] a certain distance from the topic is helpful, because on the one hand it prevents someone being a party, and on the other hand that, the acceptance among those involved is simply higher... and the second point is that this requires a high level of competence in dealing with people, in other words this, which he always referred to as 'social work' […]. So, the whole structure comes from him, that's already a profile. So it's not enough to call an agency and say I need a moderator for a topic like this; that requires an extraordinary personality".*

A distinctive feature of the ZKL process was its commitment to stakeholder dialogue in a protected space. Interviewees noted that trust-building mechanisms and confidential discussions were essential for constructive engagement and helped minimize conflicts.

Moreover, the negotiation process itself was seen as a significant learning experience. The ability to balance competing stakeholder interests while maintaining a clear direction was described as a key success factor. One respondent emphasized:

> *"The real challenge was not just negotiating positions but also understanding the motivations behind them. This required time and trust-building, which was only possible because of the structured facilitation".*

Additionally, the use of structured scenario-building and systematic perspective shifts was highlighted as effective methodological tools that enabled participants to explore different pathways for agricultural transformation. However, it was noted that these techniques were introduced relatively late in the process. Experts suggested that earlier integration of scenario development could have improved decision-making, clarified strategic options, and facilitated implementation planning.

4.2 Optimization Potential and Implementation Challenges

Regarding possible optimization potential, the respondents primarily mentioned the implementation perspective for the recommendations elaborated, the need for better cooperation between politics and the commission, as well as a lack of monitoring the political implementation process. This also includes political responsibility and accountability. According to interviewees, political responsibility for implementation should reside at the highest political level, ideally within the Federal Chancellery, rather than being delegated to sector-specific ministries. One expert emphasized:

> *"Political support must be sustained even in the event of a government change. Otherwise, the commission's recommendations risk being abandoned".*

The interviewees set priorities for improvement in the following aspects:

- Better clarification of the mandate of each stakeholder group.
- Ensuring that all participants have clear decision-making authority within their organizations.
- Greater focus on interdisciplinary collaboration between agriculture, environmental policy, and consumer advocacy.
- Integration of international and development policy aspects to align national agricultural policies with global sustainability goals.

A major concern was the lack of binding mechanisms to ensure follow-up on the ZKL recommendations. Experts stressed that multi-stakeholder processes must be designed to withstand political changes. Additionally, the lack of structured long-term implementation pathways was identified as a barrier to sustaining transformation efforts.

> *"Without a clear political mandate and structured follow-up mechanisms, even the most well-designed recommendations risk being stalled or abandoned".*

Interviewees also emphasized the need for cross-sectoral, project-oriented policies to overcome siloed decision-making. Some noted that the working group recommendations could be refined further to improve clarity and applicability across different contexts.

All interviewees discussed the challenges of maintaining stakeholder commitment over time. Regarding potential commitment problems, one expert explained:

> *"In other words, more commitment [is needed], and this brings me back to the personalities. I believe that this commitment can be secured through key personalities, but we also face the challenge that, because the whole process takes time, some players change roles. So how do we ensure that the positions adopted within the commission are upheld and defended within their organizations, even after personnel changes?"*

Another expert added:

> *"The German Farmers' Association should take a stronger leadership role in guiding the sector through this transformation. Change is difficult and unpopular among some farmers, but without these difficult steps, we will never resolve critical challenges related to animal welfare, climate adaptation, and economic sustainability".*

Table 30.1 Results

Expectations	Fulfillment of expectations	Moderation
New, innovative solutions (1)	Surpassed (2)	Objectivity (1)
Low expectations (2)	Fulfilled (1)	Conviction/motivation (1)
Balancing of interests/compromises (2)	Not fulfilled (0)	Leadership (2)
Trust-building mechanisms (2)		Neutrality (3)
Scenario development as a method (1)		Integrity (1)
Optimization potential	**Challenges**	**Success factors**
Clarification of political responsibilities (1)	Detailed consensus (1)	Common set of values/ basic consensus (1)
Ability to reach consensus (1)	Unclear definition of tasks (1)	Confidentiality (1)
Implementation perspective (3)	(party) political vs. personal interests (4)	Public recognition/ prioritization (1)
Consideration of international/ development policy aspects (1)	External crises (1)	Personal exchange/ appreciation (3)
Ensuring participants have decision-making authority (2)	Stakeholder replacement disrupts processes (1)	Moderation and crisis management (3)
Timing the commission's work at the start of a legislative period (2)	Lack of structured implementation pathways (2)	Scheduling flexibility (2)
		Steering committee (1)
		Interdisciplinary collaboration (5)
		Wealth of experience of those involved (2)
		Consideration of different perspectives (4)

The importance of political timing was also highlighted as a crucial factor. Two interviewees expressed that commissions like the ZKL should ideally be established at the beginning of a legislative period rather than later, when the political landscape is already shaped by other priorities.

The results are summarized in Table 30.1, grouped according to the developed super- and subcategories. The total number of mentions is shown in brackets. Repeated mentions of a subcategory in the course of the interview were counted multiple times, meaning the maximum number can be greater than three and does not necessarily represent responses from all interviewees.

5 Discussion and Conclusions, Outlook

The results underline both the achievements and the shortcomings of the ZKL process. To interpret these findings more systematically, the following section identifies key analytical insights into multi-stakeholder processes and links them to existing research on governance and transformation.

5.1 Analytical Insights into Multi-Stakeholder Processes

By including three key stakeholders of ZKL, the analysis captures a nuanced perspective on the issues at stake.

With regard to the questions posed at the beginning, four key findings can be highlighted in the conclusion. The critical role of skilled moderation, particularly the necessity of neutrality, integrity, and crisis management, was widely acknowledged as a key factor in the success of the ZKL.

Structural challenges in implementation, including a lack of political continuity and accountability mechanisms, hinder the long-term realization of commission recommendations.

The significance of interdisciplinary cooperation and mutual appreciation among stakeholders, as well as the necessity of maintaining decision-making authority within participant organizations, emerged as a further key factor.

Generally, the following five *Analytical Insights into Multi-stakeholder Processes* can be identified:

1. *Institutional Representation and Authority*
   The commission's effectiveness is closely tied to the authority of its members. Participants with long-standing institutional roles and decision-making power were better positioned to shape outcomes and ensure policy continuity. Conversely, those without formal decision-making authority struggled to secure long-term organizational commitments.
2. *Moderation and Conflict Resolution*
   Neutral and skilled moderation was essential in navigating conflicts and maintaining a balanced dialogue. While techniques like perspective shifts and "critical friends of the chair" helped mitigate power asymmetries, their inconsistent application across commission phases suggests the need for a more systematic approach to facilitation.
3. *Implementation and Political Commitment*
   The absence of binding mechanisms for implementing the commission's recommendations weakened its impact. Political transitions and shifting policy priorities further exacerbated this challenge, emphasizing the need for stronger accountability structures and clearer institutional positioning.
4. *Methodological Preliminaries*
   The delayed integration of scenario-based techniques limited strategic foresight and stakeholder engagement. Earlier adoption of these tools could have improved long-term policy planning and facilitated a clearer roadmap for implementation.
5. *Cross-Sectoral Policy Integration*
   The study highlights the need for stronger alignment between national and EU agricultural policies to avoid fragmented decision-making and enhance global coordination in agricultural sustainability efforts.

Finally, a more explicit engagement with broader society—beyond organized stakeholder groups—is needed to strengthen legitimacy and public trust. Future processes should consider how to meaningfully integrate the perspective of the general public, including consumers and rural communities. The success of future multi-stakeholder formats will also depend on their ability to anticipate and address emerging challenges, such as climate extremes, global market disruptions, and digital transformation in agriculture.

5.2 Key Recommendations for Future Multi-Stakeholder Processes

Key Recommendations for Future Multi-stakeholder Processes can be summarized by five aspects:

1. *Structural Clarity*
   - Define clear decision-making authority for each stakeholder to prevent ambiguity and enhance accountability.
   - Align commissions with political cycles to maximize adoption potential and sustain momentum despite government transitions.
   - Ensure participation of stakeholders with institutional decision-making power to enhance implementation feasibility.
2. *Enhanced Facilitation*
   - Implement structured stakeholder rotation and perspective-shifting mechanisms to mitigate power asymmetries.
   - Standardize moderation techniques to improve conflict resolution and consensus-building.
3. *Implementation Pathways*
   - Establish binding follow-up mechanisms to ensure long-term accountability.
   - Institutionalize scenario-based planning from the outset to prevent post-commission stagnation.
   - Assign clear delegated authority (Prokura) to secure binding organizational commitments.
4. *Knowledge Integration*
   - Introduce structured scenario development early to enhance evidence-based decision-making.
   - Foster interdisciplinary knowledge-sharing to ensure holistic policymaking.
5. *Transparency and Stakeholder Engagement*
   - Establish protected stakeholder dialogue spaces to build trust and facilitate constructive discussions.
   - Improve public communication strategies to enhance societal acceptance of transformation processes.
   - Ensure equal decision-making power among stakeholder groups to prevent dominance by singular interest groups.

The ZKL process demonstrates both the potential and limitations of multi-stakeholder governance in agricultural policy transformation. While it successfully facilitated consensus-building and knowledge exchange, persistent challenges in political implementation, leadership clarity, and power asymmetries highlight the need for structural refinements in future commissions.

This study underscores the necessity of stronger institutional anchoring, structured implementation strategies, and the consistent application of participatory methodologies to maximize the effectiveness of multi-stakeholder processes. Future commissions should integrate early-stage scenario-based foresight, structured follow-up mechanisms, and interdisciplinary collaboration to enhance strategic decision-making and policy execution.

To further advance the understanding of multi-stakeholder governance, the authors aim to include additional ZKL stakeholders, enabling a more comprehensive assessment of barriers and facilitators in agricultural policy implementation.

Additionally, an important avenue for future research is the examination of the EU-level agricultural dialogue initiated by European Commission President Ursula von der Leyen and moderated by Prof. Dr. Peter Strohschneider. This initiative aims to align agricultural sustainability with economic resilience and offers valuable insights into adaptive governance, financing models, and digital policy tools for agricultural transformation at the European level. Integrating lessons from both national and EU governance structures will be instrumental in refining future multi-stakeholder approaches to agricultural policy.

References

Austin JE, Seitanidi MM (2012) Collaborative value creation: a review of partnering between nonprofits and businesses. Part 2: partnership processes and outcomes. Nonprofit Volunt Sect Q 41(6):929–968. https://doi.org/10.1177/0899764012454685

Bäckstrand K (2006) Multi-stakeholder partnerships for sustainable development: rethinking legitimacy, accountability and effectiveness. Eur Environ 16(5):290–306. https://onlinelibrary.wiley.com/doi/abs/10.1002/eet.425

Bryman A (2016) Social research methods. Oxford University Press. https://books.google.de/books?hl=de&lr=&id=N2zQCgAAQBAJ

Bundesministerium für Ernährung und Landwirtschaft (2024) Zukunftskommission Landwirtschaft. https://www.bmel.de/DE/themen/landwirtschaft/zukunftskommission-landwirtschaft.html

Dentoni D, Bitzer V, Schouten G (2018) Harnessing wicked problems in multi-stakeholder partnerships. J Bus Ethics 150(2):333–356. https://doi.org/10.1007/s10551-018-3858-6

Eweje G, Sajjad A, Nath SD, Kobayashi K (2021) Multi-stakeholder partnerships: a catalyst to achieve sustainable development goals. Mark Intell Plan 39(2):186–212. https://doi.org/10.1108/MIP-04-2020-0135

Freeman RE (1984) Strategic management: a stakeholder approach. Pitman, Boston. https://books.google.de/books?hl=de&lr=&id=NpmA_qEiOpkC

Hajer M, Versteeg W (2018) Imagining the post-fossil city: why is it so difficult to think of new possible worlds? Territ Politics Gov 7(2):1–13. https://doi.org/10.1080/21622671.2018.1510339

Innes JE, Booher DE (2020) Planning with complexity: an introduction to collaborative rationality for public policy. Routledge. https://www.taylorfrancis.com/books/mono/10.4324/9780203864302

Jäger J, Brutschin E, Pianta S, Omann I, Kammerlander M, Sudharmma Vishwanathan S, Vrontisi Z, MacDonald J, van Ruijven B (2023) Stakeholder engagement and decarbonization pathways: meeting the challenges of the COVID-19 pandemic. Front Sustainability 3. https://doi.org/10.3389/frsus.2022.1063719

Kuckartz U, Rädiker S (2024) Fokussierte Interviewanalyse mit MAXQDA. Springer. ISBN: 9783658402112, 3658402113

Kuenkel P (2019) Stewarding sustainability transformations in multi-stakeholder collaboration, Stewarding Sustainability Transformations. Springer, Cham. https://doi.org/10.1007/978-3-030-03691-1_6

Kvale S, Brinkmann S (2009) Interviews: learning the craft of qualitative research interviewing. Sage. https://books.google.de/books?hl=de&lr=&id=bZGvwsP1BRwC

Münch R, Schulze Pals L (2025) Brücken bauen: Für ein besseres Miteinander von Bürgern und Bauern. Der Erfolg der Zukunftskommission Landwirtschaft – und deren Scheitern?. ISBN: 9783784392509, 3784392504

Pahl-Wostl C (2002) Participative and stakeholder-based policy design, evaluation and modeling processes. Integr Assess 3(1):3–14. https://doi.org/10.1076/iaij.3.1.3.7409

Patton MQ (2015) Qualitative research & evaluation methods. Sage. https://books.google.de/books?hl=de&lr=&id=ovAkBQAAQBAJ

von Soest C (2023) Why do we speak to experts? Reviving the strength of the expert interview method. Perspect Polit 21(1):277–287. https://doi.org/10.1017/S1537592722001116

Zukunftskommission Landwirtschaft (2021) Zukunft Landwirtschaft. Eine gesamtgesellschaftliche Aufgabe. Empfehlungen der Zukunftskommission Landwirtschaft, Berlin. https://www.bmel.de/SharedDocs/Downloads/DE/Broschueren/abschlussbericht-zukunftskommission-landwirtschaft.pdf

31 Sustainable Transformation of Intensive Livestock Production Systems Is Not a Return to "Old McDonald's Farm"

Gerald Shurson

1 Review of Our Challenges

Our current food system is inefficient, unsustainable, and incapable of feeding the future of 11 billion people without undergoing an extensive transformation. There are 690 million people who are hungry and malnourished, 340 million children suffering from micronutrient deficiencies, while at the same time, 2.1 billion adults are overweight or obese (FAO et al. 2020). In addition, 33 to 40% of food produced globally is lost or wasted annually, which has created a substantial environmental burden where 25% of freshwater resources (FAO 2023), 30% of agricultural land area, and 38% of total energy consumption for food production are wasted while contributing about 8% of total greenhouse gas (GHG) emissions from disposal in landfills (FAO 2019). In fact, if food loss and waste were its own country, it would be the third largest GHG emitter (after China and the United States) globally (FAO 2013).

"Industrial" or "intensive" agriculture and "factory farming" are terms frequently used to create negative images and perceptions that infer bad and unacceptable food production practices such as the use of toxic chemicals, monoculture crop production, confined animals, and the demise of traditional, small, "Old McDonald's-style" family farms. It is true that agriculture is a major contributor to many of our existing environmental problems including water pollution and scarcity, causes 80% of soil degradation, is largely responsible for deforestation and biodiversity loss, and contributes about 31% of total global greenhouse gas (GHG) emissions responsible for global warming and climate change. However, intensive industrial agriculture has enabled us to efficiently produce enormous amounts of food capable

G. Shurson (✉)
Department of Animal Science, College of Food, Agriculture, and Natural Resource Sciences, University of Minnesota, St. Paul, MN, USA
e-mail: shurs001@umn.edu

B. Grabkowsky, T. Blaha (eds.), *Intensive Livestock Production in Transition*,
https://doi.org/10.1007/978-3-031-97872-2_31

of meeting all of the current needs of our global society if food loss and waste are prevented and a better distribution system is in place to ensure complete food security for everyone. Because of the economies of scale and capability of adopting the newest and most productive and efficient technologies, large scale, intensive agriculture will continue to dominate the production of our global food supply as we strive to produce increasing amounts of food necessary to feed a growing global population. Therefore, the most important question that we must address is "Can we prevent further global warming and climate change, conserve natural resources, improve efficiency, and regenerate ecosystems to adequately feed a growing population and sustain life on Earth?"

Intensive food animal production systems contribute to the current inefficiency and unsustainability of our global food system, but they are also an essential part of the future of feeding the world sustainably. Although eating less meat has been proposed as a solution for reducing the environmental footprint of agriculture and food production, it is not a viable option for many reasons. Meat, milk, and eggs provide high-quality protein and essential vitamins and minerals that are not easily obtained by humans consuming only plant-based diets. Recycling valuable nutrients from food loss and waste, and agri-industrial by-products into animal feed is essential for a requisite circular economy and sustainable food system. Prudent management and application of animal manure to cropland is essential for regenerating soil health and fertility, which can substantially reduce reliance on fossil-fuel-derived synthetic nitrogen fertilizer and conserve the declining finite reserves of phosphate fertilizer in a circular food system. Therefore, we need to adopt many of the concepts, strategies, and recommendations provided by the scientific experts and authors of the chapters in this book on *Intensive livestock Production in Transition* to redesign our livestock and food production system for the future of people, ecosystems, and the planet.

In fact, the need to sustainably transform intensive livestock production systems is more urgent now than ever before because the rising incomes in China and other emerging economies are expected to create more meat demand and consumption in the future. To meet this demand, we will need to produce much more meat, milk, and eggs with fewer resources and substantially reduce the environmental burden while also improving resilience to the direct and indirect effects of climate change. This will require continued development and implementation of new technologies that enhance productivity and nutritional efficiency, and minimize mortality and morbidity caused by increasing transboundary disease transmission in these intensive livestock and poultry production systems. We will also need to improve communication with the general public and increase investment in scientific advances involving precision livestock farming to improve animal welfare.

Although there is a misconception that organic, natural, and local production of food is more environmentally sustainable than large-scale intensive livestock and crop production, these alternative production practices actually cause greater environmental burdens because they require more resources per unit of calories and protein produced for human consumption. Likewise, regenerative agriculture practices are often associated with reverting back to traditional small-scale "Old

McDonald's-style" farming practices, but they can be part of the solution for balancing the need between increased productivity and efficiency without creating more waste, pollution, and depletion of natural resources. Countries in the EU and the United States have the technology, infrastructure, and resources needed to continue efficiently providing a high proportion of food for the world, but these farming systems must be transformed to become more environmentally, socially, and economically sustainable. This approach is a more prudent option than outsourcing more production to countries in the developing world that do not use the most advanced technologies or have the infrastructure needed for the most efficient production, which would further exacerbate depletion of natural resources, and increase deforestation, biodiversity loss, and pollution. Therefore, our goal should be to transform intensive livestock, agriculture, and food production systems rather than replace them, even though it is not often viewed as the best approach for achieving sustainable food production.

2 We Need to Focus on the Guiding Principles of Sustainability

We can begin the food system transformation process by focusing on the essentials for life which include land, water, and air. These resources must be conserved by never extracting more than ecosystems can generate, and resources must be recycled and restored so that the amount of waste and pollution generated never exceeds the amount that ecosystems can safely absorb. Soil supports 95% of all food production, but agriculture is responsible for 80% of soil degradation, which leads to reduced crop yields; increased fertilizer use and crop production costs; loss of soil water-holding capacity and susceptibility to increased flooding; soil compaction, loss of structure, and nutrient degradation; erosion causing water pollution; and release of carbon dioxide and nitrous oxide into the atmosphere, contributing to climate change (Prăvălie et al. 2024; Thaler et al. 2022; Kwang et al. 2023). Water connects all living organisms on the planet, and more than 72% of global freshwater is used for irrigated agriculture and livestock production. However, more than 50% of the global population is currently affected by water scarcity, which is increasing due to climate change, ecosystem degradation caused by land use change (i.e., deforestation), and excess consumption of resources that is disrupting the global hydrological cycle (Mazzucato et al. 2024). As a result, increased frequency and severity of droughts and floods are occurring, and depletion of groundwater supplies continues to increase because of excess water consumption and use for irrigation. Emissions from livestock production play a vital role in the overall well-being and environmental stability of the planet and human health. About 31% of global GHG emissions are produced by agri-food systems. Air pollution is the fourth leading risk factor for human deaths globally and has detrimental effects on ecosystems (Afifa et al. 2024).

Excessive extraction and consumption of Earth's resources have resulted in exceeding six of nine planetary boundaries (Richardson et al. 2023; Rockström et al. 2020), which represent the Earth's carrying capacity to absorb the waste and pollution generated by human activity. The amount of land, water, energy, and material use is a good indicator of environmental damage and adverse effects on human health (Steinmann et al. 2017). Global material use continues to increase (Krausmann et al. 2009) and is greater than the level needed to achieve decent living standards (Vélez-Henao and Pauliuk 2023) and a safe operating space for human societies (Bringezu 2015). High-income countries are responsible for 74% of excess global material use (27% for United States and 25% for EU; Hickel et al. 2022), which has led to high international material inequality (Schaffartzik et al. 2019). This excessive extraction and use of natural resources is the main cause of environmental degradation that has led to climate change, loss of biodiversity, land system change, freshwater depletion, excess nitrogen (N) and phosphorus (P) losses, and pollution of novel entities (i.e., microplastics and per- and polyfluorinated alkyl substances (PFAS)—forever chemicals) and will require high-income countries to significantly transform economies using degrowth approaches (Hickel et al. 2022). Therefore, a qualitative or sustainable growth economy must be the objective for the reconstruction of agri-food systems to achieve efficiency, consistency, and sufficiency that is based on continuous innovation.

3 We Need to Understand the Interlinkages Among Biodiversity, Water, Food, Health, and Climate Change to Develop and Implement Meaningful Action Plans

One of the reasons that we have failed to overcome climate change and environmental degradation thus far, is partly because of our limited ability to identify and understand the interlinkages, interdependencies, cascading and compounding impacts, and feedback loops between sectors and systems in a holistic manner to develop integrated decisions that maximize synergies and minimize trade-offs for sustainable transformation of our global food, agricultural, and ecological systems. It is hoped that the deployment of artificial intelligence algorithms can provide this integrated complex analysis and recommendations in the near future, but until then, we should rely on a recent nexus assessment conducted by the Intergovernmental Science-Policy Platform on Biodiversity and Ecosystems Services (IPBES) to guide global environmental governance and action plans.

The IPBES is an independent intergovernmental organization that was established in 2012 by governments of 94 countries to engage the best expertise across scientific disciplines and knowledge communities to conduct assessments, provide policy support, and build capacity, knowledge, communications, and outreach regarding the benefits of biodiversity and nature to people. In its most recent thematic assessment (IPBES 2024), a team of global interdisciplinary scientists from 26 countries conducted a comprehensive nexus assessment of the interlinkages and challenges among five nexus elements that include **biodiversity**, **water**, **food**, **health**, and **climate change**. This analysis focuses on biodiversity and nature's contribution to people because biodiversity is an essential element that supports our water and food supplies, as well as our health and stability of the climate. Although ***energy***, ***land***, ***soil***, and ***air*** were not designated as independent nexus elements, they ***are embedded and accounted for across the five nexus elements***. This assessment is a remarkable achievement and an invaluable resource because it addresses the complex, interconnected causes of degradation of climate, water, food, health, and biodiversity across multiple spatial and temporal scales using a systems approach. It also qualifies the extent of scientific information and documented interactions to avoid making decisions in isolation that can lead to misalignment, unplanned trade-offs, and unintended consequences. This approach helps us understand the interconnections among these nexus elements so that we can identify opportunities for collaboration across sectors and scales and provide synergistic and holistic management and governance in the sustainable transformation of food and intensive livestock systems.

Our economic, societal, and policy decisions have continued to provide short-term benefits and high financial returns for a small number of privileged people while ignoring climate change and environmental impacts that have led to widening inequalities of human well-being (IPBES 2024). The "almighty dollar," constant pursuit of more money, and our current economic structure must be curtailed and balanced with an environmentally and socially sustainable transformation. Our dependence on the fossil fuel industry has enabled it to have enormous power and control in preventing the adoption of climate-smart and environmentally sustainable products and practices. Corporate greed and an unsatiable desire for more wealth has also led to huge barriers for sustainable transformation, which also contributes to excessive influence and control of dysfunctional governments that often follow misguided ideologies. We must also overcome the devastation caused by poorly managed exploitation of natural resources and vulnerable human populations, and find ways of reconnecting food producers with consumers regarding who, how, why, and where their food is produced. This IPBES nexus assessment is a great next step for attempting to overcome these major barriers to sustainably transform our global food, agriculture, and food animal production systems.

Indirect drivers of biodiversity and environmental degradation include economic, demographic, institutional, cultural, and technological change. Specifically, gross domestic product (GDP), material use intensity, trade, and poverty were used as indicators of economic drivers; population growth and urbanization were used as indicators of demographic changes; environmental regulations and armed conflicts were designated as institutional indicators; knowledge and literacy, along with per capita consumption, were used to measure cultural change; and solar and wind renewable energy, as well as use of information and communication technology, were used as indicators of changes in relevant technology (Table 31.1). All of these indirect drivers have increased or accelerated globally during the past 50 years (Table 31.1). As a result, they have increased or intensified trends in most of the direct drivers of biodiversity loss and environmental degradation, which include land and sea use change, unsustainable exploitation, climate change, pollution, and invasive alien species (Table 31.1). Although many of these effects and interactions are based on either evidence from a small number of studies or are uncertain or

Table 31.1 Trends in indirect drivers and impacts on trends in direct drivers of biodiversity and environmental degradation (Adapted from IPBES 2024)

Indicators of indirect driver		Indirect driver trend since 2001	Land and sea use change	Unsustainable exploitation	Climate change	Pollution	Invasive alien species
Economic	Gross domestic product	> +3%	Large increase	Large increase	Large increase	Large increase	Modest increase
	Material intensity	-0.3 to -3%	Uncertain	Small decrease	Small decrease	Small decrease	Uncertain
	Trade	> +3%	Large increase	Large increase	Modest increase	Modest increase	Large increase
	Poverty	> +3%	Minimal effect	Minimal effect	Minimal effect	Uncertain	Minimal effect
Demographic	Population	0.3 to 3%	Large increase	Large increase	Large increase	Large increase	Uncertain
	Urbanization	0.3 to 3%	Modest increase	Modest increase	Modest increase	Modest increase	Modest increase
Institutional	Regulations (environmental)	> +3%	Minimal effect	Minimal effect	Minimal effect	Minimal effect	Minimal effect
	Armed conflicts	> +3%	Variable	Variable	Modest increase	Modest increase	Uncertain
Cultural	Knowledge and literacy	0.3 to 3%	Uncertain	Uncertain	Uncertain	Uncertain	Uncertain
	Per capita consumption	0.3 to 3%	Large increase	Large increase	Large increase	Large increase	Modest increase
Technology	Renewable energy (solar and wind)	> +3%	Modest increase	Large increase	Small decrease	Variable	Uncertain
	Information and communication	> +3%	Uncertain	Large increase	Variable	Modest increase	Modest increase

Well-established

Established but incomplete

Evidence from small number of studies

Inconclusive - similar number of positive and negative responses inconclusive because a similar number of both positive and negative responses have been observed, increases in GDP economic growth, global trade, population, urbanization, and per capita consumption have had the greatest negative effects on the environment by increasing land and sea use change, unsustainable exploitation of natural resources, climate change, and pollution.

Indicators of indirect drivers of economic, demographic, institutional, cultural, and technological change were also assessed for their impacts on the nexus elements of biodiversity, water (availability and quality), food (quantity and quality), health (physical and mental), and climate change (Table 31.2). Again, increased GDP growth, global trade, population, urbanization, and per capita consumption in high-income countries had the greatest negative effects on biodiversity, water, food, health, and climate change (Table 31.2). Governments have failed to develop and implement policies to overcome these challenges even though research in technology innovation, education, poverty reduction, and some environmental regulations have led to some environmental improvements (Table 31.2). As a result, gaps in governance, along with conflicting objectives and incentives, have led to unintended consequences and inefficient use of resources (IPBES 2024).

Therefore, to provide guidance and a pathway forward for meaningful environmental, social, and economic sustainable transformation of our food, agricultural, and ecological systems, IPBES (2024) summarized 71 response options in 10 action categories for policymakers to use in developing regulations and new operating guidelines for a more sustainable world (Table 31.3). The ten action categories include: (1) conserve and stop conversion of ecosystems of high ecological integrity, (2) restore natural and seminatural ecosystems, (3) manage ecosystems in human-exploited lands and waters, (4) consume sustainably, (5) reduce pollution and waste, (6) integrate planning and governance, (7) manage risk, (8) ensure rights and equity, (9) align financing, and (10) actions for developing and implementing multilateral multilateral environmental agreements, reconnecting people with nature, and integrated health education (IPBES 2024). These response options are based on scientific evidence from multiple knowledge systems in different sectors and spatial and temporal scales that can be adapted to various local and national conditions (Table 31.3).

4 Useful Models and Behavioral Changes Can Lead to Successful Food and Intensive Livestock System Transformation

The previous section on understanding the interlinkages among biodiversity, water, food, health, and climate change is necessary for developing and implementing meaningful action plans to begin synthesizing this complex, interconnected, and transdisciplinary information into redesigned food system pathways that are balanced, equitable, and just while also addressing the many trade-offs and needs for customizing food animal production systems that optimize resource use under varying local conditions. We also need to redesign and build new food and livestock production systems that not only prevent further environmental and social damage

Table 31.2 Trends in indirect drivers and impacts on trends in nexus elements of climate change, biodiversity, water, food, and health (Adapted from IPBES 2024)

Indicators of indirect driver		Indirect driver trend since 2001	Nexus Elements							
			Climate change	Biodiversity	Water		Food		Health	
					Availability	Quality	Quantity	Quality	Physical	Mental
Economic	Gross domestic product	> +3%	Large increase	– –	–	– –	+	–	–	–
	Material intensity	-0.3 to -3%	Small decrease	Minimal Δ	Minimal Δ	Minimal Δ	+	Uncertain	Minimal Δ	Uncertain
	Trade	> +3%	Modest increase	– –	–	– –	+	–	–	–
	Poverty	< -3%	Minimal effect	Minimal Δ	Variable	Minimal Δ	Variable	Variable	Variable	Variable
Demographic	Population	0.3 to 3%	Large increase	– –	–	– –	+	–	–	–
	Urbanization	0.3 to 3%	Modest increase	–	–	–	+	Variable	–	–
Institutional	Regulations (environmental)	> +3%	Minimal effect	+	+	Minimal Δ	Minimal Δ	Minimal Δ	Minimal Δ	Minimal Δ
	Armed conflicts	> +3%	Modest increase	Variable	Minimal Δ	Minimal Δ	Minimal Δ	– –	– –	– –
Cultural	Knowledge and literacy	0.3 to 3%	Uncertain	Uncertain	Uncertain	Uncertain	Uncertain	Uncertain	Uncertain	Uncertain
	Per capita consumption	0.3 to 3%	Large increase	– –	–	– –	+	–	–	Minimal Δ
Technology	Renewable energy (solar and wind)	> +3%	Small decrease	–	Minimal Δ	Minimal Δ	Variable	Minimal Δ	Variable	Minimal Δ
	Information and communication	> +3%	Variable effect	–	Minimal Δ	Minimal Δ	Minimal Δ	Uncertain	–	Minimal Δ

Well-established

Established but incomplete

Evidence from small number of studies

Inconclusive - similar number of positive and negative responses

Table 31.3 Response options, actions, or policies that can advance governance and sustainable management of the complex and interconnected nexus elements to overcome food insecurity, inadequate water availability and quality, climate change, health risks, and biodiversity loss (adapted from IPBES 2024)

Action category	Response option	Nexus element affected
Conserve ecosystems	Area-based conservation	Biodiversity
	Halt conversion of intact ecosystems	Food
	Forest conservation for health	Health
Restore ecosystems	Restoration of blue carbon ecosystems	Climate change
	Wetland conservation and restoration	Climate change
	Mangrove conservation and restoration for health	Health
	Restoration of soil health	Food
	Rewilding	Biodiversity
	Restoration of inland water systems	Biodiversity
	Restoration of coastal and marine systems	Biodiversity
	Forest landscape restoration	Biodivesity
Manage ecosystem functions	Forest-based practices to address climate change	Climate change
	Integrated multi-trophic aquaculture	Climate change
	Increase soil organic carbon	Climate change
	Ecological intensification—aquatic foods	Food
	Ecological intensification—rangelands	Food
	Ecological intensification—croplands	Food
	Manage alien species	Water
	Sustainable inland fisheries	Water
	Agroecology	Biodiversity and climate change
Consume sustainably	Efficient water use in agriculture	Water
	Reduce food loss and waste	Food
	Sustainable healthy diets	Food and climate change
	Sustainable use of medicinal plants	Health
	Reduce meat overconsumption	Health
	Offshore wind power	Climate change
	Solar photovoltaics on land	Climate change
	Circular bioeconomy	Climate change
Reduce pollution	Manage wastewater	Water
	Reduce nutrient pollution	Food
	Reduce pesticide pollution	Food
	Reduce plastic pollution	Food
	Pollution prevention	Health
	Sustainable intensification	Climate change and food
	Reduce short-lived climate pollutants	Climate change

(continued)

Table 31.3 (continued)

Action category	Response option	Nexus element affected
Integrate planning and governance	Integrated landscape and seascape approaches	Biodiversity
	Land and sea planning	Biodiversity
	Integrated water infrastructure	Water
	Transboundary water cooperation	Water
	Groundwater governance	Water
	Water-sensitive urban infrastructure	Water
	Community water management	Water
	City region food systems	Food
	Integrated watershed-health interventions	Health
Manage risk	Urban nature-based solutions	Climate change and biodiversity
	Ecosystem-based adaptation in rural landscapes	Biodiversity
	Dam operation	Water
	Net-zero sustainable healthcare	Health
	Urban green infrastructure	Health
	Biodiversity management for zoonoses	Health
	Health impact assessments	Health
	The One Health approach	Health
	Multi-hazard early warning systems	Climate change
Ensure rights and equity	Intercultural health services	Health
	Universal health coverage	Health
	Access to natural resources and land	Food
	Indigenous food systems	Food
	Foster gender transformative approaches	Food
	Address gender burdens of water collection	Water
	Rights of nature	Water
	Inclusive water management	Water
	Inclusive water education	Water
	Rights-based approaches	Biodiversity
Align financing	Global cooperation for finance and technology	Climate change
	Repurpose public spending	Food
	Finance for water infrastructure	Water
	Natural capital accounting	Biodiversity
Others	Integrated health education	Health
	Nature on prescription	Health
	Reconnecting people with nature	Biodiversity
	Multilateral environmental agreements	Biodiversity

but also begin conserving, repairing, and restoring environmental resources to overcome the many consequences of the damage that has occurred so that we can create more resilient and equitable food security for all. Many sustainability, circular, regenerative agriculture, and One Health models and recommendations have been developed by leading experts from multiple disciplines around the world (Hickel 2020; Halpern et al. 2024; ISO 2024; Zhang et al. 2024; Ferreira 2024; McLennon et al. 2021). Leading experts have also authored the many chapters in this book that summarize most of the key issues that need to be addressed in transforming our current intensive food animal production systems to become sustainable and resilient. These chapters provide a wealth of science-based information that provide meaningful insights into the many benefits and limitations, along with challenges and opportunities for sustainable transformation of intensive livestock and poultry production systems.

Successful food system transformation also involves fundamental changes in values, culture, regulations, policies, markets, and governance along with changes in technologies, infrastructure, and knowledge, which Thornton et al. (2024) summarized as the eight key drivers of food system transformation:

1. Build trust among all entities in the food system.
2. Transform mindsets to adopt new approaches for producing and distributing food.
3. Engage stakeholders across society to ensure transparency and responsible innovation.
4. Ensure stable public and private financing to promote innovation without contributing to existing inequalities.
5. Redesign markets that spread risks and costs associated with innovations in supply chains.
6. Revise policies and regulations to support new technologies and industries and reduce economic and bureaucratic constraints.
7. Develop, monitor, and enforce standards and regulations of innovations using supervisory entities for adequate governance.
8. Develop transition pathways that demonstrate how desired outcomes can be achieved.

While these are reasonable ideas and recommendations, they fail to address several underlying causes that are limiting the food system transformation. We have several pervasive self-destructive practices that must be overcome.

1. We are challenged with a vast amount of inaccurate and misleading information through a wide variety of media platforms that influence individual thoughts, beliefs, and decisions about governance, the economy, and various social issues. Unfortunately, we do not have a system for helping people differentiate between facts and fiction, which often leads to poor decision-making. Furthermore, we are constantly reminded of the environmental degradation of the world but are provided with little information on how to restore and overcome this destruction.
2. Our current educational system has not prepared society to think critically, nor has it enabled people to develop the knowledge and skills required for develop-

ing and successfully implementing sustainable business models in food supply chains. Much of this is because most researchers and educators have been trained to be reductionist scientists in single disciplines, rather than being trained to become systems thinking, transdisciplinary scientists. As a result, all of us struggle, especially those in governance and business decision-making roles, to redesign food supply chains to manage trade-offs and achieve a balance of economic, social, and environmental sustainability.

3. We have allowed the economy and markets to define what we should value, which has led to fabricated demand and compulsive overconsumption behavior requiring the need to make more money to buy more things with the hope of achieving greater satisfaction and quality of life. But we are never satisfied. This consumption-driven mindset, especially in high-income countries, has led to recklessly extravagant and wasteful lifestyles that make the sellers rich, degrade the Earth's resources, and falsely lead us to believe that it provides a better quality of life. Governments have performed as though human consumption has no consequences even though perpetual economic growth is not compatible with sustainability of the Earth's resources.
4. Despite our dependence on food to survive and thrive as a global human society, the evolution of agriculture toward more intensive and efficient production systems has created a disconnect for consumers in understanding how their food is produced, and has also resulted in a lost connection for farmers with nature and the environment. The participatory role of people in protecting the well-being of land, water, air, and ecosystems has been lost and must be recovered. Restoration, recycling, and regeneration can be effective ways for humans to reconnect with nature and food production in positive ways.
5. Political views have become barriers for enacting regulations and policies to alter economies to curb waste and reduce excess material use, minimize environmentally degrading practices, and eliminate societal inequalities. Unfortunately, politicians are heavily influenced by large corporations whose primary focus is to generate economic growth and profits to appease shareholders. Corporate greed has led to a system of inequitable profit distribution in agriculture and food supply chains at the expense of low profit margins for farmers and high food prices for consumers. Selfishness has overpowered self-control to the extent that satisfaction has become no longer possible. Agriculture and food production should not be an electoral campaign issue. Historically, in most EU countries and in the United States, promising programs have been funded and initiated but subsequently stalled during election cycles and discarded by succeeding governments simply because they originated from their predecessors. Therefore, it is necessary to develop an overall societal vision and mission statement that can be used as a binding agreement to transform the current agri-food system. This will also require mandatory collaboration across government departments and agencies to avoid conflicts in meeting goals because the transformation of the agri-food system is much broader than that of agricultural ministries.

6. Climate change, political unrest, war, poverty, and other factors have caused mass migrations of people from low-income countries to higher-income countries in search of survival and a better life. Citizens in destination countries for migrants worry about competition for jobs, housing, and other essentials of life. However, these underappreciated migrant workers are necessary for agriculture and food production because native workers do not want these low-paying, dangerous, and unpleasant jobs on farms and in slaughter plants. Unfortunately, most countries provide inadequate support for these essential workers in our food system. This leads us to the question of "Who is going to produce our food in the future?"

5 We Need to Build a System That Protects and Adequately Supports Those Who Produce Our Food

Like the Earth's natural resources, farmers are also essential resources for producing food. Unfortunately, the sustainability of our agricultural workforce is rarely considered to be a vital part of our global food and livestock production system transformation. Although there are many types of farmers in our ever-evolving agricultural and food production system (Bock et al. 2020; see Table 31.3 in Chap. 2), independent intensive farmers and corporate farmers who manage production for large food corporations generally use the best available technologies for efficient production of commodity food products needed to feed the majority of people in our global society, compared with smaller-scale lifestyle, urban, community provisioning, hobby, and social care farmers who fulfill important roles and niches in the food system but do not produce food with the efficiency and scale needed to feed the world. Lack of enough workers to meet demand in the meat and dairy industries in the United States is a significant sustainability challenge, with about a third approaching retirement age, 20 to 50% of the workforce comprised of migrant workers, and emerging immigration policies that will severely deplete this vital part of the workforce on livestock farms, meat-packing facilities, and in the trucking industry (American Immigration Council 2022; Adcock et al. 2015). This situation is similar in the EU, where the decline in the number of farmers and workers on farms and other stages in the supply chain (e.g., seasonal workers for crop and produce harvesting, meat processing plants) has been partially offset by migrants (Nori 2017; Palumbo et al. 2022).

In the United States, ensuring an adequate number of well-trained farmers is in jeopardy because a limited number of young people are becoming farmers, and because of restrictive government immigration policies and inadequate support for migrant workers. Similarly in Europe, the number of farmers is also declining, primarily due to increasing bureaucracy, changing farm working conditions, and a lack of intergenerational transfer. The younger generation of prospective farmers often sees limited prospects in agriculture and is less willing to take on the responsibility of managing the family farm. Some of the social sustainability factors that contribute to difficulties of recruiting and retaining farmers involve often having a suboptimal quality of life, barely surviving financially, and expectations to constantly adapt to increasing environmental and societal demands without getting paid to implement these changes. Historically, farmers were paid for the foods they produce without being scrutinized for how it was produced. Now that society is demanding changes in how their food is produced (i.e., animal welfare, environmental impacts), farmers have not been compensated for the cost of making these changes, which often increase costs without improving productivity and efficiency incurred by farmers. Consumers are unwilling to pay more than an extra 10 to 15% for the extra costs to implement these practices. Farmers are the ultimate systems thinkers who must manage soil fertility, crop production, animal health and welfare, manure, water, land, ecosystems, biosecurity, resource conservation, and relationships with neighbors and the community while also managing finances, capital investments, contracts, and marketing decisions. They need to be supported by providing clear, practical, science-based transition pathways that allow adequate financial and political support to make the necessary sustainability transitions. Farmers will implement these practices if they get paid for them. We (society) need to pay the difference in costs of applying new sustainable technologies on farms and take this burden off farmers because we all benefit from the production of abundant, safe, nutritious, and affordable food, and because we all have a responsibility to conserve, protect, and regenerate Earth's resources as part of the global community living on this planet.

Furthermore, farmers worldwide are experiencing increasing amounts of stress, which is leading to numerous threatening mental and physical health conditions (King et al. 2023; Jones et al. 2024). In a recent US survey (Sussell et al. 2023), the risk of occupational suicide was more than 50% greater for male farmers, ranchers, and other agricultural managers than the overall suicide rates of men from all other occupations. Male farmers have increased risk of cardiovascular disease associated with chronic stress and hardships from work compared with non-farmer males in the same demographic (Desai et al. 2022). Unfortunately, the social science component of sustainability has generally been ignored, but equity, justice, and quality of life for people are equally important as economic and environmental sustainability. External (Guptill 2021) and internal (Brennan et al. 2023) social sustainability requirements of farmers were summarized in Chap. 2, Tables 2.4 and 2.5), which include relationships with communities, satisfaction with work, safety and security, reasonable income, and quality of life. Public policies are desperately needed to improve working conditions and living standards among farmers to enable greater job satisfaction and mental and physical health.

Immigrants are also essential workers on livestock farms and in trucking, meat-packing, and dairy processing operations, but these workers are often exploited and poorly treated in many countries around the world. Food system consolidation has resulted in lower wages for meat-packing workers, anti-union actions, and increased injury rates due to fewer regulatory inspections. Many legal, social, economic, and cultural factors also lead worker exploitation and low wages, and create vulnerable conditions for migrant workers in the agri-food sector in Europe (Palumbo et al. 2022). As a result, these workers are in an impossible situation of demanding their rights while risking loss of access to legal documentation and earnings they need to survive. Several actions are needed to address this critical issue including: (1) developing policies to balance immigration policy with labor policy to improve wages and labor conditions of immigrant workers in agriculture, (2) increasing government funding for enforcement of existing labor regulations and rights, (3) strengthening farm worker protections and standards, (4) providing worker autonomy to unionize and exercise their labor rights, and (5) providing a pathway for legalizing status and citizenship (Smolski 2019).

Even more concerning is the urgent need to eliminate forced labor (modern slavery) in agriculture and food production systems in many countries, which continues to be a significant problem involving more than 50 million people around the world who work in labor-intensive sectors such as agriculture in low-, middle-, and high-income countries (Kunz et al. 2023). People subjected to forced labor are coerced to work using violence or intimidation, accumulated debt, retention of identity papers, or threats of being reported to immigration authorities (International Labour Organization 2014). Unfortunately, illegal labor trafficking operations continue to thrive in the United States despite the passage of laws intended to prevent them (Childress et al. 2023). Domestic meat, poultry, and egg production and processing sectors in the United States had the highest risk (62%) of forced labor among all food commodities (Blackstone et al. 2023). The use of forced labor in global food production and processing industries is a hidden issue that rarely is exposed when consumers make food purchasing decisions about how their food was produced. Although law enforcement continues to be the primary mechanism to identify offenders and businesses involved in human trafficking, it is ineffective, and other criminology concepts and methods of disrupting labor trafficking in agriculture must also be implemented (Childress et al. 2023).

There will continue to be many types of farmers owning and managing livestock and poultry farms in the future, but those involved in large-scale intensive animal production will meet the majority of the meat, milk, and egg supplies needed to feed growing urban populations. The diverse beliefs and desires among consumers and the variable price points associated with how food is produced will continue to support niche products and markets (e.g., products from welfare-oriented husbandry systems, organic, free-range, cage-free, no antibiotics ever, climate-neutral), but the lack of economies of scale and ability to produce mass quantities of high-quality, safe, and nutritious food will depend on intensive livestock production systems, even though they have sustainability flaws that must be overcome. However, consolidation in the animal agriculture sector that supports intensive livestock production systems without focusing on sustainability, quality, and safety is a barrier for sustainable transformation that needs to be addressed.

6 We Must Eliminate Barriers for Change

The challenges of achieving systemic transformation of our food system are dependent on technological innovations and especially on good governance, political-economic conditions, and policy trade-offs. Governments were created for the purpose of developing structural guidance for societies to thrive, and achieving food system sustainability is an essential part of this responsibility that affects all of us. Unfortunately, government regulations have not been effective in meeting the urgent needs for redesigning our food system to become more sustainable and resilient. De Miranda Robeiro and Kruglianskas (De Miranda Ribeiro and Kruglianskas 2015) conducted an extensive literature review and concluded that for environmental regulations to be effective, they must be participatory, decentralized, flexible, simple and clear, preventive, inducers of innovation, multi-instrumental, rigorous on enforcement, performance-based, planned and gradual, supported by adequate resources, measured and communicated, and reflexive. Additionally, a cross-departmental coordination of different strategies and measures is crucial (e.g., land utilization).

Béné (2022) concluded that it will be our political will that determines if we can overcome self-reinforcing dynamics that have locked food systems into our current unsustainable status and are preventing their transformation toward sustainability. He identified four major actions that are needed to eliminate these barriers (Béné 2022).

6.1 Destabilize the Status Quo and Disempower Transnational Agri-food Corporations

Unprecedented concentration of resources among a few large transnational retailers and agri-food companies has led to enormous control, influence, wealth, and governing power of our food system. Because these corporations are driven by demand for continuous profits and high returns on investment, which are determined by global financial market priorities, they are currently not compatible with the objectives of a sustainable food system. As a result, these agri-food corporations are incapable of facilitating transformation into more sustainable outcomes because they are not dedicated to achieving sustainability goals. So far, these companies have invested in defensive research and development to protect their assets and business activities by preventing disruptive innovations from reaching the market. Breaking up the oligopoly of the dominant agri-food corporations in our food system can enable true competitiveness again and allow new players to destabilize the status quo and participate in the sustainable transformations needed. If these dominant agri-food corporations acknowledge and embrace the necessity for change, they can exert the same amount of power and influence for implementing all essential sustainability measures using a top-down approach. However, this will require a new code of conduct within these large-scale agri-food companies compared with current behaviors.

6.2 Align New Technology Innovations with Sustainability

Innovation is driven by profit, not by sustainability. We currently lack the guidance of purposeful, goal-oriented processes that ensure that new innovations in the agri-food system are not only economically viable but are also aligned with sustainability goals. This requires establishing a new robust governance system that prevents false promises of corporate responsibility, "self-regulation," and "license-to-operate," which have prevented sustainable transformation. Transforming our food system requires transforming the governance of food systems to proactively develop public policies that drive the sustainable innovation process and enable smaller innovators and entrepreneurs to provide sustainable solutions.

6.3 International Governance and Accountability

Most governments claim to support an urgent transformation to more sustainable and healthier food systems but continue to provide subsidies and financial support to national and transnational corporations in the meat industry (e.g., Mercosur in South America) that are not aligned with achieving sustainability. Although some monitoring initiatives exist to track countries' political commitments and progress (or lack thereof) toward meeting sustainability goals, they are not capable of monitoring the broad scope of transformations needed in multiple sectors of the food system. Therefore, a global intergovernmental organization that has the power to enforce compliance and resolve international disputes related to compliance with policies of global food system sustainability goals is needed to achieve accountability.

6.4 Create and Adopt a New Scientific Approach

We need different types of scientific data and evidence, not more of it, that connects science, policy, and society to increase our understanding of how they interact with each other to change environmental, economic, social, health, and cultural outcomes of food systems that are aligned with long-term sustainability. Scientists must move beyond being "knowledge brokers," focused only on generating more data and evidence, and become more engaged in the contested political and policy-making processes by acknowledging the trade-offs and avoiding scientific neutrality. Basic research must remain a cornerstone, but we also need applied research and a research system that recognizes and rewards transdisciplinary approaches.

We need a unified, transparent evaluation system for the external effects of agri-food systems that can serve as a foundation to jointly develop action strategies that are tailored to the specific characteristics and conditions of each region. This will require scientists to participate in identifying unsustainable actions by certain actors in food supply chains, documenting government and private sectors' deceitful communications and incoherent policies that continue to support the status quo, and explaining the trade-offs and winners and losers resulting from the transformation process. This will require a major paradigm shift in behavior and actions compared with the current generally nonparticipatory practices of academics in the sustainable transformation process. However, university experts must avoid obstructing unfavorable environmental policies and minimizing the role of food animal production on the environment by aligning with livestock commodity groups to distort the facts (Morris and Jacquet 2024).

These actions, along with providing incentives for sustainable intensification of agriculture for human prosperity and global sustainability, that have been proposed by Rockström et al. (2017), are needed. This proposal involves an operational framework for agricultural development at all scales to stop and potentially reverse environmental damage while focusing on eradicating hunger and poverty to improve well-being for all. This involves a major paradigm shift away from agriculture being the major driver of global environmental change toward becoming a key contributor in the transition to a sustainable world where human society can thrive within a safe environmental operating space.

7 Sustainable Transformation of Intensive Food and Livestock Production Has Multiple Trade-Offs That Will Result in Winners and Losers

There are trade-offs with winners and losers, but we must stay true to the guiding principles of sustainability that if we take care of nature, it will take care of us. This requires **a new paradigm and change in behavior,** especially in high-income countries that consume and waste in excess, so that the remaining finite resources can be shared and enable the efficient production of enough safe, affordable, and nutritious food needed to feed the projected 11 billion people who will live on this planet in the future. Policymakers and citizens in high-income countries have a moral and ethical responsibility to create a more sustainable livestock system because (1) they have the financial resources to do it, (2) they have the most technologically advanced livestock production systems to more efficiently use scarce resources, and (3) they have reciprocal dependence on exports. Subsidy models for sustainable agricultural and food systems need to be established in the EU. These models should incentivize sustainable practices (e.g., promoting biodiversity, regenerating soils, and improving animal welfare) rather than maintaining conventional systems that would not be viable without subsidies.

Therefore, public funds need to be diverted away from subsidizing industries (i.e., fossil fuels) that have detrimental effects on the environment and toward farmers, farming, and food system transition pathways. In low- and moderate-income countries, train-the-trainer programs should be promoted to empower local communities to effectively address climate change and its consequences. By fostering local expertise and capacity, these initiatives can enhance sustainable and efficient production practices and ultimately contribute to long-term resilience and reduce the drivers of involuntary migration. This approach emphasizes local empowerment as a cornerstone for sustainable development, enabling communities to build the skills and knowledge needed to implement practical solutions tailored to their specific environmental and socioeconomic challenges.

8 The Nexus for a Sustainable Future

Successfully transforming our global food system to become sustainable will require a multidimensional, transdisciplinary, big-picture view of a multitude of key interrelated components. We need to holistically embrace all Sustainable Development Goals (SDGs) to guide us to a different economy that is more equal, inclusive, just, sustainable, and resilient where businesses, ecosystems, and society can simultaneously succeed. Climate change should be a priority because it affects all SDGs and actions against climate change create jobs, reduce poverty, improve human health, reduce carbon emissions, and support sustainable production, energy usage, and allocation of resources. Achieving sustainable transformation of our food system is a global challenge that requires international cooperation. Alliances, partnerships, and multi-stakeholder collaborations among sectors need to be developed with a goal of collectively acting and causing necessary changes for the shared prosperity of all for future generations. Measurable sustainability objectives and strategies need to be defined, which can align businesses, organizations, government, and society with our natural environment. Transparent and effective communication of information with markets is needed at all stages of supply chains. We must ensure a sustainable labor market and quality of life for farmers, develop an adequate and socially sustainable support system for migrant workers on farms and meat-processing facilities, and eliminate forced labor in food animal production systems where it exists.

We also need to eliminate government subsidies that support environmentally and ecologically damaging industries and practices, especially those that promote the continued use of fossil fuels, and redirect this government investment toward developing a comprehensive renewable energy infrastructure. Government subsidies should be diverted from supporting water irrigation practices and toward improving water quality and sanitation, along with improving land management, biodiversity, and regeneration of natural ecosystems. A new approach to education is needed that involves learning about sustainable development in higher education institutions, and life-long learning needs to be promoted that involves interdisciplinary, problem-solving-based education. Scientists must play a key role in informing policymakers about the trade-offs and benefits and limitations of potential pathways of achieving sustainable food systems, using their expertise to provide evidence-based solutions for the transition to more sustainable agricultural practices. This aligns with the principles of Education for Sustainable Development, which emphasizes the need for interdisciplinary learning, collaboration, and the integration of scientific knowledge into policy and practice. By bridging the gap between research and implementation, scientists can help shape policies that foster sustainable food systems and ensure long-term resilience. Academic researchers and educators also need to teach farmers about the trade-offs, transition pathways, and how to implement sustainable practices. New innovations for sustainable development must address social, environmental, and economic concerns. Precision agriculture and livestock farming, artificial intelligence, and synthetic biology can enhance pathways to greater sustainability and resilience of our global food system, but they need to be aligned with sustainable development goals and adequately governed to prevent harm that can result from misuse. We need to take away the dominance of transnational agri-food corporations to reduce their control, influence, wealth, and governing power of our food system. Only then can we make meaningful progress in the sustainable and resilient transformation of intensive livestock production systems to feed future generations.

References

Adcock F, Anderson D, Rosson P (2015) The economic impacts of immigrant labor on U.S. dairy farms. National Milk Federation, Texas A & M AgriLife Research, College Station, TX, pp 1–21. https://www.nmpf.org/wp-content/uploads/immigration-survey-090915.pdf

Afifa K, Arshad, Hussain N, Ashraf MH, Saleem MZ (2024) Air pollution and climate change as grand challenges to sustainability. Sci Total Environ 928:172370. https://doi.org/10.1016/j.scitotenv.2024.172370

American Immigration Council (2022) The essential role of immigrants in America's meat and dairy industries. Labor Market Series Factsheet, Tending to America's Food Supply, p 11. https://www.americanimmigrationcouncil.org/sites/default/files/research/tending_to_americas_food_supply_meat_dairy_industries.pdf

Béné C (2022) Why the great food transformation may not happen—a deep-dive into our food systems' political economy, controversies and politics of evidence. World Dev 154:105881. https://doi.org/10.1016/j.worlddev.2022.105881

Blackstone NT, Rodriguez-Huerta E, Battaglia K, Jackson B, Jackson E, Norris CB, Sparks JLD (2023) Forced labour risk is pervasive in the US land-based food supply. Nat Food 4:596–606. https://doi.org/10.1038/s43016-023-00794-x

Bock AK, Krzysztofowicz M, Rudkin J, Winthagen V (2020) Farmers of the future. In: EUR 30464 EN. Publications Office of the, European Union, Luxembourg. https://doi.org/10.2760/680650

Brennan M, Hennessy T, Dillon E, Meredith D (2023) Putting social into agricultural sustainability: integrating assessments of quality of life and wellbeing into farm sustainability indicators. Sociol Rural 63:629–660. https://doi.org/10.1111/soru.12417

Bringezu S (2015) Possible target corridor for sustainable use of global material resources. Resources 4:25–54. https://doi.org/10.3390/resources4010025

Childress C, Farrell A, Bhimani S, Maass KL (2023) Disrupting labor trafficking in the agricultural sector: looking at opportunities beyond law enforcement interventions. Vict Offenders 18:473–511. https://doi.org/10.1080/15564886.2022.2133036

De Miranda Ribeiro F, Kruglianskas I (2015) Principles of environmental regulatory quality: a synthesis from literature review. J Clean Prod 96:58–76. https://doi.org/10.1016/j.jclepro.2014.03.047

Desai B, Sahni S, Jordan H, Sahni R, Reinbeau R, Nguyen A, Babalola O (2022) Risk of cardiovascular disease in male farmers over the age of 45: a review of literature. Cureus 14:e24642. https://doi.org/10.7759/cureus.24642

FAO (2013) Food wastage footprint—impacts on natural resources summary report.. https://www.fao.org/4/i3347e/i3347e.pdf

FAO (2019) The state of food and agriculture 2019. Moving forward on food loss and waste reduction. Food and Agriculture Organization of the United Nations, Rome. https://www.fao.org/3/ca6030en/ca6030en.pdf

FAO (2023) The state of food and agriculture 2023—revealing the true cost of food to transform agrifood systems. Food and Agriculture Organization of the United Nations, Rome. https://doi.org/10.4060/cc7724en

FAO, IFAD, WFP, UNICEF, WHO (2020) The state of food security and nutrition in the world 2020. Transforming food systems for affordable healthy diets. FAO, Rome. http://www.fao.org/publications/sofi/2020/en/

Ferreira MN (2024) Conservation priorities mapping—a first step toward building area-based strategies. Frontiers in Science 2:1440501. https://doi.org/10.3389/fsci.2024.1440501

Guptill A (2021) Understanding and measuring social sustainability. SARE quality-of-life working group, sustainable agriculture research and education. University of Maryland, College Park. https://www.sare.org/resources/understanding-and-measuring-social-sustainability/

Halpern C, Freeman KK, Barrett CB, van Dijk M, Mason-D'Croz D, Simons A, van Veen B, Herrero M, van Zanten HHE (2024) Perspective paper: framework to assess the potential of circular food system technologies. Glob Food Secur 43:100814. https://doi.org/10.1016/j.gfs.2024.100814

Hickel J (2020) The sustainable development index: measuring the ecological efficiency of human development in the anthropocene. Ecol Econ 167:106331. https://doi.org/10.1016/j.ecolecon.2019.05.011

Hickel J, O'Neill DW, Fanning AL, Zoomkawala H (2022) National responsibility for ecological breakdown: a fair-shares assessment of resource use, 1970-2017. Lancet Planet Health 6:e342–e349

International Labour Organization (2014) The meanings of forced labour. https://www.ilo.org/global/topics/forced-labour/news/WCMS_237569/lang-en/index.htm

IPBES (2024) The thematic assessment report on interlinkages among biodiversity, water, food and health—summary for policymakers. Intergovernmental Science-Policy Platform on Biodiversity and Ecosystem Services, https://biodiversiteit.org/wp-content/uploads/2024/12/Nexus-Assessment-Report-16-Dec-2024.pdf

ISO (2024) International standard ISO 59004—circular economy—vocabulary, principles and guidance for implementation. ISO, Geneva

Jones AQ, Sawatzky A, Thompson R, Hagen BNM (2024) Poor mental health negatively impacts farmers personally, interpersonally, cognitively, and professionally. FACETS 9:1–16. https://doi.org/10.1139/facets-2023-0230

King E, Lamont K, Wendelboe-Nelson C, Williams C, Stark C, van Woerden HC, Maxwell M (2023) Engaging the agricultural community in the development of mental health interventions: a qualitative research study. BMC Psychiatry 23:399. https://doi.org/10.1186/s12888-023-04806-9

Krausmann F, Gingrich S, Eisenmenger N, Erb K-H, Haberl H, Fischer-Kowalski M (2009) Growth in global materials use, GDP and population during the 20th century. Ecol Econ 68:2696–2705. https://doi.org/10.1016/j.ecolecon.2009.05.007

Kunz N, Chesney T, Trautrims A, Gold S (2023) Adoption and transferability of joint interventions to fight modern slavery in food supply chains. Int J Prod Econ 258:108809. https://doi.org/10.1016/j.ijpe.2023.108809

Kwang JS, Thaler EA, Larsen IJ (2023) The future of soils in the Midwestern United States. Earths Future 11:e2022EF003104. https://doi.org/10.1029/2022EF003104

Mazzucato M, Okonjo-Iweala N, Rockström J, Shanmugaratnam T (2024) The economics of water: valuing the hydrological cycle as a global common good. Global Commission on the Economics of Water, Paris

McLennon E, Dari B, Jha G, Sihi D, Kankarla V (2021) Regenerative agriculture and integrative permaculture for sustainable and technology driven global food production and security. Agron J 113:4541–4559. https://doi.org/10.1002/agj2.20814

Morris V, Jacquet J (2024) The animal agriculture industry, US universities, and the obstruction of climate change understanding and policy. Clim Change 177:41. https://doi.org/10.1007/s10584-024-03690-w

Nori M (2017) The shades of green: migrants' contribution to EU agriculture. In: Context, trends, opportunities, challenges. European University Institute in Florence: Migration Policy Centre

Palumbo L, Corrado A, Triandafyllidou A (2022) Migrant labour in the Agri-food system in Europe: unpacking the social and legal factors of exploitation. Eur J Migr Law 24:179–192. https://doi.org/10.1163/15718166-12340125

Prăvălie R, Borrelli P, Panagos P, Ballabio C, Lugato E, Chappell A, Miguez-Macho G, Maggi F, Peng J, Niculiţă M, Roşca B, Patriche C, Dumitraşcu M, Bandoc G, Nita I-A, Birsan M-V (2024) A unifying modelling of multiple land degradation pathways in Europe. Nat Commun 15:3826. https://doi.org/10.1038/s41467-024-48252-x

Richardson K, Steffen W, Lucht W, Bendtsen J, Cornell SE, Donges JF, Drüke M, Fetzer I, Bala G, von Bloh W, Feulner G, Fiedler S, Gerten D, Gleeson T, Hofmann M, Huiskamp W, Kummu M, Mohan C, Nogués-Bravo D, Petri S, Porkka M, Rahmstorf S, Schaphoff S, Thonicke K, Tobian A, Virkki V, Wang-Erlandsson L, Weber L, Rockström J (2023) Earth beyond six of nine planetary boundaries. Sci Adv 9:eadh2458. https://doi.org/10.1126/sciadv.adh2458

Rockström J, Williams J, Daily G, Noble A, Matthews N, Gordon L, Wetterstrand H, DeClerck F, Shah M, Steduto P, de Fraiture C, Hatibu N, Unve O, Bird J, Sibanda L, Smith J (2017) Sustainable intensification of agriculture for human prosperity and global sustainability. Ambio 46:4–17. https://doi.org/10.1007/s13280-016-0793-6

Rockström J, Edenhofer O, Gaertner J, DeClerck F (2020) Planet-proofing the global food system. Nat Food 1:3–5. https://doi.org/10.1038/s43016-019-0010-4

Schaffartzik A, Duro JA, Krausmann F (2019) Global appropriation of resources causes high international material inequality—growth is not the solution. Ecol Econ 163:9–19. https://doi.org/10.1016/j.ecolecon.2019.05.008

Smolski AR (2019) Stemming the exploitation of immigrant farm labor. Contexts 18:70–71. https://doi.org/10.1177/1536504219854727

Steinmann ZJN, Schipper AM, Hauck M, Giljum S, Wernet G, Huijbregts MAJ (2017) Resource footprints are good proxies of environmental damage. Environ Sci Technol 51:6360–6366. https://doi.org/10.1021/acs.est.7b00698

Sussell A, Peterson C, Li J, Miniño A, Scott KA, Stone DM (2023) Suicide rates by industry and occupation – National Vital Statistics System, United States, 2021. Morbidity and MMWR Morb Mortal Wkly Rep, US Department of Health and Human Services, Centers for Disease Control and Prevention 72:1346–1350. https://doi.org/10.15585/mmwr.mm7250a2

Thaler EA, Kwang JS, Quirk BJ, Quarrier CL, Larsen IJ (2022) Rates of historical anthropogenic soil erosion in the Midwestern United States. Earths Future 10:e2021EF002396. https://doi.org/10.1029/2021EF002396

Thornton P, D'Croz DM, Kugler C, Remans R, Zornetzer H, Herrero M (2024) Enabling food system innovation: accelerators for change. Glob Food Secur 40:100738. https://doi.org/10.1016/j.gfs.2023.100738

Vélez-Henao JA, Pauliuk S (2023) Material requirements of decent living standards. Environ Sci Technol 57:14206–14217. https://doi.org/10.1021/acs.est.3c03957

Zhang Y, Summers S, Jones JW, Reid JF (2024) A scalable index for quantifying circularity of bioeconomy systems. Resour Conserv Recycling 201:107821. https://doi.org/10.1016/j.resconrec.2024.107821

MIX
Papier aus verantwortungsvollen Quellen
Paper from responsible sources
FSC® C105338

If you have any concerns about our products,
you can contact us on
ProductSafety@springernature.com

In case Publisher is established outside the EU,
the EU authorized representative is:
Springer Nature Customer Service Center GmbH
Europaplatz 3, 69115 Heidelberg, Germany

Printed by Libri Plureos GmbH
in Hamburg, Germany